W9-BMD-520

FINITE ELEMENT METHODS IN MECHANICS

The author would like to express
his extreme indebtedness to
Professor Richard A. Scott
and
the late Professor Roger D. Low,
who contributed a great deal of time and effort
to help bring this book to fruition.

Finite element methods in mechanics

NOBORU KIKUCHI

Department of Mechanical Engineering and Applied Mechanics
The University of Michigan, Ann Arbor

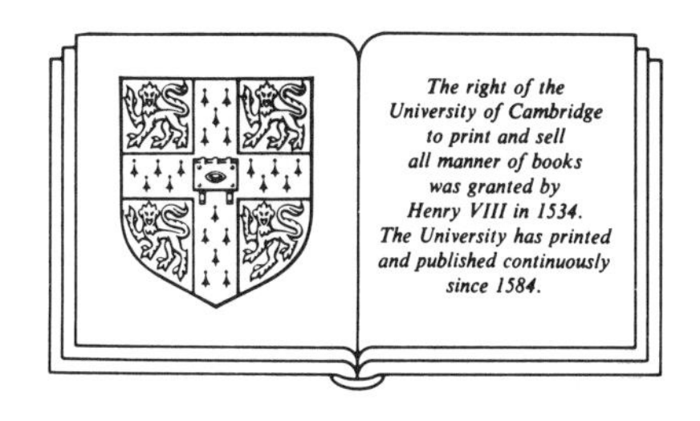

CAMBRIDGE UNIVERSITY PRESS

Cambridge

London New York New Rochelle

Melbourne Sydney

Published by the Press Syndicate of the University of Cambridge
The Pitt Building, Trumpington Street, Cambridge CB2 1RP
32 East 57th Street, New York, NY 10022, USA
10 Stamford Road, Oakleigh, Melbourne 3166, Australia

First published 1986

Printed in the United States of America

Library of Congress Cataloging in Publication Data

Kikuchi, Noboru.

Finite element methods in mechanics.

Bibliography: p.

1. Finite element method. 2. Heat – Conduction.
3. Structures, Theory of. I. Title.
TA347.F5K55 1985 620'.001'515353 85-2508
ISBN 0 521 30262 5 hard covers
ISBN 0 521 33972 3 paperback

ISBN 0 521 30953 0 program diskette
ISBN 0 521 30952 2 program manual

TO NANAE AND AYAKO

CONTENTS

PREFACE

This textbook was initially developed for the introductory course in finite element methods at the Department of Mechanical Engineering and Applied Mechanics, the University of Michigan, Ann Arbor, Michigan. It is based on four years of teaching experience of first-year graduate students and some senior undergraduate students in the engineering college. Because of the mechanical engineering environment, heat conduction problems are covered, as are standard stress analysis of solids and structures. Many small-size BASIC and FORTRAN programs are given so that readers can apply them to solve exercises using either microcomputers such as IBM PCs and compatibles or mainframes of computer networks that support FORTRAN IV. BASIC programs are available in separate volumes with a diskette for the IBM PC and compatibles by reader's request (see back of book for ordering instructions). FORTRAN programs are also available to readers who specially request them from the author. At present, BASIC compilers are available for some microcomputers in order to improve speed of computation. In the author's opinion, they are very impressive and encourage the use of microcomputers even for finite element methods. It is noted, however, that these programs are primarily designed for educational purposes – to teach the theory of finite element methods. Programs listed in this book are the following:

BASIC programs
- Jacobi's method for solving eigenvalue problems (p. 13)
- BASIC-FEM1 for heat conduction (p. 61)
- generalized Jacobi's method (p. 140)
- inverse iteration method for solving eigenvalue problems (p. 146)
- BASIC-FEM2 for plane linear elasticity (p. 265)

FORTRAN programs
- FEM1 for heat conduction (p. 328)
- PRE-FEM1 for generating finite element models (p. 345)
- POST-FEM1 for plotting computed results (p. 360)
- FEM2 for plane linear elasticity (p. 372)

The emphasis of this book is the development of finite element methods based on applied mechanics background. Starting from very fundamental formulations of heat conduction and linear elasticity, the weak form (i.e., the principle of virtual work in elasticity) is derived from a boundary value problem that represents mechanics behavior of solids and fluids. Finite element approximations are then derived from this weak form, applying adequate interpolation of functions even for simple truss and beam structures without using other conventional "engineering" approaches. Although the book is intended to be self-contained, it may be necessary for readers to review some background of heat conduction and linear elasticity. This book also contains many mathematical aspects of finite element methods. It is recommended that readers examine these during a second reading and follow all the derivations of equations and inequalities: An advanced calculus background would suffice for this. The mathematics given in this book should provide the fundamental structure of finite element methods as well as the basic nature of mechanics. Furthermore, this information is key to the design of finite element grids that give fairly accurate approximations.

Materials covered in this book are the minimum needed for an understanding of introductory finite element methods and provide a good foundation for more advanced nonlinear mechanics and dynamics. Accurate finite element approximations can be obtained only by clear insight into mechanics and an appreciation of mathematical consequence. Note, therefore, that the exercises in this book are not trivial; that is, mere mechanical substitution of numbers will not provide answers. They are designed both to extend the scope of and to provide more advanced knowledge of finite element methods. Thus, it is recommended that all of them be *read* whether or not they are to be solved. Many can be topics for term projects.

Chapter 1 is a quick review of background necessary for the study of introductory finite element methods. It reviews index notations (for coding of programs), the Gaussian elimination method for solving systems of linear equations, Jacobi's method for eigenvalue problems, and variational methods. Readers sufficiently familiar with this material can skip this chapter.

Chapter 2 describes basic structure and the procedure for obtaining finite element approximations for heat conduction problems using the three-node triangular element – the simplest finite element for plane problems. One notable section is 2.10, in which many discussions on finite element approximations are given. For example, section 2.10.2 extensively discusses the idea of adaptive finite element methods for the improvement of the quality of approximations and the design of better finite element grids. Although its importance is widely accepted, this subject has not been treated in textbooks published so far.

Chapter 3 treats upwind methods for dealing with convection-dominated problems in heat transfer and time-dependent problems basically using the θ-method. Brief stability consideration is also given. This chapter may be skipped in a one-semester course.

Chapter 4 deals with structural mechanics for trusses, beams, and frames. Not only static but also dynamic problems are treated here. Rather detailed discussion is given for eigenvalue problems using a generalized Jacobi method and inverse iteration method for free vibration of structures. Newmark's β-method and Wilson's θ-method are also introduced to solve forced vibration problems as well as Gear's stiff integration schemes. As an application of finite element methods to structural optimization, three optimal design problems are solved using the optimality criteria method.

Chapter 5 is for plane linear elasticity and considers finite element approximations by the four-node isoparametric element. As in preceding chapters, after deriving all necessary formulations for linear elasticity, the form of the principle of virtual work is discretized by four-node elements. The concept of isoparametric elements is discussed. Section 5.4, where details of characteristics of the four-node element are discussed, must be read very carefully. Eigenvalue structure, a necessity of the selective reduced integration technique for bending problems and for nearly incompressible materials and convergence property are especially discussed. Using the FORTRAN program FEM2, several typical applied elasticity problems are solved and the results compared to those of analytical solutions, as well as those of photoelasticity. BASIC-FEM2 is then developed by applying the reduced integration method with hourglass control to reduce computing time in microcomputers. Interpolation theory is discussed in Section 5.9 for isoparametric, serendipity, and singular elements. In Section 5.10, modification of FEM2 is discussed to solve Navier–Stokes flow problems. Using the similarity of two governing equations, this modification is rather straightforward.

Chapter 6 discusses finite element approximations of Mindlin plates and non-conforming approximations of Love–Kirchhoff plates. This chapter may be skipped for a one-semester course.

The author would like to express his sincere appreciation to all the students who took the course "Introduction to Finite Element Methods" at the University of Michigan. Their comments were valuable in the writing of this book. Among these students, Mr. Byeong Cheon Koh and Mr. Toshikazu Torigaki must be specially mentioned for helping to edit the course notes and for completing the programs listed here. The author is also grateful to all the secretaries in the Department of Mechanical Engineering and Applied Mechanics for their professional typing job, particularly Ms. Susan Martin, who devoted large amounts of time.

Ann Arbor, Michigan Noboru Kikuchi

1 REVIEW OF BACKGROUND MATERIALS

We shall briefly review index notation, the Gauss elimination method to solve a system of linear equations, Jacobi's method for an eigenvalue problem, and variational methods. These comprise the minimum background required in order to understand the materials given in this book. Readers with a good grasp of these topics can skip this chapter and start with Chapter 2.

1.1 Index notation

The most covenient notation for the study of finite element methods is *index notation,* since equations written using it can be translated to FORTRAN statements directly. For example, let us consider the dot product of two vectors,

$$\mathbf{u} = \sum_{i=1}^{N} u_i \mathbf{i}_i = u_1 \mathbf{i}_1 + u_2 \mathbf{i}_2 + \cdots + u_N \mathbf{i}_N \tag{1.1}$$

and

$$\mathbf{v} = \sum_{i=1}^{N} v_i \mathbf{i}_i = v_1 \mathbf{i}_1 + v_2 \mathbf{i}_2 + \cdots + v_N \mathbf{i}_N \tag{1.2}$$

Here the numbers $\{u_i\}$ are the components of the vector in the $\mathbb{R}^N$ space. For simplicity, let $N = 3$; we are then in the three-dimensional space that we use in mechanics. In most cases, x-, y-, and z-coordinate axes are set up in the space $\mathbb{R}^3$, and the unit vectors $\mathbf{i}$, $\mathbf{j}$, and $\mathbf{k}$ are introduced along each axis x, y, and z, respectively (see Figure 1.1). In index notation, we change these as follows:

$$\begin{array}{cccccc} x & y & z & \mathbf{i} & \mathbf{j} & \mathbf{k} \\ \downarrow & \downarrow & \downarrow & \downarrow & \downarrow & \downarrow \\ x_1 & x_2 & x_3 & \mathbf{i}_1 & \mathbf{i}_2 & \mathbf{i}_3 \end{array}$$

Then the position vector $\mathbf{r} = x\mathbf{i} + y\mathbf{j} + z\mathbf{k}$ is written as $\mathbf{r} = x_i \mathbf{i}_i$, instead of $\mathbf{r} = \sum_{i=1}^{3} x_i \mathbf{i}_i$. The rule involved here is that summation is taken over the index i, which is repeated exactly once (i.e. appears exactly twice) in a term. If clarity is necessary on the range of summation, we may write

$$\mathbf{r} = x_i \mathbf{i}_i, \quad i = 1, 2, \ldots, N \tag{1.3}$$

The unit vectors $\mathbf{i}_1$, $\mathbf{i}_2$, and $\mathbf{i}_3$ (i.e., $\mathbf{i}$, $\mathbf{j}$, and $\mathbf{k}$) are called the *base vectors* for the $\mathbb{R}^3$ space, and the numbers x_1, x_2, and x_3 are components of the vector $\mathbf{r}$ with respect to the base vectors $\mathbf{i}_1$, $\mathbf{i}_2$, and $\mathbf{i}_3$. Generalization to the $\mathbb{R}^N$ space is straightforward. Thus, the vectors $\mathbf{u}$ and $\mathbf{v}$ in (1.1) and (1.2) are represented by

$$\mathbf{u} = u_i \mathbf{i}_i \qquad \mathbf{v} = v_i \mathbf{i}_i, \quad i = 1, 2, \ldots, N \tag{1.4}$$

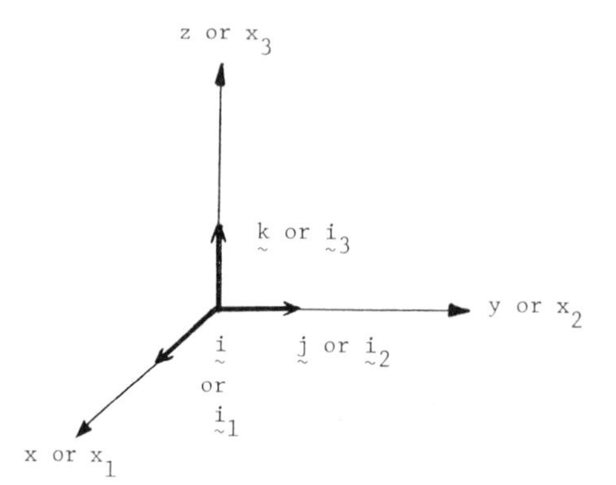

Figure 1.1 Coordinate system for $\mathbb{R}^3$.

and their dot product can be written as

$$\mathbf{u} \cdot \mathbf{v} = u_i v_i \left(\triangleq \sum_{i=1}^{N} u_i v_i \right) \tag{1.5}$$

For the dot product of two vectors we have the FORTRAN statements

```
    DOT = 0
    DO 100 I = 1,N
100 DOT = DOT + U(I) * V(I)
```
(1.6)

The index in index notation directly becomes the one that indicates the entry of the array in FORTRAN.

By the summation convention, we do not sum over i in the term $a_i b_i c_i$, since the index i is repeated three times; however, we take summation over j in $a_j c_i d_j$. A remark on the summation convention is that the letter used for the repeated index is immaterial in the sense that $a_j c_i d_j$ is exactly the same as the expression $a_k c_i d_k$ since summation is taken over j in the first and over k in the second. Thus, repeated indices are called *dummy indices*. The index i in the expression $a_j c_i d_j$ above is called a free index that takes any number from $i = 1, 2, \ldots, N$. A FORTRAN statement for this is given as

```
    SUM = 0
    DO 100 J = 1,N
100 SUM = SUM + A(J) * D(J)
    ACD(I) = SUM * C(I)
```
(1.7)

Similarly, let us introduce the basis for a matrix (or tensor) $\mathbf{T}$ defined in the space $\mathbb{R}^N \mathbb{R}^M$ as the set $\{\mathbf{i}_i \mathbf{e}_I\}$, $i = 1, 2, \ldots, N$, $I = 1, 2, \ldots, M$, where $\mathbf{i}_i \mathbf{e}_I$ is the *dyadic*, or outer product, and is best defined by the following two operations:

$$\mathbf{i}_i \mathbf{e}_I \circ \mathbf{v}_J = \mathbf{i}_i (\mathbf{e}_I \circ \mathbf{v}_J) \qquad \mathbf{v}_j \circ (\mathbf{i}_i \mathbf{e}_I) = (\mathbf{v}_j \circ \mathbf{i}_i) \mathbf{e}_I \tag{1.8}$$

where the product indicated by $\circ$ denotes ordinary, dot, or cross multiplication. Using the components T_{iI} of the tensor $\mathbf{T}$ with respect to $\mathbf{i}_i \mathbf{e}_I$, $\mathbf{T}$ is represented as

$$\mathbf{T} = T_{iI} \mathbf{i}_i \mathbf{e}_I \tag{1.9}$$

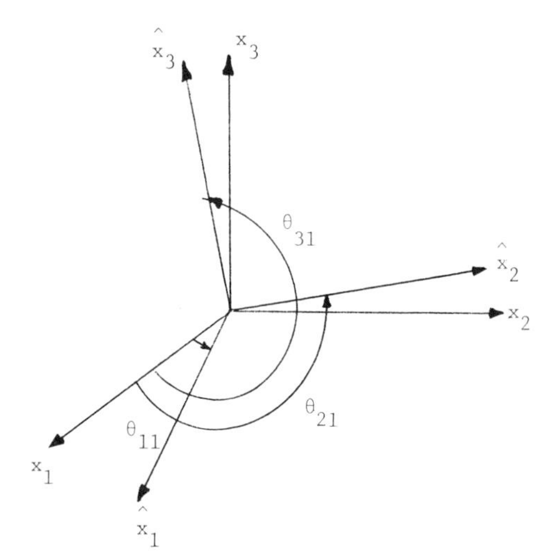

Figure 1.2 Rotation of the coordinate system.

We sometimes describe the tensor $\mathbf{T}$ by the matrix form $[T_{iI}]$ by arraying its components T_{iI}, $i = 1, 2, \ldots, N$ and $I = 1, 2, \ldots, M$. The operation of multiplying a matrix and a vector can be given as

$$\mathbf{v} = \mathbf{T} \cdot \mathbf{u} \qquad \text{or} \qquad v_i = T_{iJ} u_J \tag{1.10}$$

where $\mathbf{v} = v_i \mathbf{i}_i \in \mathbb{R}^N$ and $\mathbf{u} = u_J \mathbf{e}_J \in \mathbb{R}^M$. Note that a matrix transforms one vector into another vector. One very typical example of a matrix is the coordinate rotation matrix $\boldsymbol{\beta} = \beta_{Ii} \mathbf{e}_I \mathbf{i}_i$ (see Figure 1.2) defined by

$$\beta_{Ii} = \cos(\theta_{Ii}) \tag{1.11}$$

where θ_{Ii} is the angle between the I axis and the i axis. Then the unit base vectors $\mathbf{e}_I$ are related to the vectors $\{\mathbf{i}_i\}$:

$$\mathbf{e}_I = \beta_{Ii} \mathbf{i}_i \tag{1.12}$$

Using these, we can obtain the components of $\mathbf{v}$ in the rotated coordinate system $(\hat{x}_1, \hat{x}_2, \hat{x}_3)$:

$$\mathbf{v} = v_i \mathbf{i}_i = \hat{v}_I \mathbf{e}_I \tag{1.13a}$$

where

$$\begin{aligned} \hat{v}_I &\triangleq \mathbf{v} \cdot \mathbf{e}_I = (\mathbf{v}_i \mathbf{i}_i) \cdot (\beta_{Ij} \mathbf{i}_j) \\ &= v_i \beta_{Ij} \mathbf{i}_i \cdot \mathbf{i}_j = v_i \beta_{Ij} \delta_{ij} = \beta_{Ii} v_i \end{aligned} \tag{1.13b}$$

Here we have used the fact that

$$\mathbf{i}_i \cdot \mathbf{i}_j = \delta_{ij} = \begin{cases} 1 & \text{if } i = j \\ 0 & \text{if } i \neq j \end{cases} \tag{1.14}$$

by using the Kronecker delta. Similarly, we can find the new components of a matrix after a coordinate rotation has been performed. Indeed, for a given matrix

$$\mathbf{A} = A_{ij} \mathbf{i}_i \mathbf{i}_j \tag{1.15}$$

we also have

$$\mathbf{A} = \hat{A}_{IJ}\mathbf{e}_I\mathbf{e}_J \tag{1.16a}$$

where

$$\begin{aligned}\hat{A}_{IJ} \triangleq \mathbf{e}_I \cdot \mathbf{A}\mathbf{e}_J &= (\beta_{Ii}\mathbf{i}_i) \cdot (A_{jk}\mathbf{i}_j\mathbf{i}_k)(\beta_{Jl}\mathbf{i}_l) \\ &= \beta_{Ii}A_{jk}\beta_{Jl}\mathbf{i}_i \cdot (\mathbf{i}_j\mathbf{i}_k)\mathbf{i}_l \\ &= \beta_{Ii}A_{jk}\beta_{Jl}\underbrace{(\mathbf{i}_i \cdot \mathbf{i}_j)}_{\delta_{ij}}\underbrace{(\mathbf{i}_k \cdot \mathbf{i}_l)}_{\delta_{kl}} \\ &= \beta_{Ii}A_{ik}\beta_{Jk}\end{aligned} \tag{1.16b}$$

The transformation (1.16) is sometimes represented in *matrix* form as

$$\hat{\mathbf{A}} = \boldsymbol{\beta}\mathbf{A}\boldsymbol{\beta}^{\mathrm{T}} \tag{1.17}$$

by using the transpose of the matrix. A translation of equation (1.17) to FORTRAN is

```
      DO 100 IT = 1,3
      DO 100 JT = 1,3
      ATIJ = 0
      DO 102 I = 1,3
      DO 102 J = 1,3
  102 ATIJ = ATIJ + B(IT,I) * A(I,J) * B(JT,J)
  100 AT(IT,JT) = ATIJ
```
(1.18)

We shall now look at index notation for the gradient and divergence operators. In the (x, y, z) coordinate system, the gradient of a scalar function ϕ is given by

$$\nabla\phi = \text{grad } \phi \triangleq \mathbf{i}\frac{\partial\phi}{\partial x} + \mathbf{j}\frac{\partial\phi}{\partial y} + \mathbf{k}\frac{\partial\phi}{\partial z} \tag{1.19}$$

If we introduce the notation $\phi_{,i}$ for the partial derivative with respect to x_i, that is,

$$\phi_{,i} \triangleq \frac{\partial\phi}{\partial x_i} \tag{1.20}$$

then the gradient of ϕ becomes

$$\nabla\phi = \mathbf{i}_i\phi_{,i} \qquad \left(\text{i.e.,} \quad \nabla = \mathbf{i}_i\frac{\partial}{\partial x_i}\right) \tag{1.21}$$

The gradient of a vector $\mathbf{v} = v_i\mathbf{i}_i$ is a tensor $\nabla\mathbf{v}$ represented by

$$\nabla\mathbf{v} \triangleq v_{i,j}\mathbf{i}_j\mathbf{i}_i \tag{1.22}$$

The divergence of a vector $\mathbf{v} = v_x\mathbf{i} + v_y\mathbf{j} + v_z\mathbf{k}$ is defined as

$$\nabla \cdot \mathbf{v} = \text{div } \mathbf{v} = \frac{\partial v_x}{\partial x} + \frac{\partial v_y}{\partial y} + \frac{\partial v_z}{\partial z} \tag{1.23}$$

Using index notation, we have

$$\nabla \cdot \mathbf{v} = \left(\mathbf{i}_i \frac{\partial}{\partial x_i}\right) \cdot (v_j \mathbf{i}_j)$$

$$= \frac{\partial v_j}{\partial x_i} \mathbf{i}_i \cdot \mathbf{i}_j = (v_{j,i})\delta_{ij} = v_{i,i} \tag{1.24}$$

The divergence of a tensor $\mathbf{T} = T_{ij}\mathbf{i}_i\mathbf{i}_j$ is given by

$$\nabla \cdot \mathbf{T} = \operatorname{div} \mathbf{T} = \left(\mathbf{i}_i \frac{\partial}{\partial x_i}\right) \cdot (T_{jk}\mathbf{i}_j\mathbf{i}_k)$$

$$= T_{jk,i} \underbrace{\mathbf{i}_i \cdot (\mathbf{i}_j \mathbf{i}_k)}_{\underbrace{(\mathbf{i}_i \cdot \mathbf{i}_j)}_{\delta_{ij}}\mathbf{i}_k}$$

$$= T_{jk,j}\mathbf{i}_k \tag{1.25}$$

In the usual notation, we have

$$\nabla \cdot \mathbf{T} = \left(\frac{\partial T_{xx}}{\partial x} + \frac{\partial T_{yx}}{\partial y} + \frac{\partial T_{zx}}{\partial z}\right)\mathbf{i}$$

$$+ \left(\frac{\partial T_{xy}}{\partial x} + \frac{\partial T_{yy}}{\partial y} + \frac{\partial T_{zy}}{\partial z}\right)\mathbf{j}$$

$$+ \left(\frac{\partial T_{xz}}{\partial x} + \frac{\partial T_{yz}}{\partial y} + \frac{\partial T_{zz}}{\partial z}\right)\mathbf{k} \tag{1.26}$$

Using the gradient and divergence operators, the Laplacian is given as

$$\Delta\phi \triangleq \nabla \cdot \nabla\phi = \frac{\partial^2 \phi}{\partial x^2} + \frac{\partial^2 \phi}{\partial y^2} + \frac{\partial^2 \phi}{\partial z^2}$$

$$= \left(\mathbf{i}_i \frac{\partial}{\partial x_i}\right) \cdot \left(\mathbf{i}_j \frac{\partial}{\partial x_j}\right)\phi$$

$$= \phi_{,ij}\mathbf{i}_i \cdot \mathbf{i}_j = \phi_{,ii} \tag{1.27}$$

More generally,

$$\operatorname{div}(\mathbf{k} \operatorname{grad} \phi) = \nabla \cdot (\mathbf{k}\, \nabla\phi) = (k_{ij}\phi_{,j})_{,i} \tag{1.28}$$

Another useful convention using index notation can be obtained from the permutation symbol

$$e_{ijk} = \begin{cases} 1 & \text{if } i, j, k \text{ is an even permutation of } 1, 2, 3 \\ -1 & \text{if } i, j, k \text{ is an odd permutation of } 1, 2, 3 \\ 0 & \text{otherwise} \end{cases} \tag{1.29}$$

More precisely, $e_{112} = 0$, $e_{231} = 1$, $e_{321} = -1$, $e_{132} = -1$, $e_{133} = 0$, $e_{312} = 1$, and others. If we use this symbol, the cross product $\mathbf{w} = \mathbf{u} \times \mathbf{v}$ of two vectors $\mathbf{u}$ and $\mathbf{v}$

can be written as

$$w_i = e_{ijk} u_j v_k \tag{1.30}$$

if $\mathbf{w} = w_i \mathbf{i}_i$, $\mathbf{u} = u_j \mathbf{i}_j$, and $\mathbf{v} = v_k \mathbf{i}_k$, since

$$\mathbf{w} = \mathbf{u} \times \mathbf{v} = \begin{vmatrix} \mathbf{i} & \mathbf{j} & \mathbf{k} \\ u_x & u_y & u_z \\ v_x & v_y & v_z \end{vmatrix} \tag{1.31}$$

Exercise 1.1: Suppose that the range of all indices is from 1 to 3 in the following.

1. Show that (a) $\delta_{ij}\delta_{ij} = 3$, (b) $e_{ijk}e_{kji} = -6$, (c) $e_{kki} = 0$, (d) $\delta_{ij}\delta_{jk} = \delta_{ik}$, and (e) $e_{ijk}A_jA_k = 0$.
2. If $b_i = a_i/\sqrt{a_j a_j}$, show that $\mathbf{b} = b_i \mathbf{i}_i$ is a unit vector.
3. Use index notation to prove that

$$\mathbf{a} \cdot (\mathbf{b} \times \mathbf{c}) = (\mathbf{a} \times \mathbf{b}) \cdot \mathbf{c}$$

4. Using the definition of a determinant, show that

$$\det \begin{pmatrix} a_{11} & a_{12} & a_{13} \\ a_{21} & a_{22} & a_{23} \\ a_{31} & a_{32} & a_{33} \end{pmatrix} = e_{ijk} a_{1i} a_{2j} a_{3k}$$

5. Show that (a) $e_{ijk}e_{imn} = \delta_{jm}\delta_{kn} - \delta_{jn}\delta_{km}$, (b) $e_{ijk}e_{ijn} = 2\delta_{kn}$, and

(c) $$\det \begin{pmatrix} a_{11} & a_{12} & a_{13} \\ a_{21} & a_{22} & a_{23} \\ a_{31} & a_{32} & a_{33} \end{pmatrix} = \tfrac{1}{6} e_{ijk} e_{lmn} a_{il} a_{jm} a_{kn} \tag{1.32}$$

6. Develop a FORTRAN program to normalize the vector $\mathbf{a} = a_i \mathbf{i}_i$ whose components are stored in the one-dimensional array $A(I)$, $I = 1, \ldots, N$.
7. Suppose that a 3×3 matrix array $A(I, J)$, $I, J = 1, 2, 3$, is given. Develop a FORTRAN program to compute its determinant.

1.2 Gauss elimination method for solving a system of linear equations

In the subsequent chapters, we shall solve the system of linear equations obtained by finite element approximations of problems in mechanics. Roughly speaking, a finite element method is a process by which a continuous problem in mechanics is reduced to a discrete problem, whose solution leads to a system of linear equations symbolically represented by

$$\mathbf{K}\mathbf{u} = \mathbf{f} \qquad \text{or} \qquad K_{ij} u_j = f_i \tag{1.33}$$

where $\mathbf{K}$ is the stiffness matrix, $\mathbf{u}$ the generalized displacement vector, and $\mathbf{f}$ the generalized load vector. Thus, in order to obtain the generalized displacement $\mathbf{u}$, we have to solve the system of linear equation (1.33).

One of the methods used to solve (1.33) is the Gauss elimination method discussed below. Suppose that we are given the system of linear equations

$$a_{ij} x_j = b_i, \quad i = 1, \ldots, N \tag{1.34a}$$

that is,

$$
\begin{aligned}
a_{11}x_1 + a_{12}x_2 + a_{13}x_3 + \cdots + a_{1N}x_N &= b_1 \\
a_{21}x_1 + a_{22}x_2 + a_{23}x_3 + \cdots + a_{2N}x_N &= b_2 \\
&\ \vdots \\
a_{N1}x_1 + a_{N2}x_2 + a_{N3}x_3 + \cdots + a_{NN}x_N &= b_N
\end{aligned}
\tag{1.34b}
$$

If matrix notation is used, (1.34a) and (1.34b) are also expressed by the form

$$
\begin{bmatrix}
a_{11} & a_{12} & a_{13} & \cdots & a_{1N} \\
a_{21} & a_{22} & a_{23} & \cdots & a_{2N} \\
 & & & \vdots & \\
a_{N1} & a_{N2} & a_{N3} & \cdots & a_{NN}
\end{bmatrix}
\begin{Bmatrix} x_1 \\ x_2 \\ \vdots \\ x_N \end{Bmatrix}
=
\begin{Bmatrix} b_1 \\ b_2 \\ \vdots \\ b_N \end{Bmatrix}
\tag{1.34c}
$$

A standard Gaussian elimination process is divided into two parts, the forward elimination and the back substitution. We shall describe in detail the forward elimination for the first two steps and shall generalize the forward elimination process using index notation.

The first step is to eliminate the terms $a_{21}x_1, a_{31}x_1, \ldots, a_{N1}x_1$ from the system of linear equations (1.34) as indicated below:

$$
\begin{array}{lllll}
a_{11}x_1 + & a_{12}x_2 + & a_{13}x_3 + \cdots + & a_{1N}x_N = b_1 \\
\left(a_{21} - \dfrac{a_{21}}{a_{11}} a_{11}\right) x_1 + & \left(a_{22} - \dfrac{a_{21}}{a_{11}} a_{12}\right) x_2 + & \left(a_{23} - \dfrac{a_{21}}{a_{11}} a_{13}\right) x_3 + \cdots + & \left(a_{2N} - \dfrac{a_{21}}{a_{11}} a_{1N}\right) x_N = b_2 - \dfrac{a_{21}}{a_{11}} b_1 \\
\left(a_{31} - \dfrac{a_{31}}{a_{11}} a_{11}\right) x_1 + & \left(a_{32} - \dfrac{a_{31}}{a_{11}} a_{12}\right) x_2 + & \left(a_{33} - \dfrac{a_{31}}{a_{11}} a_{13}\right) x_3 + \cdots + & \left(a_{3N} - \dfrac{a_{31}}{a_{11}} a_{1N}\right) x_N = b_3 - \dfrac{a_{31}}{a_{11}} b_1 \\
 & & & \qquad \vdots \\
\left(a_{N1} - \dfrac{a_{N1}}{a_{11}} a_{11}\right) x_1 + & \left(a_{N2} - \dfrac{a_{N1}}{a_{11}} a_{12}\right) x_2 + & \left(a_{N3} - \dfrac{a_{N1}}{a_{11}} a_{13}\right) x_3 + \cdots + & \left(a_{NN} - \dfrac{a_{N1}}{a_{11}} a_{1N}\right) x_N = b_N - \dfrac{a_{N1}}{a_{11}} b_1
\end{array}
$$

Denoting the coefficients of the new equations by $\tilde{a}_{ij}$ where

$$
\tilde{a}_{11} = a_{11}, \qquad \tilde{a}_{12} = a_{12}
$$

$$
\tilde{a}_{22} = a_{22} - \frac{a_{21}}{a_{11}} a_{12}, \qquad \tilde{a}_{23} = a_{23} - \frac{a_{21}}{a_{11}} a_{13}
$$

$$
\tilde{b}_2 = b_2 - \frac{a_{21}}{a_{11}} b_1, \text{ etc.}
$$

we have

$$
\begin{aligned}
\tilde{a}_{11}x_1 + \tilde{a}_{12}x_2 + \tilde{a}_{13}x_3 + \cdots + \tilde{a}_{1N}x_N &= \tilde{b}_1 \\
\tilde{a}_{22}x_2 + \tilde{a}_{23}x_3 + \cdots + \tilde{a}_{2N}x_N &= \tilde{b}_2 \\
\tilde{a}_{32}x_2 + \tilde{a}_{33}x_3 + \cdots + \tilde{a}_{3N}x_N &= \tilde{b}_3 \\
&\ \vdots \\
\tilde{a}_{N2}x_2 + \tilde{a}_{N3}x_3 + \cdots + \tilde{a}_{NN}x_N &= \tilde{b}_N
\end{aligned}
$$

The second step is to eliminate the terms $\tilde{a}_{32}x_2, \tilde{a}_{42}x_2, \ldots, \tilde{a}_{N2}x_2$ using the second equation as follows:

$$
\begin{aligned}
&\tilde{a}_{11}x_1 + \tilde{a}_{12}x_2 + \tilde{a}_{13}x_3 + \cdots + \tilde{a}_{1N}x_N = \tilde{b}_1 \\
&\tilde{a}_{22}x_2 + \tilde{a}_{23}x_3 + \cdots + \tilde{a}_{2N}x_N = \tilde{b}_2 \\
&\left(\tilde{a}_{32} - \frac{\tilde{a}_{32}}{\tilde{a}_{22}}\tilde{a}_{22}\right)x_2 + \left(\tilde{a}_{33} - \frac{\tilde{a}_{32}}{\tilde{a}_{22}}\tilde{a}_{23}\right)x_3 + \cdots + \left(\tilde{a}_{3N} - \frac{\tilde{a}_{32}}{\tilde{a}_{22}}\tilde{a}_{2N}\right)x_N = \tilde{b}_3 - \frac{\tilde{a}_{32}}{\tilde{a}_{22}}\tilde{b}_2 \\
&\vdots \\
&\left(\tilde{a}_{N2} - \frac{\tilde{a}_{N2}}{\tilde{a}_{22}}\tilde{a}_{22}\right)x_2 + \left(\tilde{a}_{N3} - \frac{\tilde{a}_{N2}}{\tilde{a}_{22}}\tilde{a}_{23}\right)x_3 + \cdots + \left(\tilde{a}_{NN} - \frac{\tilde{a}_{N2}}{\tilde{a}_{22}}\tilde{a}_{2N}\right)x_N = \tilde{b}_N - \frac{\tilde{a}_{N2}}{\tilde{a}_{22}}\tilde{b}_2
\end{aligned}
$$

Continuing the above two steps, we can generate the forward elimination procedure for the kth step:

$$
\begin{aligned}
&\tilde{a}_{ij} = \tilde{a}_{ij} - \frac{\tilde{a}_{ik}}{\tilde{a}_{kk}}\tilde{a}_{kj} \quad \text{(no sum on } k) \\
&i = k+1, k+2, \ldots, N \quad j = k+1, k+2, \ldots, N
\end{aligned} \tag{1.35}
$$

$$
\begin{aligned}
&\tilde{b}_i = \tilde{b}_i - \frac{\tilde{a}_{ik}}{\tilde{a}_{kk}}\tilde{b}_k \quad \text{(no sum on } k) \\
&i = k+1, k+2, \ldots, N
\end{aligned} \tag{1.36}
$$

for given $k = 1, 2, \ldots, N-1$. The above index expressions suggest a FORTRAN program for the forward elimination by the Gauss method:

```
C
C      (FORWARD ELIMINATION)
C
       N1 = N - 1
       DO 100 K = 1,N1
       K1 = K + 1
       DO 102 L = K,N
   102 C(L) = A(K,L)
       AKK = 1./C(K)
       BK = B(K)
       DO 108 I = K1,N                                        (1.37)
       AIK = A(I,K) * AKK
       B(I) = B(I) - AIK * BK
       DO 108 J = K,N
   108 A(I,J) = A(I,J) - AIK * C(J)
C
       WRITE(6,600) K                                  ]*
   600 FORMAT(///10X, 'STEP = ',I3,/)                  ]
       WRITE(6,602) ((A(I,J),J = 1,4),B(I),I = 1,N)    ]
   602 FORMAT(4(E10.3,1X),5X,E10.3)                    ]
   100 CONTINUE
```

* These four steps of the program are prepared only for the purpose of checking if the program for the forward elimination is working correctly.

A routine for the back substitution can be obtained by

$$\tilde{a}_{kk}x_k + \tilde{a}_{kk+1}x_{k+1} + \cdots + \tilde{a}_{kN}x_N = \tilde{b}_k$$

that is,

$$x_k = \left(\tilde{b}_k - \sum_{j=k+1}^{N} \tilde{a}_{kj}x_j\right)\Big/\tilde{a}_{kk} \quad \text{(no sum over } k\text{)} \tag{1.38}$$

This can be carried out by the program

```
C
C     (BACK SUBSTITUTION)
C
      K = N
      B(K) = B(K)/A(K,K)
  104 K = K - 1
      IF(K.LE.0) RETURN
      K1 = K + 1                                            (1.39)
      SUM = 0.
      DO 106 J = K1,N
  106 SUM = SUM + A(K,J) * B(J)
      B(K) = (B(K) - SUM)/A(K,K)
      GOTO 104
```

We now present an example illustrating the above two routines.

$$\begin{bmatrix} 0.200E+01 & 0.300E+01 & -0.100E+01 & 0.500E+01 \\ 0.400E+01 & 0.400E+01 & -0.300E+01 & 0.300E+01 \\ -0.200E+01 & 0.300E+01 & -0.100E+01 & 0.100E+01 \\ -0.300E+01 & 0.200E+01 & -0.100E+01 & 0.500E+01 \end{bmatrix} \begin{Bmatrix} x_1 \\ x_2 \\ x_3 \\ x_4 \end{Bmatrix} = \begin{Bmatrix} 0.150E+02 \\ 0.100E+02 \\ -0.500E+01 \\ -0.100E+01 \end{Bmatrix}$$

Since the number of equations is 4, three steps are necessary in the forward elimination as shown in Table 1.1. Then the back substitution yields the solution

```
1    3.00000
2    1.00000
3    4.00000
4    2.00000
```

Exercise 1.2: It is inconvenient to transfer two-dimensional arrays, such as the coefficient matrix **A** in the above example, from one subroutine to another. Modify the above programs so that the coefficient matrix **A** is stored in a one-dimensional manner as

$$\mathbf{A} = (a_{11}, a_{12}, a_{13}, \ldots, a_{1n}, a_{21}, a_{22}, \ldots, a_{2n}, \ldots, a_{n1}, a_{n2}, \ldots, a_{nn})$$

Exercise 1.3: If the property of symmetry $a_{ij} = a_{ji}$ is assumed in a system of linear equations, it is possible to save storage space, that is, the array for the coefficient matrix **A**. Modify the programs to accomplish this.

Table 1.1

STEP = 1				
0.200E + 01	0.300E + 01	−0.100E + 01	0.500E + 01	0.150E + 02
0.0	−0.200E + 01	−0.100E + 01	−0.700E + 01	−0.200E + 02
0.0	0.600E + 01	−0.200E + 01	0.600E + 01	0.100E + 02
0.0	0.650E + 01	−0.250E + 01	0.125E + 02	0.215E + 02
STEP = 2				
0.200E + 01	0.300E + 01	−0.100E + 01	0.500E + 01	0.150E + 02
0.0	−0.200E + 01	−0.100E + 01	−0.700E + 01	−0.200E + 02
0.0	−0.0	−0.500E + 01	−0.150E + 02	−0.500E + 02
0.0	−0.0	−0.575E + 01	−0.103E + 02	−0.435E + 02
STEP = 3				
0.200E + 01	0.300E + 01	−0.100E + 01	0.500E + 01	0.150E + 02
0.0	−0.200E + 01	−0.100E + 01	−0.700E + 01	−0.200E + 02
0.0	−0.0	−0.500E + 01	−0.150E + 02	−0.500E + 02
0.0	−0.0	−0.191E − 05	0.700E + 01	0.140E + 02

Exercise 1.4: If the coefficient matrix is banded – that is, if

$$a_{ij} = \begin{cases} a_{ij} & \text{if } |j-i| \leq M \\ 0 & \text{if } |j-i| > M \end{cases} \qquad M < n$$

we need not compute and store the zero elements. Modify the programs in order to exploit this property.

Exercise 1.5: Develop a BASIC program that is equivalent to the programs (1.37) and (1.39).

1.3 Jacobi's method for solving an eigenvalue problem

Another typical discrete form obtained by finite element approximations is the eigenvalue problem

$$\mathbf{Ku} = \lambda\mathbf{Mu} \quad \text{or} \quad K_{ij}u_j = \lambda M_{ij}u_j \tag{1.40}$$

where $\mathbf{M}$ is called the mass matrix in the area of finite element methods. In this case, the problem is to find λ and $\mathbf{u}$ satisfying (1.40). If the matrix is invertible, then (1.40) becomes

$$\mathbf{M}^{-1}\mathbf{Ku} = \lambda\mathbf{u} \qquad \text{or} \qquad M_{ik}^{-1}K_{kj}u_j = \lambda u_i \tag{1.41}$$

If the matrix $\mathbf{S} = \mathbf{M}^{-1}\mathbf{K}$ is a symmetric $N \times N$ matrix (in most vibration problems it is not!), there exist N pairs of solutions to the eigenvalue problem (1.41); that is, there exist the solutions $(\lambda_1, \mathbf{u}_1)$, $(\lambda_2, \mathbf{u}_2)$, . . . , $(\lambda_n, \mathbf{u}_n)$ to (1.41). One method for finding the solutions to (1.41), that is,

$$\mathbf{Su} = \lambda\mathbf{u} \qquad \text{or} \qquad S_{ij}u_j = \lambda u_i \tag{1.42}$$

is the Jacobi method, which is based on the similarity transformation of a square matrix.

Suppose that $\mathbf{P}$ is an orthogonal matrix such that det $\mathbf{P} = 1$ and $\mathbf{P}^{-1} = \mathbf{P}^{\mathrm{T}}$, where $\mathbf{P}^{\mathrm{T}}$ is the transpose of the matrix $\mathbf{P}$. Then the similarity transformation of $\mathbf{S}$ by $\mathbf{P}$ is given by

$$\hat{\mathbf{S}} = \mathbf{P}^{\mathrm{T}}\mathbf{S}\mathbf{P} \qquad \text{or} \qquad \hat{S}_{ij} = P_{ki}S_{kl}P_{lj} \tag{1.43}$$

Suppose that λ and $\mathbf{u}$ are an eigenvalue and the corresponding eigenvector of $\mathbf{S}$, that is, λ and $\mathbf{u}$ satisfy the system (1.42). Premultiplying by $\mathbf{P}^{\mathrm{T}}$ yields

$$\mathbf{P}^{\mathrm{T}}\mathbf{S}\mathbf{u} = \lambda\mathbf{P}^{\mathrm{T}}\mathbf{u}$$

Define

$$\mathbf{v} = \mathbf{P}^{\mathrm{T}}\mathbf{u}, \qquad \text{i.e.,} \qquad \mathbf{u} = \mathbf{P}\mathbf{v} \tag{1.44}$$

Then we have

$$\mathbf{P}^{\mathrm{T}}\mathbf{S}\mathbf{P}\mathbf{v} = \lambda\mathbf{v}, \qquad \text{i.e.,} \quad \hat{\mathbf{S}}\mathbf{v} = \lambda\mathbf{v} \tag{1.45}$$

This means that if $(\lambda, \mathbf{u})$ is a pair of eigenvalue and eigenvector of $\mathbf{S}$, then $(\lambda, \mathbf{v})$ is a pair of eigenvalue and eigenvector of $\hat{\mathbf{S}}$. The converse is also true. Therefore, solving problem (1.42) is equivalent to solving problem (1.45).

The Jacobi method is based on the above similarity transformation of a matrix and the idea that one of the off-diagonal terms of the new matrix can be forced to be zero after applying the transformation (as will be seen below). For example, if S_{ij}, $i \neq j$, is the element of $\mathbf{S}$ whose magnitude is the largest among S_{kl}, $k < l$, $l = 1, 2, \ldots, N$, the $\hat{S}_{ij}$ can be made zero by choosing properly the orthogonal matrix $\mathbf{P}$. Repeating this process until all off-diagonal terms become zero, we have

$$\mathbf{S}^*\mathbf{u}^* = \lambda\mathbf{u}^* \tag{1.46}$$

where

$$\begin{aligned} \mathbf{S}^* &= \cdots \mathbf{P}_3^{\mathrm{T}}\mathbf{P}_2^{\mathrm{T}}\mathbf{P}_1^{\mathrm{T}}\mathbf{S}\mathbf{P}_1\mathbf{P}_2\mathbf{P}_3 \cdots \\ \mathbf{u}^* &= \cdots \mathbf{P}_3^{\mathrm{T}}\mathbf{P}_2^{\mathrm{T}}\mathbf{P}_1^{\mathrm{T}}\mathbf{u} \triangleq \mathbf{T}^{\mathrm{T}}\mathbf{u} \end{aligned} \tag{1.47}$$

and $\mathbf{P}_i^{\mathrm{T}}$ is the orthogonal matrix used at the ith process of forcing an off-diagonal term to be zero. Since $\mathbf{S}^*$ is supposed to be a diagonal matrix, its diagonal elements are then the eigenvalues of $\mathbf{S}$. Using (1.47), the corresponding eigenvectors are obtained from

$$\mathbf{u} = \mathbf{P}_1\mathbf{P}_2\mathbf{P}_3 \cdots \mathbf{u}^* = \mathbf{T}\mathbf{u}^* \tag{1.48}$$

In general, there is no guarantee that the matrix $\mathbf{S}^*$ will be obtained after a finite number of applications of the similarity transformation. However, we shall assume this is obtainable for the matrix $\mathbf{S}$ considered here.

Let us denote by $\mathbf{P}^{(ij)}$ the orthogonal matrix that yields $\hat{S}_{ij} = 0$ by the similarity transformation of the matrix $\mathbf{S}$, where the indices i and j are defined so that the absolute value of S_{ij} is the largest among the off-diagonal terms S_{kl}, $k < l$, $l = 1, 2, \ldots, N$. To do this, we seek the angle ϕ to be used in the orthogonal matrix

$\mathbf{P}^{(ij)}$, whose elements are given by

$$[P_{kl}^{(ij)}] = \begin{bmatrix} 1 & & \vdots & & \vdots & & \\ & 1 & \vdots & & \vdots & & \\ \cdots & \cdots & \cos\phi & \cdots & \sin\phi & \cdots & \\ & & \vdots & & \vdots & & \\ \cdots & \cdots & -\sin\phi & \cdots & \cos\phi & \cdots & \\ & & \vdots & & \vdots & & \\ & & \vdots & & \vdots & & 1 \end{bmatrix} \begin{matrix} \\ \\ i \\ \\ j \\ \\ \\ \end{matrix} \tag{1.49}$$

(column i contains $\cos\phi$, $-\sin\phi$; column j contains $\sin\phi$, $\cos\phi$)

that is,

$$\begin{aligned} &P_{ii}^{(ij)} = P_{jj}^{(ij)} = \cos\phi, \qquad && P_{ij}^{(ij)} = -P_{ji}^{(ij)} = \sin\phi \quad \text{(no sum)} \\ &P_{kl}^{(ij)} = 0, \quad k \neq i, \quad l \neq j, \qquad && P_{kk}^{(ij)} = 1, \quad k \neq i, \quad k \neq j \end{aligned} \tag{1.50}$$

Using this rotation matrix, the matrix $[S_{ij}]$ becomes $[\hat{S}_{kl}]$;

$$\hat{S}_{kl} = P_{mk}^{(ij)} S_{mn} P_{nl}^{(ij)} \tag{1.51}$$

As a result of the rotation, we want

$$\hat{S}_{ij} = 0 \tag{1.52}$$

Hence, the angle ϕ is determined by $\hat{S}_{ij} = 0$.

$$\hat{S}_{ij} = \hat{S}_{ji} = \tfrac{1}{2}(S_{ii} - S_{jj}) \sin 2\phi + S_{ij} \cos 2\phi = 0$$

that is,

$$\tan 2\phi = \frac{-2S_{ij}}{S_{ii} - S_{jj}}, \qquad S_{ii} \neq S_{jj} \quad \text{(no sum)} \tag{1.53}$$

Once the angle ϕ has been determined, $\hat{S}_{kl}$ in (1.51) is obtained as

$$\begin{aligned} &\hat{S}_{kl} = S_{kl}, \quad k, l \neq i, j \\ &\hat{S}_{ik} = \hat{S}_{ki} = S_{ik} \cos\phi - S_{jk} \sin\phi, \quad k \neq i, j \\ &\hat{S}_{jk} = \hat{S}_{kj} = S_{ik} \sin\phi + S_{jk} \cos\phi, \quad k \neq i, j \\ &\hat{S}_{ii} = \tfrac{1}{2}(S_{ii} + S_{jj}) + \tfrac{1}{2}(S_{ii} - S_{jj}) \cos 2\phi - S_{ij} \sin 2\phi \\ &\hat{S}_{jj} = \tfrac{1}{2}(S_{ii} + S_{jj}) + \tfrac{1}{2}(S_{ii} - S_{jj}) \cos 2\phi + S_{ij} \sin 2\phi \\ &\qquad \text{(no sum)} \end{aligned} \tag{1.54}$$

Convergence of the iterations may be checked by the quantity $|S_{ij}|$; that is, if

$$|S_{ij}| < \text{TOLE} \tag{1.55}$$

is satisfied for a given tolerance TOLE, the process of applying similarity transformations will be terminated. Thus, after a certain number of applications of the transformation, the matrix **S** is transformed to a diagonal matrix **S*** within errors whose order of magnitude is less than the given tolerance.

Let us now write a program of Jacobi's method to find the eigenvalues and eigenvectors for a given symmetric matrix. A flowchart for the program is shown in Figure 1.3. The program shown below is written in the BASIC language of the system of IBM PC-XT.

```
1000 PRINT "*******************************"
1010 PRINT "   JACOBI'S METHOD    (S)   "
1020 PRINT "*******************************"
1030 PRINT : PRINT :PRINT
1040 INPUT "Size of the symmetric matrix      NX = ";NX
1050 DIM S(NX,NX),T(NX,NX)
1060 PRINT : PRINT "SYMMETRIC MATRIX  S " : PRINT
1070 FOR I=1 TO NX : FOR J=I TO NX : PRINT "    S(";I;",";J;") = ";: INPUT S(I,J) :
     S(J,I)=S(I,J) : NEXT J : NEXT I
1080 GOSUB 1100
1090 GOTO 1580
1100 REM  Subroutine Jacobi's Method  ( Standard Eigenvalue Problems  )
1110 REM ------------------------------------------------------------------------
1120 PRINT
1130 INPUT "Tolerance   EP = ";EP
1140 PRINT :
     PRINT "ITERATION PROCESS ............................................." : PRINT
1150 FOR I=1 TO NX : FOR J=I TO NX : T(I,J)=0 : T(J,I)=0 : NEXT J : T(I,I)=1 : NEXT I
1160 N=0 : NP=1
1170 N=N+1
1180 REM  < FIND THE MAXIMUM ABSOLUTE VALUE  >
1190 SM=0
1200 FOR I=1 TO NX-1 : FOR J=I+1 TO NX : IF ABS(S(I,J))<SM THEN GOTO 1220
1210 SM=ABS(S(I,J)) : IM=I : JM=J
1220 NEXT J  : NEXT I
1230 REM  < FIND THE TWICE VALUE OF THE ANGLE  >
1240 SI=S(IM,IM) : SJ=S(JM,JM) : SK=S(IM,JM)
1250 IF SI=SJ THEN GOTO 1280
1260 SL=-2*SK/(SI-SJ)   :   T2=ATN(SL)
1270 GOTO 1310
1280 IF SK>0 THEN T2=-3.141592654#/2
1290 IF SK=0 THEN T2=0
1300 IF SK<0 THEN T2=3.141592654#/2
1310 T1=.5*T2
1320 C1=COS(T1)   :   S1= SIN(T1)   :   C2=COS(T2)   :   S2=SIN(T2)
1330 REM  <  MODIFY THE MATRIX S  >
1340 FOR K=1 TO NX   :   SY=S(IM,K)   :   SZ=S(JM,K)
1350 S3=SY*C1-SZ*S1   :   S(IM,K)=S3   :   S(K,IM)=S3
1360 S4=SY*S1+SZ*C1   :   S(JM,K)=S4   :   S(K,JM)=S4
1370 NEXT K
1380 S(IM,IM)=.5*((SI+SJ)+(SI-SJ)*C2)-SK*S2
1390 S(JM,JM)=.5*((SI+SJ)-(SI-SJ)*C2)+SK*S2
1400 S(IM,JM)=0   :   S(JM,IM)=0
1410 REM   <  MODIFY  T  >
1420 FOR K=1 TO NX   :   T3=T(K,IM)   :   T4=T(K,JM)
1430 T(K,IM)=T3*C1-T4*S1   :   T(K,JM)=T3*S1+T4*C1
1440 NEXT K
1450 IF N<NP THEN GOTO 1480
1460 PRINT N;TAB(10);SM
1470 NP=NP+10
1480 IF SM>EP THEN GOTO 1170
1490 PRINT :
     PRINT "RESULTS/EIGENVALUES & EIGENVECTORS ........................" : PRINT
1500 PRINT "NUMBER OF ITERATION = ";N
1510 PRINT "TOLERANCE   = ";SM
1520 PRINT :
     INPUT "Which eigenvalue and eigenvector will be output? (1,....,NX) = ";I
1530 IF I<=0 OR I>NX THEN GOTO 1570
1540 PRINT : PRINT " EIGENVALUE = ";S(I,I) : PRINT : PRINT "EIGENVECTOR" :  PRINT
1550 FOR J=1 TO NX : PRINT J;TAB(10);T(J,I) : NEXT J
1560 GOTO 1520
1570 RETURN
1580 END
```

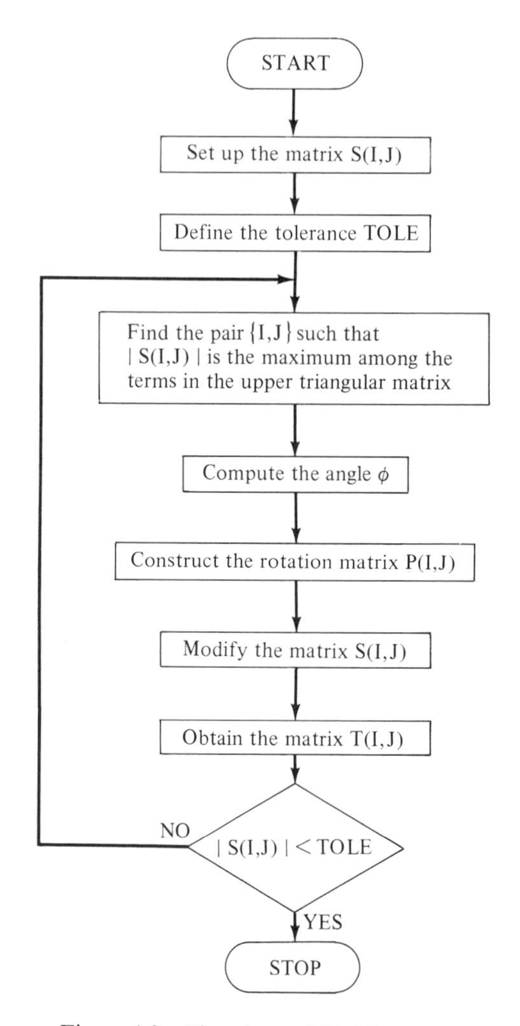

Figure 1.3 Flowchart of JACOBI.BAS.

Exercise 1.6: Write a program of Jacobi's method to obtain eigenvalues and eigenvectors by using the FORTRAN language.

1.4 Variational methods

We shall briefly review variational methods to solve an algebraic and a boundary value problem. For simplicity, we first study the problem: Find a vector $\mathbf{u} = u_i\mathbf{i}_i$ satisfying a system of linear equations

$$\mathbf{Ku} = \mathbf{f} \quad (\text{i.e.,} \quad K_{ij}u_j = f_i) \tag{1.56}$$

for a given vector $\mathbf{f} = f_i\mathbf{i}_i$ and a symmetric matrix $\mathbf{K} = K_{ij}\mathbf{i}_i\mathbf{i}_j$ such that

$$K_{ij} = K_{ji}, \qquad i, j = 1, 2, \ldots, N \tag{1.57}$$

Using the symmetry of $\mathbf{K}$, it can be shown that if $\mathbf{u}$ is a *minimizer* of a functional

$$F(\mathbf{v}) = \tfrac{1}{2}\mathbf{v} \cdot \mathbf{K}\mathbf{v} - \mathbf{v} \cdot \mathbf{f} \tag{1.58}$$

among all vectors $\mathbf{v} \in \mathbb{R}^n$, that is, if $\mathbf{u}$ satisfies

$$F(\mathbf{v}) \geq F(\mathbf{u}), \quad \forall \mathbf{v} \in \mathbb{R}^n \tag{1.59}$$

then $\mathbf{u}$ is also a solution of equation (1.56). Here $\mathbf{v} \in \mathbb{R}^n$ means "a vector $\mathbf{v}$ is an element of an N-dimensional Euclidean space $\mathbb{R}^n$," and $\forall$ means "for every." A functional is a special function whose range is a scalar field such as a real line $\mathbb{R}$. We now show that (1.59) yields (1.56). Taking $\mathbf{v} = \mathbf{u} + \varepsilon\mathbf{w}$ in (1.59) for an arbitrary vector $\mathbf{w}$ and a positive number $\varepsilon > 0$, we have

$$F(\mathbf{u} + \varepsilon\mathbf{w}) \geq F(\mathbf{u}), \quad \forall \mathbf{w} \in \mathbb{R}^n$$

Expanding the left side using (1.58) yields

$$\varepsilon\{\tfrac{1}{2}(\mathbf{w} \cdot \mathbf{K}\mathbf{u} + \mathbf{u} \cdot \mathbf{K}\mathbf{w}) - \mathbf{w} \cdot \mathbf{f}\} + \tfrac{1}{2}\varepsilon^2\mathbf{w} \cdot \mathbf{K}\mathbf{w} \geq 0$$

Dividing by $\varepsilon > 0$ and passing through the limit $\varepsilon \to 0$, we have

$$\tfrac{1}{2}(\mathbf{w} \cdot \mathbf{K}\mathbf{u} + \mathbf{u} \cdot \mathbf{K}\mathbf{w}) - \mathbf{w} \cdot \mathbf{f} \geq 0, \quad \forall \mathbf{w} \in \mathbb{R}^n$$

Using the symmetry of $\mathbf{K}$, this implies

$$\mathbf{w} \cdot (\mathbf{K}\mathbf{u} - \mathbf{f}) \geq 0, \quad \forall \mathbf{w} \in \mathbb{R}^n \tag{1.60}$$

Since $\mathbf{w}$ is an arbitrary vector in $\mathbb{R}^n$, (1.60) has to be satisfied for the choice $-\mathbf{w}$ instead of $\mathbf{w}$. That is,

$$-\mathbf{w} \cdot (\mathbf{K}\mathbf{u} - \mathbf{f}) \geq 0, \quad \forall \mathbf{w} \in \mathbb{R}^n \tag{1.61}$$

Inequalities (1.60) and (1.61) yield

$$\mathbf{w} \cdot (\mathbf{K}\mathbf{u} - \mathbf{f}) = 0, \quad \forall \mathbf{w} \in \mathbb{R}^n \tag{1.62}$$

That is, $\mathbf{u}$ has to be a solution of (1.56):

$$\mathbf{K}\mathbf{u} - \mathbf{f} = 0$$

On the other hand, if an additional condition

$$\mathbf{w} \cdot \mathbf{K}\mathbf{w} \geq 0, \quad \forall \mathbf{w} \in \mathbb{R}^n \tag{1.63}$$

is imposed, the inverse relation to the above can be verified. Indeed, suppose that $\mathbf{u}$ is a solution of (1.56). Then applying the relation

$$F(\mathbf{v}) - F(\mathbf{u}) = (\mathbf{v} - \mathbf{u}) \cdot (\mathbf{K}\mathbf{u} - \mathbf{f}) + \tfrac{1}{2}(\mathbf{v} - \mathbf{u}) \cdot \mathbf{K}(\mathbf{v} - \mathbf{u})$$

the nonnegativeness condition (1.63) of the matrix $\mathbf{K}$ yields inequality (1.59):

$$F(\mathbf{v}) - F(\mathbf{u}) \geq 0, \quad \forall \mathbf{v} \in \mathbb{R}^n$$

Therefore, under the condition

$$K_{ij} = K_{ji} \qquad \text{and} \qquad \mathbf{w} \cdot \mathbf{K}\mathbf{w} \geq 0, \quad \forall \mathbf{w} \in \mathbb{R}^n \tag{1.64}$$

the following three are *equivalent* to each other:

$$\begin{aligned} &(\text{P1}) && \mathbf{Ku} - \mathbf{f} = 0 \\ &(\text{P2}) && \mathbf{w} \cdot (\mathbf{Ku} - \mathbf{f}) = 0, \quad \forall \mathbf{w} \in \mathbb{R}^n \\ &(\text{P3}) && F(\mathbf{v}) - F(\mathbf{u}) \geq 0, \quad \forall \mathbf{v} \in \mathbb{R}^n \end{aligned} \tag{1.65}$$

where F is the functional defined by (1.58). Very roughly speaking, (P1), (P2), and (P3) correspond to the equilibrium equation, the principle of virtual work, and the principle of minimum potential energy, respectively, in mechanics.

Let us extend the above to the boundary value problem:

$$-\frac{d}{dx} k \frac{du}{dx} = f \text{ in } (0, 1), \qquad u(0) = g, \qquad k\frac{du}{dx}(1) = h \tag{1.66}$$

Since the differential equation is defined on an interval (0, 1), we can obtain a functional F through integrating some quantity over the interval,

$$F(v) = \frac{1}{2} \int_0^1 k\left(\frac{dv}{dx}\right)^2 dx - \int_0^1 fv \, dx - hv(1) \tag{1.67}$$

The manner of defining a functional F for a given boundary value problem (1.66) is that the boundary value problem (1.66) is obtained as the *Euler equation* of the functional F as follows. Suppose that u is a minimizer of F such that

$$u(0) = g, \qquad F(v) \geq F(u), \quad \forall v \text{ with } v(0) = g \tag{1.68}$$

In other words, u is a minimizer of F among the functions defined on the interval (0, 1) that satisfy the "essential" boundary condition $v(0) = g$ at $x = 0$. Then, for every w such that $w(0) = 0$, taking $v = u \pm \varepsilon w$ in (1.68) yields

$$F(u \pm \varepsilon w) - F(u) \geq 0$$

which reduces to

$$\pm\varepsilon\left\{\int_0^1 \left(k\frac{du}{dx}\frac{dw}{dx} - fw\right) dx - hw(1)\right\} + \frac{1}{2}\varepsilon^2 \int_0^1 k\left(\frac{dw}{dx}\right)^2 dx \geq 0$$

Dividing by $\varepsilon > 0$ and passing to the limit $\varepsilon \to 0$, we have

$$\int_0^1 \left(k\frac{du}{dx}\frac{dw}{dx} - fw\right) dx - hw(1) = 0 \tag{1.69}$$

for every w such that $w(0) = 0$. It it clear that (1.69) is exactly the same as the first variation of the functional F by the variation $\delta u = w$ from the minimizer u.* Indeed, the first variation of F is

$$\delta F(u) = \int_0^1 k\frac{du}{dx}\frac{d(\delta u)}{dx} dx - \int_0^1 f\,\delta u \, dx - h\,\delta u(1)$$

* The reader should note that for contact problems involving inequality constraints the current approach is applicable, whereas the traditional "See approach" is not.

Now, the Euler equation follows from the application of integration by parts in (1.69):

$$\int_0^1 \left[-\frac{d}{dx}\left(k\frac{du}{dx}\right) - f \right] w \, dx + k\frac{du}{dx}(1)w(1) - hw(1) = 0 \tag{1.70}$$

since $w(0) = 0$. Noting that w and $w(1)$ are arbitrary, equation (1.70) yields

$$-\frac{d}{dx}\left(k\frac{du}{dx}\right) = f \text{ in } (0, 1), \qquad k\frac{du}{dx}(1) = h \tag{1.71}$$

Since the boundary condition at $x = 1$ for the first derivative of the minimizer u is obtained as a part of the Euler equation, this is called a "natural" boundary condition. If the integrand of a functional F consists of the function v and its derivatives up to mth order, applying integration by parts m times yields boundary terms involving derivatives of $m, \ldots, (2m - 1)$th order. Then we can define natural boundary conditions for such boundary terms derived from the process of integration by parts. On the other hand, if boundary conditions are written in terms of derivatives of $0, \ldots, (m - 1)$th order, these are essential boundary conditions.

In the above, we have shown that the minimizer u of the minimization problem (1.68) is also a solution of the boundary value problem (1.66). As for the discrete system (1.56), it can be shown that a solution u of (1.66) is also a minimizer of the functional F under the condition

$$k \geq 0 \text{ in } (0, 1) \tag{1.72}$$

To see this, it suffices to note the following relation:

$$F(v) - F(u) = \int_0^1 \left[k\frac{du}{dx}\frac{d}{dx}(v - u) - f(v - u) \right] dx$$
$$- h(v - u)(1) + \tfrac{1}{2}\int_0^1 k\left[\frac{d}{dx}(v - u)\right]^2 dx \tag{1.73}$$

for every v and u. Therefore, including the intermediate step (1.69) of (1.68) and (1.71), we have the equivalent relation among the following three forms:

(P1) $$-\frac{d}{dx}\left(k\frac{du}{dx}\right) = f \text{ in } (0, 1), \qquad k\frac{du}{dx}(1) = h, \qquad u(0) = g$$

(P2) $$\int_0^1 \left(k\frac{du}{dx}\frac{dw}{dx} - fw\right) dz - hw(1) = 0, \quad \forall w \text{ with } w(0) = 0$$

(P3) $$u(0) = g, \qquad F(v) \geq F(u), \quad \forall v \text{ with } v(0) = g$$

We shall call (P1), (P2), and (P3) the *local, weak,* and *variational forms,* respectively.

In the above, the functional form F is chosen so that (1.66) can be obtained as the Euler equation. However, for a given boundary value problem, it might be difficult to find the corresponding functional for the variational formulation (P3). To avoid this difficulty, the weak form (P2), an intermediate step of the originally given boundary value problem and the variational form, could be a basis for the *variational method* to solve (P1) instead of using (P3), since the form

(P2) is easily obtained from the local form (P1). Indeed, multiplying an arbitrary function w to the differential equation of (1.66) and applying integration by parts after integrating them over the domain (0, 1), we have

$$\int_0^1 \left(-\frac{d}{dx} k \frac{du}{dx} - f \right) w \, dx = 0$$

and

$$\int_0^1 \left(k \frac{du}{dx} \frac{dw}{dx} - fw \right) dx - \left[k \frac{du}{dx} w \right]_0^1 = 0$$

Assuming $w(0) = 0$ and using the boundary condition of (1.66) yields

$$\int_0^1 \left(k \frac{du}{dx} \frac{dw}{dx} - fw \right) dx - hw(1) = 0$$

which is nothing but the form (P2). The derivation of (P2) from (P1) is very straightforward, as shown. Thus, if it is possible to use the weak form (P2) for approximation methods to solve the original boundary value problem (P1), we need not go further up to the variational form (P3), which requires the deduction of a functional.

1.4.1 Approximation (direct methods)

Before discussing finite element methods, we shall briefly review approximation methods based on (P2) and (P3) that are precursors to the finite element methods developed during the 1960s.

Let the minimizer u of F be assumed to be a polynomial

$$u(x) = g + \sum_{j=1}^{N} u_j x^j, \quad u_j \in \mathbb{R}, \quad j = 1, \ldots, N \tag{1.74}$$

that satisfies the essential boundary condition at $x = 0$. Substitution of (1.74) into the functional F yields

$$F(u) = \tfrac{1}{2} \sum_{i=1}^{N} \sum_{j=1}^{N} k \frac{ij}{i+j-1} u_i u_j - f_0 g - \sum_{i=1}^{N} b_i u_i - h\left(g + \sum_{i=1}^{N} u_i \right) \tag{1.75}$$

where $b_i = \int_0^1 f x^i \, dx$, $f_0 = \int_0^1 f \, dx$, and k is assumed to be a constant. If u is a minimizer of F, we have to satisfy

$$\frac{\partial}{\partial u_i} F(u) = 0, \quad i = 1, \ldots, N \tag{1.76}$$

that is,

$$\sum_{j=1}^{N} k \frac{ij}{i+j-1} u_j - b_i - h = 0, \quad i = 1, \ldots, N$$

Defining

$$K_{ij} = \frac{kij}{i+j-1} \qquad \text{and} \qquad f_i = b_i + h \tag{1.77}$$

we have the system of linear equations

$$K_{ij}u_j = f_i, \quad i = 1, \ldots, N \tag{1.78}$$

Solving (1.78) yields an approximation to the minimizer u of F. By taking $N \to \infty$, we may have a minimizer u. This procedure to obtain u is called the *Ritz method.*

(P2) can also be used to find an approximation of u. Suppose that

$$w(x) = \sum_{i=1}^{N} w_i x^i, \quad w(0) = 0 \tag{1.79}$$

Substitution of (1.74) and (1.79) yields

$$\sum_{i=1}^{N} \sum_{j=1}^{N} k \frac{ij}{i+j-1} w_i u_j - \sum_{i=1}^{N} b_i w_i - \sum_{i=1}^{N} h w_i = 0$$

for every w, that is, for every w_i, $i = 1, \ldots, N$. Using (1.77), we have

$$w_i K_{ij} u_j = w_i f_i, \quad \forall w_i, \quad i = 1, \ldots, N \tag{1.80}$$

that is,

$$K_{ij}u_j = f_i, \quad i = 1, \ldots, N$$

Thus, the same system of linear equations as (1.78) is obtained from (P2). We shall call this *Galerkin's method* to find an approximation of a solution to (P1).

If the function w in (P2) is assumed to be

$$w(x) = \sum_{i=1}^{N} w_i \sin\left(\frac{i\pi}{2} x\right)$$

a similar system of linear equations to (1.78) can be obtained. In this case, since different representations to u and w are assumed in (P2), we shall call this the *generalized Galerkin* (or *weighted residual*) *method.*

Exercise 1.7: Solve (1.78) for $N = 3$, and $f(x) = 1 + x + x^2$.

Exercise 1.8: Assume that

$$u(x) = g + \sum_{i=1}^{N} u_i \sin\left(\frac{i\pi}{2} x\right), \qquad w(x) = \sum_{i=1}^{N} w_i \sin\left(\frac{i\pi}{2} x\right)$$

and apply the Ritz and Galerkin methods to solve (1.66).

Exercise 1.9: Find the (P2) and (P3) forms corresponding to the local formulation.

$$\text{(P1)} \quad \begin{cases} -\dfrac{d}{dx}\left(k \dfrac{du}{dx}\right) + \lambda u = f & \text{in } (0, 1) \\[2ex] k \dfrac{du}{dx} = -k_0(u - g) + h & \text{at } x = 0 \text{ and } 1 \end{cases}$$

Exercise 1.10: Suppose that a functional F is defined as

$$F(v) = \int_a^b f(v, v^{(1)}, \ldots, v^{(m)})\, dx$$

where $v^{(i)}$ is the ith derivative of a function v defined on the interval (0, 1). Assuming that the integrand f is infinitely many times differentiable in its arguments, derive the Euler equation by taking the *first variation* of F and by applying integration by parts m times. Note that Taylor's expansion of the integrand f is given as follows:

$$
\begin{aligned}
f(u + \Delta u, u^{(1)} + \Delta u^{(1)}, \ldots, u^{(m)} + \Delta u^{(m)}) \\
= f(u, u^{(1)}, \ldots, u^{(m)}) + \frac{\partial f}{\partial u}(u, u^{(1)}, \ldots, u^{(m)})\,\Delta u \\
+ \cdots + \frac{\partial f}{\partial u^{(m)}}(u, u^{(1)}, \ldots, u^{(m)})\,\Delta u^{(m)} \\
+ \frac{1}{2}\frac{\partial^2 f}{\partial u^2}(u, u^{(1)}, \ldots, u^{(m)})\,\Delta u^2 + \cdots
\end{aligned}
$$

Exercise 1.11: Find the Euler equation for the functional

$$F(w) = \tfrac{1}{2}\int_0^1 EI\left(\frac{d^2w}{dx^2}\right)^2 dx - \int_0^1 fw\,dx + \left[M\frac{dw}{dx}\right]_0^1 - \left[Pw\right]_0^1$$

where $[g]_0^1 = g(1) - g(0)$.

1.4.2 Lagrange multiplier methods

In the above, the minimization problem has the constraint $u = g$ on the boundary point $x = 0$. This yields the condition $w(0) = 0$ for a variation w from the solution u. Now, if a Lagrange multiplier p is introduced to release the constraint $u = g$ at $x = 0$, the original minimization problem can be reformulated as a saddle point problem of a corresponding Lagrangian

$$L(v, q) = F(v) - qv(0) - g \tag{1.81}$$

where q is an arbitrary *admissible* Lagrange multiplier. That is, we shall seek a *saddle point* (u, p) such that

$$L(u, q) \leq L(u, p) \leq L(v, p), \quad \forall (v, q) \tag{1.82}$$

where v does not have any restriction. The first inequality of (1.82) yields the minimization problem

$$qu(0) - g \geq pu(0) - g, \quad \forall q \tag{1.83}$$

Taking $q = p \pm \varepsilon r$, $\varepsilon > 0$, $\forall r$, in (1.83) and dividing by ε, we have

$$ru(0) - g = 0, \quad \forall r, \qquad \text{i.e.,} \quad u = g \text{ at } x = 0 \tag{1.84}$$

On the other hand, the second inequality in (1.82) yields the minimization problem

$$F(v) - pv(0) \geq F(u) - pu(0) \tag{1.85}$$

Taking $v = u \pm \varepsilon w$, $\varepsilon > 0$ in (1.85), that is, taking the first variation, we have

$$\int_0^1 k\frac{du}{dx}\frac{dw}{dx}\,dx - pw(0) = \int_0^1 fw\,dx, \quad \forall w \tag{1.86}$$

Applying integration by parts, we have Euler's equation

$$-\frac{d}{dx}\left(k\frac{du}{dx}\right) = f \text{ in } (0, 1)$$

$$-k\frac{du}{dx} = p \text{ at } x = 0, \qquad k\frac{du}{dx} = 0 \text{ at } x = 1$$

Thus, the Lagrange multiplier p is the *heat flux* at the boundary $x = 0$.

If an inequality constraint $u - \bar{u} \leq 0$ is assumed in (0, 1) for a given function $\bar{u}$ such that $g \leq \bar{u}$ at $x = 0$, the Lagrange multiplier p has to be restricted by $p < 0$ and the Lagrangian is defined by

$$L(v, q) = F(v) - \int_0^1 q(v - \bar{u})\, dx, \quad \forall q \leq 0$$

Note that v and u are free from the constraint $v - \bar{u} \leq 0$ and $u - \bar{u} \leq 0$. Thus, there are no restrictions on the variation with respect to u, although the Lagrange multiplier is restricted by $p \leq 0$. The first inequality of (1.82) yields

$$\int_0^1 (q - p)(u - \bar{u})\, dx \geq 0, \quad q \leq 0$$

that is,

$$p \leq 0, \quad u - \bar{u} \leq 0 \qquad p(u - \bar{u}) = 0 \quad \text{in } (0, 1) \tag{1.87}$$

The second inequality of (1.82) implies

$$\int_0^1 \left(k\frac{du}{dx}\frac{dw}{dx} - pw\right) dx = \int_0^1 fw\, dx, \quad \forall w, w(0) = 0$$

The local form of this integral identity can be obtained as

$$-\frac{d}{dx}\left(k\frac{du}{dx}\right) - p = f \quad \text{in } (0, 1)$$
$$u = g \quad \text{at } x = 0, \quad k\frac{du}{dx} = 0 \quad \text{at } x = 1 \tag{1.88}$$

It follows from (1.87) that $p = 0$ if $u < \bar{u}$ and p needs not be zero if $u = \bar{u}$. That is, if the solution reaches to the upper bound $\bar{u}$ at a point x, the Lagrange multiplier p becomes *active*. If the strict inequality $u < \bar{u}$ is satisfied, p is *nonactive*.

Exercise 1.12: Let us consider a minimization problem for a functional

$$F(w) = \tfrac{1}{2}\int_0^1 \left[EI\left(\frac{d^2w}{dx^2}\right)^2 + kw^2\right] dx - \int_0^1 fw\, dx$$

If a new function M is defined by

$$M = EI\frac{d^2w}{dx^2} \tag{1.89}$$

the original functional $F(w)$ is written as

$$\hat{F}(w, M) = \tfrac{1}{2}\int_0^1 \left(\frac{M^2}{EI} + kw^2\right) dx - \int_0^1 fw\, dx$$

and the minimization problem becomes a problem with respect to two variables w and M under a subsidiary condition (1.89). If a Lagrange multiplier p is introduced to make an unconstrained problem, a corresponding Lagrangian becomes

$$L(w, M, p) = \hat{F}(w, M) - \int_0^1 p\left(\frac{M}{EI} - \frac{d^2w}{dx^2}\right) dx$$

Assuming the boundary condition on the Lagrange multiplier p such that

$$p(0) = p(1) = 0$$

integration by parts on the last term of the Lagrangian yields

$$L(w, M, p) = \hat{F}(w, M) - \int_0^1 \left(p\frac{M}{EI} + \frac{dp}{dx}\frac{dw}{dx}\right) dx$$

Obtain similar expressions to (1.84) and (1.86) for the Lagrangian $L(w, M, p)$ starting from a saddle point problem similar to (1.82):

$$L(w, M, q) \leq L(w, M, p) \leq L(v, N, p), \quad \forall(v, N, q)$$

Using the relation $p = M$, which represents one of three equations obtained by the saddle point problem, rewrite the other two equations in terms of w and M.

1.4.3 Penalty methods

There are other ways to derive an unconstrained minimization problem to a constrained problem, which is subject to subsidiary conditions such as essential boundary conditions. The exterior penalty method is such a method. We shall briefly explain this using the minimization problem (1.68), which is constrained by the boundary condition $v(0) = g$.

The first step is to introduce a penalty functional $P(v)$ such that:

i. $P(v) = 0$ if and only if subsidiary conditions are exactly satisfied;
ii. $P(v) \geq 0$ and P is a continuous convex functional.

If a functional P satisfies inequality $P((1-\theta)u + \theta v) \leq (1-\theta)P(u) + \theta P(v)$ for every $\theta \in [0, 1]$, P is said to be convex. Now, if the condition $v(0) = g$ is concerned, a functional defined by

$$P(v) = \tfrac{1}{2}[v(0) - g]^2 \tag{1.90}$$

is clearly a penalty functional.

The second step is to define an unconstrained minimization problem to find u_ε such that

$$F_\varepsilon(u_\varepsilon) \leq F_\varepsilon(v), \quad \forall v \tag{1.91}$$

where

$$F_\varepsilon(v) = F(v) + \frac{1}{\varepsilon} P(v) \tag{1.92}$$

for a sufficiently small positive number ε. The weak form derived from (1.91) is

$$\int_0^1 \left(k \frac{du_\varepsilon}{dx} \frac{dw}{dx} - fw \right) dx - hw(1) + \frac{1}{\varepsilon} [u(0) - g] w(0) = 0, \quad \forall w \tag{1.93}$$

This then yields Euler's equation

$$-\frac{d}{dx}\left(k \frac{du_\varepsilon}{dx} \right) = f \text{ in } (0, 1)$$

$$k \frac{du_\varepsilon}{dx}(1) = h, \qquad k \frac{du_\varepsilon}{dx}(0) = \frac{1}{\varepsilon} [u_\varepsilon(0) - g] \tag{1.94}$$

Since the heat flux at the boundary is finite, it can be expected by the third equation in (1.94) that

$$u_\varepsilon(0) - g = \varepsilon \left(k \frac{du_\varepsilon}{dx}(0) \right) \to 0 \text{ as } \varepsilon \to 0$$

That is, $u_\varepsilon(0) - g = O(\varepsilon)$. If a constant ε is sufficiently small, the boundary condition $u(0) = g$ is approximately satisfied in the unconstrained minimization problem (1.91). A formal proof for the above consequence is as follows:

$$F_\varepsilon(u_\varepsilon) = \min_v F_\varepsilon(v) \leq \min_{v, v(0) = g} F_\varepsilon(v) = F(u) \tag{1.95}$$

that is,

$$P(u_\varepsilon) \leq \varepsilon [F(u) - F(u_\varepsilon)]$$

Since $F(u)$ and $F(u_\varepsilon)$ are finite values for any ε, the right side goes to zero as $\varepsilon \to 0$. Because of $P(v) \geq 0$, we have

$$\lim_{\varepsilon \to 0} P(u_\varepsilon) = 0$$

If a sequence $\{u_\varepsilon\}$ converges to a function $\hat{u}$, then $P(\hat{u}) = 0$; that is, $\hat{u}(0) = g$, since P is assumed to be continuous. The remaining question is to show that $\hat{u} = u$, that is, that $\hat{u}$ is a minimizer of the original problem. Noting that $P(v) \geq 0$ and $\varepsilon \geq 0$, (1.95) yields

$$F(u_\varepsilon) \leq F(u)$$

Since u_ε is assumed to converge to $\hat{u}$, we have $F(\hat{u}) \leq F(u)$; that is, $\hat{u}$ is also a minimizer of a functional F under the condition $\hat{u}(0) = g$.

The last remark is a relation between the Lagrange multiplier and penalty methods. If

$$p_\varepsilon = -\frac{1}{\varepsilon} [u_\varepsilon(0) - g] \tag{1.96}$$

is defined, the weak form (1.93) suggests that p_ε is an approximation of the Lagrange multiplier. Indeed, if p_ε converges to a function $\hat{p}$ as $\varepsilon \to 0$, it can be shown that $\hat{p}$ is the Lagrange multiplier to the constraint $u(0) - g = 0$.

Exercise 1.13: For a functional

$$F(w) = \frac{1}{2} \int_0^L \left[EI \left(\frac{d^2 w}{dx^2} \right)^2 + kw^2 \right] dx - \int_0^L fw \, dx$$

Let us consider that the relation

$$\theta = \frac{dw}{dx}$$

is a subsidiary condition. Then the original functional becomes

$$\hat{F}(w, \theta) = \frac{1}{2} \int_0^L \left[EI \left(\frac{d\theta}{dx} \right)^2 + kw^2 \right] dx - \int_0^L fw \, dx$$

Applying the penalty method for a penalty functional

$$P(w, \theta) = \frac{1}{2} \int_0^L \left(\theta - \frac{dw}{dx} \right)^2 dx$$

derive the weak form and Euler's equation.

2
FINITE ELEMENT ANALYSIS OF HEAT CONDUCTION PROBLEMS

A simple, but nontrivial, application of finite element methods is to a class of two-dimensional heat transfer problems. In this chapter, we shall derive a finite element approximation to a heat conduction problem in a plane domain by using the simplest finite element, the three-node triangular element known as the constant-strain triangular (CST) element.

2.1 Heat conduction problems

Let Ω be a bounded domain in which the temperature field T is defined. This may be a solid body or the bounded flow domain of a fluid. Let us consider an infinitesimal element Δa in a plane domain Ω (i.e., $\mathbb{R}^2$) as shown in Figure 2.1.

Let $\mathbf{q}(x, y) = q_x(x, y)\mathbf{i} + q_y(x, y)\mathbf{j}$ be the heat flux vector at an arbitrary point P of Ω. Then, during an infinitesimal time increment Δt, the balance of heat flow for the differential element $\Delta a = \Delta x\, \Delta y$ is given by

$$\begin{aligned} Q\,\Delta t = \{&[q_x(x - \tfrac{1}{2}\Delta x, \bar{y}) - q_x(x + \tfrac{1}{2}\Delta x, \bar{\bar{y}})]\,\Delta y \\ &+ [q_y(\bar{x}, y - \tfrac{1}{2}\Delta y) - q_y(\bar{\bar{x}}, y + \tfrac{1}{2}\Delta y)]\,\Delta x + f\,\Delta x\,\Delta y\}\,\Delta t \end{aligned} \tag{2.1}$$

where f is the heat source (if any) and Q is the increase in the amount of heat in the element Δa during a unit time interval. If we divide by $\Delta x\, \Delta y\, \Delta t$ and pass to the limit as Δx and $\Delta y \to 0$, we obtain the relation

$$\lim_{\Delta x, \Delta y \to 0} \frac{Q}{\Delta x\, \Delta y} = -\frac{\partial q_x}{\partial x} - \frac{\partial q_y}{\partial y} + f \tag{2.2}$$

If we introduce *Fourier's law* of heat conduction:

$$\begin{aligned} q_x &= -k_{xx}\frac{\partial T}{\partial x} - k_{xy}\frac{\partial T}{\partial y} \\ q_y &= -k_{yx}\frac{\partial T}{\partial x} - k_{yy}\frac{\partial T}{\partial y} \end{aligned} \qquad (\text{or } q_i = -k_{ij}T_{,j}) \tag{2.3}$$

where the tensor $\mathbf{k} = k_{ij}\mathbf{i}_i\mathbf{i}_j$ represents the heat conductivity of the body. Using the relationship

$$Q = \rho c_p \dot{T}\,\Delta x\,\Delta y \tag{2.4}$$

where c_p is the heat capacity at constant pressure and ρ is the density, we obtain the equation of heat conduction

$$\rho c_p \dot{T} = \frac{\partial}{\partial x}\left(k_{xx}\frac{\partial T}{\partial x} + k_{xy}\frac{\partial T}{\partial y}\right) + \frac{\partial}{\partial y}\left(k_{yx}\frac{\partial T}{\partial x} + k_{yy}\frac{\partial T}{\partial y}\right) + f \quad \text{in } \Omega \tag{2.5}$$

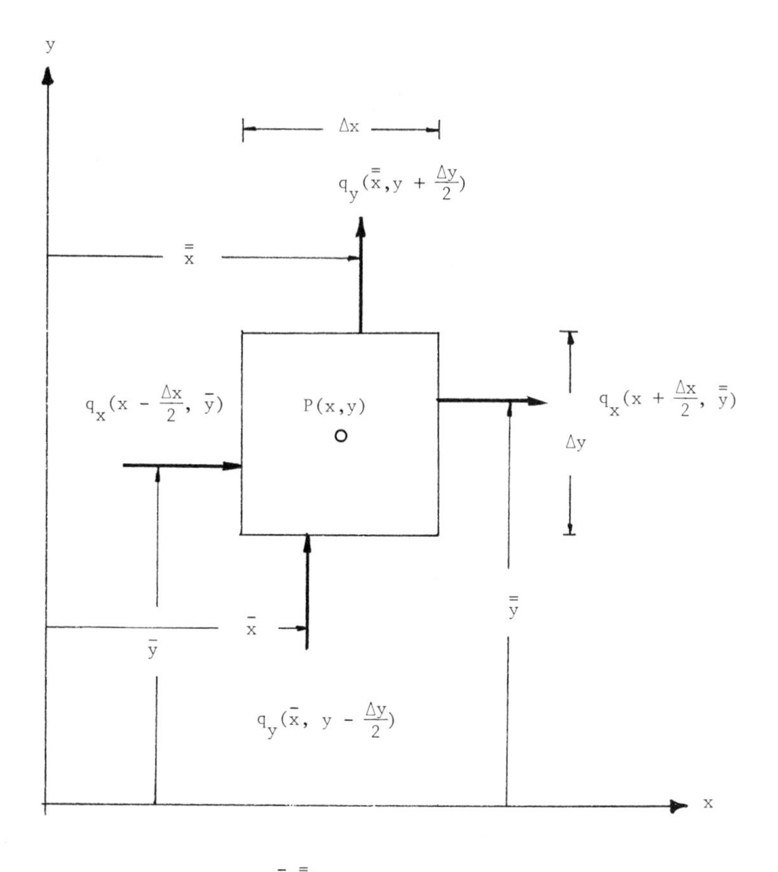

Figure 2.1 Balance of heat flux in a differential element.

Here $\dot{T}$ is the material derivative of the temperature field defined by

$$\dot{T} = \frac{\partial T}{\partial t} + u_x \frac{\partial T}{\partial x} + u_y \frac{\partial T}{\partial y} \tag{2.6}$$

if fluid bodies are considered (velocity components, u_x, u_y), and

$$\dot{T} = \frac{\partial T}{\partial t} \tag{2.7}$$

if solids are considered. If index notation is used, (2.5) becomes

$$\rho c_p \dot{T} = (k_{ij} T_{,j})_{,i} + f \quad \text{in } \Omega \tag{2.8}$$

whereas in *coordinate-free* notation, it reads:

$$\rho c_p \dot{T} = \text{div}(\mathbf{k} \text{ grad } T) + f \quad \text{in } \Omega \tag{2.9}$$

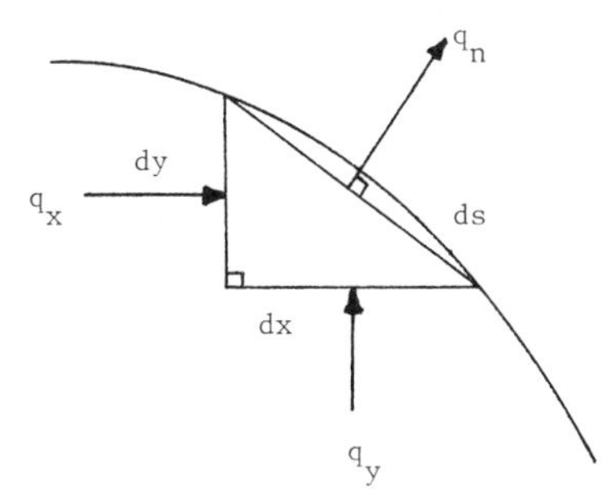

Figure 2.2 Flux along the boundary.

In analyses of heat conduction problems, we usually assume symmetry of **k**; that is,

$$k_{ij} = k_{ji} \quad \text{(i.e., } k_{xy} = k_{yx}\text{)} \tag{2.10}$$

and that **k** is positive definite:

$$k_{ij}X_jX_i \geq kX_iX_i \quad \text{in } \Omega \tag{2.11}$$

for some positive constant $k > 0$. Conditions (2.10) and (2.11) ensure well-posedness of the problem of heat conduction.

Equation (2.5), or equivalently (2.9), governs the conduction of heat at interior points of the domain. We now derive a condition to be satisfied at points on the boundary where the heat flux is specified. To do this, consider the differential element on the boundary as shown in Figure 2.2. Let q_n be the heat flux from the boundary per unit length, and let ds be the differential arc length corresponding to dx and dy. Then the balance of heat flow yields

$$q_n\, ds = q_x\, dy + q_y\, dx \tag{2.12}$$

Noting that if $\mathbf{n} = n_x\mathbf{i} + n_y\mathbf{j}$ is the unit vector outward normal to the boundary, so that

$$n_x = \frac{dy}{ds}, \qquad n_y = \frac{dx}{ds} \tag{2.13}$$

we obtain

$$q_n = q_x n_x + q_y n_y \tag{2.14}$$

Using index notation,

$$q_n = q_i n_i = -(k_{ij}T_{,j})n_i \tag{2.15}$$

Suppose that the inflow of heat is specified on some part Γ_2 of the boundary by the function h. Then we must satisfy

$$(k_{ij}T_{,j})n_i = h \quad \text{on } \Gamma_2 \tag{2.16}$$

Another commonly used condition on the boundary is that the outflow of heat is proportional to the difference between the body's surface temperature and the room temperature, T_∞, say. If we impose this condition on a part Γ_3 of the boundary, we have

$$k_0(T - T_\infty) = -(k_{ij}T_{,j})n_i \quad \text{on } \Gamma_3 \tag{2.17}$$

We can also specify the surface temperature of the body on a part Γ_1 of the boundary as

$$T = g \quad \text{on } \Gamma_1 \tag{2.18}$$

where g is the given temperature.

Thus, a mathematical description of the heat conduction problem is

$$\begin{aligned} \rho c_p \dot{T} &= (k_{ij}T_{,j})_{,i} + f \quad \text{in } \Omega \\ T &= g \quad \text{on } \Gamma_1 \\ (k_{ij}T_{,j})n_i &= h \quad \text{on } \Gamma_2 \\ (k_{ij}T_{,j})n_i &= -k_0(T - T_\infty) \quad \text{on } \Gamma_3 \end{aligned} \tag{2.19}$$

under the assumption that the boundary Γ of the domain Ω consists of three disjoint segments Γ_1, Γ_2, and Γ_3. Finally, if the *initial condition*

$$T = T_0 \quad \text{at } t = 0 \tag{2.20}$$

is given, the problem [(2.19), (2.20)] is called the *initial boundary value problem* of heat conduction. The *steady-state solution* (if there is one) of (2.19) and (2.20) is governed by the *boundary value problem:*

$$\begin{aligned} -(k_{ij}T_{,j})_{,i} &= f \quad \text{in } \Omega \\ T &= g \quad \text{on } \Gamma_1 \\ (k_{ij}T_{,j})n_i &= h \quad \text{on } \Gamma_2 \\ (k_{ij}T_{,j})n_i &= -k_0(T - T_\infty) \quad \text{on } \Gamma_3 \end{aligned} \tag{2.21}$$

Exercise 2.1: Suppose that $k_{ij} = k\delta_{ij}$. If $k_0 \to +\infty$ while k is held fixed in (2.17), we have

$$T - T_\infty \to 0$$

Explain either mathematically or physically (or both) why this happens. Note that this suggests a form of a generalized third type of boundary condition,

$$(k_{ij}T_{,j})n_i = -k_0(T - g) + h$$

that contains both the first and second types of boundary conditions. Show that if $h = 0$ and $k_0 \to \infty$, we have the first type of boundary condition $T = g$. Further, if $k_0 = 0$, then the second type of boundary condition is obtained.

2.2 Weak form for finite element approximations

We illustrate with the steady-state problem (2.21). A standard procedure of finite element methods in the solution of physical problems described by a system of differential equations is as follows: The first step is to change the boundary value problem (2.21) to a more general and simpler form for finite element approximations. If finite difference methods are used, the first step is not necessary since finite difference schemes are obtainable from the boundary value problems themselves.

This first step, which leads to the *weak form* (or the *principle of virtual work* in structural mechanics), converts the local description (or pointwise description) to a global integral form. To do this, we multiply both sides of the first equation of (2.21) by an arbitrary function $\bar{T}$, which is akin to a virtual displacement in structural mechanics, and integrate over the domain Ω. We get

$$\int_\Omega [-(k_{ij}T_{,j})_{,i}]\bar{T}\, d\Omega = \int_\Omega f\bar{T}\, d\Omega \tag{2.22}$$

Note the results, obtained by integration by parts:

$$\begin{aligned} \int_\Omega \frac{\partial \phi}{\partial x}\psi\, d\Omega &= \int_\Omega \phi\left(-\frac{\partial \psi}{\partial x}\right) d\Omega + \int_\Gamma (\phi n_x)\psi\, d\Gamma \\ \int_\Omega \frac{\partial \phi}{\partial y}\psi\, d\Omega &= \int_\Omega \phi\left(-\frac{\partial \psi}{\partial y}\right) d\Omega + \int_\Gamma (\phi n_y)\psi\, d\Gamma \end{aligned} \tag{2.23a}$$

for *continuous* functions ϕ and ψ, defined on Ω and its boundary Γ. If the index notation is applied, one can write

$$\int_\Omega \phi_{,i}\psi\, d\Omega = \int_\Omega \phi(-\psi_{,i})\, d\Omega + \int_\Gamma (\phi n_i)\psi\, d\Gamma, \quad i = 1, 2 \tag{2.23b}$$

for *continuous* functions ϕ and ψ. If continuity is not assumed, (2.23) is no longer valid. In this treatment, we shall assume continuity of the functions discussed. We sometimes say that C^0-continuity is assumed since continuity of the derivatives is not assumed.

Applying (2.23) to (2.22) yields

$$\int_\Omega k_{ij}T_{,j}\bar{T}_{,i}\, d\Omega - \int_\Gamma k_{ij}T_{,j}n_i\bar{T}\, d\Gamma = \int_\Omega f\bar{T}\, d\Omega \tag{2.24}$$

Note that

$$\int_\Gamma \eta\, d\Gamma = \int_{\Gamma_1} \eta\, d\Gamma + \int_{\Gamma_2} \eta\, d\Gamma + \int_{\Gamma_3} \eta\, d\Gamma$$

for the case that Γ consists of three disjoint parts Γ_1, Γ_2, and Γ_3. Now if we look at the boundary conditions in (2.21), we notice that the second term of (2.24) can be changed as follows:

$$\int_\Gamma k_{ij}T_{,j}n_i\bar{T}\, d\Gamma = \int_{\Gamma_1} (k_{ij}T_{,j})n_i\bar{T}\, d\Gamma + \int_{\Gamma_2} h\bar{T}\, d\Gamma - \int_{\Gamma_3} k_0(T - T_\infty)\bar{T}\, d\Gamma$$

Remember that the temperature T is specified only on the portion Γ_1 of the boundary. We call the boundary condition $T = g$ on Γ_1 the *essential* (or *kinematic*, in structural mechanics) condition. Although the function $\bar{T}$ multiplying the differential equation is arbitrary in Ω, it is not arbitrary on (all of) the boundary Γ and hence is required to satisfy some conditions there. It must be realized that on Γ_1 the nature of the condition differs from those on the other parts Γ_2 and Γ_3 where the temperature T is unknown. What restrictions are put on a virtual displacement in structural mechanics? If a portion of the boundary is fixed, the virtual displacement must be zero on this portion, since a virtual displacement is a variation of the real displacement and is arbitrary except that it must not violate the kinematic constraints. The same idea holds in heat conduction problems. Since $\bar{T}$ is a virtual temperature field, it is a variation of the true temperature field T. Since $T = g$ on Γ_1, $\bar{T}$ is arbitrary except in Γ_1, where it must be zero. Hence, we have to impose on $\bar{T}$ the condition

$$\bar{T} = 0 \quad \text{on } \Gamma_1 \tag{2.25}$$

Thus, the boundary term becomes

$$\int_\Gamma k_{ij}T_{,j}n_i\bar{T}\,d\Gamma = \int_{\Gamma_2} h\bar{T}\,d\Gamma + \int_{\Gamma_3} k_0 T_\infty \bar{T}\,d\Gamma - \int_{\Gamma_3} k_0 T\bar{T}\,d\Gamma$$

Substitution of this into (2.24) yields the form

$$\int_\Omega k_{ij}T_{,j}\bar{T}_{,i}\,d\Omega + \int_{\Gamma_3} k_0 T\bar{T}\,d\Gamma = \int_\Omega f\bar{T}\,d\Omega + \int_{\Gamma_2} h\bar{T}\,d\Gamma + \int_{\Gamma_3} k_0 T_\infty \bar{T}\,d\Gamma \tag{2.26}$$

for an *arbitrary continuous function* $\bar{T}$ such that $\bar{T} = 0$ on Γ_1.

Thus far, we have changed the boundary value problem (2.21) of heat conduction to the *weak form*

$$T = g \quad \text{on } \Gamma_1$$

$$\int_\Omega k_{ij}T_{,j}\bar{T}_{,i}\,d\Omega + \int_{\Gamma_3} k_0 T\bar{T}\,d\Gamma = \int_\Omega f\bar{T}\,d\Omega + \int_{\Gamma_2} h\bar{T}\,d\Gamma + \int_{\Gamma_3} k_0 T_\infty \bar{T}\,d\Gamma, \quad \forall \bar{T} \ni \bar{T} = 0 \text{ on } \Gamma_1 \tag{2.27}$$

Note that all heat flux conditions in (2.21) are included in the integral form (2.27) and that only first derivatives of the temperature are involved in (2.27) whereas second derivatives of T are present in the differential equation (2.21). This weak form has many advantages, when approximations are considered, compared to finite difference methods in one dimension. The first derivative $d\phi/dx$ is approximated by $(\phi_n - \phi_{n-1})/\Delta x$, the second derivative $d^2\phi/dx^2$ by $(\phi_{n+1} - 2\phi_n + \phi_{n-1})/\Delta x^2$, and so on, and as the order of the derivative increases, more nodal points need to be considered. It is known that boundary conditions involving the heat flux are difficult to deal with in the finite difference methods usually studied. It turns out that the form (2.27) avoids these difficulties. We base our finite element approximations on the weak form (2.27).

Exercise 2.2: Derive the weak form (2.27) without using index notation.

For comparison purposes, we comment here on the relation between the weak form (2.27) and a variational form based on the *principle of minimum "potential energy."* If the heat conductivity is symmetric, that is, if

$$k_{ij} = k_{ji}$$

it is possible to define a *total potential energy* for the heat conduction problem (2.21) by

$$F(T) = \tfrac{1}{2}\int_\Omega k_{ij}T_{,j}T_{,i}\,d\Omega + \tfrac{1}{2}\int_{\Gamma_3} k_0 T^2\,d\Gamma - \int_\Omega fT\,d\Omega - \int_{\Gamma_2} hT\,d\Gamma - \int_{\Gamma_3} k_0 T_\infty T\,d\Gamma$$

The principle of minimum potential energy is then stated as

$$F(T) = \min_{\tilde{T}} F(\tilde{T}), \quad \text{subjected to } \tilde{T} = g \text{ on } \Gamma_1 \tag{2.28}$$

and then,

$$F(T) \le F(T \pm \theta(\hat{T} - T)), \quad \forall \hat{T} \ni \hat{T} = g \text{ on } \Gamma_1, \text{ and } \forall \theta \in (0, 1)$$

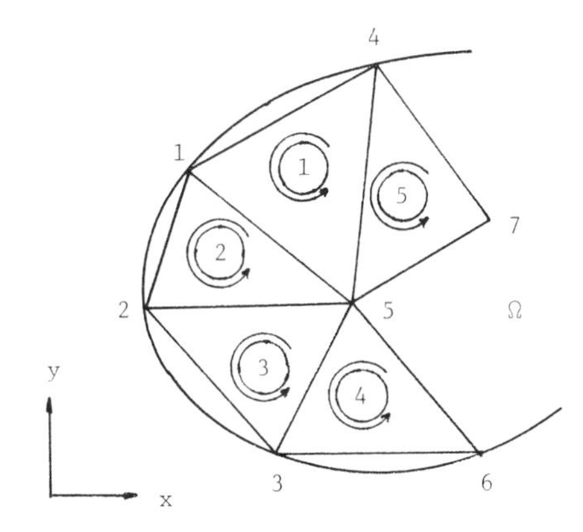

Figure 2.3 Discretization of the domain Ω by three-node triangular elements.

Expanding this inequality and passing to the limit $\theta \to 0$, we have

$$\pm \int_{\Omega} k_{ij} T_{,j} (\hat{T} - T)_{,i} \, d\Omega \pm \int_{\Gamma_3} k_0 T (\hat{T} - T) \, d\Gamma$$
$$\geq \pm \int_{\Omega} f (\hat{T} - T) \, d\Omega \pm \int_{\Gamma_2} h (\hat{T} - T) \, d\Gamma \pm \int_{\Gamma_3} k_0 T_{\infty} (\hat{T} - T) \, d\Gamma$$

Using the upper and then the lower signs allows us to conclude that

$$\int_{\Omega} k_{ij} T_{,j} (\hat{T} - T)_{,i} \, d\Omega + \int_{\Gamma_3} k_0 T (\hat{T} - T) \, d\Gamma$$
$$= \int_{\Omega} f (\hat{T} - T) \, d\Omega + \int_{\Gamma_2} h (\hat{T} - T) \, d\Gamma + \int_{\Gamma_3} k_0 T_{\infty} (\hat{T} - T) \, d\Gamma \quad (2.29)$$

Since $\hat{T}$ is arbitrary, $\bar{T} = \hat{T} - T$ is also arbitrary, but $\bar{T} = 0$ on Γ_1. Thus, (2.29) is the same as the weak form (2.27). The derivation of (2.29) from (2.28) is obviously equivalent to taking the first variation of F at the solution T, and $\bar{T}$ is, in general, denoted by δT in the variational calculus.

2.3 Finite element discretization

The second step of the finite element approximation is the discretization of the domain Ω and the functions T and $\bar{T}$ in the weak form (2.27). We shall use the three-node triangular element, which is known as the CST element in the area of structural mechanics.

We first consider the discretization of the domain Ω. Let Ω be divided into E three-node triangular elements, and let Ω_e be a typical one of these elements. It is preferable to use smaller elements in those portions of Ω where the gradient of the temperature field is expected to be large. After this discretization of the domain, we have to identify each element in some way. To do this, we number each element as shown in Figure 2.3. The numbering must be sequential so that each element can be distinguished from the others by its unique number. We call these numbers the *element numbers*. The next job is to be able to identify an element by giving the numerical labels of its three vertices, which are called its nodal points. For example, in Figure 2.3, the element 5 (or Ω_5) consists of three vertices 4, 5, and 7. Note that nodal points (or vertices of triangular elements) also have sequential numbering. We call these the *node numbers*. Thus, if we know the coordinates of each of its nodal points, the geometry of a particular element is defined. The *element connectivity* of a finite element is simply the sequence consisting of its (three) node numbers. Hence, for example, the three node numbers 4, 5, and 7 constitute the element connectivity of the finite element 5. The element connectivity of the element 3 is 5, 2, and 3. We store the connectivities of the elements in the array IJK(J,NEL), J = 1, 2, 3, NEL = 1, . . . , E, where $J = 1, 2, 3$ indicates the ordering of the connectivity and NEL is the element number. For convenience, the ordering of the node numbers in the connectivity is defined by starting with (any) one and proceeding around the element in the counterclockwise direction. The contents of the array IJK for the example shown in Figure 2.3 are as follows:

NEL	IJK(1,NEL)	IJK(2,NEL)	IJK(3,NEL)
1	1	5	4
2	5	1	2
3	5	2	3
4	6	5	3
⋮	⋮	⋮	⋮
E			

If the coordinates of each nodal point are stored in the array X(NODE), Y(NODE), NODE = 1, 2, . . . , N, where N is the number of the last nodal point, the *geometry* of an element Ω_e for e = NEL can be obtained by the coordinates of the three nodal points that belong to the element. Let the arrays XE(IA), YE(IA), IA = 1, 2, 3, store the coordinates of these three nodal points.

Thus, for a particular element Ω_e, e = NEL, the geometry of the element is given by the routine

```
      DO 100 IA = 1, NNODE
      IJKIA = IJK(IA,NEL)
      XE(IA) = X(IJKIA)
  100 YE(IA) = Y(IJKIA)
```
(2.30)

for the total number NNODE = 3 of nodal points of an element.

We call the arrays XE and YE the *element coordinates*. Once we know XE(IA) and YE(IA), for example, the area of the element is computed by

$$\text{AREA} = \tfrac{1}{2}\det\begin{bmatrix} 1 & 1 & 1 \\ \text{XE(1)} & \text{XE(2)} & \text{XE(3)} \\ \text{YE(1)} & \text{YE(2)} & \text{YE(3)} \end{bmatrix}$$

It must be realized that the index I refers to coordinate axis: IA to the numbering of three nodal points of the element connectivity; and NODE indicates the node number in the global system $\Omega = \bigcup_{e=1}^{E} \Omega_e$.

Exercise 2.3: Consider the L-shaped domain Ω shown in Figure 2.4, and divide the domain into the 16 elements shown. Define the element connectivity IJK of each finite element Ω_e, $e = 1, 2, \ldots, 16$. Also define the coordinates X and Y of each nodal point. After specifying IJK, X, and Y, determine the element coordinates XE and YE by using the routine (2.30) described above.

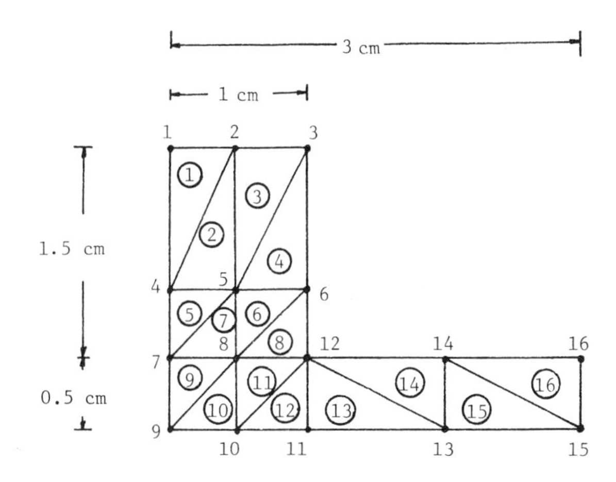

Figure 2.4 L-shaped domain for exercise 2.3.

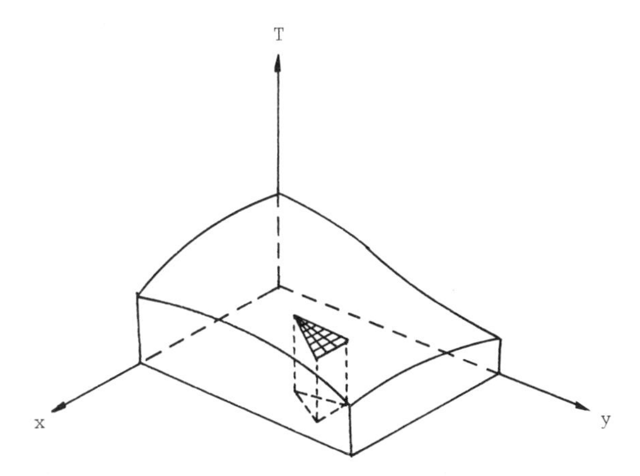

Figure 2.5 Surface of the function T.

We have so far considered the discretization of the domain into finite elements. We now discuss the discretization of the functions T and $\bar{T}$ that appear in the weak form (2.27). This part is known as the *interpolation* of a function. Suppose that T and $\bar{T}$ are *continuous* functions of x and y (or x_1 and x_2 in index notation). Then, if the function T is plotted in a three-dimensional manner, T forms a smooth curved surface on the domain Ω, as shown in Figure 2.5. Let us consider an element Ω_e of the domain Ω and approximate the portion of the surface above Ω_e by a linear function of x and y [for (x, y) in Ω_e] in the sense that the values of the linear function coincide with those of the original function T at each of the three vertices of the triangular element Ω_e. Thus, if we denote the linear function by T_h on Ω_e, we have

$$T_h(x_\alpha, y_\alpha) = T(x_\alpha, y_\alpha), \quad \alpha = 1, 2, 3 \tag{2.31}$$

where (x_α, y_α) are the coordinates of the α-th nodal point in Ω_e. If we use the notation employed in the discretization of the domain Ω,

$$x_1 = \mathrm{XE}(1), \qquad y_1 = \mathrm{YE}(1)$$
$$x_2 = \mathrm{XE}(2), \qquad y_2 = \mathrm{YE}(2)$$
$$x_3 = \mathrm{XE}(3), \qquad y_3 = \mathrm{YE}(3)$$

Since T_h is assumed to be linear on Ω_e, T_h has the form

$$T_h = d_0 + d_1 x + d_2 y \tag{2.32}$$

where the constants d_0, d_1, and d_2 are chosen so that the conditions (2.31) are satisfied. Thus, if the values of the function T at the three vertices are given as T_1, T_2, and T_3, that is,

$$T_\alpha = T(x_\alpha, y_\alpha), \quad \alpha = 1, 2, 3 \tag{2.33}$$

we have

$$T_h(x, y) = \left(\sum_{\alpha=1}^{3} T_\alpha N_\alpha(x, y)\right) = T_\alpha N_\alpha(x, y) \tag{2.34}$$

on the element Ω_e, where

$$N_\alpha(x, y) = a_\alpha + b_\alpha x + c_\alpha y \tag{2.35}$$

and

$$\{a_\alpha\} = \frac{1}{J}\begin{Bmatrix} x_2 y_3 - x_3 y_2 \\ x_3 y_1 - x_1 y_3 \\ x_1 y_2 - x_2 y_1 \end{Bmatrix}, \qquad \{b_\alpha\} = \frac{1}{J}\begin{Bmatrix} y_2 - y_3 \\ y_3 - y_1 \\ y_1 - y_2 \end{Bmatrix}$$

$$\{c_\alpha\} = \frac{1}{J}\begin{Bmatrix} x_3 - x_2 \\ x_1 - x_3 \\ x_2 - x_1 \end{Bmatrix}, \qquad J = 2 * \mathrm{AREA}$$

To see this, note that (2.32) and (2.33) yield the relations

$$T_\alpha = d_0 + d_1 x_\alpha + d_2 y_\alpha, \quad \alpha = 1, 2, 3$$

which can be written as

$$\begin{Bmatrix} T_1 \\ T_2 \\ T_3 \end{Bmatrix} = \begin{bmatrix} 1 & x_1 & y_1 \\ 1 & x_2 & y_2 \\ 1 & x_3 & y_3 \end{bmatrix} \begin{Bmatrix} d_0 \\ d_1 \\ d_2 \end{Bmatrix}$$

Solving this system of equations, we have

$$\begin{Bmatrix} d_0 \\ d_1 \\ d_2 \end{Bmatrix} = \begin{bmatrix} a_1 & a_2 & a_3 \\ b_1 & b_2 & b_3 \\ c_1 & c_2 & c_3 \end{bmatrix} \begin{Bmatrix} T_1 \\ T_2 \\ T_3 \end{Bmatrix}$$

Hence we find

$$T_h(x, y) = (a_1 + b_1 x + c_1 y)T_1 + (a_2 + b_2 x + c_2 y)T_2 + (a_3 + b_3 x + c_3 y)T_3$$

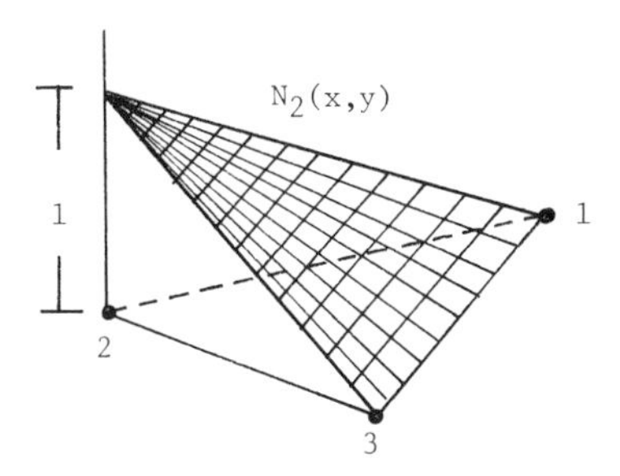

Figure 2.6 Example of a shape function.

which is equation (2.34). It is clear that

$$\sum_{\alpha=1}^{3} N_\alpha(x, y) = 1, \qquad N_\alpha(x_\beta, y_\beta) = \delta_{\alpha\beta} \tag{2.36}$$

A plot of $N_2(x, y)$ is given in Figure 2.6. We call the functions N_α the shape functions for the finite element Ω_e. Conditions (2.36) also imply

$$x = x_\alpha N_\alpha(x, y), \qquad y = y_\alpha N_\alpha(x, y) \tag{2.37}$$

From (2.35), we have also the relations

$$\frac{\partial N_\alpha}{\partial x} = b_\alpha, \qquad \frac{\partial N_\alpha}{\partial y} = c_\alpha \tag{2.38a}$$

Using index notation, these become

$$N_{\alpha,1} = b_\alpha \qquad \text{and} \qquad N_{\alpha,2} = c_\alpha \tag{2.38b}$$

If the array DN(IA,I), IA = 1, 2, 3, I = 1, 2, is introduced for the first derivatives of the shape functions $\{N_\alpha(x, y)\}$, we can construct DN(IA,I) by using the geometry of the element Ω_e. Indeed, we have

```
DN(1,1) = (YE(2) − YE(3))/(2. * AREA)
DN(1,2) = (XE(3) − XE(2))/(2. * AREA)
DN(2,1) = (YE(3) − YE(1))/(2. * AREA)
DN(2,2) = (XE(1) − XE(3))/(2. * AREA)          (2.39)
DN(3,1) = (YE(1) − YE(2))/(2. * AREA)
DN(3,2) = (XE(2) − XE(1))/(2. * AREA)
```

Note that the first derivatives of the shape functions are constants if three-node triangular elements are considered. This yields constant-gradient (strain in stress analysis) fields in each element. Thus, we call such an element a *CST element*.

The first derivatives of $T_h(x, y)$ are piecewise constant in Ω and can be computed by

$$\frac{\partial T_h}{\partial x} = T_\alpha \frac{\partial N_\alpha}{\partial x} \qquad (\text{or } T_{h,1} = T_\alpha N_{\alpha,1})$$

$$\frac{\partial T_h}{\partial y} = T_\alpha \frac{\partial N_\alpha}{\partial y} \qquad (\text{or } T_{h,2} = T_\alpha N_{\alpha,2}) \tag{2.40a}$$

If full index notation is used, we have

$$T_{h,i} = T_\alpha N_{\alpha,i}, \quad i = 1, 2 \tag{2.40b}$$

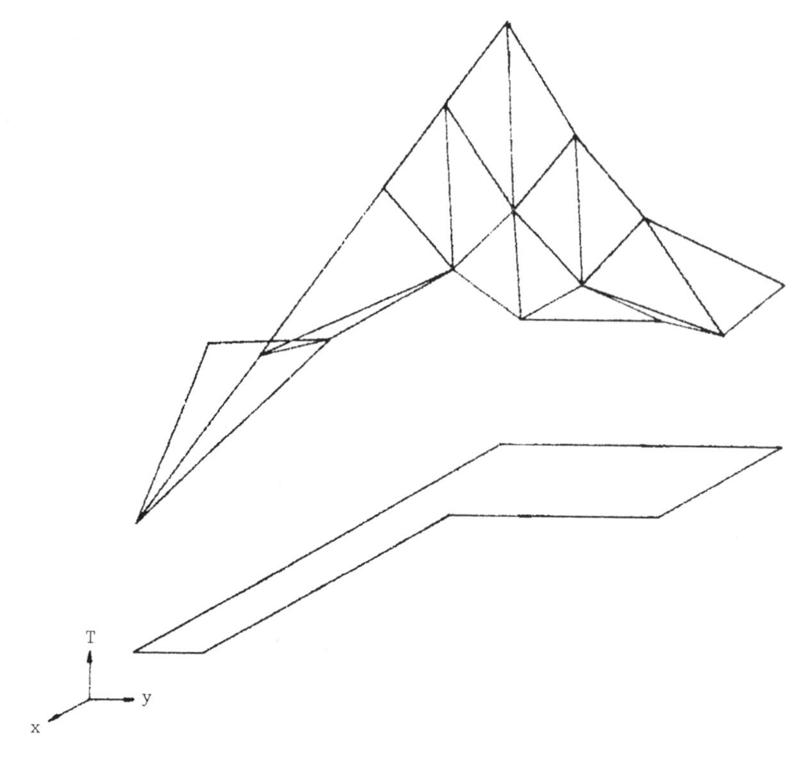

Figure 2.7 Description of a function $T(x, y)$.

in each finite element. The FORTRAN statements for (2.40) are as follows:

```
      DO 100 I = 1,2
      DTI = 0
      DO 102 IA = 1,3                                    (2.41)
  102 DTI = DTI + TE(IA) * DN(IA,I)
  100 DT(I) = DTI
```

Here DT is the array for the first derivatives of the interpolated function T_h in each element, and TE is the array containing T_α, $\alpha = 1, 2, 3$.

Example 2.1: Let us consider interpolation of a function

$$T(x, y) = 2 + 2x^2y + 3(1 - x) + 2(2 - y)^2$$

on the L-shaped domain Ω shown in Figure 2.4. Sixteen three-node triangular elements are used to discretize the given domain and the function. The plot in Figure 2.7 is a three-dimensional representation of the result of the interpolation.

2.4 Stiffness matrix and load vector

The third step is to derive the stiffness matrix and the generalized load vector from the weak form (2.27) by using the discretization discussed in the previous section.

To this end, we first note that for $\Omega = \bigcup_{e=1}^{E} \Omega_e$

$$\int_{\Omega} k_{ij} T_{,j} \bar{T}_{,i} \, d\Omega = \sum_{e=1}^{E} \int_{\Omega_e} k_{ij} T_{,j} \bar{T}_{,i} \, d\Omega$$

and

$$\int_{\Omega} f \bar{T} \, d\Omega = \sum_{e=1}^{E} \int_{\Omega_e} f \bar{T} \, d\Omega$$

Apply the interpolation (2.33) (obtained from the three-node triangular element) to the functions T and $\bar{T}$ in Ω_e:

$$T(x, y) = T_\beta N_\beta(x, y) \qquad \text{and} \qquad \bar{T}(x, y) = \bar{T}_\alpha N_\alpha(x, y)$$

Then the first term of the weak form (2.27) becomes

$$\begin{aligned} \int_{\Omega} k_{ij} T_{,j} \bar{T}_{,i} \, d\Omega &= \sum_{e=1}^{E} \int_{\Omega_e} k_{ij} T_{,j} \bar{T}_{,i} \, d\Omega \\ &= \sum_{e=1}^{E} \int_{\Omega_e} k_{ij} T_\beta N_{\beta,j} \bar{T}_\alpha N_{\alpha,i} \, d\Omega \\ &= \sum_{e=1}^{E} \bar{T}_\alpha \left(\int_{\Omega_e} k_{ij} N_{\alpha,i} N_{\beta,j} \, d\Omega \right) T_\beta \end{aligned} \tag{2.42}$$

Using (2.40b), this becomes

$$\int_{\Omega} k_{ij} T_{,j} \bar{T}_{,i} \, d\Omega = \sum_{e=1}^{E} \bar{T}_\alpha K^e_{\alpha\beta} T_\beta \tag{2.43}$$

where

$$K^e_{\alpha\beta} = \left(\int_{\Omega_e} k_{ij} \, d\Omega \right) N_{\alpha,i} N_{\beta,j} \tag{2.44}$$

Suppose that k_{ij} is *piecewise constant*, that is, the conductivity coefficients k_{ij} are constants on each element, but values may be different from those on other elements. Then the evaluation of (2.44) becomes trivial and we have

$$K^e_{\alpha\beta} = (k_{ij} N_{\alpha,i} N_{\beta,j}) * \text{AREA} \tag{2.45}$$

The matrix $[K^e_{\alpha\beta}]$ is called the *element stiffness matrix* in finite element analysis.

Let SKE(IA,JB) be the array for the element stiffness matrix $[K^e_{\alpha\beta}]$, and let XK(I,J) be the array for the conductivity coefficients, which are assumed to be piecewise constant. Then the form (2.45) suggests the following FORTRAN routine for computing the element stiffness matrix:

```
    DO 100 IA = 1,3
    DO 100 JB = 1,3
    SKEAB = 0.
    DO 102 I = 1,2                                              (2.46)
    DO 102 J = 1,2
102 SKEAB = SKEAB + XK(I,J) * DN(IA,I) * DN(JB,J)
100 SKE(IA,JB) = SKEAB * AREA
```

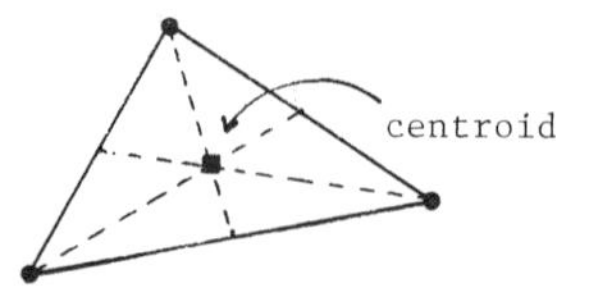

Figure 2.8 One-point integration rule.

Here the indices α and β in index notation have been replaced by IA and JB in the FORTRAN routine. If the stiffness matrix is known to be symmetric, we need not evaluate all of its elements. We may change the statement DO 100 JB = 1,3 to DO 100 JB = IA,3.

If the conductivity k_{ij} is not piecewise constant, (2.45) must be modified. To do this, we introduce a technique of numerical integration over a triangular element. Suppose that the integration

$$\int_{\Omega_e} \hat{f}\, d\Omega \tag{2.47}$$

is to be performed for a given function $\hat{f}$ on Ω_e. If $\hat{f}$ is a constant function of Ω_e,

$$\int_{\Omega_e} \hat{f}\, d\Omega = \hat{f} * \text{AREA} \tag{2.48}$$

If $\hat{f}$ is not constant, the *one-point integration rule* (or *mean value theorem*) on the triangular domain shown in Figure 2.8 gives

$$\int_{\Omega_e} \hat{f}\, d\Omega = \hat{f}(\hat{x}) * \text{AREA} \tag{2.49}$$

where $\hat{x}$ is the centroid of the triangle. Recall that the coordinates $\hat{x}$, $\hat{y}$ of the centroid of a triangle are given by

$$\hat{x} = \frac{1}{\text{AREA}} \int_{\Omega_e} x\, d\Omega, \qquad \hat{y} = \frac{1}{\text{AREA}} \int_{\Omega_e} y\, d\Omega \tag{2.50}$$

This furnishes the integration rule for a linear function $\hat{f}$. If $\hat{f} = \hat{f}_0 + \hat{f}_1 x + \hat{f}_2 y$, then

$$\int_{\Omega_e} \hat{f}\, d\Omega = \hat{f}(\hat{\mathbf{x}}) * \text{AREA} \tag{2.51}$$

which is the one-point integration rule. From this we see that the one-point rule is exact for linear functions. This suggests that if the values of the components k_{ij} of heat conduction are given at the centroid of the element Ω_e and stored in the array XK(I,J), the form (2.45) is valid for piecewise linear approximations of the conductivity coefficients k_{ij}. Note that almost any function can be reasonably approximated over a small region by a piecewise linear function. Therefore, the only modification needed in the routine (2.46) is that the array XK(I,J) should contain the values of k_{ij} at the centroid of each triangle. Hence there are no significant changes in the routine (2.46) for computing the element stiffness matrix, even for nonconstant conductivity k_{ij}.

Similar ideas can be used for the evaluation of the first term on the right side of the weak form (2.27). The heat supply term f is first approximated by a piecewise constant function so that the one-point integration rule (2.49) is applicable. Then, we have

$$\int_{\Omega} f\bar{T}\, d\Omega = \sum_{e=1}^{E} \int_{\Omega_e} f\bar{T}_\alpha N_\alpha(x, y)\, d\Omega = \sum_{e=1}^{E} \bar{T}_\alpha f_\alpha^e \tag{2.52}$$

where

$$f_\alpha^e = \int_{\Omega_e} f N_\alpha \, d\Omega$$
$$\doteq [f(\hat{\mathbf{x}}) N_\alpha(\hat{\mathbf{x}})] * \text{AREA}$$
$$= \tfrac{1}{3} f(\hat{\mathbf{x}}) * \text{AREA} \tag{2.53}$$

since the value of the shape function N_α at the centroid $\hat{\mathbf{x}}$ is $\frac{1}{3}$.

Thus, if the array FE(IA) is to contain the terms f_α^e, and if FEC denotes the value of the heat supply at the centroid, then the *element generalized load vector* FE(IA) is computed by the routine

```
    DO 100 IA = 1,3
100 FE(IA) = FEC * AREA/3                                        (2.54)
```

Example 2.2: For simplicity let us consider the evaluation of the integral

$$\int_\Omega f \bar{T} \, d\Omega \tag{2.55}$$

in the one-dimensional domain $\Omega = (0, 1)$ under the assumption that f and $\bar{T}$ are functions of x. First, f is approximated by a piecewise constant function, and $\bar{T}$ is interpolated by a linear function within each element. Then, the forms equivalent to (2.52) and (2.53) become

$$\int_\Omega f \bar{T} \, d\Omega = \sum_{e=1}^{E} \bar{T}_\alpha f_\alpha^e \quad (\alpha = 1, 2) \tag{2.56}$$

where

$$f_\alpha^e = \int_\Omega f N_\alpha \, d\Omega \doteq \tfrac{1}{2} f(\hat{x}) * \text{LENGTH} \tag{2.57}$$

$\hat{x}$ is the centroid of the interval $\Omega_e = (x_e, x_{e+1})$, and LENGTH is the length of Ω_e. Note that

$$N_1(x) = \frac{1}{\text{LENGTH}}(-x + x_{e+1}), \qquad N_2(x) = \frac{1}{\text{LENGTH}}(x - x_e)$$

Now let us perform the integration by using the forms (2.56) and (2.57), and compare the result with the exact evaluation for the case in which

$$f = e^{-x}, \qquad \bar{T} = x^2$$

If we compute analytically,

$$\int_\Omega f \bar{T} \, d\Omega = \int_0^1 e^{-x} x^2 \, dx = 2 - 5e^{-1} \doteq 0.1606028$$

If the finite element approach is used, the results of (2.56) and (2.57) are as follows:

NE	FEM	Error	NE	FEM	Error
3	0.17646	0.01583	30	0.16076	0.00013
5	0.16631	0.00568	40	0.16069	0.00006
10	0.16203	0.00140	50	0.16066	0.00003
20	0.16096	0.00033	∞	0.16063	—

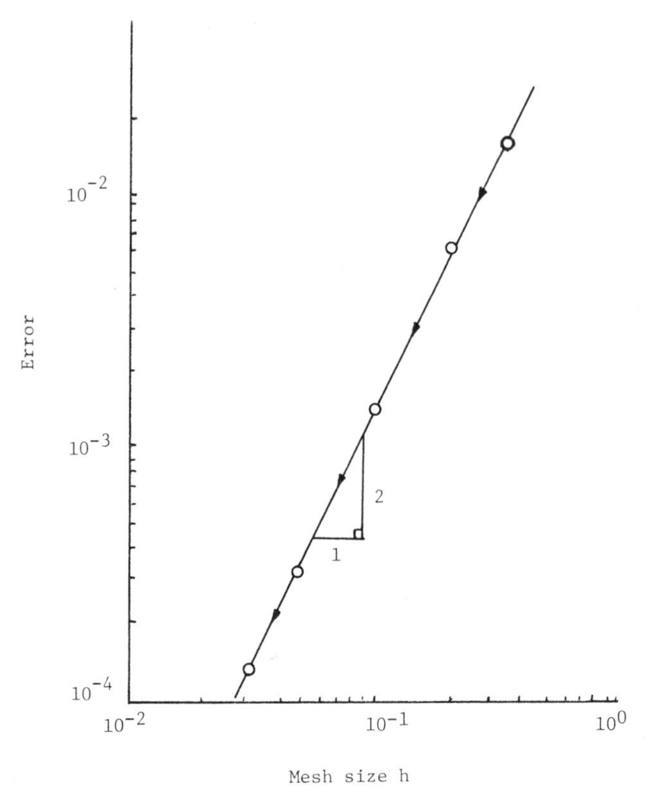

Figure 2.9 Convergence of the numerical integration using finite element methods.

The number NE is the total number of finite elements used for the evaluation. Even for 3 elements, the error is only 9.9‰. If 10 elements are used, the error is reduced to 0.9‰. The graph shown in Figure 2.9 indicates the rate of convergence as the mesh size $h = 1/\text{NE}$ tends to zero. This rate of convergence to the correct value is 2 for the approximation (2.57).

2.5 Treatment of boundary terms

In the previous section, the two terms of the weak form (2.27) that involve interior points of the domain Ω have been approximated by using the three-node triangular element. We shall now discretize the remaining three terms, which involve points on the boundaries Γ_2 and Γ_3.

Let us consider a part of the domain whose boundary is a part of, for example, Γ_3 as shown in Figure 2.10. Since the domain Ω has been divided into a set of three-node triangular elements, we divide the boundary Γ_3 into two-node line elements Γ_e^3, which are edges of the triangular elements. As was done for the geometry of three-node elements Ω_e, let us define two-node line elements on the boundary by introducing their element connectivities IJB(IA,NEL), IA = 1, 2 and NEL = 1, . . . , NEB, where NEB is the number of two-node line elements on the bound-

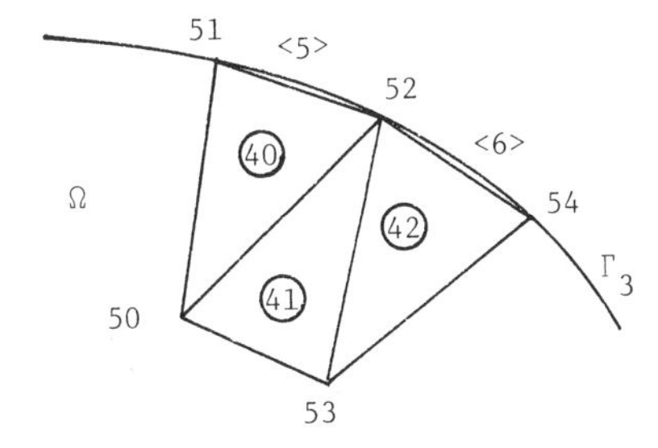

Figure 2.10 Boundary element Γ_e:⟨·⟩, line elements; ⊙, triangular elements.

ary Γ_3. For the situation shown in Figure 2.10, we have the following array IJB for the boundary elements:

NEL	IJB (1,NEL)	IJB (2,NEL)
⋮	⋮	⋮
5	52	51
6	54	52
⋮	⋮	⋮

The numbering order in IJB is the same as that used for interior elements (i.e., counterclockwise).

Let XE(IA), YE(IA), IA = 1, 2, be the array in which the coordinates of the end points of line elements are stored. For the line element NEL = 5 we have

$$\text{XE}(1) = \text{X}(52), \qquad \text{YE}(1) = \text{Y}(52)$$
$$\text{XE}(2) = \text{X}(51), \qquad \text{YE}(2) = \text{Y}(51)$$

This identification is obtained by the routine

```
      DO 100 IA = 1,2
      IJBIA = IJB(IA,NEL)
      XE(IA) = X(IJBIA)
  100 YE(IA) = Y(IJBIA)
```
(2.58)

It is now clear that the geometry of any line element Γ_e^3 is determined by the connectivity IJB.

The next step is the interpolation of the functions T and $\bar{T}$ on the boundary elements. Since T and $\bar{T}$ are approximated by piecewise linear functions, the trace of these functions on the boundary must be piecewise linear. That is, if a specific line element Γ_e^3 is considered, the functions T and $\bar{T}$ are both approximated by linear functions. Figure 2.11 depicts the situation for the function T. If an arc coordinate s is used along the line element, the first node can be assigned the coordinate $s = 0$ and the second node the coordinate $s = H$. Here H is the length of the element computed by the routine

```
SX = XE(2) - XE(1)
SY = YE(2) - YE(1)
H = SQRT(SX**2 + SY**2)
```
(2.59)

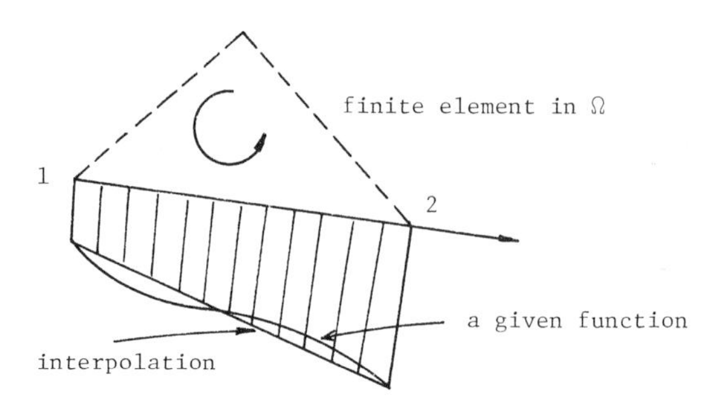

Figure 2.11 Interpolation on Γ_e.

Let T_1 and T_2 be the values of the function T at the first and second nodal points, respectively. Then the coefficients c_0 and c_1 in the polynomial

$$T(s) = c_0 + c_1 s \tag{2.60}$$

are found to be $c_0 = T_1$ and $c_1 = (T_2 - T_1)/H$. This yields the expression

$$T(s) = T_1\left(1 - \frac{s}{H}\right) + T_2\left(\frac{s}{H}\right) = T_\alpha N_\alpha(s)$$

$$N_1(s) = 1 - \frac{s}{H}, \qquad N_2(s) = \frac{s}{H} \tag{2.61}$$

The functions $N_1(s)$ and $N_2(s)$ are called the *shape functions* for the two-node line element.

Now we are ready to evaluate the second term of the weak form (2.27):

$$\int_{\Gamma_3} k_0 T\bar{T}\, d\Gamma = \sum_{e=1}^{E'_3} \int_{\Gamma_e^3} k_0 T\bar{T}\, d\Gamma$$

where E'_3 is the number of line elements on Γ_3. Substitution of

$$T = T_\beta N_\beta \qquad \text{and} \qquad \bar{T} = \bar{T}_\alpha N_\alpha$$

yields

$$\int_{\Gamma_e^3} k_0\, T\bar{T}\, d\Gamma = \bar{T}_\alpha\left(\int_{\Gamma_e^3} k_0 N_\alpha N_\beta\, d\Gamma\right) T_\beta \tag{2.62a}$$

If k_0 is assumed to be constant on Γ_e^3, we have

$$\int_{\Gamma_e^3} k_0\, T\bar{T}\, d\Gamma = \{\bar{T}_1, \bar{T}_2\} \underbrace{\frac{k_0 H}{6}\begin{bmatrix} 2 & 1 \\ 1 & 2 \end{bmatrix}}_{[K_{\alpha\beta}^{e,3}]} \begin{Bmatrix} T_1 \\ T_2 \end{Bmatrix} \tag{2.62b}$$

where

$$K_{\alpha\beta}^{e,3} = k_0 \int_0^H N_\alpha(s) N_\beta(s)\, ds \tag{2.63}$$

The matrix $[K_{\alpha\beta}^{e,3}]$ is the *element stiffness matrix* on the boundary Γ_3.

The remaining two terms

$$\int_{\Gamma_3} k_0 T_\infty \bar{T}\, d\Gamma \qquad \text{and} \qquad \int_{\Gamma_2} h\bar{T}\, d\Gamma$$

in the weak form (2.27) can be approximated as follows: Let h_e and T_e be, respectively, the values of h and $k_0 T_\infty$ at the center of a line element. Then we have

$$\int_{\Gamma_2} h\bar{T}\, d\Gamma = \sum_{e=1}^{E'_2} \int_{\Gamma_e^2} h\bar{T}\, d\Gamma = \sum_{e=1}^{E'_2} \bar{T}_\alpha f_\alpha^{e,2}$$

$$\int_{\Gamma_3} k_0 T_\infty \bar{T}\, d\Gamma = \sum_{e=1}^{E'_3} \int_{\Gamma_e^3} k_0 T_\infty \bar{T}\, d\Gamma = \sum_{e=1}^{E'_3} \bar{T}_\alpha f_\alpha^{e,3}$$

$$\left\{\int_{\Gamma_e^2} hN_\alpha(s)\, d\Gamma\right\} = \frac{h_e H}{2}\begin{Bmatrix}1\\1\end{Bmatrix}, \qquad f_\alpha^{e,2} = \int_{\Gamma_e^2} hN_\alpha(s)\, d\Gamma \tag{2.64}$$

$$\left\{\int_{\Gamma_e^3} k_0 T_\infty N_\alpha(s)\, d\Gamma\right\} = \frac{T_e H}{2}\begin{Bmatrix}1\\1\end{Bmatrix}, \qquad f_\alpha^{e,3} = \int_{\Gamma_e^3} k_0 T_\infty N_\alpha(s)\, d\Gamma$$

where E'_2 is the number of elements on Γ_2. The last two vectors of (2.64) are called the *generalized load vectors* from the boundary conditions.

A FORTRAN program that computes the stiffness matrix and the generalized load vector from the boundary condition is as follows.

On the boundary Γ_e^3:

```
CONST = XKO * H/6.
SKE(1,1) = 2. * CONST
SKE(1,2) = 1. * CONST
SKE(2,1) = 1. * CONST
SKE(2.2) = 2. * CONST

FE(1) = 0.5 * TE * H
FE(2) = 0.5 * TE * H
```

(2.65)

On the boundary Γ_e^2:

```
FE(1) = 0.5 * HE * H
FE(2) = 0.5 * HE * H
```

(2.66)

Here HE = h_e, TE = T_e, and XKO = k_0 at the center of the line element Γ_e^3. The element length H is computed by the geometry of the line element using (2.59).

Thus far, we have obtained the finite element approximation for the weak form (2.27):

$$T_\alpha = g_\alpha \quad \text{on } \Gamma_1$$

$$\sum_{e=1}^{E} \bar{T}_\alpha K_{\alpha\beta}^e T_\beta + \sum_{e=1}^{E'_3} \bar{T}_\alpha K_{\alpha\beta}^{e,3} T_\beta \tag{2.67}$$

$$= \sum_{e=1}^{E} \bar{T}_\alpha f_\alpha^e + \sum_{e=1}^{E'_2} \bar{T}_\alpha f_\alpha^{e,2} + \sum_{e=1}^{E'_3} \bar{T}_\alpha f_\alpha^{e,3}, \quad \forall \bar{T} \ni \bar{T} = 0 \text{ on } \Gamma_1$$

2.6 Assembling of element stiffness matrices and generalized load vectors

We have so far shown how to approximate the weak formulation (2.27) for the steady-state problem of heat conduction. That is, the element stiffness matrices $[K^e_{\alpha\beta}]$ and $[K^{e,3}_{\alpha\beta}]$ and the generalized load vectors $\{f^e_\alpha\}$, $\{f^{e,2}_\alpha\}$, and $\{f^{e,3}_\alpha\}$ are obtained for each finite element Ω_e or Γ_e. One of the remaining tasks is *to assemble* these "element level" matrices and vectors to form the global stiffness matrix and the global generalized load vector by performing the indicated summations in the discretized weak form (2.67). Assemblage of those element stiffness matrices and load vectors is based on the assumption that the interpolation of the temperature T and its virtual temperature $\bar{T}$ *must be continuous on the whole domain* $\Omega \cup \Gamma$. The assumption of continuity ensures the validity of the integrations by parts of the weighted equation (2.22). If T and $\bar{T}$ were allowed to be discontinuous, the form (2.24) is no longer correct. That is, the weak form (2.27), which forms our basis of the finite element approximation, is incorrect. Thus, we must have continuity of the functions and their interpolations if the weak form (2.27) is assumed to be the basis of the approximation. There are other formulations that do not use the continuity assumption. These formulations are known as *mixed/hybrid finite element methods.* We shall not cover these methods in our introductory study of finite element methods.

What is a method of discretization that maintains continuity of T and $\bar{T}$? Recall that T and $\bar{T}$ are linear functions in each finite element, and hence very linear on the edges of these elements. If adjacent elements have common nodes, and if T and $\bar{T}$ each have common nodal values, then continuity will prevail. For example, consider the adjacent elements shown in Figure 2.12. The element connectivities for the finite elements Ω_9 and Ω_{12} are tabulated as follows:

NEL	IJK(1,NEL)	IJK(2,NEL)	IJK(3,NEL)
9	13	16	15
12	16	14	15

The nodes 15 and 16 are common to Ω_9 and Ω_{12}. Thus, in order to maintain continuity of the function T along the common boundary (line 15 ~ 16), we have to assign

$$T_2 \text{ for } \Omega_9 \quad \text{and} \quad T_1 \text{ for } \Omega_{12} \text{ at node } 16$$
$$T_3 \text{ for } \Omega_9 \quad \text{and} \quad T_3 \text{ for } \Omega_{12} \text{ at node } 15$$

If T^N_α denotes the value of T at the α local node of the Nth element, the identification may be described as

$$T^9_2 = T^{12}_1 = T(\mathbf{x}^{16})$$
$$T^9_3 = T^{12}_3 = T(\mathbf{x}^{15})$$

The relation between the local numbering α and the global numbering of the nodal

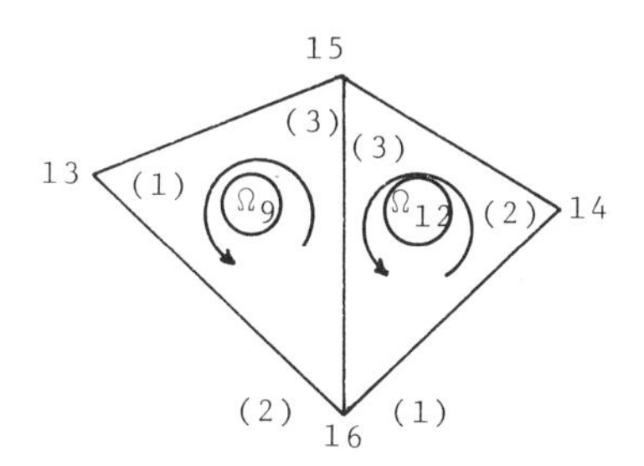

Figure 2.12 Continuity.

points can be expressed by the routine

```
      DO 100 IA = 1,3
      IJKIA = IJK(IA,NEL)                                    (2.68)
  100 TE(IA) = T(IJKIA)
```

for the element NEL. The connectivity IJK takes care of all transferring from global to elements and from elements to global. The array T(NOD), NOD = 1, . . . , NX, is used to store the nodal values of the temperature field T; NX is the total number of nodal points and NOD indicates the global number of the node.

Within each element Ω_e, the terms in (2.67) can be expressed in FORTRAN as follows:

Term C	FORTRAN
$\bar{T}_\alpha K^e_{\alpha\beta} T_\beta$	VTE(IA) * SKE(IA,IB) * TE(IB)
$\bar{T}_\alpha K^{e,3}_{\alpha\beta} T_\beta$	VTE(IA) * SKE(IA,IB) * TE(IB)
$\bar{T}_\alpha f^e_\alpha$	VTE(IA) * FE(IA)
$\bar{T}_\alpha f^{e,2}_\alpha$	VTE(IA) * FE(IA)
$\bar{T}_\alpha f^{e,3}_\alpha$	VTE(IA) * FE(IA)

Note that the contents of the arrays SKE and FE are defined by the forms (2.45), (2.53), and (2.64). The array VTE(IA) contains the values of the virtual temperature.

Now we assemble these to form their global counterparts. That is, the summations in (2.67) must be performed. We first look at the parts related to interior points of the domain. To this end, three arrays SK(N,M), F(N), and VT(N) are introduced and referred to as the *global stiffness matrix*, the *global generalized load vector*, and the *virtual temperature,* respectively. Indices N and M coincide with the global numbering of nodal points. For example, SK (5,12) is the element of the global stiffness matrix for nodes 5 and 12. Similarly, F(6) is the generalized load at node 6. Since a nodal point is, in general, common to several elements, contributions to the stiffness matrix comes from several elements. Thus, we have to define the global stiffness matrix SK by performing the summation. For the "domain

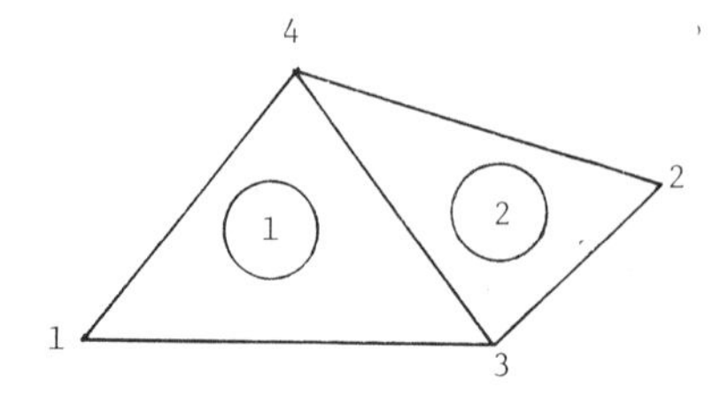

Figure 2.13 Example of global stiffness matrix assembly.

part" of the form (2.67), we have a routine

```
      DO 100 NEL = 1,NELX
      ———Form SKE and FE———
      DO 200 IA = 1,3
      IJKIA = IJK(IA,NEL)
      F(IJKIA) = F(IJKIA) + FE(IA)                                    (2.69)
      DO 200 JB = 1,3
      IJKJB = IJK(JB,NEL)
      SK(IJKIA,IJKJB) = SK(IJKIA,IJKJB) + SKE(IA,JB)
  200 CONTINUE
  100 CONTINUE
```

Here NELX is the total number of triangular elements inside the domain. This total number was denoted by E in (2.67). The procedure (2.69) is called the *assembling* of element stiffness matrices and load vectors.

In order to see the idea behind the algorithm (2.69) for the assembly of element stiffness matrices and load vectors into global ones, let us consider an example of this assembly for the two elements shown in Figure 2.13. Suppose that the element connectivities are defined as follows:

NEL	IJK(1,NEL)	IJK(2,NEL)	IJK(3,NEL)
1	3	4	1
2	2	4	3

Suppose that the element stiffness matrices are

$$\begin{bmatrix} k^1_{11} & k^1_{12} & k^1_{13} \\ k^1_{21} & k^1_{22} & k^1_{23} \\ k^1_{31} & k^1_{32} & k^1_{33} \end{bmatrix} \quad \text{and} \quad \begin{bmatrix} k^2_{11} & k^2_{12} & k^2_{13} \\ k^2_{21} & k^2_{22} & k^2_{23} \\ k^2_{31} & k^2_{32} & k^2_{33} \end{bmatrix}$$

Since the total number of nodes in the domain is 4, and since only the temperature is unknown at each node, the total degrees of freedom in the finite element model is 4. Thus, the size of the global stiffness matrix is 4×4 and has the form

$$\begin{bmatrix} k_{11} & k_{12} & k_{13} & k_{14} \\ k_{21} & k_{22} & k_{23} & k_{24} \\ k_{31} & k_{32} & k_{33} & k_{34} \\ k_{41} & k_{42} & k_{43} & k_{44} \end{bmatrix}$$

The term k_{ij} is the component of the global stiffness matrix related to the virtual temperature $\bar{T}_i$ at the ith node and the temperature T_j at the jth node in the global numbering system. Now, our question is how the global stiffness matrix is assembled from the two element stiffness matrices.

Because of the element connectivity for the first element, the correspondence between the local and global node numbering is as follows:

Local	Global
1 ⟷	3
2 ⟷	4
3 ⟷	1

Thus, for example, the first row and first column of the element stiffness matrix contributes to form a part of the third row and third column of the global stiffness matrix. Schematically, this correspondence could be represented by

$$\begin{array}{c} \\ 3 \\ 4 \\ 1 \\ \uparrow \\ \text{IJK(I,1)} \end{array} \begin{array}{l} \quad 3 \qquad 4 \qquad 1 \quad \leftarrow \text{IJK(J,1)} \\ \begin{bmatrix} k_{11}^1 & k_{12}^1 & k_{13}^1 \\ k_{21}^1 & k_{22}^1 & k_{23}^1 \\ k_{31}^1 & k_{32}^1 & k_{33}^1 \end{bmatrix} \\ \\ \\ \end{array}$$

Then the insertion of the first element stiffness matrix to the global scheme becomes the following:

$$\begin{array}{c} \\ 1 \\ 2 \\ 3 \\ 4 \\ \uparrow \\ \text{Global} \end{array} \begin{array}{l} \quad 1 \qquad 2 \qquad 3 \qquad 4 \quad \leftarrow \text{Global} \\ \begin{bmatrix} k_{33}^1 & 0 & k_{31}^1 & k_{32}^1 \\ 0 & 0 & 0 & 0 \\ k_{13}^1 & 0 & k_{11}^1 & k_{12}^1 \\ k_{23}^1 & 0 & k_{21}^1 & k_{22}^1 \end{bmatrix} \\ \\ \\ \end{array}$$

Similarly, the second element whose element stiffness matrix is

$$\begin{array}{c} \\ 2 \\ 4 \\ 3 \\ \uparrow \\ \text{IJK(I,2)} \end{array} \begin{array}{l} \quad 2 \qquad 4 \qquad 3 \quad \leftarrow \text{IJK(J,2)} \\ \begin{bmatrix} k_{11}^2 & k_{12}^2 & k_{13}^2 \\ k_{21}^2 & k_{22}^2 & k_{23}^2 \\ k_{31}^2 & k_{32}^2 & k_{33}^2 \end{bmatrix} \\ \\ \\ \end{array}$$

is inserted into the global array as follows:

$$\begin{array}{c} \\ 1 \\ 2 \\ 3 \\ 4 \\ \uparrow \\ \text{Global} \end{array} \begin{array}{l} \quad 1 \qquad 2 \qquad 3 \qquad 4 \quad \leftarrow \text{Global} \\ \begin{bmatrix} 0 & 0 & 0 & 0 \\ 0 & k_{11}^2 & k_{13}^2 & k_{12}^2 \\ 0 & k_{31}^2 & k_{33}^2 & k_{32}^2 \\ 0 & k_{21}^2 & k_{23}^2 & k_{22}^2 \end{bmatrix} \\ \\ \\ \end{array}$$

Therefore, the complete assembling of the two elements yields the global stiffness matrix:

$$\begin{array}{c} \\ 1 \\ 2 \\ 3 \\ 4 \end{array}\begin{array}{c} \begin{array}{cccc} 1 & 2 & 3 & 4 \end{array} \\ \begin{bmatrix} k_{33}^1 & 0 & k_{31}^1 & k_{32}^1 \\ 0 & k_{11}^2 & k_{13}^2 & k_{12}^2 \\ k_{13}^1 & k_{31}^2 & k_{11}^1 + k_{33}^2 & k_{12}^1 + k_{32}^2 \\ k_{23}^1 & k_{21}^2 & k_{21}^1 + k_{23}^2 & k_{22}^1 + k_{22}^2 \end{bmatrix} \end{array}$$

From this example, it should be clear that the algorithm (2.69) is merely a FORTRAN translation of the above processes of assembling.

For the boundary Γ_3, a routine similar to (2.69) can be developed. Let NEB3 be the total number of line elements involved in the discretization of Γ_3. Then the routine is

```
      DO 100 NEL = 1,NEB3
      ———Form SKE and FE———
      DO 200 IA = 1,2
      IJBIA = IJB(IA,NEL)
      F(IJBIA) = F(IJBIA) + FE(IA)                                  (2.70)
      DO 200 JB = 1,2
      IJBJB = IJB(JB,NEL)
      SK(IJBIA,IJBJB) = SK(IJBIA,IJBJB) + SKE(IA,JB)
  200 CONTINUE
  100 CONTINUE
```

For the remaining term in (2.67), there are several ways, but we shall use the following:

```
      DO 300 NEL = NEB3 + 1,NEB3 + NEB2
      ———Form FE———
      DO 400 IA = 1,2                                               (2.71)
      IJBIA = IJB(IA,NEL)
  400 F(IJBIA) = F(IJBIA) + FE(IA)
  300 CONTINUE
```

Note that the element connectivity on the boundary Γ_2 is also stored in the array IJB, which was introduced earlier for the boundary Γ_3, and that NEB2 is the total number of line elements on Γ_2.

After assembling by the routines (2.69)–(2.71), the discretized weak form (2.67) becomes

$$\begin{gathered} T_I = g_I \quad \text{on } \Gamma_1 \\ \bar{T}_I K_{IJ} T_J = \bar{T}_I F_I, \quad \forall \bar{T}_I \ni \bar{T}_I = 0 \text{ on } \Gamma_1 \end{gathered} \tag{2.72a}$$

or

$$\begin{gathered} \mathrm{T(I) = G(I)} \quad \text{on } \Gamma_1 \\ \mathrm{VT(I) * SK(I,J) * T(J) = VT(I) * F(I)}, \quad \forall \mathrm{VT(I)} \ni \mathrm{VT(I)} = 0 \text{ on } \Gamma_1 \end{gathered} \tag{2.72b}$$

We here simply assign the boundary values of g at the nodal points of the boundary Γ_1, so that G(I) is the value of g at the Ith node on the boundary Γ_1.

Noting that

$$a = b \quad \text{if } ax = bx, \quad \forall x \tag{2.73}$$

(2.72) is reduced to the *finite element version* of the boundary value problem (2.21):

$$\begin{aligned} &T_I = g_I \text{ on } \Gamma_1 \\ &K_{IJ}T_J = F_I \end{aligned} \quad \text{or} \quad \begin{aligned} &\text{T(I)} = \text{G(I) on } \Gamma_1 \\ &\text{SK(I,J)} * \text{T(J)} = \text{F(I)} \end{aligned} \tag{2.74}$$

It is clear that (2.74) is a system of algebraic (linear) equations that can be solved by, for example, taking the inverse of the global stiffness matrix $[K_{IJ}]$ stored in SK(I,J).

2.7 Boundary conditions and the band solver

As shown above, we have obtained a system of linear equations whose solution represents the finite element approximation of the solution of the heat conduction problem (2.21). The temperature field T is determined by solving the system of equations along with the boundary condition

$$T_I = g_I \quad \text{on} \quad \Gamma_1 \tag{2.75}$$

If the boundary consists of Γ_1 and Γ_2 only, the stiffness matrix $[K_{IJ}]$ is singular. That is, the matrix $[K_{IJ}]$ has a zero eigenvalue together with a constant eigenvector $T_I = 1/\sqrt{N}, I = 1, \ldots, N$, where N is the total number of nodes in the finite element model.

One way to impose the boundary condition (2.75) on the system of linear equations is to use the penalty method described in section 1.4.3. To explain this method, suppose that node 8 is on Γ_1 so that $T_8 = g_8$ must be satisfied. From the system of linear equations (2.74), we already have

$$K_{81}T_1 + K_{82}T_2 + \cdots + K_{87}T_7 + K_{88}T_8 + \cdots + K_{8N}T_N = F_8 \tag{2.76}$$

in addition to the boundary condition. If (2.76) is rewritten as

$$\begin{aligned} K_{81}T_1 + K_{82}T_2 + \cdots + K_{87}T_7 + (K_{88} + \hat{K})T_8 + K_{89}T_9 + \cdots + K_{89}T_N \\ = F_8 + \hat{K}g_8 \end{aligned} \tag{2.77}$$

by adding the term $\hat{K}T_8$ on the left side and the terms $\hat{K}g_8$ on the right side, and if the number $\hat{K}$ is considerably larger than $K_{81}, K_{82}, \ldots, K_{8N}$ and F_8, then the relation (2.77) is approximately equivalent to the boundary condition (2.75). For example, if $\hat{K}$ is chosen to be

$$\hat{K} = N * 10^p * \text{MAX } K_{IJ}, \quad I, J = 1, \ldots, N \tag{2.78}$$

equation (2.77) yields

$$T_8 = g_8 + \text{O}(10^{-p}) \tag{2.79}$$

Thus, $T_8 = g_8$ is satisfied to within an error 10^{-p}. Due to the intrinsic error introduced by the problem modification, the technique has been termed the *penalty method*. Thus, if the parameter p is properly chosen (e.g., $p = 3 \sim 5$), the boundary condition (2.75) can be realized by the above modification of (2.77). Hence we add a large number to each of those diagonal elements of the global stiffness matrix that correspond to a boundary node on Γ_1 and add products of boundary values with these same large numbers to the global generalized load vector. A FORTRAN routine that performs the modification mentioned above is the following:

```
      DO 100 I = 1, NEB1
      READ(5,500) NODE, VALUE
      SK(NODE,NODE) = SK(NODE,NODE) + PENALT
      F(NODE) = F(NODE) + PENALT * VALUE                         (2.80)
  100 CONTINUE
  500 FORMAT (I5,F10.4)
```

Here NEB1 $= E_1'$ is the total number of nodal points on the boundary Γ_1, PENALT $= \hat{K}$ (the large number) and VALUE $= g_1$ (the boundary value at the node I).

Once the modification of the global stiffness matrix is made in conjunction with the boundary condition (2.75), the matrix $[K_{IJ}]$ becomes nonsingular, and then the system of linear equations is solvable. To solve it, we may use the Gaussian elimination method.

It is to be noted that for most problems the stiffness matrix is a square matrix with many zero elements, and the nonzero elements are near its main diagonal. Moreover, if problem (2.21) is symmetric, that is, if the conductivity satisfies

$$k_{ij} = k_{ji}$$

then the stiffness matrix is symmetric too. Thus, we need to store only the shaded parts of the matrix, as shown in Figure 2.14. The rectangular matrix obtained by the shaded part of the stiffness matrix is called the *banded stiffness matrix*. The typical situation shown in Figure 2.14 also suggests the manner in which element stiffness matrices should be assembled in order to make the banded stiffness matrix instead of the square one. A typical routine for this purpose is obtained by a slight modification of (2.69):

```
      DO 100 NEL = 1,NELX
      ___________________
      DO 200 IA = 1,3
      IJKIA = IJK(IA,NEL)
      F(IJKIA) = F(IJKIA) + FE(IA)
      DO 200 JB = 1,3                                            (2.81)
      IJKJB = IJK(JB,NEL)
      JBIA1 = IJKJB - IJKIA + 1
      IF(JBIA1.LE.O) GO TO 200
      SK(IJKIA,JBIA1) = SK(IJKIA,JBIA1) + SKE(IA,JB)
  200 CONTINUE
  100 CONTINUE
```

It is noted that the banded structure of the global stiffness matrix strongly depends on the nodal numbering in the finite element model. For example, for the numbering shown in Figure 2.15, the bandwidth is 12, and there is no banded structure in the stiffness matrix. However, if the numbering is changed as indicated

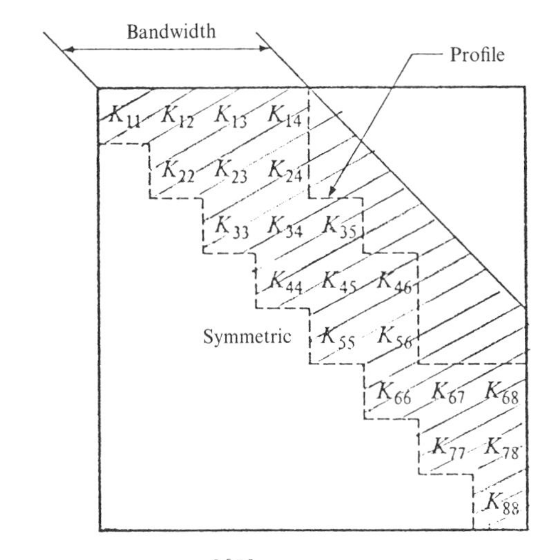

Figure 2.14 Banded matrix.

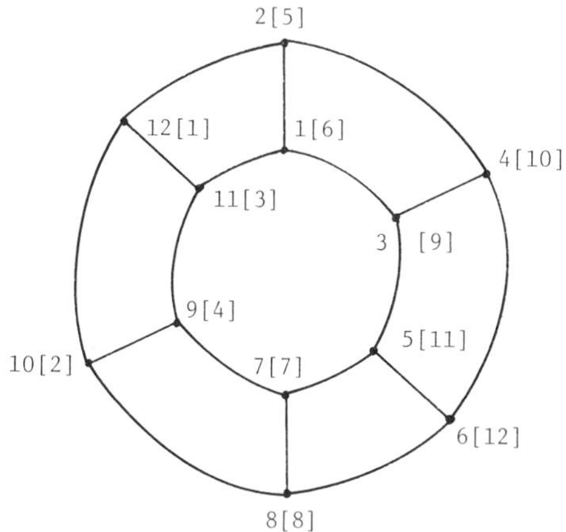

Figure 2.15 Nonbanded structure of the stiffness matrix.

by the bracketed numbers, the bandwidth becomes 7 and the stiffness matrix could be stored as a 12×7 rectangular matrix.

A method that determines the bandwidth of the global stiffness matrix is obtained using the element connectivities:

```
    MBAND = 0
    DO 100 NEL = 1,NELX
    DO 100 I = 1,3
    IJKI = IJK(I,NEL)
    DO 100 J = 1,3                                          (2.82)
    IJKJ = IJK(J,NEL)
    IJKIJ = IJKJ – IJKI + 1
    IF (IJKIJ.GT.MBAND) MBAND = IJKIJ
100 CONTINUE
```

Since the bandwidth MBAND is computed by using only the element connectivities, this routine should be performed at the *preprocessing* of a finite element program and must be used to determine the size of the rectangular banded global stiffness matrix.

Because of the method of storing the stiffness matrix into a band (i.e., rectangular) matrix, the Gauss elimination method (1.37) described in section 1.2 must be adequately modified. For example, the part of the forward elimination may be modified as follows:

```
        N1 = N - 1
        DO 100 K = 1,N1
        K1 = K + 1
        KMAX = K + MBAND - 1
        IF(KMAX.GT.N) KMAX = N
        DO 102 L = K,KMAX
    102 C(L) = A(K,L - K + 1)
        AKK = 1./C(K)                                  (2.83)
        BK = B(K)
        DO 104 I = K1, KMAX
        AIK = A(K,I - K + 1) * AKK
        B(I) = B(I) - AIK * BK
        DO 104 J = I,KMAX
    104 A(I,J - I + 1) = A(I,J - I + 1) - AIK * C(J)
    100 CONTINUE
```

2.8 Program FEM1 and numerical examples

We are now ready to make a program that will solve the boundary value problem (2.21) by using finite element methods. The program described below consists of three "processings": preprocessing, finite element processing, and postprocessing. The main objective of our present study is to learn how to develop the finite element (FE) processing. The preprocessing provides all necessary information to the FE processing, and the postprocessing may be developed to retrieve the information necessary to design/analyze a structure by using the results from the FE processing. If fairly complex structures are considered, some sophistication in preprocessing might be necessary. One of the common "sophistications" in the first stage is to introduce a mesh generator that automatically provides the coordinates of nodal points and the element connectivities. Plotting of the computed quantities may be very helpful in the analysis of a problem. This is a part of the postprocessing. However, these are not the most essential points in the development of finite element codes.

The program FEM1, presented in Appendix 1, consists of 13 subroutines:

Preprocessing
HRMESH: Read nodal coordinates and element connectivities
MATERL: Read heat conductivity
HBANDW: Compute the (half-) bandwidth

Finite element processing
ASEMBO: Assembling element stiffness matrices and load vectors
  ESTIF3: Construct element stiffness matrices
  GRADI3: Compute global gradient of shape functions
HEATGE: Construct the load vector due to external heat generation
BOUND1: Input the first type of boundary condition

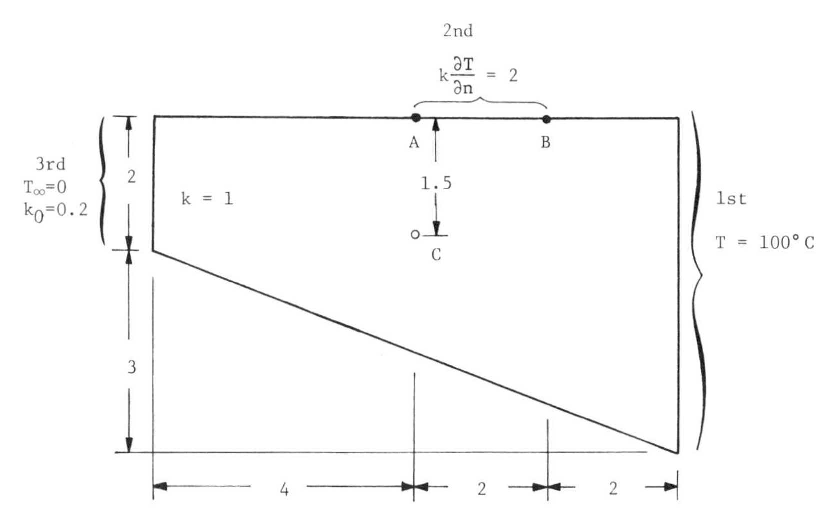

Figure 2.16 Domain for example 2.3.

BOUND2: Input the nonhomogeneous second type of boundary condition
BOUND3: Input the third type of boundary condition
BANDSL: Solve the system of linear equations obtained

Postprocessing
OUTPUT: Output the results, such as temperature and heat flux
QUALIT: Output the quality indices of the finite element approximations

To provide easy access to the program FEM1, an automatic mesh generator, which provides input data for the subroutine HRMESH, is introduced in Appendix 2. The basic idea of the mesh generator PRE-FEM1 is that a given domain is decomposed into possibly curved quadrilaterals, which are further discretized into many three-node finite elements. After providing coordinates of nodes characterizing quadrilaterals and connectivities of these quadrilaterals, we specify number of meshes in the x and y directions, mesh gradient, material property, and so forth. PRE-FEM1 then automatically generates element connectivities IJK of three-node triangular elements, nodal coordinates X and Y, material group number MPE, and LMN, which stores element numbers adjacent to a finite element. Details can be found in Appendix 2.

Another program is given in Appendix 3: POST-FEM1, which draws the finite element model applied in FEM1, isothermal lines, distribution of the temperature, heat flux vectors, and isoerror lines for the temperature and the heat flux. Almost all necessary data to this program are automatically generated in FEM1.

Example 2.3: In order to demonstrate the use of the program, let us consider the problem of heat conduction shown in Figure 2.16. A quadrilateral domain is divided into 16 triangular elements consisting of 15 nodal points. Suppose that the material in the quadrilateral is homogeneous and isotropic, so that

$$k_{ij} = k\delta_{ij}, \quad k = 1 \text{ kcal/m}^2 \text{ h °C}$$

The temperature is kept constant at the right surface, say at $T = 100°C$. There is a natural convection by air flow at the left surface. Suppose that the convection coefficient k_0 is given as $k_0 = 0.2$ kcal/m^2 h °C and the outside temperature is specified as $T_\infty = 0°C$, that is, the third type of boundary condition is assumed at the left surface:

$$k\frac{\partial T}{\partial n} = -k_0(T - T_\infty)$$

Table 2.1 *Input data for example 2.3*

Line					
1	TITLE OF THE PROBLEM (MARCH, 1984)				
2	EXAMPLE 2.3: HEAT CONDUCTION				
3	CONTROL NUMBERS				
4	1	1ST BOUNDARY ON			
5	1	NONHOMOGENEOUS 2ND BOUNDARY ON			
6	1	3RD BOUNDARY ON			
7	1	HEAT GENERATION ON			
8	FINITE ELEMENT MODEL				
9	15	16			
10	(COORDINATES: NODES)				
11	1	0.	3.		
12	2	0.	4.		
13	3	0.	5.		
14	4	2.	2.25		
15	5	2.	3.625		
16	6	2.	5.		
17	7	4.	1.5		
18	8	4.	3.25		
19	9	4.	5.		
20	10	6.	0.75		
21	11	6.	2.875		
22	12	6.	5.		
23	13	8.	0.		
24	14	8.	2.5		
25	15	8.	5.		
26	(CONNECTIVITIES: ELEMENTS)				
27	1	1	1	4	5
28	2	1	1	5	2
29	3	1	2	5	3
30	4	1	3	5	6
31	5	1	4	7	8
32	6	1	4	8	5
33	7	1	5	8	6
34	8	1	6	8	9
35	9	1	7	10	11
36	10	1	7	11	8
37	11	1	8	11	9
38	12	1	9	11	12
39	13	1	10	13	14
40	14	1	10	14	11
41	15	1	11	14	12
42	16	1	12	14	15
43	MATERIAL CONSTANTS				
44	1				
45	1	1.		1.	0.
46	HEAT SINK (POINTWISE)				
47	−1				
48	8	−50.			
49	1ST TYPE BOUNDARY CONDITIONS				
50	3	10000000000.			
51	13	100.			
52	14	100.			
53	15	100.			
54	NONHOMOGENEOUS 2ND TYPE BOUNDARY CONDITIONS				
55	1				
56	9	12	0.2		
57	3RD TYPE BOUNDARY CONDITIONS				
58	2	1			
59	1	1	1	2	
60	2	1	2	3	
61	1	0.2		0.	

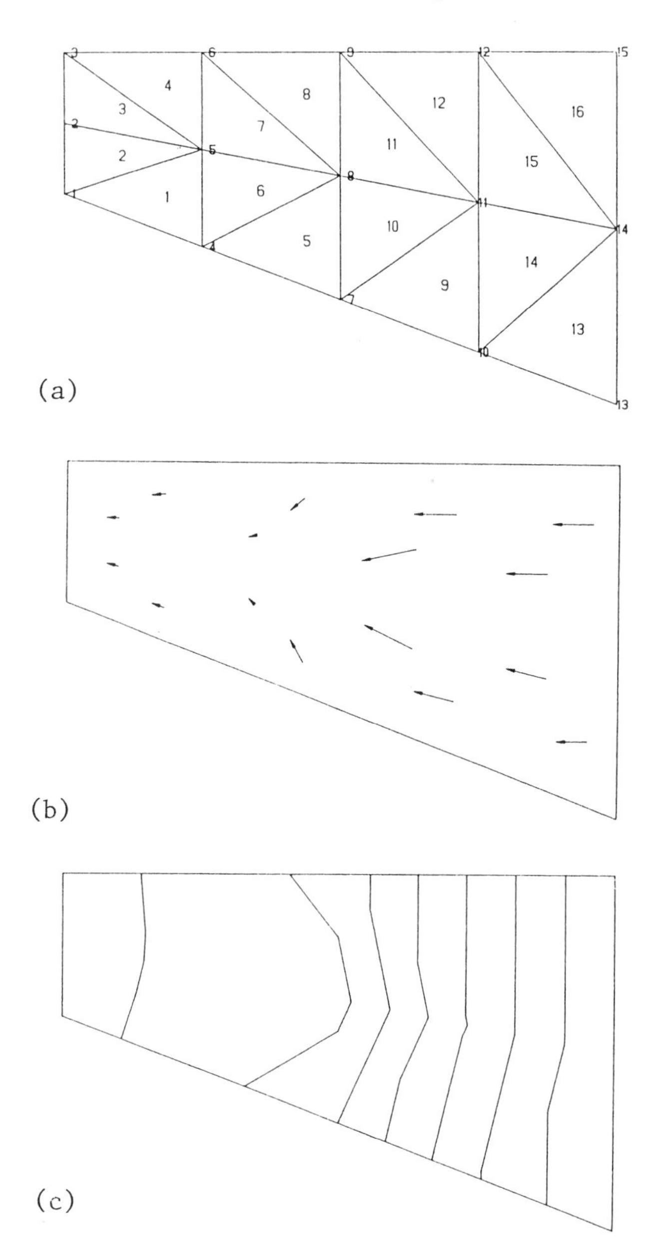

Figure 2.17 Results of example 2.3: (a) finite element model; (b) distribution of heat flux; (c) isotemperature lines.

The bottom surface is completely insulated, that is, $k\,\partial T/\partial n = 0$. The top surface is also insulated except the portion specified by AB in Figure 2.16. Suppose that there is heat flux from this portion, say $k\partial T/\partial n = 0.2\ \text{kcal/m}^2\ \text{h}$. At last, we assume a heat sink at the point C with the strength 50 kcal/h.

The input data to the program FEM1 is prepared in Table 2.1. Figures 2.17(a)–(c) are the plots by the program POST-FEM1 given in Appendix 3. These are the equitemperature distribution, and the heat flux vector.

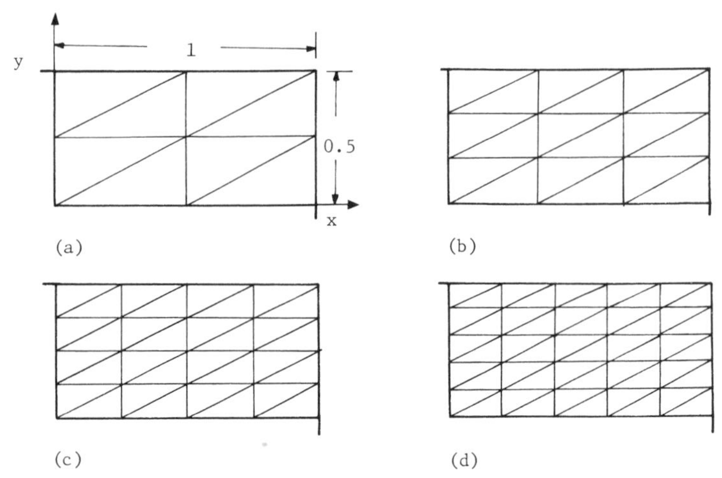

Figure 2.18 Uniform meshes for a convergence test: (a) 2 × 2, (b) 3 × 3, (c) 4 × 4, (d) 5 × 5.

Example 2.4 (a convergence test): Using the program FEM1, let us solve a problem, the analytical solution of which is known, in order to verify the code. To this end, a rectangular domain $(a, 0) \times (0, b)$ is considered in the xy coordinate system. Let heat conductivity be given by $k_{ij} = k\delta_{ij}$ and $k = 1$ kcal/m^2 h °C. Assuming the homogeneous boundary condition $T = 0$°C and the distribution heat source $f =$ 1 kcal/m^2 h over the entire domain, the analytical solution T is obtained as

$$T(x, y) = \sum_{m,n=1,3,5,\ldots} \frac{16}{\pi^2 mn\left[\left(\frac{m\pi}{a}\right)^2 + \left(\frac{n\pi}{b}\right)^2\right]} \sin m\pi\frac{x}{a} \sin n\pi\frac{y}{b} \tag{2.84}$$

For $a = 2$ and $b = 1$, let us solve this over the right upper quarter domain using uniform 2 × 2, 3 × 3, 4 × 4, and 5 × 5 meshes shown in Figure 2.18. If we plot the amount of error of T at the centroid of the rectangular domain, it can be realized that error decreases rapidly as the number of elements increases, as shown in Figure 2.19. It is also noted that similar results can be obtained even for nonuniform finite element models. Indeed, for the finite element models given in Figure 2.20, convergence is obtained as shown in Figure 2.19. Although the amount of error is larger than that by uniform models, the rate of convergence is almost the same.

Example 2.5: Let us consider a heat conduction problem, the domain of which has a hole inside. Assuming complete radiation in the hole, we have constant temperature on the surface of the hole and the balance of in and out heat flux through the inner surface Γ_S:

$$T = \text{const} \quad \text{(unknown a priori)}$$
$$\int_{\Gamma_S} k\frac{\partial T}{\partial n}\, d\Gamma = 0 \tag{2.85}$$

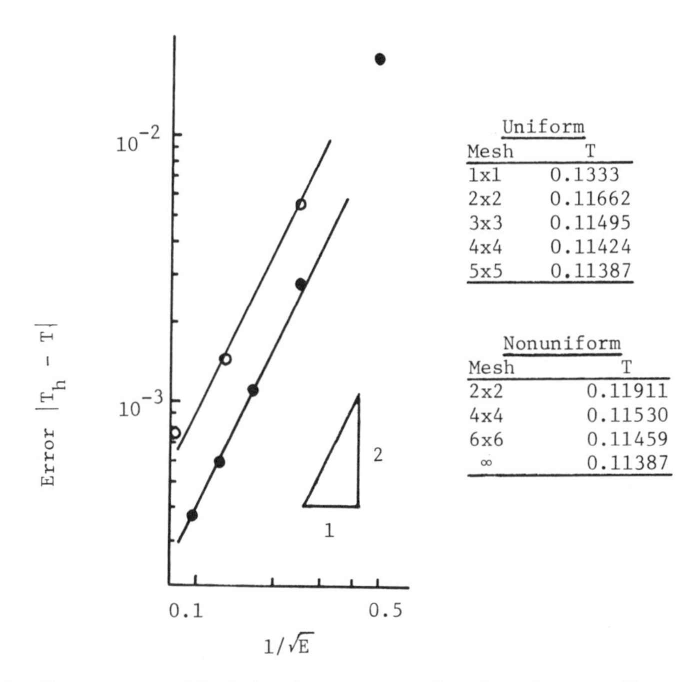

Figure 2.19 Convergence of the finite element approximation: ○, nonuniform; ●, uniform.

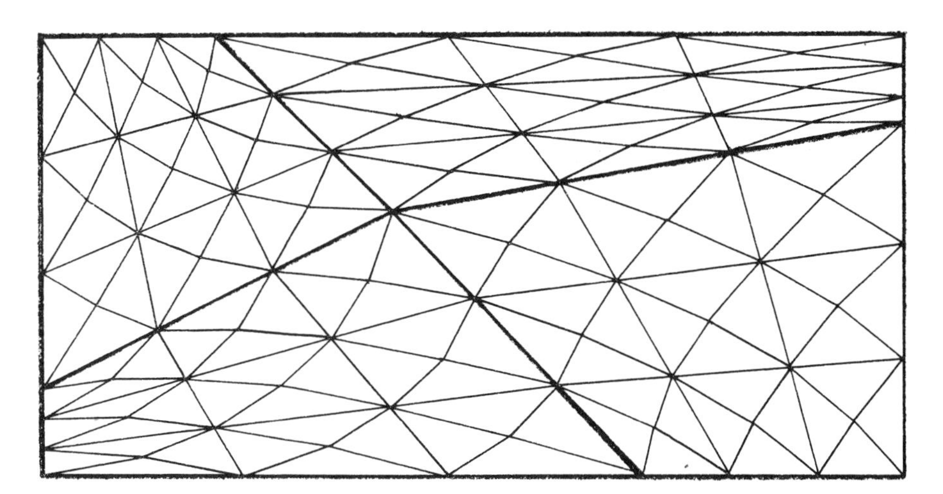

Figure 2.20 Irregular finite element discretization for a convergence test.

This problem can be solved by a similar technique to penalty methods, which have been applied to resolve the first type of boundary condition. Let us first review this. If the condition $T = g$ must be satisfied on Γ_1, this can be approximated by the third type of boundary condition

$$k\frac{\partial T}{\partial n} = -k_0(T - g) \quad \text{on } \Gamma_1 \tag{2.86}$$

applying a very large heat transfer coefficient k_0. Physically, if the heat transfer coefficient k_0 is extremely large, the surface temperature of the

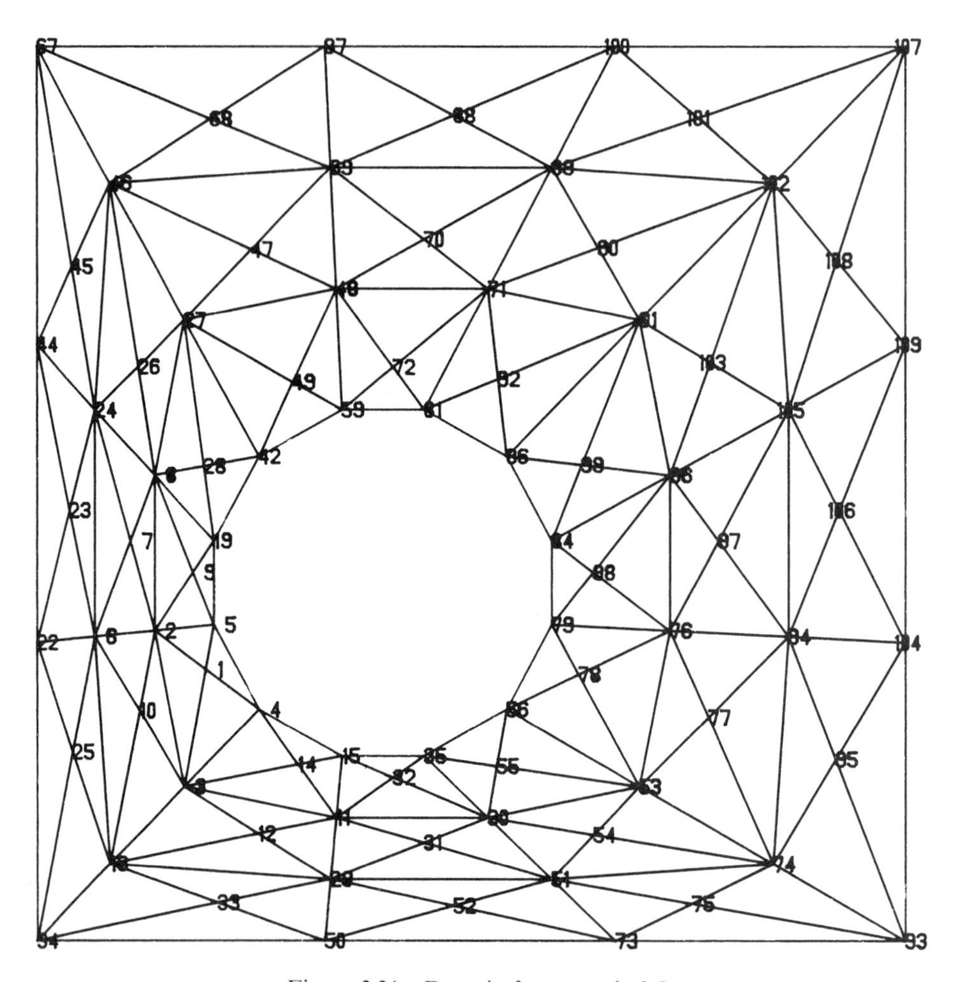

Figure 2.21 Domain for example 2.5.

body becomes almost the same temperature g on the outside of the boundary. Indeed, if the heat flux $k\,\partial T/\partial n$ is finite, it follows from (2.86) that $T - g = O(1/k_0)$.

Now noting that (2.85) can be equivalently represented by

$$T_{,i} = 0, \quad i = 1, 2 \text{ in } \Omega_0 \tag{2.87}$$

where Ω_0 is the inside hole, addition of the term

$$\int_{\Omega_0} \hat{k}\delta_{ij} T_{,j} \bar{T}_{,i} \, d\Omega \tag{2.88}$$

to the left side of the weak form (2.27) implies the condition (2.85) in an approximate sense. This means that the hole Ω_0 is identified with a material, the conductivity of which is extremely large. It is physically obvious that the temperature in Ω_0 is almost the same everywhere. Mathematically, it is possible to show that

$$T_{,i} T_{,i} = O(1/\hat{k})$$

that is, $T_{,i} = O(1/\sqrt{\hat{k}})$ for $i = 1$ and 2.

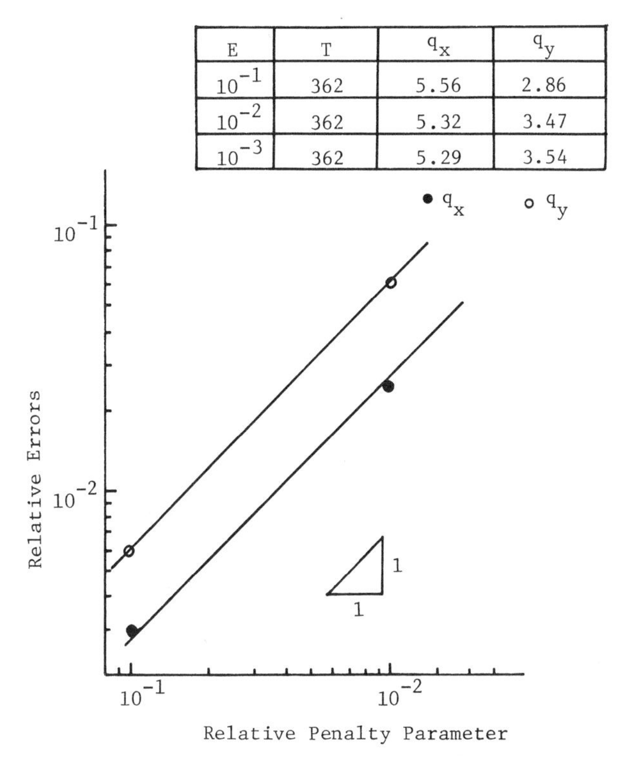

E	T	q_x	q_y
10^{-1}	362	5.56	2.86
10^{-2}	362	5.32	3.47
10^{-3}	362	5.29	3.54

Figure 2.22 Convergence of the penalty method ($1/k \to 0$).

We shall solve the problem shown in Figure 2.21 for $\hat{k} = 10k$, $100k$, and $1000k$, where k is the heat conductivity for Ω. It is clear that even for $\hat{k} = 10k$ constant temperature is almost achieved in the "hole" Ω_0. Convergence of the solutions for varying the constant $\hat{k}$ is given in Figure 2.22. Figure 2.23 shows the heat flux in Ω.

Exercise 2.4: Suppose that a rectangular domain consists of two different materials, say aluminum and copper, as shown in Figure 2.24. Find the matrix of heat conductivity $\mathbf{k}$ in each block.

Exercise 2.5: Suppose that the heat conductivity $\mathbf{k}$ depends on temperature. In this case, the weak form (2.27) becomes

$$T = g \quad \text{on } \Gamma_1$$

$$\int_\Omega k_{ij}(T) T_{,j} \bar{T}_{,i}\, d\Omega + \int_{\Gamma_3} k_0 T \bar{T}\, d\Gamma = \int_\Omega f \bar{T}\, d\Omega + \int_{\Gamma_2} h \bar{T}\, d\Gamma + \int_{\Gamma_3} k_0 T_\infty \bar{T}\, d\Gamma, \quad \forall \bar{T} \ni \bar{T} = 0 \text{ on } \Gamma_1 \tag{2.89}$$

This problem can be solved by the *successive iterative method.* Indeed, we first assume the temperature field T_0 to evaluate the heat conductivity $\mathbf{k}(T_0)$. Using this, we solve the problem based on the weak

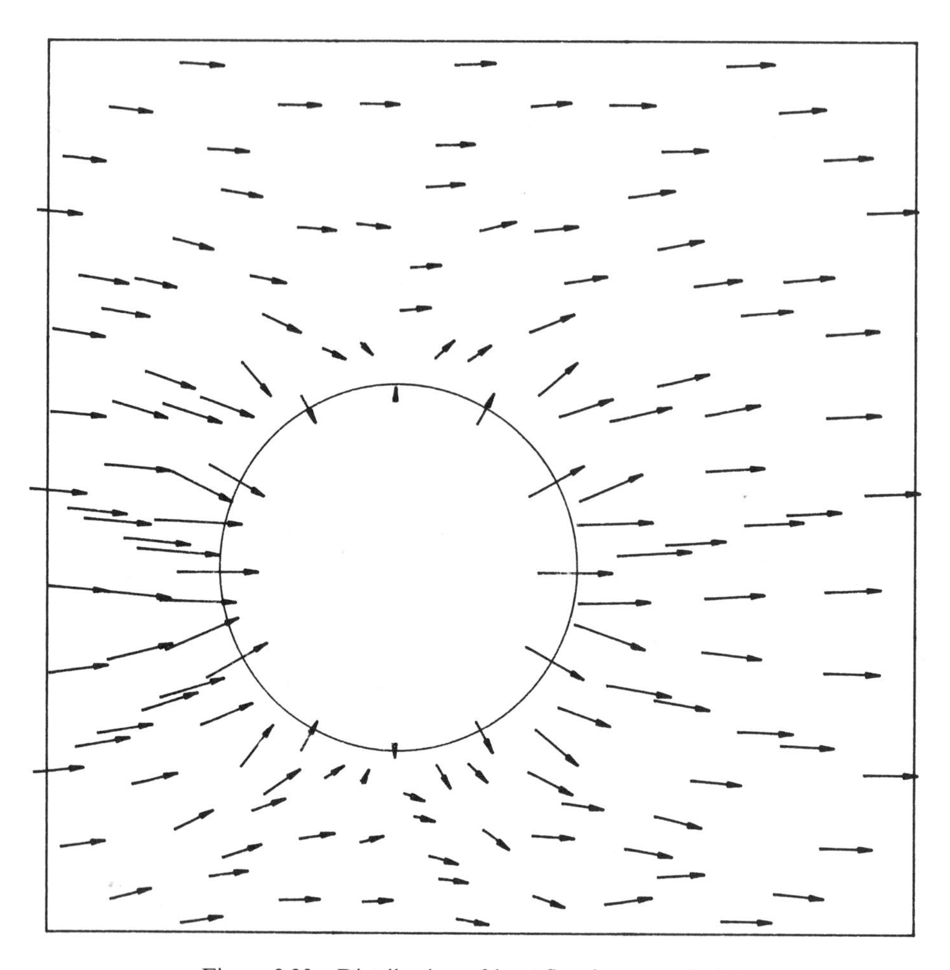

Figure 2.23 Distribution of heat flux for example 2.5.

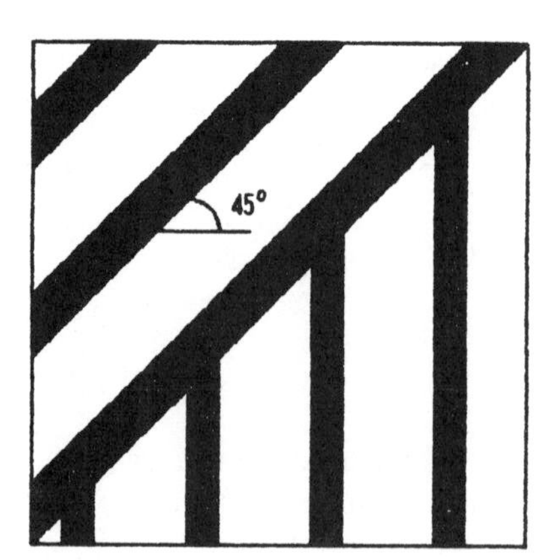

Figure 2.24 A composite material: ■ copper, □ aluminum.

form (2.27) and obtain a solution, say T_1. Then the heat conductivity is updated by the newly obtained temperature T_1. Using $\mathbf{k}(T_1)$, we solve (2.27) and obtain the solution T_2. If the difference between T_1 and T_2 is large, repeat the above process until this becomes small enough. After a certain number of iterations, it is expected to have convergence.

Now, modify the program FEM1 so that the successive iterative method is applicable for the case of dependent heat conductivity $\mathbf{k}(T)$ on temperature.

2.9 BASIC-FEM1 for a microcomputer

The development of electronics during the 1970s enables us to use microcomputers such as TRS80, APPLE II+, IBM PC, and others to develop small-size special-purpose finite element codes. Microcomputers in the early 1980s are no longer "micro" if we remember the situation in the 1960s and 1970s. Indeed, in the price range of $3000–$5000 in 1984, at least 256K–512K memory was available in microcomputers. This amount of memory is often sufficient for a finite element program that solves a specific small-size problem, especially if the program is separated into three processings: pre-, FE, and postprocessing. Using the disk drive, all input data created in the preprocessing can be written on a disk and then retrieved at the FE processing. Computed results in the FE processing are also written in a disk so that the last postprocessing has access to all data necessary. It is certainly true that the speed of computation cannot compare with the supercomputers, for example, CYBER, CRAY, and others. But if the purpose of the program is restricted to small-size problems, or is for education, then microcomputers have a place even in the area of finite element methods that has always been closely tied with large-size, superfast computers.

To exploit fully the new development of microcomputers, let us transfer the program FEM1 to a microcomputer. Here we shall use an IBM PC with 256K memory and two disk drives. The memory size could be 64K if the size of the finite element model is small. A characteristic of microcomputers with BASIC is the fully interactive style. Thus, it is easy to make a program to which we can talk. It will ask us for the input data, and it will serve only one person at a time.

The BASIC language is standard. However, each microcomputer has its own way of interacting with the disk drive. Thus, the parts related to the disk drive have to be modified if other microcomputers are used. The programs listed below are programmed according to Pode and Schneider (1983).

```
1000 CLS : SCREEN 0
1010 PRINT "   ****************"
1020 PRINT "       F E M  1     "
1030 PRINT "   ****************"
1040 PRINT "                        Version 1.0 / 1985"
1050 PRINT : PRINT
1060 PRINT "                  THIS PROGRAM IS ALSO APPLICABLE TO "
1070 PRINT "                        POTENTIAL/IRROTATIONAL/SEEPAGE FLOW"
1080 PRINT "                        LUBRICATION PROBLEMS"
1090 PRINT "                        ELASTIC TORSION PROBLEMS"
1100 PRINT "                        ELASTIC MEMBRANE PROBLEMS ETC."
```

```
1110 PRINT : PRINT
1120 Y$ = "YES":N$ = "NO"
1130 PRINT "PRE-PROCESSING"
1140 PRINT "--------------------------------------------------------------------"
1150 PRINT
1160 INPUT "IS THIS PROBLEM PLANE OR AXISYMMETRIC ?   (P/A) = ";PA$
1170 PRINT : PRINT
1180 PRINT "WHICH FUNCTION IS ASSUMED IN THIS PROBLEM ?"
1190 PRINT
1200 PRINT "                 0 = VELOCITY POTENTIAL"
1210 PRINT "                     TEMPERATURE"
1220 PRINT "                     PRESSURE"
1230 PRINT "                     TOTAL HEAD"
1240 PRINT "                 1 = STREAM FUNCTION"
1250 PRINT "                     STRESS FUNCTION"
1260 PRINT
1270 INPUT "YOUR SELECTION (0/1) = ";NF
1280 PRINT : PRINT
1290 PRINT : PRINT
1300 INPUT "ARE THE DATA (COORDINATES AND ELEMENT CONNECTIVITIES) IN A FILE ?
     (YES/NO)  = ";L$
1310 IF L$<>"YES" THEN GOTO 1390
1320 PRINT : INPUT "FILE NAME CONTAINING THE DATA = ";FD$
1330 OPEN FD$ FOR INPUT AS #1
1340 INPUT#1,NX,NE
1350 DIM X(NX),Y(NX),MP(NE),IK(3,NE),F(NX+1),QX(NE),QY(NE)
1360 FOR I=1 TO NX : INPUT#1,X(I),Y(I) : NEXT I
1370 FOR I=1 TO NE : INPUT#1,MP(I),IK(1,I),IK(2,I),IK(3,I) : NEXT I
1380 GOTO 1510
1390 PRINT
1400 INPUT "TOTAL NUMBER OF NODES    NX = ";NX
1410 INPUT "TOTAL NUMBER OF ELEMENTS    NE = ";NE
1420 DIM X(NX),Y(NX),MP(NE),IK(3,NE),F(NX+1),QX(NE),QY(NE)
1430 PRINT " PRINT "< COORDINATES OF NODES >" : PRINT
1440 FOR I=1 TO NX
1450 PRINT I;TAB(10);:INPUT "(X,Y) = ";X(I),Y(I)
1460 NEXT I
1470 PRINT : PRINT : PRINT "< ELEMENT CONNECTIVITIES >" : PRINT
1480 FOR N=1 TO NE
1490 PRINT N;TAB(7);:INPUT "<MP>,<CONNECTIVITY;I,J,K> = ";
     MP(N),IK(1,N),IK(2,N),IK(3,N)
1500 NEXT N
1510 PRINT : PRINT : PRINT "< MATERIAL CONSTANTS ; HEAT CONDUCTIVITY > " : PRINT
1520 INPUT "TOTAL NUMBER OF DIFFERENT MATERIALS  MX = ";MX : PRINT
1530 DIM HC(2,MX)
1540 FOR I=1 TO MX
1550 PRINT "GROUP ";I;TAB(10);:INPUT "(KXX,KYY) = ";HC(1,I),HC(2,I)
1560 NEXT I
1570 PRINT : PRINT : PRINT
1580 PRINT "DISTRIBUTED SOURCE/SINK xxxxxxxxxxxxxxx"
1590 PRINT
1600 INPUT "ARE THERE ANY SOURCES/SINKS ?  (YES/NO) = ";L$
1610 IF L$ < > Y$ THEN GOTO 1650
1620 PRINT : PRINT " ASSUME A POLYNOMIAL FORM : ":
     PRINT " F = FO+F1*X+F2*Y+F3*X*X+F4*X*Y+F5*Y*Y": PRINT
1630 INPUT "? FO,F1,F2,F3,F4,F5 = ";FO,F1,F2,F3,F4,F5
1640 GOTO 1750
1650 FO = 0:F1 = 0:F2 = 0:F3 = 0:F4 = 0:F5 = 0
1660 PRINT : PRINT : PRINT
1670 PRINT "COMPUTE THE BAND WIDTH xxxxxxxxxxxxxxx"
1680 PRINT
1690 MB=0
1700 FOR N=1 TO NE
1710 FOR IA=1 TO 3 : I1=IK(IA,N)
1720 FOR JB=1 TO 3 : I2=IK(JB,N)
1730 IF MB<I2-I1+1 THEN MB=I2-I1+1
1740 NEXT JB : NEXT IA : NEXT N
1750 PRINT "BAND WIDTH   MB = ";MB
1760 PRINT : PRINT :PRINT
1770 PRINT "FINITE ELEMENT PROCESSING"
1780 PRINT "--------------------------------------------------------------------"
```

```
1790 PRINT
1800 DIM SK(NX,MB),SE(3,3),FE(3)
1810 REM < INITIAL CLEARANCE OF THE STIFFNESS MATRIX SK AND THE LOAD VECTOR F >
1820 FOR I = 1 TO NX
1830 F(I) = 0
1840 FOR J = 1 TO MB
1850 SK(I,J) = 0
1860 NEXT J : NEXT I
1870 PRINT "STIFFNESS MATRIX xxxxxxxxxxxxxxx" : PRINT
1880 N = 0
1890 N=N+1
1900 PRINT N;" ";
1910 I1=IK(1,N) : I2=IK(2,N) : I3=IK(3,N)
1920 X1=X(I1) : Y1=Y(I1) : X2=X(I2) : Y2=Y(I2) : X3=X(I3) : Y3=Y(I3)
1930 KX=HC(1,MP(N)) : KY=HC(2,MP(N))
1940 JC = X1 * (Y2 - Y3) + X2 * (Y3 - Y1) + X3 * (Y1 - Y2)
1950 AR = .5 * JC
1960 B(1) = (Y2 - Y3) / JC:C(1) = (X3 - X2) / JC
1970 B(2) = (Y3 - Y1) / JC:C(2) = (X1 - X3) / JC
1980 B(3) = (Y1 - Y2) / JC:C(3) = (X2 - X1) / JC
1990 XE = (X1 + X2 + X3) / 3:YE = (Y1 + Y2 + Y3) / 3
2000 F9 = (F0 + F1 * XE + F2 * YE +F3 * XE * XE + F4 * XE * YE
          + F5 * YE * YE) * AR / 3
2010 FE(1) = F9:FE(2) = F9:FE(3) = F9
2020 FOR I = 1 TO 3: FOR J = 1 TO 3
2030 SE(I,J) = (KX * B(I) * B(J) + KY * C(I) * C(J)) * AR
2040 NEXT J : NEXT I
2050 CS = 1: IF PA$ = "A" THEN CS = XE
2060 FOR I=1 TO 3 : I1=IK(I,N)
2070 F(I1) = F(I1) + CS * FE(I)
2080 FOR J=1 TO 3 : JJ=IK(J,N)
2090 MM = JJ - I1 + 1
2100 IF MM < = 0 THEN GOTO 2120
2110 SK(I1,MM) = SK(I1,MM) + CS * SE(I,J)
2120 NEXT J : NEXT I
2130 IF N < NE THEN GOTO 1890
2140 IF NF = 1 THEN GOTO 2290
2150 PRINT : PRINT
2160 PRINT "POINT SOURCES/SINKS xxxxxxxxxxxxxxx"
2170 PRINT
2180 INPUT "ARE THERE ANY POINT SOURCES/SINKS ?  (YES/NO) = ";L$
2190 IF L$ < > Y$ THEN GOTO 2290
2200 PRINT " INPUT THEIR POSITIONS BY NODE NUMBERS AND THEIR STRENGTH."
2210 PRINT "              (+) = SOURCE     (-) = SINK"
2220 INPUT "HOW MANY POINT SOURCES/SINKS ?    NS = ";NS : PRINT
2230 FOR I = 1 TO NS
2240 PRINT I;TAB(10);:INPUT "(NODE NUMBER, STRENGTH) = ";N1,V1
2250 CS = 1: IF PA$ = "A" THEN CS = X(N1)
2260 IF CS = 0 THEN CS = 1
2270 F(N1) =  CS * V1
2280 NEXT I
2290 PRINT : PRINT
2300 PRINT "1-ST BOUNDARY CONDITION xxxxxxxxxxxxxxx"
2310 PRINT
2320 INPUT "TOTAL NUMBER OF NODES AT WHICH TEMPERATURE IS SPECIFIED = ";NS
2330 IF NS < = 0 THEN GOTO 2220
2340 PRINT
2350 INPUT "PENALTY PARAMETER (THAT IS A VERY LARGE NUMBER IN COMPARISON
     WITH KXX AND KYY.) = ";SP
2360 EP = 0: PRINT
2370 IF NS = 0 THEN GOTO 2460
2380 FOR I = 1 TO NS
2390 INPUT "(NODE,TEMPERATURE) = ";N1,V1
2400 CS = 1: IF PA$ = "A" THEN CS = X(N1)
2410 IF CS = 0 THEN CS = 1
2420 F(N1) = F(N1) + CS * SP * V1
2430 SK(N1,1) = SK(N1,1) + CS * SP
2440 NEXT I
2450 GOTO 2470
2460 EP=.0001
2470 PRINT : PRINT
```

```
2480 PRINT "2-ND BOUNDARY CONDITIONS xxxxxxxxxxxxxxx"
2490 PRINT
2500 INPUT "HOW MANY LINE ELEMENTS ARE THERE ON THE 2-ND BOUNDARY ?  NS = ";NS
2510 IF NS < = 0 THEN GOTO 2610
2520 PRINT
2530 PRINT " INPUT LINE ELEMENT CONNECTIVITIES AND VALUES OF HEAT FLUX." : PRINT
2540 FOR I = 1 TO NS
2550 PRINT I;TAB(10);: INPUT "(NODE1,NODE2),FLUX = ";N1,N2,V1
2560 LE = SQR((X(N1)-X(N2))*(X(N1)-X(N2)) + (Y(N1)-Y(N2))*(Y(N1)-Y(N2)))
2570 CS = 1: IF PA$ = "A" THEN CS = .5 * (X(N1) + X(N2))
2580 F(N1) = F(N1) + CS * V1 * LE / 2
2590 F(N2) = F(N2) + CS * V1 * LE / 2
2600 NEXT I
2610 PRINT : PRINT
2620 PRINT "3-RD BOUNDARY CONDITION xxxxxxxxxxxxxxx"
2630 PRINT
2640 INPUT "HOW MANY LINE ELEMENTS ARE THERE ON THE 3-RD BOUNDARY ?  NY = ";NY
2650 IF NY < = 0 THEN GOTO 2770
2660 FOR I = 1 TO NY
2670 PRINT I;TAB(10);: INPUT "(NODE1,NODE2),KO,GO = ";N1,N2,KO,GO
2680 CS = 1: IF PA$ = "A" THEN CS = .5 * (X(N1) + X(N2))
2690 LE = SQR((X(N2)-X(N1))*(X(N2)-X(N1)) + (Y(N2)-Y(N1))*(Y(N2)-Y(N1)))
2700 K1 = CS * KO * LE / 3:G1 = CS * KO * GO * LE / 2
2710 N3 = N2 - N1 + 1
2720 SK(N1,1) = SK(N1,1) + K1:F(N1) = F(N1) +G1
2730 SK(N2,1) = SK(N2,1) + K2:F(N2) = F(N2) +G1
2740 IF N3 > 0 THEN SK(N1,N3) = SK(N1,N3) + .5 * K1
2750 IF N3 < = 0 THEN SK(N2,N1 - N2 + 1) = SK(N2,N1 - N2 + 1) + .5 * K1
2760 NEXT I
2770 PRINT : PRINT
2780 PRINT "SOLVE THE FINITE ELEMENT EQUATION"
2790 PRINT "------------------------------------------------------------------------"
2800 PRINT
2810 PRINT "< FORWARD ELIMINATION >"
2820 N1=NX-1
2830 FOR N=1 TO N1
2840 PRINT N;" ";
2850 M = N - 1:MR = MB:NM = NX - M
2860 IF MR > NM THEN MR = NM
2870 PI = SK(N,1)
2880 FOR L=2 TO MR
2890 CP = SK(N,L) / PI
2900 IF CP = 0 THEN GOTO 2960
2910 I = M + L:J = 0
2920 FOR K = L TO MR
2930 J = J + 1
2940 SK(I,J) = SK(I,J) - CP * SK(N,K)
2950 NEXT K
2960 SK(N,L) = CP
2970 NEXT L
2980 NEXT N
2990 PRINT
3000 PRINT "< REDUCTION OF THE LOAD VECTOR >"
3010 FOR N=1 TO N1
3020 M = N - 1:MR = MB:NM = NX - M
3030 IF MR > NM THEN MR = NM
3040 CP = F(N)
3050 IF CP = 0 THEN GOTO 3110
3060 F(N) = CP / SK(N,1)
3070 FOR L = 2 TO MR
3080 I = M + L
3090 F(I) = F(I) - SK(N,L) * CP
3100 NEXT L
3110 NEXT N
3120 PRINT "<  BACK SUBSTITUTION  >"
3130 F(NX) = F(NX) / SK(NX,1)
3140 FOR I=1 TO N1
3150 N = NX - I:M = N - 1:MR = MB:NM = NX - M
3160 IF MR > NM THEN MR = NM
3170 FOR K=2 TO MR
```

```
3180 L = M + K
3190 F(N) = F(N) - SK(N,K) * F(L)
3200 NEXT K
3210 NEXT I
3220 PRINT : PRINT : PRINT : PRINT
3230 PRINT "POST-PROCESSING"
3240 PRINT "------------------------------------------------------------------------"
3250 PRINT
3260 PRINT "     --------------------------------------------------------"
3270 PRINT "                 TEMPERATURE    ( NODES )"
3280 PRINT "     --------------------------------------------------------"
3290 PRINT
3300 FOR I=1 TO NX STEP 2
3310 PRINT I;TAB(10);INT(100000!*F(I))/100000!;TAB(40);I+1;TAB(50);
     INT(100000!*F(I+1))/100000!
3320 NEXT L
3330 PRINT : PRINT : PRINT : PRINT
3340 PRINT "     --------------------------------------------------------"
3350 PRINT "          TEMPERATURE AND HEAT FLUX     ( ELEMENTS ) "
3360 PRINT "     --------------------------------------------------------"
3370 PRINT : PRINT
3380 FOR N=1 TO NE
3390 I1=IK(1,N) : I2=IK(2,N) : I3=IK(3,N)
3400 X1=X(I1) : Y1=Y(I1) : X2=X(I2) : Y2=Y(I2) : X3=X(I3) : Y3=Y(I3)
3410 KX=HC(1,MP(N)) : KY=HC(2,MP(N))
3420 JC = X1 * (Y2 - Y3) + X2 * (Y3 - Y1) + X3 * (Y1 - Y2)
3430 AR = JC / 2
3440 B1 = (Y2 - Y3) / JC:C1 = (X3 - X2) / JC
3450 B2 = (Y3 - Y1) / JC:C2 = (X1 - X3) / JC
3460 B3 = (Y1 - Y2) / JC:C3 = (X2 - X1) / JC
3470 V1 = F(I1) * B1 + F(I2) * B2 + F(I3) * B3
3480 V2 = F(I1) * C1 + F(I2) * C2 + F(I3) * C3
3490 F3=(F(I1)+F(I2)+F(I3))/3
3500 XN=(X(I1)+X(I2)+X(I3))/3 : YN=(Y(I1)+Y(I2)+Y(I3))/3
3510 VX=-KX*V1 : IF NF=1 THEN VX=KX*V2
3520 VY=-KY*V2 : IF NF=1 THEN VY=-KY*V1
3530 PRINT N;TAB(8);"X ";INT(10000*XN)/10000;TAB(20);"Y ";INT(10000*YN)/10000;
     TAB(32);"T ";INT(10000000#*F3)/10000000#,TAB(47);"QX ";
     INT(10000000#*VX)/10000000#;TAB(62);"QY ";INT(10000000#*VY)/10000000#
3540 QX(N)=VX : QY(N)=VY
3550 NEXT N
3560 PRINT : PRINT
3570 PRINT "STORE THE FE-MODEL AND RESULTS xxxxxxxxxxxxxxx"
3580 REM     < THE FOLLOWING IS SPECIFICALLY FOR IBM PC  >
3590 PRINT
3600 INPUT "WHAT IS THE FILE NAME ?   (e.g. A:RESULT)   = ";F$
3610 OPEN F$ FOR OUTPUT AS #2
3620 PRINT#2,NX : PRINT#2,NE : PRINT#2,MB
3630 FOR I=1 TO NX : PRINT#2,X(I) : PRINT#2,Y(I) : PRINT#2,F(I) : NEXT I
3640 FOR N=1 TO NE
3650 PRINT#2,IK(1,N) : PRINT#2,IK(2,N) : PRINT#2,IK(3,N) :
     PRINT#2,QX(N) : PRINT#2,QY(N)
3660 NEXT N
3670 CLOSE #2
3680 PRINT :
     PRINT "END OF COMPUTATION xxxxxxxxxxxxxxxxxxxxxxxxxxxxxxxxxxxxxxxxxxxxxxxxxxxxxx"
3690 END
```

Example 2.6: We shall solve the heat conduction problem shown in Figure 2.25 using BASIC-FEM1. For simplicity, the material in the domain is assumed to be isotropic and homogeneous, say $k_{ij} = k\delta_{ij}$ and $k = 1$. No heat source and sink exist in the domain. The first type of boundary condition is assumed on the top surface, while the second type is specified on the rest of the boundary. The domain is discretized by 74 three-node triangular elements.

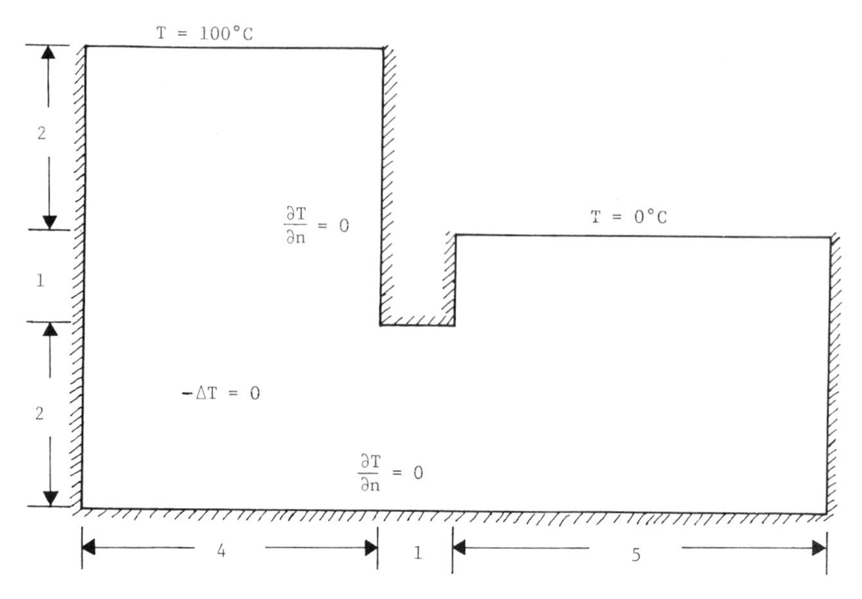

Figure 2.25 Domain for example 2.6 for BASIC-FEM1.

Computational results are obtained within 600 s and are visualized in Figure 2.26. These figures of the finite element model, heat flux, and isotemperature lines are plotted by IBM PC using a postprocessing program for BASIC-FEM1. This example shows a possibility of using microcomputers for the study of finite element methods.

2.10 General remarks

We have so far explained a finite element method that solves heat conduction problems and have developed computer codes using both FORTRAN and BASIC. In this section, we shall attempt to obtain a deeper understanding of the method introduced.

2.10.1 Other field problems

Many problems in mechanics and physics are governed by boundary value problems similar to that which governs the heat conduction problem studied above. For example, if we are interested in the transverse deflection of a soap film (or a thin elastic membrane), the boundary value problem becomes

$$-Tu_{,ii} = f \quad \text{in } \Omega$$

$$u = g \quad \text{on } \Gamma_1, \qquad T\frac{\partial u}{\partial n} = h \quad \text{on } \Gamma_2$$

where T is the surface tension, u the deflection, f the applied load on the surface, g the prescribed boundary deflection, and h the normal traction on the boundary. If T is constant in Ω, then the above problem is exactly the same as the heat conduction problem. Similarly, many other problems in mechanics have this form. Some of these problems are listed in Table 2.2.

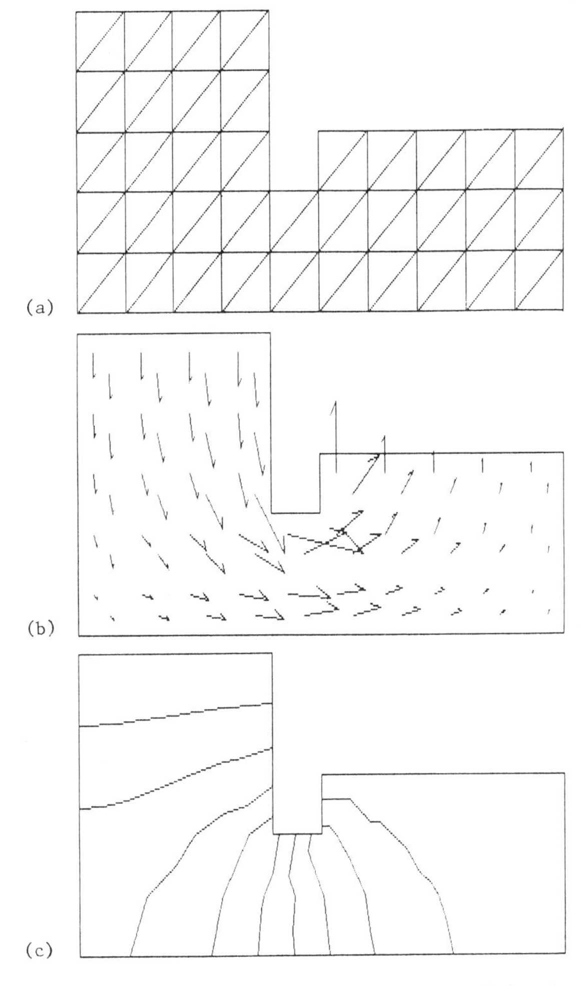

Figure 2.26 Results of example 2.6 for BASIC-FEM1: (a) finite element model; (b) distribution of heat flux; (c) isotemperature lines.

Table 2.2 *List of field problems*

Field problem	Unknown	k_x, k_y (or $k_x = k_y = k$)	f
Heat conduction	Temperature (T)	Thermal conductivity	Internal heat
Seepage flow	Hydraulic head (h)	Permeability	None
Incompressible flow	Stream function (ψ)	Unity	Twice the vorticity
Soap film	Deflection (u)	Surface tension	Pressure
Elastic torsion	Stress function (ϕ)	(Shear modulus)$^{-1}$	Twice the rate of twist
Elastic torsion	Warping function (ψ)	Unity	None
Electrical conduction	Voltage (u)	Electrical conductivity	Internal current source
Magnetostatics	Magnetic potential (u)	Reluctivity	Current density

The following exercises are introduced in order to extend the finite element method for heat conduction to a variety of problems in mechanics, some of which involve nonlinearity. These may be considered as stepping stones to study advanced problems using finite element methods. Problems are obtained from potential flow with free boundary, seepage flow through porous media, lubrication of oil films, torsion of elastic bars, and deflection of elastic membranes. Details of the mechanics of these problems are not discussed here, but fundamental references are given at the end of each exercise.

Exercise 2.6 (flow through an open channel): Let the average angular velocity of a flow be defined by

$$\boldsymbol{\omega} = \omega_x \mathbf{i} + \omega_y \mathbf{j} + \omega_z \mathbf{k}$$

where

$$\omega_x = \frac{1}{2}\left(\frac{\partial w}{\partial y} - \frac{\partial v}{\partial z}\right), \qquad \omega_y = \frac{1}{2}\left(\frac{\partial u}{\partial z} - \frac{\partial w}{\partial x}\right), \qquad \omega_z = \frac{1}{2}\left(\frac{\partial v}{\partial x} - \frac{\partial u}{\partial y}\right)$$

and $\mathbf{v} = u\mathbf{i} + v\mathbf{j} + w\mathbf{k}$ is the velocity vector of the flow. If the problem is two dimensional so that $u = u(x, y)$, $v = v(x, y)$, and $w = 0$ are assumed, then only one component remains:

$$\omega_z = \frac{1}{2}\left(\frac{\partial v}{\partial x} - \frac{\partial u}{\partial y}\right).$$

If a stream function ψ is assumed to exist for the flow field, the velocity is related to it by

$$u = \frac{\partial \psi}{\partial y} \quad \text{and} \quad v = -\frac{\partial \psi}{\partial x} \tag{2.90}$$

and then the equation governing the average angular velocity becomes

$$-\left(\frac{\partial^2 \psi}{\partial x^2} + \frac{\partial^2 \psi}{\partial y^2}\right) = 2\omega_z \quad \text{in } \Omega \tag{2.91}$$

where Ω is the flow domain.

Find the velocity field of the flow in the domain shown in Figure 2.27 using FEM1 or BASIC-FEM1. If some modification of the program is necessary, please do so. The boundary conditions are

$$\frac{\partial \psi}{\partial n} = 0 \text{ on } \Gamma_1 \text{ and } \Gamma_3$$

$$\psi = 5 \text{ on } \Gamma_2 \text{ and } \psi = 0 \text{ on } \Gamma_4$$

and the function ω_z is defined as

$$\omega_z = \begin{cases} 1 & \text{at A} \\ 0 & \text{elsewhere} \end{cases}$$

Since

$$\frac{\partial \psi}{\partial n} = n_x \frac{\partial \psi}{\partial x} + n_y \frac{\partial \psi}{\partial y} = \frac{\partial \psi}{\partial y}$$

on Γ_1 and Γ_3, we have assumed $u = 0$ on these boundaries.

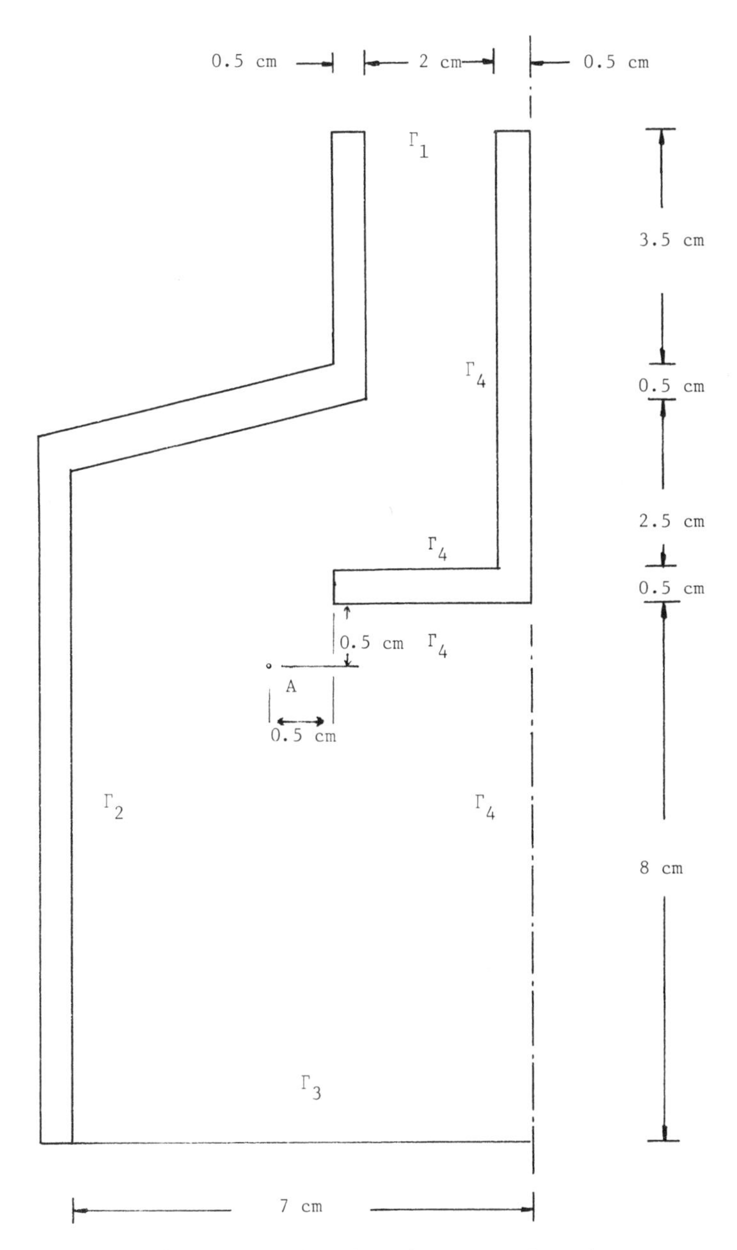

Figure 2.27 Flow through an open channel.

Compute the velocity fields for the cases $\psi = 4, 3, 2$, and 1 on Γ_2, assuming that the other conditions remain the same. Compute the total amount of fluid flowing through the boundaries Γ_1 and Γ_3 for the cases $\psi = 5, 4, 3, 2$, and 1. Plot the relation between the value of the stream function on Γ_2 and the amount of flow.

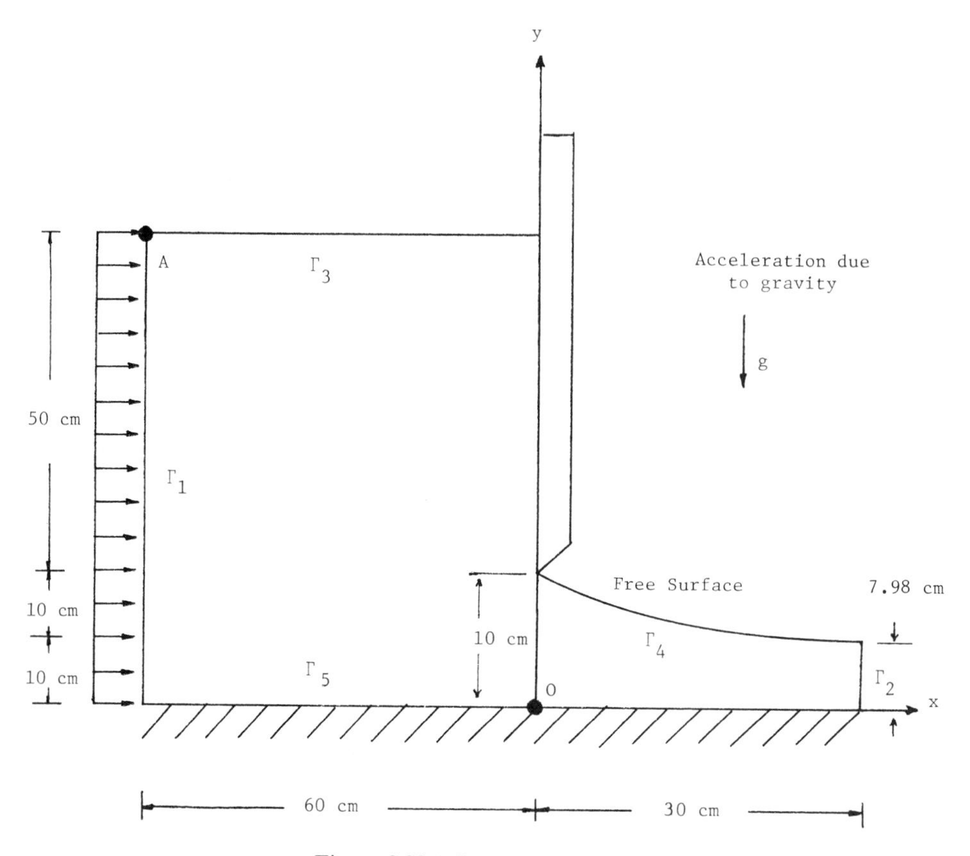

Figure 2.28 Open-gate problem.

Exercise 2.7 (open-gate problem): Suppose that there exists a velocity potential ϕ such that

$$\mathbf{v} = -\nabla\phi \tag{2.92}$$

where $\mathbf{v}$ is the velocity vector of the flow. It follows from the principle of mass conservation that the equation

$$\nabla \cdot \mathbf{v} = -\nabla \cdot \nabla\phi = 0 \tag{2.93}$$

holds for the steady flow of incompressible fluids in the flow domain Ω.

Find the velocity field of the flow shown in Figure 2.28. Assume that the upstream velocity is specified as $u = 40$ cm/s and $v = 0$ cm/s, that is,

$$\frac{\partial\phi}{\partial n} = \mathbf{n} \cdot \nabla\phi = 40 \text{ cm/s} \quad \text{on } \Gamma_1$$

Under the assumption that the downstream value of the potential ϕ is zero,

$$\phi = 0 \quad \text{on } \Gamma_2$$

the flow problem must be solved. The boundaries Γ_3 and Γ_4 are free-stream lines, and the slope of Γ_4 is assumed to be quadratic, that is,

$$y = \tfrac{1}{250}(x - 30)^2 + 7.98$$

The boundary Γ_5 is the bottom line.

After finding the velocity field of the flow, compute the quantity

$$E = \tfrac{1}{2}|\nabla\phi|^2 + gy \tag{2.94}$$

on the boundary Γ_4, where g = acceleration due to gravity, and then find the difference between E and E_0 defined by

$$E_0 = \tfrac{1}{2}u^2 + gH \tag{2.95}$$

where u and H are the velocity and the height of the point A shown in the figure. If the Bernoulli equation is to be satisfied along the free-stream lines Γ_3 and Γ_4, we must have

$$E = E_0 \text{ along } \Gamma_4 \tag{2.96}$$

Adjust the y coordinates of the nodal points on the line Γ_4 so that $E = E_0$ is satisfied, and recompute the velocity field. Check once more whether or not the condition $E = E_0$ is satisfied on Γ_4. If E differs from E_0, then readjust the boundary Γ_4, and compute the velocity field again. If we repeat this process until $E = E_0$ is satisfied, the exact flow domain and the velocity field will have been obtained.

It is noted that the iterative scheme based on

$$y_{\text{new}} = \frac{E_0 - \tfrac{1}{2}u_{\text{old}}^2}{g}$$

would not imply convergence to a desirable solution, where u_{old} is the speed of the flow on the free boundary Γ_4 computed on the domain defined by the coordinate y_{old} along Γ_4.

To solve the problem, we may need to state the weak form to the free boundary value problem in terms of both the velocity potential and the y coordinate of the boundary Γ_4. That is,

$$\begin{aligned} &\int_{\Omega(y)} \nabla\phi \cdot \nabla\bar{\phi}\, d\Omega = \int_{\Gamma_1} h\bar{\phi}\, d\Gamma, \quad \forall\bar{\phi} \ni \bar{\phi} = 0 \text{ on } \Gamma_2 \\ &\int_0^a \left(\tfrac{1}{2}|\nabla\phi|^2 + gy\right)\bar{y}\, dx = \int_0^a E_0\bar{y}\, dx, \quad \forall\bar{y} \end{aligned} \tag{2.97}$$

Here h is the flux of the inflow through the left side Γ_1 and a is the x coordinate of the location of the boundary Γ_2. These are assumed as $h = 40$ cm/s and $a = 30$ cm in the present example. The domain Ω is a function of y. The second equation is a weak form to the Bernoulli equation along the free boundary Γ_4.

It is clear that the system of equations (2.97) is nonlinear. Solve this using the Newton–Raphson method discussed in section 5.10. An interesting reference to the present problem is Ikegawa and Washizu (1973).

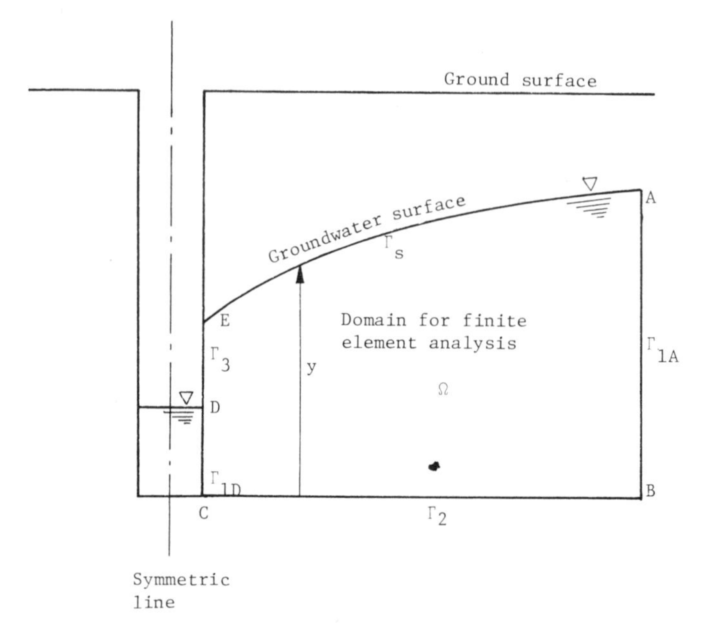

Figure 2.29 Seepage flow problem.

Exercise 2.8 (seepage flow problem): Let us consider a flow-through porous media so slow that the Darcy law

$$\mathbf{v} = -\mathbf{k}\ \nabla h \tag{2.98}$$

holds, where $\mathbf{v}$ is the velocity of flow, $\mathbf{k} = k_{ij}\mathbf{e}_i\mathbf{e}_j$ is the permeability in the rectangular coordinate system, and h is the hydraulic head. If the conservation of mass is considered in the steady state, we have

$$-\nabla \cdot \mathbf{v} = q$$

where q is a distributed source/sink of discharge per unit volume. Combining two equations yields the governing equation

$$-\nabla \cdot (\mathbf{k}\ \nabla h) = q \quad \text{in } \Omega \tag{2.99}$$

It is clear that this has the exact same form to heat conduction problems.

Now suppose that a well is constructed in a ground, as shown in Figure 2.29. Let us obtain the steady-state flow configuration and the flow field toward the well using finite element methods. Since the well only affects groundwater near to its location, a point A far from the well can be assumed on the level of groundwater surface before its construction. After defining a flow domain for finite element analysis as shown in Figure 2.29, let us assume the following boundary condition:

$$\begin{aligned}
&h = h_A \quad \text{on } \Gamma_{1A}, \quad h = h_D \quad \text{on } \Gamma_{1D} \\
&\mathbf{n} \cdot (\mathbf{k}\ \nabla h) = 0 \quad \text{on } \Gamma_2 \\
&\mathbf{n} \cdot (\mathbf{k}\ \nabla h) = -k_0(h - h_D) \quad \text{on } \Gamma_3 \\
&\mathbf{n} \cdot (\mathbf{k}\ \nabla h) = 0 \qquad \text{and} \qquad h = y \quad \text{on } \Gamma_S
\end{aligned} \tag{2.100}$$

where h_A and h_D are the "heights" from the horizontal line BC (the datum line for hydraulic head) and y is the location of the free boundary Γ_S, which is unknown a priori. For simplicity, let us assume the ground is isotropic and homogeneous, that is, $k_{ij} = k\delta_{ij}$ and k is a constant. It is noted that the seepage point E may coincide with D if DC is deep enough. That is, the third type of boundary Γ_3 may not exist for certain choices of configuration. Along the free boundary Γ_S, two boundary conditions are imposed. Since its location is unknown, two boundary conditions on Γ_S are not contradictory to the nature of physics.

How can we solve this problem? One of many ways is the following successive iterative method. We first assume the flow domain, say Ω_0, that is, the location of the free surface Γ_S, by specifying the y coordinate, say y_0. Since the location is assumed, two boundary conditions on Γ_S are now inadequate. Let us take only the no-flux condition $\mathbf{n} \cdot \mathbf{k}\, \nabla h = 0$ across the free boundary Γ_S. Solving the problem on Ω_0, the hydraulic head h_1 is obtained. If the location of the free boundary Γ_S assumed is close to the true one, h_1 must be almost the same as y_0 as Γ_S. If h_1 is not close to y_0 on Γ_S, we set up the new location of the free boundary y_1 by, for example,

$$y_1 = y_0 + \alpha(h_1 - y_0) \tag{2.101}$$

where α is a given constant that defines the amount of location change. Repeating this iteration procedure, we expect to obtain the location of the free boundary Γ_S that satisfies two boundary conditions.

Solve the problem using finite element methods for the case shown in Figure 2.29, and compare the result with Neumann and Witherspoon (1970) and Taylor and Brown (1967).

Exercise 2.9 (lubrication problem): After many assumptions about the flow of a lubricant, some lubrication problems are modeled by the Reynolds equation

$$-\nabla \cdot (h^3\, \nabla p) = -6U\eta \frac{\partial h}{\partial x} \tag{2.102}$$

where $\nabla = \mathbf{i}\, \partial/\partial x + \mathbf{j}\, \partial/\partial y$, $h = h(\mathbf{x})$ is the thickness of the lubricant film, p is the pressure, U is twice the value of the average velocity of the two surfaces that contain the lubricant film, and η is the coefficient of viscosity.

Solve the Reynolds equation with the boundary conditions

$$p = 0 \quad \text{on } \Gamma_1 \qquad \text{and} \qquad h^3 \nabla p \cdot \mathbf{n} = 0 \quad \text{on } \Gamma_2$$

for properly chosen constants U and η for a machine oil and for

$$h = R - \sqrt{R^2 - x^2}$$

where $R = 5$ cm. The domain Ω is defined as in Figure 2.30.

For certain thickness of the lubricant film, the pressure obtained by solving the boundary value problem based on the Reynolds equation may be negative in a portion of the domain where cavities are generated. Since the Reynolds equation holds only in the domain that does not contain cavities, we have a contradiction in such a case. To overcome

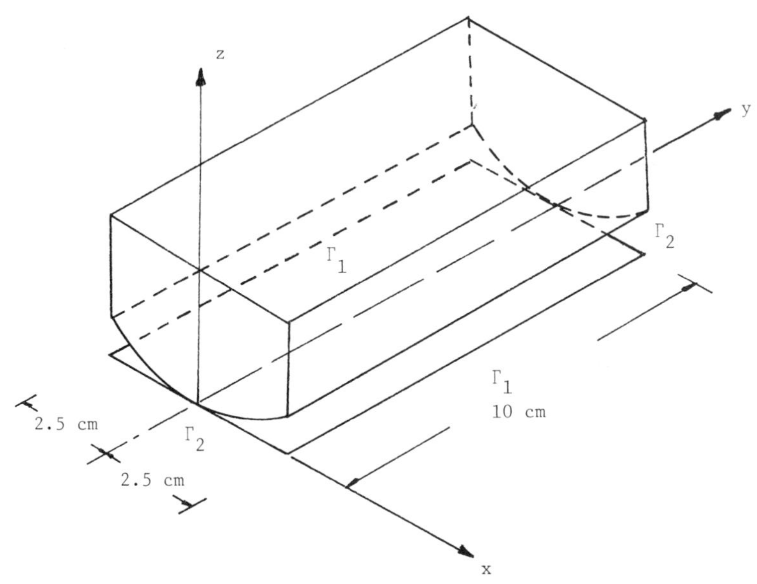

Figure 2.30 Lubrication problem.

this contradiction, we restrict the domain where boundary value problems are defined to the subdomain Ω_R in which the pressure p is nonnegative:

$$\Omega_R = \{\mathbf{x} \in \Omega: p(\mathbf{x}) \geq 0\} \tag{2.103}$$

The subdomain Ω_R may be the same as the original domain Ω for many cases. However, it is noted that Ω_R is unknown a priori because the pressure p is a primal variable to be obtained. A method to solve this problem is again application of penalty methods. That is, the Reynolds equation (2.102) is modified as

$$-\nabla \cdot (h^3 \nabla p) + kp^- = -6U\eta \frac{\partial h}{\partial x} \tag{2.104}$$

where $p^- = 0$ if $p \geq 0$, and $p^- = p$ if $p < 0$. In the domain Ω_R where $p \geq 0$ holds, the Reynolds equation is satisfied. On the other hand, if k is a very large number,

$$p = \frac{\nabla \cdot (h^3 \nabla p) - 6U\eta \, \partial h/\partial x}{k}$$

yields $|p|$ very small in the domain where the pressure p is negative. That is, modification (2.104) of the Reynolds equation violates the condition $p \geq 0$ to a very small degree in the whole domain Ω, while (2.104) yields the Reynolds equation in Ω_R. Since (2.104) violates the condition $p \geq 0$, and since a "penalty" kp^- is "paid," the above is called a *penalty approximation* of the Reynolds equation with respect to the nonnegativeness of the pressure. Solving the nonlinear equation (2.104) is straightforward using a successive iteration method.

Details of lubrication theory can be found in Dowson and Higginson (1977). An extensive study on finite element methods for lubrication is

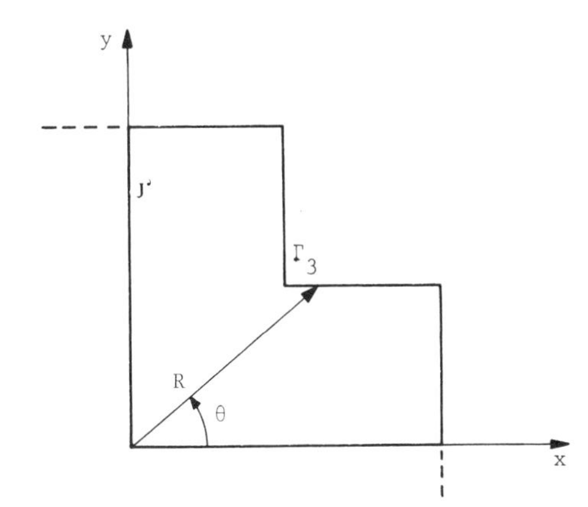

Figure 2.31 A quarter of the domain for an elastic torsion problem.

given by Huebner (1975), while mathematical structure of the Reynolds equation is analyzed by Kinderlehrer and Stampaccia (1980).

Exercise 2.10 (elastic torsion problem): Let us consider the torsion of a prismatic bar using the stress function

$$\sigma_{zx} = \frac{\partial \phi}{\partial y}, \quad \sigma_{zy} = -\frac{\partial \phi}{\partial x} \tag{2.105}$$

where the z axis coincides with the bar axis, and the xy plane contains the cross section of the bar. If the angle of twist per unit length, α, is specified, the problem is represented by the equation

$$-\nabla \cdot \left(\frac{1}{G} \nabla \phi\right) = 2\alpha \quad \text{in } \Omega \tag{2.106}$$

where G is the shear modulus of the bar.

If the domain (i.e., the cross section of the bar) is singly connected, the boundary condition of the stress function is given by

$$\phi = 0 \quad \text{on } \Gamma$$

Let us consider only the quarter of the cross section of the bar shown in Figure 2.31. Maintaining symmetry of the geometry along the x and y axes, remodel the boundary Γ_3 by adding material so that the condition

$$\sqrt{\tfrac{3}{2}(\sigma_{zx}^2 + \sigma_{zy}^2)} \le \bar{\sigma}$$

is satisfied on the remodeled boundary Γ_3, where $\bar{\sigma}$ is the yield stress of the bar for a uniaxial tension test. It is noted that there are infinitely many remodelings if the amount of material added is not specified. Find the remodeling by the minimum amount of material added using the iterative scheme described below. Let $R(\theta)$ be the distance of a point on Γ_3 from the origin, where θ is the angle from the x axis. Using the initial design, we solve the boundary value problem using finite element methods and obtain the quantity

$$J_2 = \sqrt{\tfrac{3}{2}(\sigma_{zx}^2 + \sigma_{zy}^2)}$$

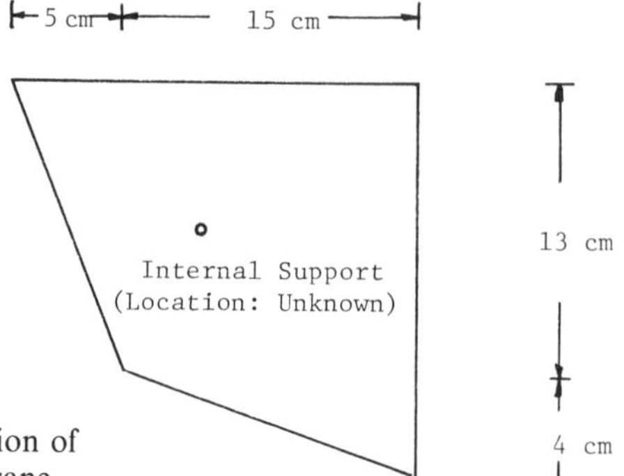

Figure 2.32 Design of location of internal support for a membrane.

on the boundary Γ_3. The location of Γ_3 is modified by

$$R_{\text{new}}(\theta) = \max\{R_{\text{init}}(\theta), R_{\text{old}}(\theta) + \alpha\, \Delta R_{\max}(J_2 - \bar{\sigma})/\bar{\sigma}\} \quad (2.107)$$

where R_{init} is the distance for the initial design, R_{old} is the present design, α is a sufficiently small magnification factor, and $\Delta R_{\max}$ is the maximum design change at each iteration. The maximum is taken for two values in (2.107), since the remodeling by adding material is considered in this example. If R_{new} differs from R_{old}, we shall solve the boundary value problem for the new design R_{new} and shall obtain the quantity J_2 along Γ_3. The design is then updated by (2.107). This process will be repeated until the design change will be very small.

Exercise 2.11 (elastic membrane problem): In terms of its transverse deflection $w = w(x, y)$, an elastic membrane is governed by the differential equation

$$-h \frac{\partial}{\partial x_i}\left(\sigma_{ji} \frac{\partial w}{\partial x_j}\right) = f \quad \text{in } \Omega \quad (2.108)$$

where h is the thickness of the membrane, σ_{ji} the stress tensor, and f is the transverse applied load. Although we often assume $h\sigma_{ji} = T\delta_{ji}$ in many cases, where T is the absolute magnitude of the stretching force per unit length of the edge of the membrane, we can consider the case of a general σ_{ji}.

Determine the location of an internal support so that the maximum deflection will be as small as possible for a uniform transverse load f. Use $\sigma_{ji} = T\delta_{ji}$, $T =$ constant, and $h = 1$ mm. Assume that the boundary of the membrane shown in Figure 2.32 is supported, that is, $w = 0$ on Γ.

2.10.2 Interpolation error and design of finite element models

We shall now discuss discretization of a domain into finite elements. As briefly mentioned in section 2.3, we should use smaller finite elements in those parts of Ω where "the temperature gradient is likely to be large" so that piecewise linear functions will reasonably approximate the temperature distribution. We shall elaborate the above statement using the theory of interpolation of functions. To do this, let us first consider the interpolation of a function u on an interval (x_1, x_{n+1}) using a piecewise linear function, where n is the total number of subintervals (x_i, x_{i+1}), $i = 1, 2, \ldots, n$, in which the function u is to be interpolated. More specifically, in a subinterval (x_i, x_{i+1}), the function u is approximated by

$$u_h^I(x) = \{u_i(x_{i+1} - x) + u_{i+1}(x - x_i)\}/h_i \tag{2.109}$$

where $u_i = u(x_i)$, $u_{i+1} = u(x_{i+1})$, $h_i = x_{i+1} - x_i$, and the summation convention is not applied. Noting the relation

$$u(x) = u(y) + \int_y^x u'(t)\,dt$$

we have, in the subinterval (x_i, x_{i+1}),

$$u(x) - u_h^I(x) = \left\{(x_{i+1} - x)\int_{x_i}^x u'\,dt + (x - x_i)\int_{x_{i+1}}^x u'\,dt\right\}\Big/ h_i \tag{2.110}$$

Similarly, applying the relation

$$u(y) = u(x) + u'(x)(y - x) + \int_x^y \int_x^t u''\,ds\,dt \tag{2.111}$$

we have

$$u(x_{i+1}) = u(x) + u'(x)(x_{i+1} - x) + \int_x^{x_{i+1}} \int_x^t u''\,ds\,dt \tag{2.112}$$

and

$$u(x_i) = u(x) + u'(x)(x_i - x) + \int_x^{x_i} \int_x^t u''\,ds\,dt \tag{2.113}$$

Subtraction of (2.112) from (2.113) yields the relation

$$u'(x) - (u_h^I)'(x) = \left\{\int_x^{x_i} \int_x^t u''\,ds\,dt - \int_x^{x_{i+1}} \int_x^t u''\,ds\,dt\right\}\Big/ h_i \tag{2.114}$$

If u'' is *assumed to be constant in the subinterval* (x_i, x_{i+1}), we have

$$u'(x) - (u_h^I)'(x) = \{\tfrac{1}{2}(2x - x_i - x_{i+1})\}u'' \tag{2.115}$$

A weighted interpolation error e_{i1}^I is defined

$$e_{i1}^I = \left\{\max_{x \in (x_i, x_{i+1})} |ku'(x) - k(u_h^I)'(x)|\right\}\sqrt{h_i} \tag{2.116}$$

for a constant k given in the interval $\Omega_i = (x_i, x_{i+1})$. Then it is estimated as follows:

$$e_{i1}^I \le \tfrac{1}{2}(h_i^{3/2}|(ku')'|) \tag{2.117}$$

On the other hand, for another interpolation error e_{i0}^I defined by

$$e_{i0}^I = \left\{\max_{x \in (x_i, x_{i+1})} |u(x) - u_h^I(x)|\right\}\sqrt{h_i}$$

we have an estimate

$$e_{i0}^I \le \left(\int_{x_i}^{x_{i+1}} |u'|\,dt\right)\sqrt{h_i} \tag{2.118}$$

The term $\sqrt{h_i}$ is multiplied in (2.116) and (2.118) so that the interpolation error is considered in the average sense. More precisely, if the mean square norm of interpolation error is defined in a subinterval (x_i, x_{i+1}) by

$$\hat{e}_{i1}^I = \left\{\int_{x_i}^{x_{i+1}} |ku'(x) - k(u_h^I)'(x)|^2\,dx\right\}^{1/2}$$

then the following estimate holds:

$$\hat{e}_{i1}^{I} \leq \max_{x \in (x_i, x_{i+1})} |ku'(x) - k(u_h^I)'(x)| \sqrt{h_i}$$

The weighted interpolation error e_{i1}^{I} is defined as the upper bound of the mean square norm of interpolation error $\hat{e}_{i1}^{I}$. Similarly, (2.118) can be regarded as an upper bound of the mean square norm

$$\hat{e}_{i0}^{I} = \left\{ \int_{x_i}^{x_{i+1}} |u(x) - u_h^I(x)|^2 \, dx \right\}^{1/2}$$

It is also noted that although the flux ku' has the singularity at a point x_0 such that

$$ku' \sim O(r^{-1/2}), \qquad r = |x - x_0|$$

the upper bound of the weighted interpolation error e_{i1}^{I} still enables the maintenance of finite value [see (2.117)]. However, if the weight $\sqrt{h_i}$ is not multiplied to define the error measure related to the flux, the error measure becomes infinity no matter how small are the finite elements introduced near the singular point.

Now let us consider a design problem of subdivision of a given interval (x_1, x_{n+1}) into subintervals (x_i, x_{i+1}), $i = 1, \ldots, n$, in which a given function u is interpolated by a piecewise linear polynomial. How shall we design the subdivision? If the total number of subintervals is fixed, a natural way is to subdivide the whole interval so that an interpolation error in the global sense is the minimum. That is, for the global interpolation errors defined by

$$e_1^I = \max_{i=1,\ldots,n} e_{i1}^I \qquad \text{or} \qquad e_0^I = \max_{i=1,\ldots,n} e_{i0}^I \tag{2.119}$$

the design problem of subintervals is stated as

$$\min e_\alpha^I = \min(\max_{i=1,\ldots,n} e_{i\alpha}^I), \quad \alpha = 0 \text{ or } 1 \tag{2.120}$$

with respect to the location of nodes $x_2, x_3, \ldots, x_n$. A necessary condition for optimality to the problem (2.120) is obtained as

$$e_{i\alpha}^I = \text{const}, \quad i = 1, \ldots, n, \alpha = 0, 1 \tag{2.121}$$

If condition (2.121) is not met, it is possible in general to reduce the maximum value of error by making an adjustment of the subdivision. Condition (2.121) means that the error has to be equally distributed in intervals.

Applying the above idea, design of subintervals may be obtained approximately by requiring

$$\bar{e}_{i\alpha}^I = \text{const}, \quad i = 1, \ldots, n, \alpha = 0, 1 \tag{2.122}$$

using upper bounds of interpolation errors, where $\bar{e}_{i\alpha}^I$, $\alpha = 0, 1$, are given by

$$\bar{e}_{i0}^I = \int_{x_i}^{x_{i+1}} |u'| dt \sqrt{h_i} \qquad \text{and} \qquad \bar{e}_{i1}^I = \tfrac{1}{2}[h_i^{3/2}|(ku')'|]$$

Thus, for e_0^I subintervals should satisfy the condition

$$\bar{e}_{i0}^I = \int_{x_i}^{x_{i+1}} |u'| \, dt \sqrt{h_i} = \text{const}, \quad i = 1, \ldots, n \tag{2.123}$$

If u' is assumed to be constant in each subinterval, then, roughly speaking, the weighted increment Δu by $\sqrt{h_i}$ is the same in each subinterval. Therefore, subdivi-

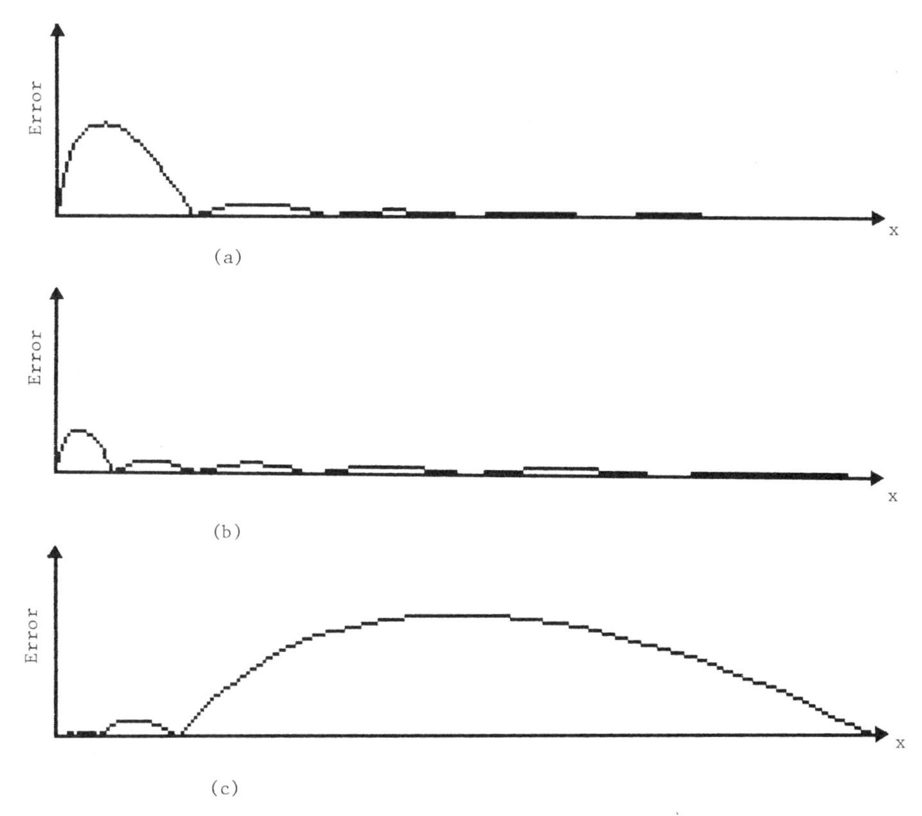

Figure 2.33 Distribution of interpolation error $u - u_h^I$ for three discretizations: (a) uniform, (b) contour lines of u, (c) contour lines of u'.

sion of the interval (x_1, x_{n+1}) can be obtained by plotting the contour lines of the function $u\sqrt{h_i}$.

For the interpolation error e_1^I, subintervals satisfy

$$\bar{e}_{i1}^I = \tfrac{1}{2}[h_i^{3/2}\,|(ku')'|] = \text{const}, \quad i = 1, \ldots, n \tag{2.124}$$

If quantity $q = ku'$ is referred to as *flux*, this means that the weighted increment Δq of the flux q is the same in each subinterval.

Example 2.7: Let us interpolate a function $u(x) = \sqrt{x + 0.01}$ in the interval (0, 1) using five subintervals. We shall subdivide the interval in the following three ways:

(i) uniform subdivision,
(ii) using the contour lines of the function u, and
(iii) using the contour lines of the flux $q = u'$.

It is clear that (ii) and (iii) are closely related to (2.123) and (2.124), respectively. Figures 2.33 and 2.34 show the distributions of interpolation errors $u - u_h^I$ and $u' - (u_h^I)'$.

For the error $u - u_h^I$, the subdivision obtained by using the contour lines of the function u provides an almost equally distributed error in

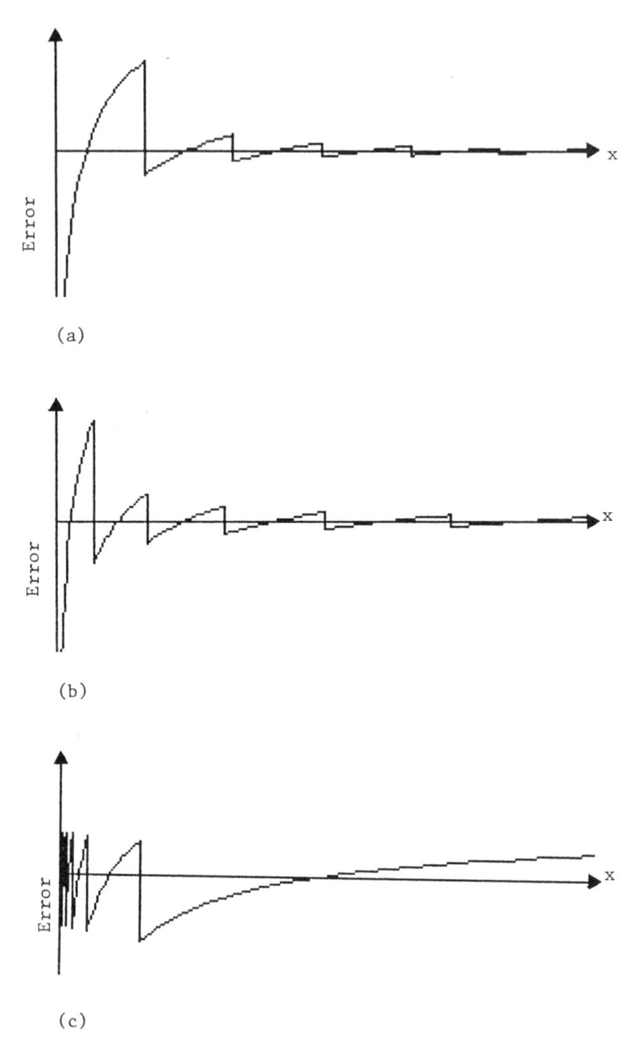

Figure 2.34 Distribution of interpolation error $u' - (u_h^I)'$ for three discretizations: (a) uniform, (b) contour lines of u, (c) contour lines of u'.

the subintervals. The error is large near the origin for the uniform subdivision, whereas it becomes large near to the right end for the one obtained by using the contour lines of the flux. If very fine accuracy of interpolation is required near the singular point where the flux becomes infinite, the subdivision based on the contour lines of the flux gives the smallest error.

For the error $u' - (u_h^I)'$, it is clear that the subdivision by the contour lines of the flux gives the smallest error that is almost equally distributed in subintervals. The subdivision obtained by the contour lines of the function, however, generates a significantly large error near to the singular point, although it fits the function u very well. This indicates that discretization of a given interval strongly depends on its purpose.

If flux needs be approximated well, discretization must be based on the design (iii), which yields the smallest interpolation error in flux.

Another important observation is that quantity $\bar{e}^I_{i\alpha}$ indicates the amount of error in the ith subinterval (element). Similarly, if the amount of deviation from the average $\bar{e}^I_{A\alpha}$, $\alpha = 0$ or 1, is defined by

$$d_\alpha = \sum_{i=1}^{n} \left(\bar{e}^I_{i\alpha} - \bar{e}^I_{A\alpha}\right)^{1/2} \Big/ \bar{e}^I_{A\alpha} \tag{2.125}$$

$$\bar{e}^I_{A\alpha} = \sum_{i=1}^{n} \bar{e}^I_{i\alpha} \Big/ n$$

then it can be an indicator of quality of discretization. If d_α is close to zero, *optimality* condition (2.122) is almost satisfied.

Now, how can we construct a discretization of a given interval (domain) into subintervals (finite elements) in which optimality condition (2.122) is approximately satisfied? If only piecewise linear interpolation is allowed to be used, two methods can be considered. The first is the *node relocation* (r-) *method*, which adjusts locations of internal nodes so that condition (2.122) can be satisfied. Another way is to use the *h-method*, which refines elements in which the values $\bar{e}^I_{i\alpha}$ are "too large." The r-method contains a fixed number of elements, whereas the h-method increases the total number of elements. Both methods need adaptation, that is, starting from an initial discretization, we adapt either node relocation or additional number of elements in order to satisfy 'optimality' condition (2.122). If other kinds of interpolation are used, adaptation by changing degrees of interpolation is also possible. This approach is called the *p-method*, since we apply different degrees of polynomials to interpolate a given function.

Example 2.8 (r-method): Starting from a discretization of an interval (0, 1) into five uniform subintervals, let us interpolate a function $u(x) = \sqrt{x + 0.01}$ using the r-method to satisfy optimality condition

$$\bar{e}^I_{i0} = \left(\int_{x_i}^{x_{i+1}} |u'|\, dt \right) \sqrt{h_i} = \text{const}, \quad i = 1, \ldots, 5$$

Assuming u' is constant in each subinterval, say, $u' = u'[(x_i + x_{i+1})/2]$, we shall obtain the location of internal nodes so that the following holds:

$$\bar{e}^I_{i0} = h_i^{3/2} u'\left(\frac{x_i + x_{i+1}}{2}\right) = \text{const}, \quad i = 1, \ldots, 5 \tag{2.126}$$

An adaptive scheme is defined by

$$x_i \leftarrow \frac{\dfrac{x_{i-1} + x_i}{2} \dfrac{\bar{e}^I_{i-1,0}}{x_i - x_{i-1}} + \dfrac{x_i + x_{i+1}}{2} \dfrac{\bar{e}^I_{i,0}}{x_{i+1} - x_i}}{\dfrac{\bar{e}^I_{i-1,0}}{x_i - x_{i-1}} + \dfrac{\bar{e}^I_{i,0}}{x_{i+1} - x_i}} \tag{2.127}$$

for this example. This scheme yields no relocation of nodes if condition (2.122) is satisfied.

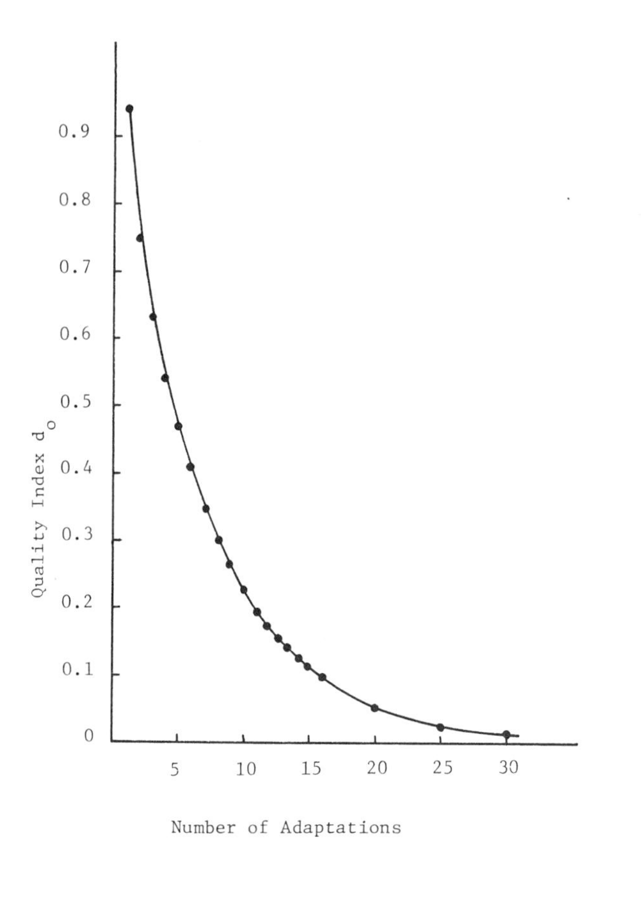

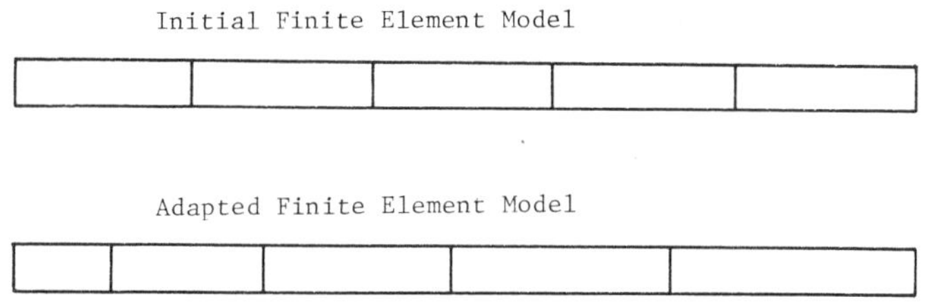

Figure 2.35 Convergence of r-method and adaptive finite element model.

Figure 2.35 shows convergence of index d_0 defined by (2.125) to zero as the number of adaptations increases. At the third adaptation, 30.4% reduction is obtained for the interpolation error e_0^I. At the tenth adaptation, 71.3% reduction is obtained.

Example 2.9 (h-method): We shall solve the same problem in Example 2.8 using the h-method. Starting from five subintervals, we shall refine the subintervals in which the error is large. Here we refine subintervals whose error is greater than 75% of the maximum error. Table 2.3 shows the result of adaptation. Although index d_0 never goes to zero, the interpolation error e_0^I becomes small very rapidly. Just one adaptation

Table 2.3 *Reduction of error by the h-method*

Adaptation	Number of elements	Maximum error	Index d_0
1	6	0.16021	1.14126
2	9	0.0814	0.59462
3	13	0.0532	1.02308
4	15	0.03241	0.96809
5	20	0.02334	1.14327
6	25	0.01705	1.53084
7	30	0.01154	1.3243
8	38	0.00847	1.63972
9	48	0.00634	2.14698
10	57	0.00444	1.84936

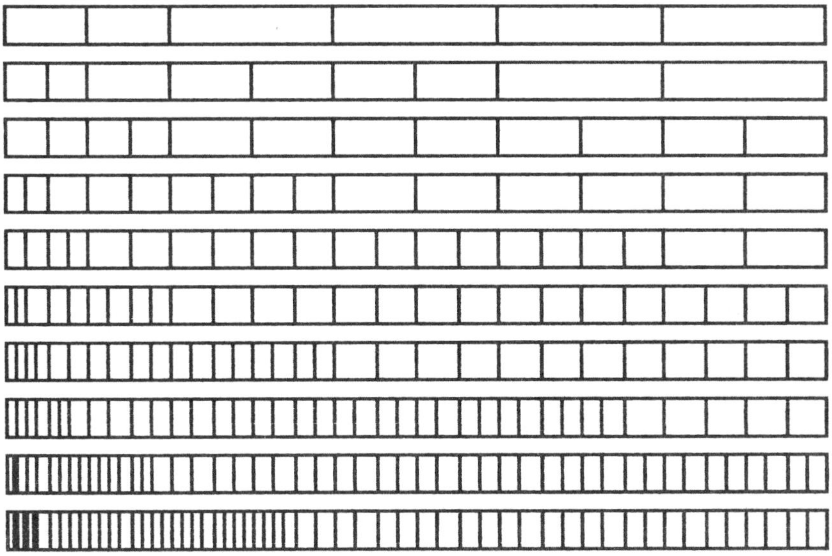

Figure 2.36 Adaptation process by the *h*-method.

implies 49% reduction of the interpolation error. At the eighth adaptation, the error becomes less than 5.3% of that of the initial discretization, while the number of elements becomes more than 6 times. The adaptation process is given in Figure 2.36.

Extension of the above adaptation scheme to two-dimensional problems may be obtained analogously.* For a triangular element Ω_e shown in Figure 2.37, the upper bounds $\bar{e}^I_{e\alpha}$, $\alpha = 0$ and 1, are defined by

$$\bar{e}^I_{e0} = \max\left[\max_{(x,y)\in\Omega_e} \int_{x_e}^{x_e+h_x} \left|\frac{\partial u}{\partial x}(s, y)\right| ds, \max_{(x,y)\in\Omega_e} \int_{y_e}^{y_e+h_y} \left|\frac{\partial u}{\partial y}(x, t)\right| dt\right] \sqrt{A_e} \quad (2.128)$$

* Details of mathematical theory of interpolation error can be found in Strang and Fix (1973).

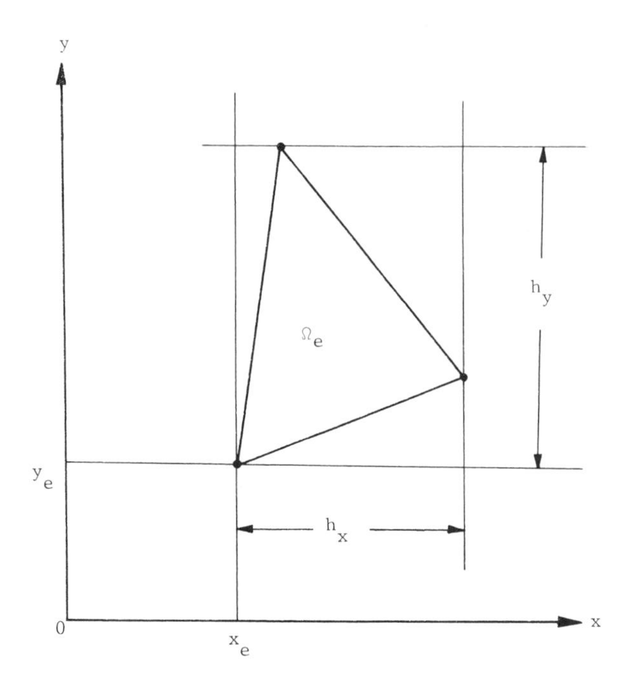

Figure 2.37 Definition of h_x and h_y.

and

$$\bar{e}_{e1}^{I} = \max\left[\frac{h_x}{2}\left|\frac{\partial q_x}{\partial x}\right|, \frac{h_x}{2}\left|\frac{\partial q_y}{\partial x}\right|, \frac{h_y}{2}\left|\frac{\partial q_x}{\partial y}\right|, \frac{h_y}{2}\left|\frac{\partial q_y}{\partial y}\right|\right] \tag{2.129}$$

where

$$x_e = \min_{(x,y)\in\Omega_e} x, \qquad y_e = \min_{(x,y)\in\Omega_e} y,$$

h_x is the size of Ω_e in the x direction, h_y is the size in the y direction, A_e is the area of Ω_e, and $\mathbf{q} = q_x\mathbf{i} + q_y\mathbf{j}$ is the flux defined by

$$q_x = k_{xx}\frac{\partial u}{\partial x} + k_{xy}\frac{\partial u}{\partial y} \quad \text{and} \quad q_y = k_{yx}\frac{\partial u}{\partial x} + k_{yy}\frac{\partial u}{\partial y} \tag{2.130}$$

Finite element models are defined so that condition (2.122) is approximately satisfied.

We have discussed design of finite element models for the *best interpolation* of a function. Now, let us extend the above concept to finite element analysis in which the solution u is unknown a priori. If heat conduction problems are solved by piecewise linear approximations using two-node line and three-node triangular elements, the heat flux is constant in elements and is discontinuous along element boundaries. Thus, replacement of u by the finite element solution u_h in (2.128) is fine, but $\mathbf{q}$ in (2.129) cannot be replaced by $\mathbf{q}_h$ obtained by a finite element solution. A way to resolve this difficulty is to use the heat flux $\hat{\mathbf{q}}_h$, which is spanned by the shape functions for temperature:

$$\hat{q}_{hi}(\mathbf{x}) = \hat{q}_{i\alpha} N_\alpha(\mathbf{x}) \quad \text{in } \Omega_e \tag{2.131}$$

where $\hat{q}_{i\alpha}$ are defined by solving the least-squares problem

$$\min_{\hat{q}_h} \tfrac{1}{2} \sum_{e=1}^{E} \int_{\Omega_e} (\hat{\mathbf{q}}_h - \mathbf{q}_h) \cdot (\hat{\mathbf{q}}_h - \mathbf{q}_h) \, d\Omega \tag{2.132}$$

The heat flux $\mathbf{q}_h$ is obtained by the finite element solution and is constant in each finite element. The least-squares problem (2.132) yields the system of linear equations

$$\sum_{e=1}^{E} M^e_{i\alpha j\beta} \hat{q}_{j\beta} = \sum_{e=1}^{E} r^e_{i\alpha} \tag{2.133}$$

where

$$M^e_{i\alpha j\beta} = \int_{\Omega_e} \delta_{ij} N_\alpha N_\beta \, d\Omega, \qquad r^e_{i\alpha} = \int_{\Omega_e} q_{ei} N_\alpha \, d\Omega \tag{2.134}$$

and $\mathbf{q}_h = q_{ei}\mathbf{i}_i$. Solving the system of linear equations, the value of the heat flux is obtained at nodes of the finite element model. Since $\hat{\mathbf{q}}_h$ is piecewise linear in an element and is continuous in the whole domain, we can differentiate it in each element. Therefore, $\mathbf{q}$ can be replaced by $\hat{\mathbf{q}}_h$ in the upper bound of the interpolation error (2.129) to define finite element models.

In summary, an adaptive design process of finite element meshes (i.e., models) is based on the strategy to reduce the interpolation error of a solution u and is described in four steps as follows:

1. Define an initial finite element model. If the r-method is applied, a *sufficiently* refined model must be introduced, since the total number of elements is fixed during adaptation. If the h-method is used, a *uniform* grid must be introduced at the initial stage.
2. Using the initial finite element model, solve a given problem by finite element methods. Compute $\bar{e}^I_{e\alpha}$ in each finite element Ω_e, $e = 1, \ldots, E$, $\alpha = 0$ or 1.
3. For the r-method, relocate nodes so that condition
$$\bar{e}^I_{e\alpha} = \text{const.} \qquad e = 1, \ldots, E, \ \alpha = 0 \text{ or } 1$$
can be achieved more closely. For the h-method, find elements in which $\bar{e}^I_{e\alpha}$ are large. Then, refine these elements. This step defines a new adapted finite element model.
4. Solve the problem using the adapted model, and compute $\bar{e}^I_{e\alpha}$ for the next adaptation.

A method to relocate nodes in the r-method may be defined by

$$\mathbf{x}_n = \sum_k \mathbf{x}^c_{nk} (\bar{e}^I_{k\alpha}/A_k) \Big/ \sum_k (\bar{e}^I_{k\alpha}/A_k) \tag{2.135}$$

where summation is taken over the elements that share the nth node, $\mathbf{x}^c_{nk}$ is the coordinate of the centroid of the kth finite element, and A_k is the area of the kth element. It is clear that (2.135) is a two-dimensional expression of (2.127). It is noted, however, that (2.135) may not be a unique way that yields $\bar{e}^I_{e\alpha}$ constant

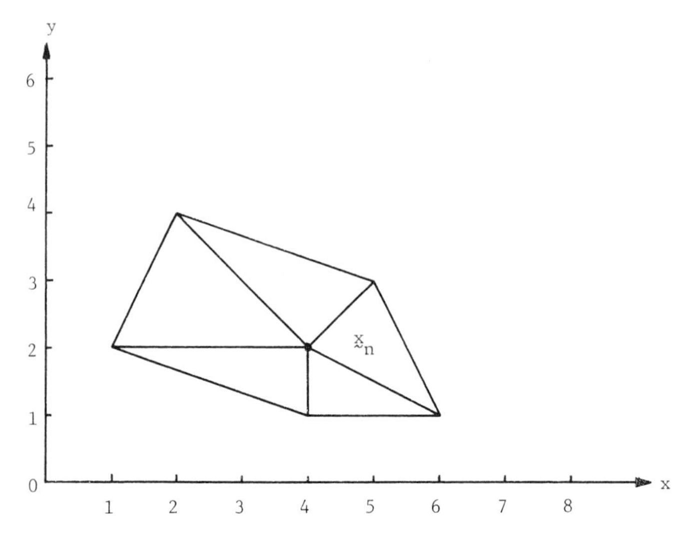

Figure 2.38 Relocation scheme.

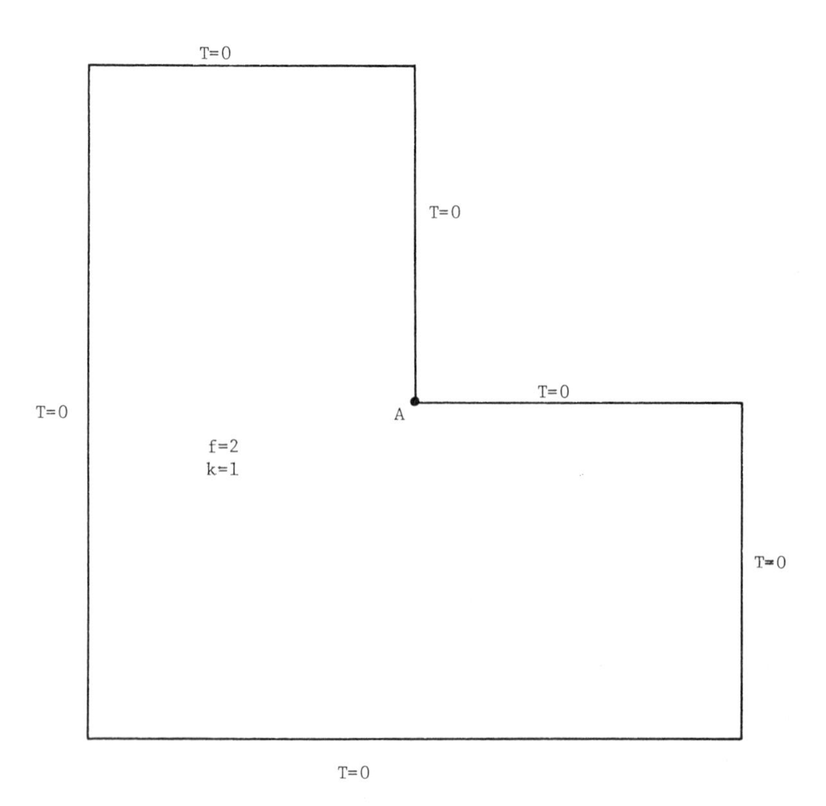

Figure 2.39 L-shaped domain for example 2.10.

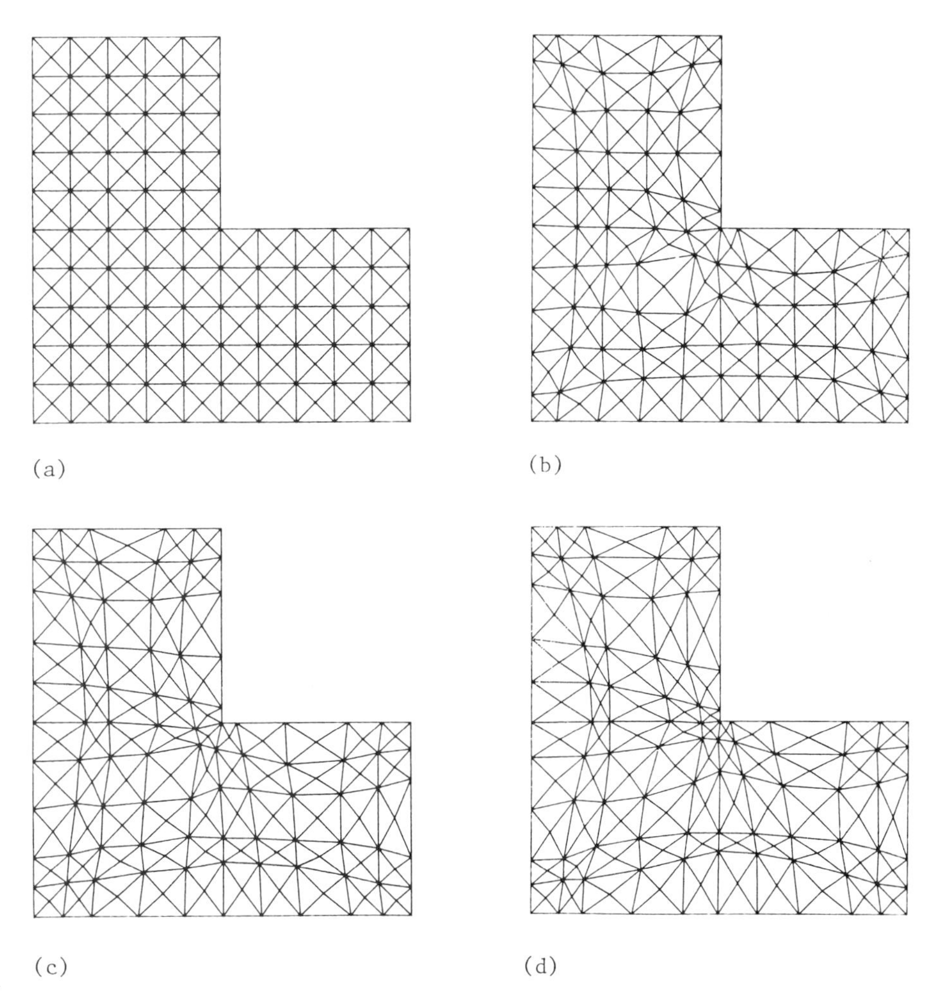

Figure 2.40 Series of r-adaptation: (a) initial model, (b) second adaptation, (c) third adaptation, (d) fourth adaptation.

for $e = 1, \ldots, E$. An important property of the relocation scheme (2.135) is that of $\bar{e}^I_{k\alpha}$ are constant, many iterations will yield the convergent location $\mathbf{x}_n$. For example, if the model shown in Figure 2.38 is considered, the limit of relocation is obtained as (3.95985, 1.79439).

Example 2.10 (r-method): Let us consider the heat conduction problem on the L-shaped domain shown in Figure 2.39. Let temperature be zero along the boundary, and let the distributed heat source have strength $f = 2$. Heat conductivity is assumed to be $k_{ij} = k\delta_{ij}$, where $k = 1$. It is clear that the rate of change of heat flux is considerably large at the corners of the L-shaped domain, especially at corner A in Figure 2.39.

Figure 2.40 gives a series of the r-adaptation of the finite element model which shows that *refinement* occurs at the corner points, especially at corner A, in order to catch rapid changes of heat flux. Figure

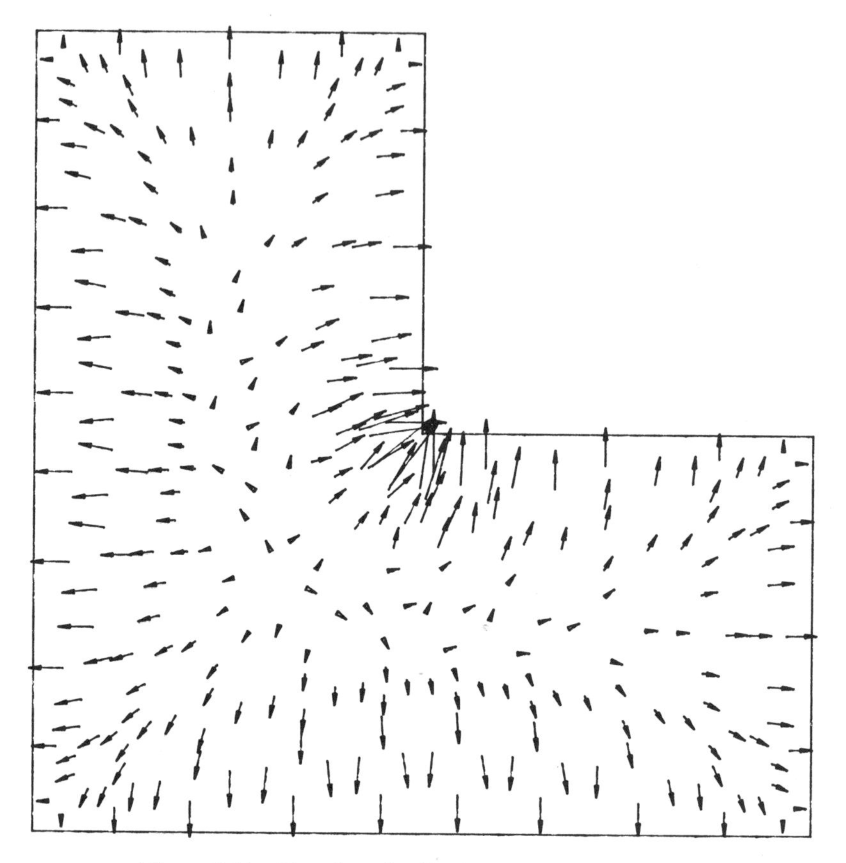

Figure 2.41 Heat flux distribution in L-shaped domain.

2.41 shows the heat flux distribution in the L-shaped domain. It is obvious that adapted finite element models provide sharp changes of heat flux at the corner A. The temperature distribution is given in Figure 2.42 and is computed at the last step of the r-adaptation.

An h-adaptation for triangular elements is described as follows. If an element Ω_e must be refined, we first find elements that share edges with Ω_e. Suppose that these elements are not assigned to be refined. But because of refinement of Ω_e, additional nodes are created at the middle of edges that must be connected to the adjacent elements. Thus, adaptation becomes as shown in Figure 2.43. An important note is that continuing application of the h-adaptation mentioned above may yield finite elements whose *aspect ratio* defined by

$$R_e = \min_{i=1,2,3} H_i/S_i \tag{2.136}$$

becomes very small, where H_i is the normal distance of the ith vertex to its opposite edge, the length of which is S_i. Monitoring of this ratio may be important in the h-adaptation for triangular elements. Refinement of elements may be determined

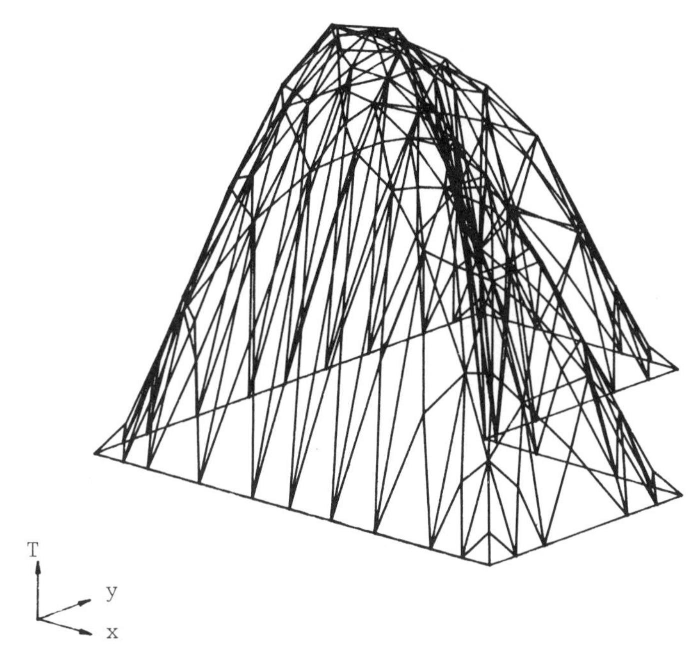

Figure 2.42 Temperature distribution.

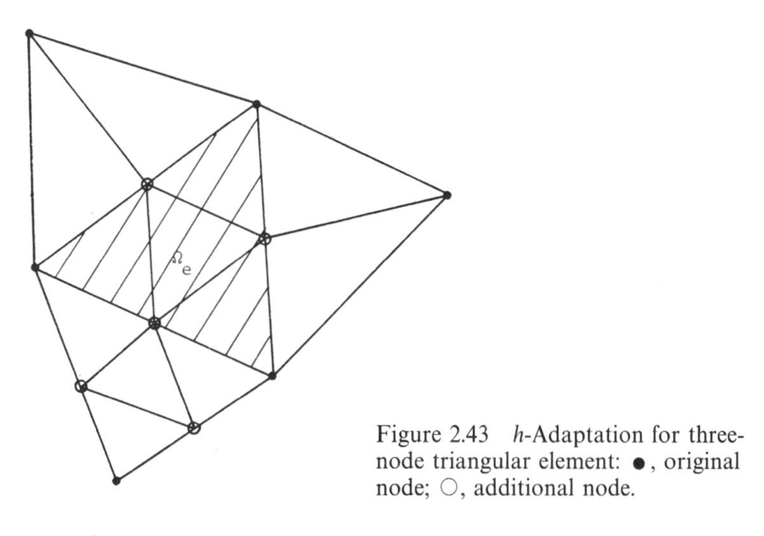

Figure 2.43 h-Adaptation for three-node triangular element: ●, original node; ○, additional node.

by values of $\bar{e}^I_{e\alpha}$. For example, if

$$\bar{e}^I_{e\alpha} \geq 0.60 \max_{i=1,\ldots,E} \bar{e}^I_{i\alpha} \tag{2.137}$$

finite element Ω_e will be refined.

Example 2.11 (h-method): Let us solve a similar problem, using the h-adaptation method for a different boundary condition. Figure 2.44 shows the process of adaptation together with the boundary condition. Decreasing of the "error" monitored by the maximum value of $\bar{e}^I_{e1}$ is

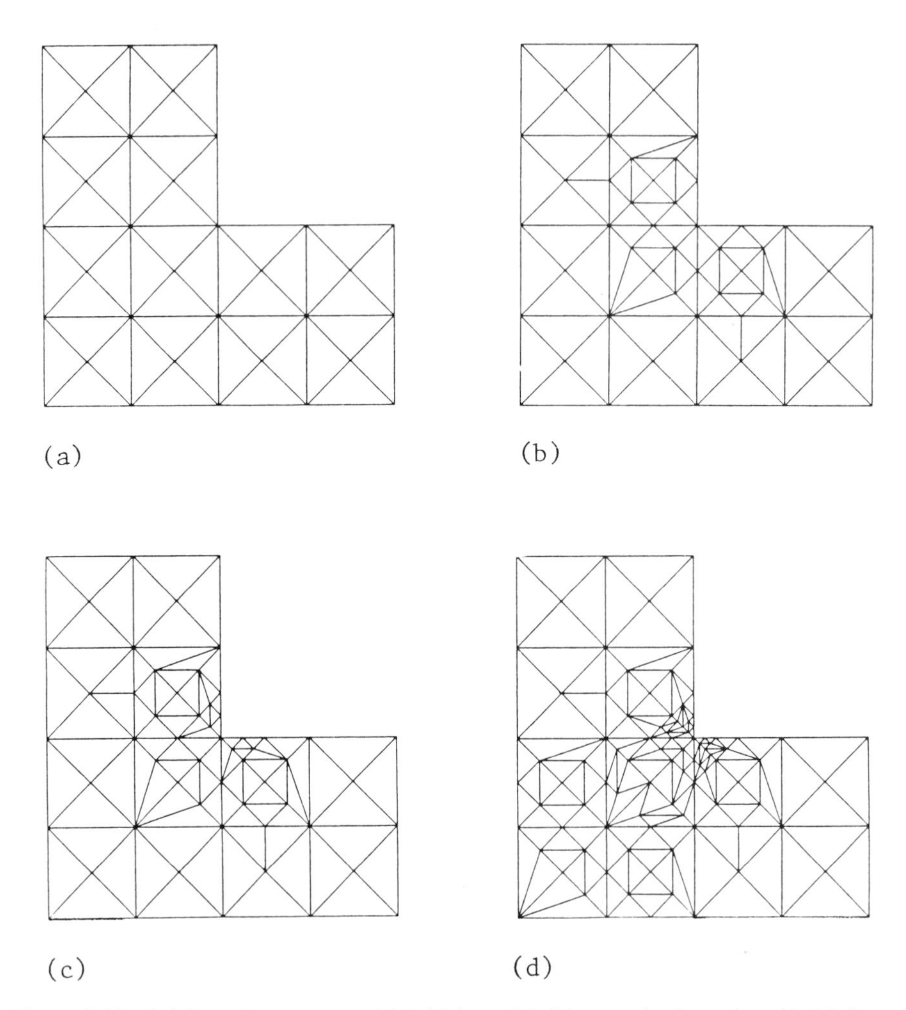

Figure 2.44 h-Adaptation, process: (a) initial model, (b) second adaptation, (c) third adaptation, (d) fourth adaptation.

shown in Figure 2.45. It is also clear that refinement is assigned in elements close to corner A.

It is noted that extensive literature on adaptive finite element methods can be found in the proceedings of the conferences held at the University of Maryland and the Technical University of Lisbon (Babuska et al., 1983, 1984). It is suggested that references of contributions in the proceedings are also good resources of information on this subject. Another interesting proceeding for optimal design of finite element grids (Shephard and Gallagher, 1979) contains a review article on this subject.

2.10.3 Convergence of finite element approximations

We have studied interpolation error and its application to design of finite element models based on the adaptive concept. Here we shall explain convergence of finite element approximation using an example from the potential flow problem.

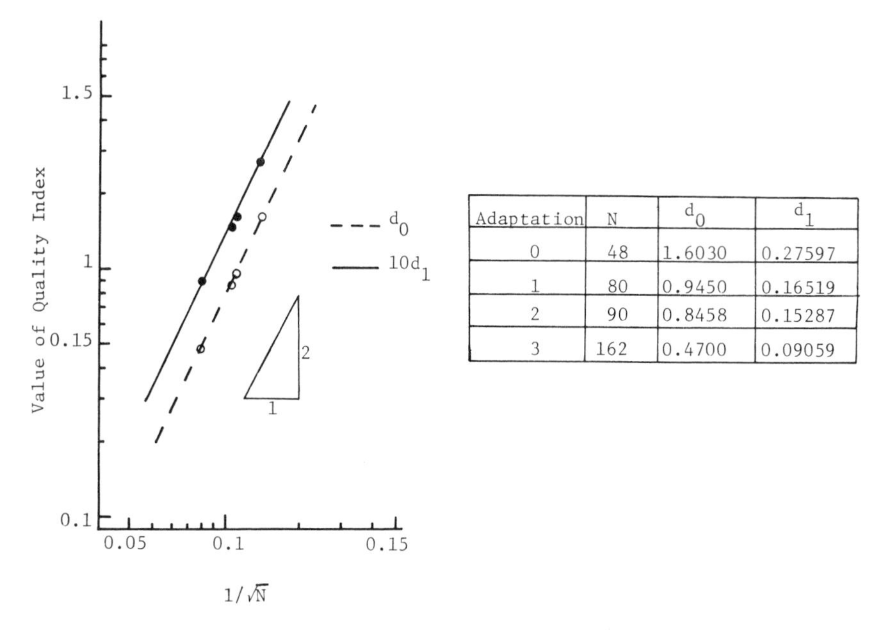

Adaptation	N	d_0	d_1
0	48	1.6030	0.27597
1	80	0.9450	0.16519
2	90	0.8458	0.15287
3	162	0.4700	0.09059

Figure 2.45 Convergence of h-adaptive method; N is total number of elements.

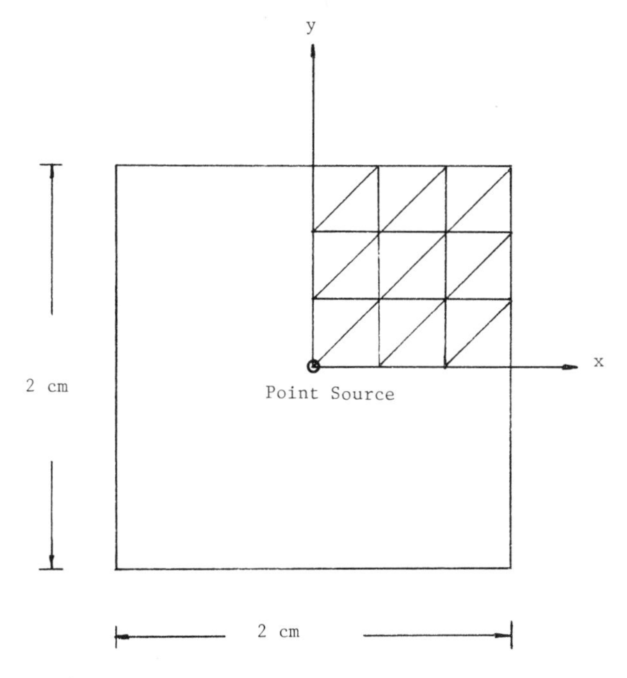

Figure 2.46 Finite element model for example 2.12.

Example 2.12: Let us consider the square domain shown in Figure 2.46. In the problem, there is a point source whose strength is 1 cm^3/s at the centroid of the square domain $(-1, 1) \times (-1, 1)$. If the value of the potential is specified to be $\phi = 0$ along the boundary, the problem

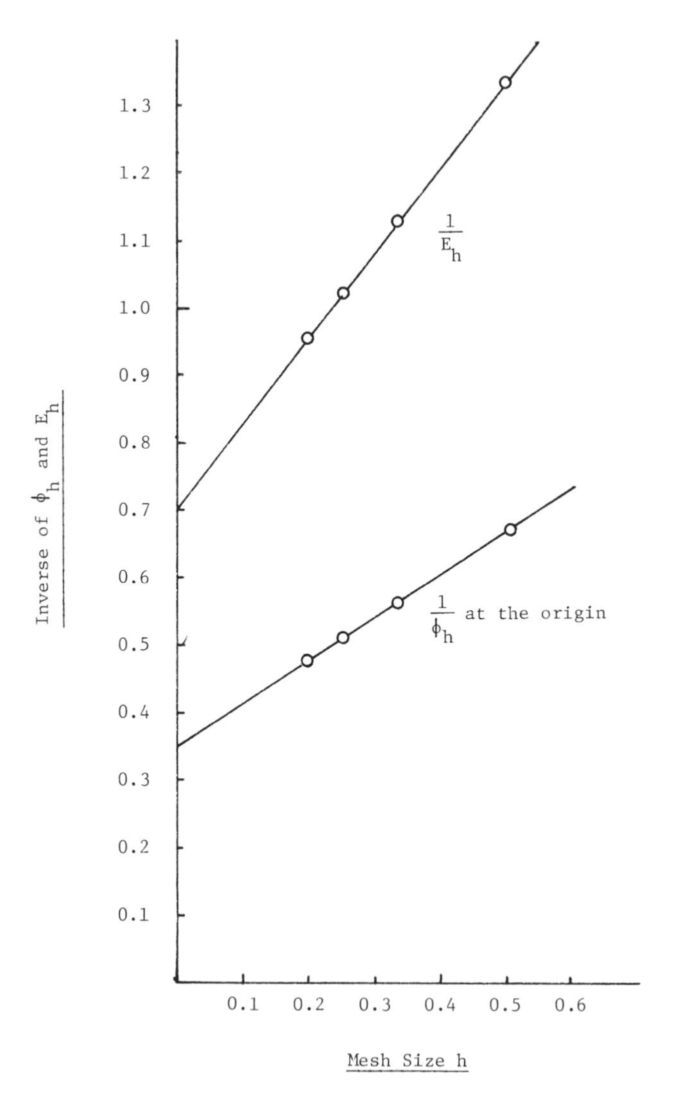

Figure 2.47 Convergence of ϕ_h and E_h as mesh size h goes to zero.

is represented by the boundary value problem

$$-\nabla^2\phi = \delta(0, 0) \quad \text{in } \Omega, \qquad \phi = 0 \quad \text{on } \Gamma$$

where $\delta(x, y)$ is the delta function at the point (x, y). Then the velocity field $\mathbf{v}$ is obtained by

$$\mathbf{v} = -\nabla\phi$$

Because of the delta function, the boundary value problem possesses a unique solution ϕ such that

$$|\nabla\phi| \to +\infty \quad \text{as } (x, y) \to (0, 0)$$

$$2\pi r|\nabla\phi| = 1 \quad \text{along any circle of radius } r$$

These meanings are quite clear in physics. Indeed, the velocity becomes large near the point source, and the discharge of fluid through a closed

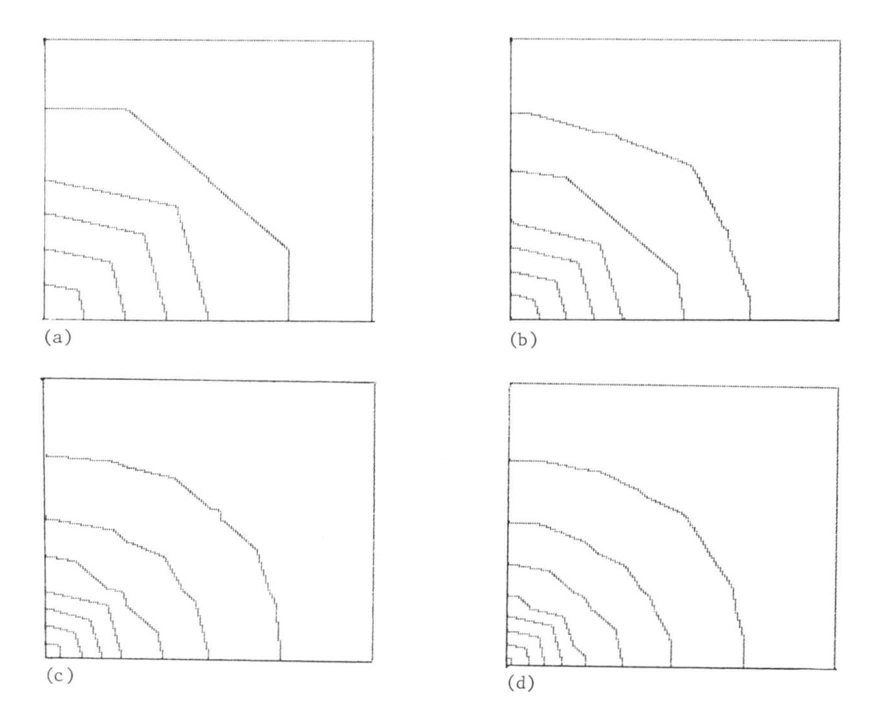

Figure 2.48 Equipotential lines by meshes: (a) 2 × 2, (b) 3 × 3, (c) 4 × 4, (d) 5 × 5.

curve that contains the point source must be the same as the amount of fluid issuing from the source.

According to the mathematical theory of finite element methods, we know that if the condition

$$r^s|\nabla\phi| < +\infty, \quad 0 \le s \tag{2.138}$$

is satisfied for a nonnegative parameter s, where r is the radial coordinate of the polar coordinate system, the following convergence property, in general, holds for uniform meshes:

$$\frac{1}{E_0} - \frac{1}{E_h} \approx ah^{2-s} \tag{2.139}$$

where a is a constant, E_h is defined by

$$E_h = \tfrac{1}{2}\int_\Omega |\nabla\phi_h|^2 \, d\Omega \tag{2.140}$$

ϕ_h is the finite element solution, and h is the mesh size. Note that the property mentioned above holds for three-node triangular elements and only for uniform meshes. Furthermore, in many cases,

$$\frac{1}{\phi_0} - \frac{1}{\phi_h} \approx bh^{2-s} \quad \text{at } \mathbf{x} \in \Omega \tag{2.141}$$

for a proper constant b. The value ϕ_0 may be the limiting value of ϕ_h as $h \to 0$.

Figures 2.47 and 2.48 show, respectively, the above convergence property and the equipotential lines (with $\Delta\phi = 0.25$) for uniform finite element meshes using 2 × 2, 3 × 3, 4 × 4, and 5 × 5 elements.

If the domain Ω has a corner whose angle θ exceeds π, then the parameter s is defined as $s = (\theta/\pi) - 1$.

In the above example, we have used the results of the mathematical theory of finite element methods without any development. Extensive works on this subject can be found in Ciarlet (1978), Oden and Reddy (1975), Oden (1983), and others. Here we shall provide a brief description of an introductory plot of the mathematical theory.

Suppose that T_h is the solution of the finite element approximation based on the *discretized* weak form

$$\int_\Omega k_{ij} T_{h,j} \bar{T}_{h,i} \, d\Omega = \int_\Omega f \bar{T}_h \, d\Omega + \int_{\Gamma_2} h \bar{T}_h \, d\Gamma \tag{2.142}$$

for every $\bar{T}_h$ such that $\bar{T}_h = 0$ on Γ_1, where $T_h = g$ on Γ_1 and g is a piecewise linear function that can be exactly reproduced by the interpolation for the finite element method. If T is the solution of the original boundary value problem

$$\begin{gathered} -(k_{ij} T_{,j})_{,i} = f \quad \text{in } \Omega \\ T = g \quad \text{on } \Gamma_1 \qquad \text{and} \qquad k_{ij} T_{,j} n_i = h \quad \text{on } \Gamma_2 \end{gathered} \tag{2.143}$$

the function T also satisfies the weak form

$$\int_\Omega k_{ij} T_{,j} \bar{T}_{,i} \, d\Omega = \int_\Omega f T \, d\Omega + \int_{\Gamma_2} h \bar{T} \, d\Gamma \tag{2.144}$$

for every $\bar{T}$ such that $\bar{T} = 0$ on Γ_1. Note that (2.144) holds for any $\bar{T}$, so it holds for $\bar{T} = \bar{T}_h$, where $\bar{T}_h$ is obtained by the finite element approximation. Hence, if $\bar{T} = \bar{T}_h$ is specifically assumed in (2.144), then subtraction of (2.142) from (2.144) yields

$$\int_\Omega k_{ij} (T - T_h)_{,j} \bar{T}_{h,i} \, d\Omega = 0 \tag{2.145}$$

for every $\bar{T}_h$ such that $\bar{T}_h = 0$ on Γ_1. The relation (2.145) defines what is meant by the *orthogonality* of the finite element solution T_h to the exact solution T with respect to the finite element approximation. If the *error of the finite element approximation* is defined by

$$e_h = T - T_h \tag{2.146}$$

then (2.145) is written by

$$\int_\Omega k_{ij} e_{h,j} \bar{T}_{h,i} \, d\Omega = 0 \tag{2.147}$$

This means that the error is orthogonal to the finite element approximation. It follows from (2.147) that

$$\int_\Omega k_{ij} e_{h,j} e_{h,i} \, d\Omega \le \int_\Omega k_{ij} e_{h,j} (T - \hat{T}_h)_{,i} \, d\Omega \tag{2.148}$$

for any $\hat{T}_h$ such that $\hat{T}_h = 0$ on Γ_1 and $\hat{T}_h$ is the interpolation of the exact solution T. Note that $\hat{T}_h$ need not be the same as the finite element solution T_h. It is certainly true that T_h is very close to $\hat{T}_h$ and that we have expected that $T_h \approx \hat{T}_h$ in the development of the finite element approximation in section 2.10.2. Applying the

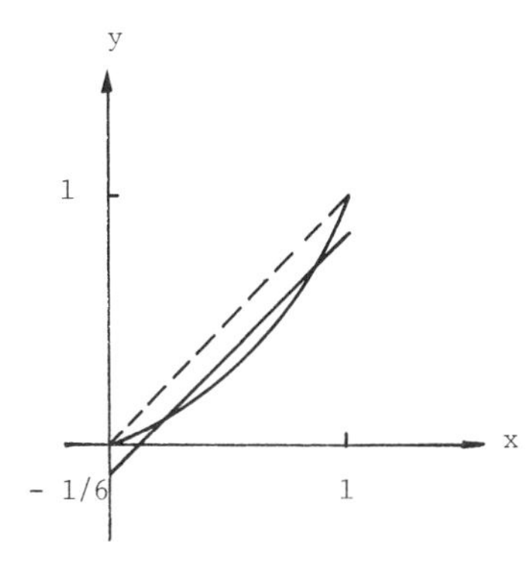

Figure 2.49 Interpolation $y = x$ (dashed) and best approximation $y = x - \frac{1}{6}$ (solid) for the function $y = x^2$.

Schwarz inequality for any two functions f and g,

$$\int_\Omega fg \, d\Omega \leq \left\{\int_\Omega f^2 \, d\Omega\right\}^{1/2} \left\{\int_\Omega g^2 \, d\Omega\right\}^{1/2} \tag{2.149}$$

(2.148) implies

$$\int_\Omega k_{ij} e_{h,j} e_{h,i} \, d\Omega \leq \int_\Omega k_{ij} e^I_{h,j} e^I_{h,i} \, d\Omega \tag{2.150}$$

where $e^I_h = T - \hat{T}_h$ is the *interpolation error*. This means that the error in the finite element approximation does not exceed the error in the interpolation of a function. Because of this, the finite element approximation is sometimes called the *best approximation,* in the sense of least squares.

The property of being the best approximation may be visualized by an example in which we approximate the function $y = x^2$ in the interval (0, 1). If the linear interpolation is applied, this is interpolated by $\hat{y}_h = x$. However, the best approximation of the function $y = x^2$ by a linear function is $y_h = x - \frac{1}{6}$ as shown in Figure 2.49.

Now suppose that $k_{ij} = k\delta_{ij}$ in (2.150) for a positive constant k. In this case, if the estimate of interpolation error is obtained as

$$\int_\Omega \nabla e^I_h \cdot \nabla e^I_h \, d\Omega \leq Ch^{2-S} \tag{2.151}$$

for a constant C and a parameter S, inequality (2.150) yields

$$E_h - E_0 \leq Ch^{2-S} \tag{2.152}$$

where

$$E_0 = \int_\Omega \nabla T \cdot \nabla T \, d\Omega \qquad \text{and} \qquad E_h = \int_\Omega \nabla T_h \cdot \nabla T_h \, d\Omega$$

Here we have used the following identity:

$$\nabla e_h \cdot \nabla e_h = \nabla T_h \cdot \nabla T_h - \nabla T \cdot \nabla T + 2\, \nabla T_h \cdot \nabla(T - T_h) + 2\nabla(T - T_h) \cdot \nabla(T - T_h)$$

Furthermore, we have the following relation:

$$\frac{1}{E_0} - \frac{1}{E_h} = \frac{E_h - E_0}{E_0 E_h} \leq \frac{E_h - E_0}{E_0^2} \leq \frac{C}{E_0^2} h^{2-S}$$

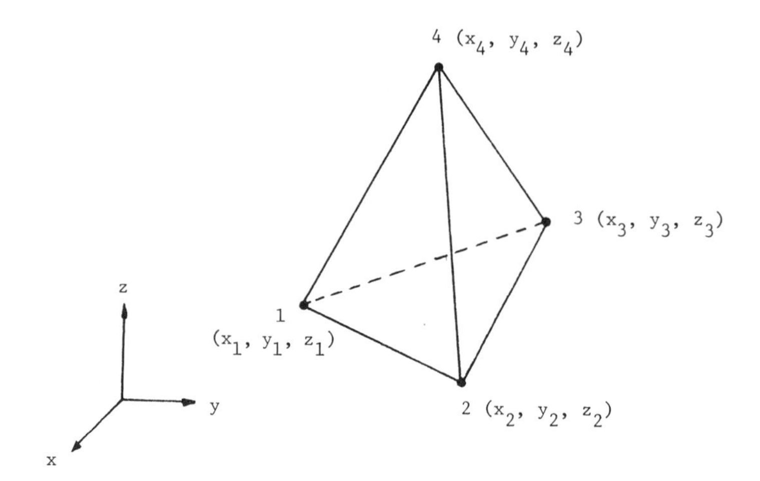

Figure 2.50 Four-node tetrahedron element in $\mathbb{R}^3$.

that is,

$$\frac{1}{E_0} - \frac{1}{E_h} \leq ah^{2-s} \tag{2.153}$$

where $a = c/E_0^2$. This is a brief explanation for the convergent property (2.139) applied in example 2.12.

2.10.4 Four-node tetrahedron element for three-dimensional problems

This section concerns a generalization of the three-node (i.e., CST) element for plane problems to the four-node tetrahedron element for three-dimensional problems. For the four-node tetrahedron element shown in Figure 2.50, the shape functions $N_\alpha(x, y, z)$, $\alpha = 1, \ldots, 4$ are defined as

$$N_\alpha(x, y, z) = a_\alpha + b_\alpha x + c_\alpha y + d_\alpha z, \quad \alpha = 1, \ldots, 4 \tag{2.154}$$

where $\{(a_\alpha, b_\alpha, c_\alpha, d_\alpha)\}$ are obtained as

$$\begin{bmatrix} a_1 & b_1 & c_1 & d_1 \\ a_2 & b_2 & c_2 & d_2 \\ a_3 & b_3 & c_3 & d_3 \\ a_4 & b_4 & c_4 & d_4 \end{bmatrix} = \begin{bmatrix} 1 & x_1 & y_1 & z_1 \\ 1 & x_2 & y_2 & z_2 \\ 1 & x_3 & y_3 & z_3 \\ 1 & x_4 & y_4 & z_4 \end{bmatrix}^{-1} \tag{2.155}$$

where $\{(x_\alpha, y_\alpha, z_\alpha)\}$ are the coordinates of the αth node of the tetrahedron. More precisely,

$$\begin{aligned} a_1 &= \frac{1}{6V}\begin{vmatrix} x_2 & y_2 & z_2 \\ x_3 & y_3 & z_3 \\ x_4 & y_4 & z_4 \end{vmatrix}, & b_1 &= -\frac{1}{6V}\begin{vmatrix} 1 & y_2 & z_2 \\ 1 & y_3 & z_3 \\ 1 & y_4 & z_4 \end{vmatrix} \\ c_1 &= -\frac{1}{6V}\begin{vmatrix} x_2 & 1 & z_2 \\ x_3 & 1 & z_3 \\ x_4 & 1 & z_4 \end{vmatrix}, & d_1 &= -\frac{1}{6V}\begin{vmatrix} x_2 & y_2 & 1 \\ x_3 & y_3 & 1 \\ x_4 & y_4 & 1 \end{vmatrix} \end{aligned} \tag{2.156}$$

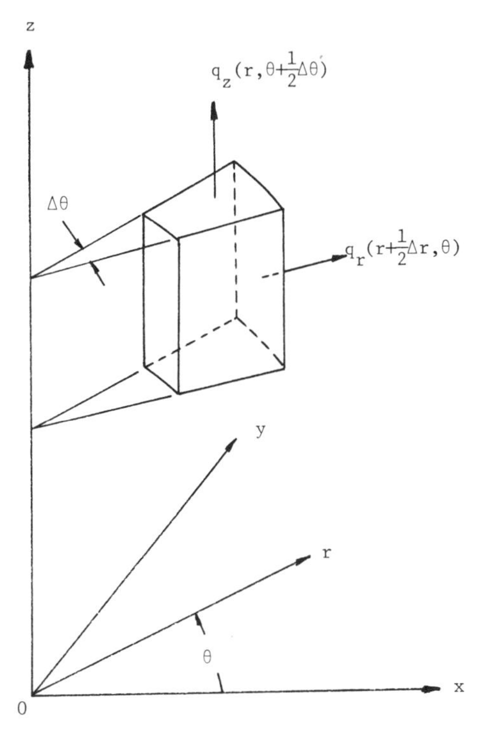

Figure 2.51 Balance of heat flux in cylindrical coordinate system ($q_\theta = 0$).

where

$$6V = \begin{vmatrix} 1 & x_1 & y_1 & z_1 \\ 1 & x_2 & y_2 & z_2 \\ 1 & x_3 & y_3 & z_3 \\ 1 & x_4 & y_4 & z_4 \end{vmatrix} = 6 \times (\text{volume of a tetrahedron}) \qquad (2.157)$$

The other constants are obtained by cyclicly permuting the subscripts 1, 2, 3, and 4.

2.10.5 Axisymmetric problems

So far we discussed the finite element approximation for the plane problem. As shown in Section 2.10.4, the extension to three-dimensional problems is straightforward. However, if axisymmetric problems are considered, we have to discuss more about finite element approximations, since this extension is less obvious compared with the case of general three-dimensional problems.

Let us first go back and derive the governing equation from a differential element. In an axisymmetric problem, the balance of out–in heat flux from/to the differential element described in Figure 2.51 is given by

$$-\rho c_p[rT]_{\bar{r},\bar{z}}\,\Delta r\,\Delta\theta\,\Delta\theta z = ([rq_r]_{r+(1/2)\Delta r,\bar{z}} - [rq_r]_{r-(1/2)\Delta r,\bar{z}})\,\Delta\theta\,\Delta z + ([rq_z]_{\bar{r},z+(1/2)\Delta z} - [rq_z]_{\bar{r},z-(1/2)\Delta z})\,\Delta r\,\Delta\theta$$

where $r - \frac{1}{2}\Delta r \le \bar{r} \le r + \frac{1}{2}\Delta r$ and $z - \frac{1}{2}\Delta z \le \bar{z} \le z + \frac{1}{2}\Delta z$. If we divide by $\Delta r\,\Delta\theta\,\Delta z$ and pass to the limit in which Δr, $\Delta\theta$, and Δz tend to zero, we get

$$\rho c_p r\dot{T} = \frac{\partial}{\partial r}(rq_r) + \frac{\partial}{\partial z}(rq_z) \qquad (2.158)$$

The next step is to write out the relation between the heat flux vector $\mathbf{q} = q_r\mathbf{e}_r + q_z\mathbf{e}_z$ and the temperature gradient ∇T. Since the gradient of a scalar function ϕ is given in the cylindrical coordinate system by

$$q_r = -\left(k_{rr}\frac{\partial T}{\partial r} + k_{rz}\frac{\partial T}{\partial z}\right)$$

$$q_z = -\left(k_{zr}\frac{\partial T}{\partial r} + k_{zz}\frac{\partial T}{\partial z}\right) \tag{2.159}$$

for the axisymmetric problem. We have used here that the temperature is a function of r and z but not of θ. Similarly, the material derivative $\dot{T}$ is given by

$$\dot{T} = \frac{\partial T}{\partial t} + \mathbf{u}\cdot\nabla T = \frac{\partial T}{\partial t} + u_r\frac{\partial T}{\partial r} + u_z\frac{\partial T}{\partial z} \tag{2.160}$$

for the axisymmetric problem.

If we introduce the indicial identifications

$$r \to 1, \qquad z \to 2$$

then Fourier's law [equation (2.159)] reads

$$q_i = -k_{ij}\frac{\partial T}{\partial x_j} \tag{2.161}$$

(2.160) becomes

$$\dot{T} = \frac{\partial T}{\partial t} + u_i\frac{\partial T}{\partial x_i} \tag{2.162}$$

and the right side of (2.158) is

$$\frac{\partial}{\partial x_i}(rq_i) \tag{2.163}$$

If we substitute (2.161), (2.162), and (2.163) into (2.158), it reads

$$\rho c_p r\left(\frac{\partial T}{\partial t} + u_i\frac{\partial T}{\partial x_i}\right) = \frac{\partial}{\partial x_i}\left(rk_{ij}\frac{\partial T}{\partial x_j}\right)$$

and if a heat source term is included, we have

$$\rho c_p r\dot{T} = \frac{\partial}{\partial x_i}\left(rk_{ij}\frac{\partial T}{\partial x_j}\right) + rf \tag{2.164}$$

where $\dot{T}$ is defined by (2.162). When heat conduction in solid bodies is considered, the material derivative becomes $\dot{T} = \partial T/\partial t$. The only difference between equation (2.164) and the equation for two-dimensional problems is the presence of the factor r in all terms of (2.164).

The corresponding weak form of (2.164) with the proper boundary and initial conditions can be obtained as plane problems. Indeed, applying the divergence theorem to the form obtained by integrating equation (2.164) after multiplying an arbitrary function $\bar{T}$ yields

$$\int_0^r \int_0^{2\pi} \int_{z_0}^z \rho c_p r \dot{T} \bar{T} \, dr \, d\theta \, dz$$
$$= \int_0^r \int_0^{2\pi} \int_{z_0}^z \left\{ \frac{\partial}{\partial x_i} \left(r k_{ij} \frac{\partial T}{\partial x_j} \right) \bar{T} + r f \bar{T} \right\} dr \, d\theta \, dz$$
$$= \int_0^{2\pi} \int_\Omega \left(-k_{ij} \frac{\partial T}{\partial x_j} \frac{\partial \bar{T}}{\partial x_i} + f \bar{T} \right) d\Omega \, d\theta$$
$$+ \int_0^{2\pi} \int_\Gamma k_{ij} \frac{\partial T}{\partial x_j} n_j \bar{T} \, d\Gamma \, d\theta$$

where $\Omega = (0, r) \times (z_0, z_1)$ is the cross section of the axisymmetric body at $\theta = 0$, and Γ is its boundary. Here $z_0 = z_0(r)$ and $z_1 = z_1(r)$ are the lower and upper bounds of the body, respectively,

$$d\Omega = r \, dr \, dz \qquad \text{and} \qquad d\Gamma = r \, ds = r\sqrt{dr^2 + dz^2}$$

Since axisymmetry is assumed, we have

$$\int_\Omega \left(\rho c_p \dot{T} \bar{T} + k_{ij} \frac{\partial T}{\partial x_j} \frac{\partial \bar{T}}{\partial x_i} \right) d\Omega = \int_\Omega f \bar{T} \, d\Omega + \int_{\Gamma_2} h \bar{T} \, d\Gamma$$

for every $\bar{T}$ such that $\bar{T} = 0$ on Γ_1 if the first type of boundary condition

$$T = g \quad \text{on } \Gamma_1$$

and the second type of boundary condition

$$q_n = k_{ij} \frac{\partial T}{\partial x_j} n_i = h \quad \text{on } \Gamma_2$$

are introduced as the boundary conditions. It is clear that the weak form obtained is very similar to that for plane problems.

2.11 Higher-order interpolation

In this last section we develop higher-order interpolation of a function defined in a two-dimensional domain. To do this, we first assume, as we have in the past, that the edges of triangular elements are straight lines. We first recall the shape functions for the three-node CST element:

$$N_\alpha(x, y) = a_\alpha + b_\alpha x + c_\alpha y, \quad \alpha = 1, 2, 3 \tag{2.165}$$

It was shown that these shape functions satisfy the relations

$$N_1 + N_2 + N_3 = 1, \qquad N_\alpha(x_\beta, y_\beta) = \delta_{\alpha\beta} \tag{2.166}$$

where (x_β, y_β) are the coordinates of the β-node of an element. Because of the first of (2.166), only two of the three shape functions are independent. Since the xy coordinate system has two independent variables x and y, we may take two shape functions for the CST element as the basis of polynomials that define the shape functions for the higher-order triangular elements. To make a clear distinction between the shape functions and the basis for higher-order interpolation, we introduce the "coordinates" ξ_i, $i = 1, 2, 3$, such that

$$\xi_i = N_i, \quad i = 1, 2, 3 \tag{2.167}$$

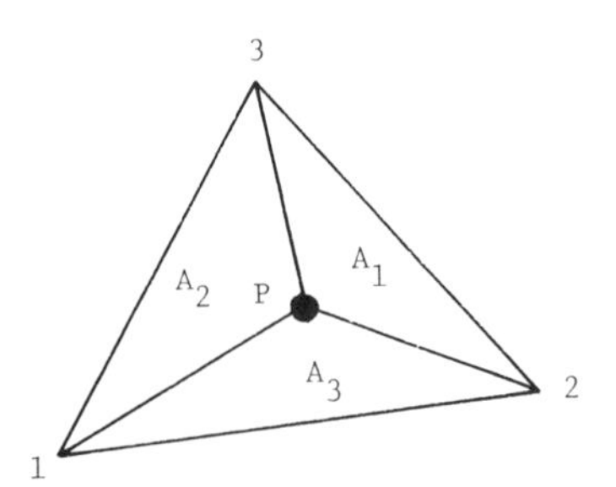

Figure 2.52 Area coordinates of P.

where $\{N_i\}$, $i = 1, 2, 3$, is the set of shape functions for the CST element. As independent variables, we shall take ξ_1 and ξ_2.

A remark concerning the basis (or coordinates ξ_i, $i = 1, 2, 3$) is that these satisfy the geometric relations

$$\xi_i = A_i/A \tag{2.168}$$

where A_i and A are, respectively, the areas of the three subtriangles having P as a common vertex and of the entire element. See Figure 2.52. Because of the relations (2.168), the coordinates ξ_i, $i = 1, 2, 3$, are sometimes called the *area coordinates.*

Another remark is that if the values of ξ_1, ξ_2, and ξ_3 are given, the corresponding xy coordinates of P are obtained by

$$x = x_\alpha \xi_\alpha \qquad \text{and} \qquad y = y_\alpha \xi_\alpha \tag{2.169}$$

where (x_α, y_α), $\alpha = 1, 2, 3$, are the coordinates of the vertices of the triangular element in the xy coordinate system. Conversely, if the xy coordinates of P are given, the corresponding values ξ_1, ξ_2, and ξ_3 are defined by (2.167). Furthermore, we have, from (2.169) and the fact that $\xi_3 = 1 - \xi_1 - \xi_2$,

$$\begin{bmatrix} \dfrac{\partial x}{\partial \xi_1} & \dfrac{\partial x}{\partial \xi_2} \\ \dfrac{\partial y}{\partial \xi_1} & \dfrac{\partial y}{\partial \xi_2} \end{bmatrix} = \begin{bmatrix} (x_1 - x_3) & (x_2 - x_3) \\ (y_1 - y_3) & (y_2 - y_3) \end{bmatrix} \tag{2.170}$$

and this yields the relation between the area elements

$$\Delta A_{xy} = \{(x_1 - x_3)(y_2 - y_3) - (x_2 - x_3)(y_1 - y_3)\}\, \Delta A_{\xi_1 \xi_2} \tag{2.171}$$

Consider now a function f expressed in terms of the coordinates ξ_1, ξ_2, and ξ_3. In view of (2.167) and the chain rules, we have

$$\begin{aligned} \frac{\partial f}{\partial x} &= \frac{\partial \xi_1}{\partial x}\frac{\partial f}{\partial \xi_1} + \frac{\partial \xi_2}{\partial x}\frac{\partial f}{\partial \xi_2} + \frac{\partial \xi_3}{\partial x}\frac{\partial f}{\partial \xi_3} \\ &= b_1 \frac{\partial f}{\partial \xi_1} + b_2 \frac{\partial f}{\partial \xi_2} + b_3 \frac{\partial f}{\partial \xi_3} \\ \frac{\partial f}{\partial y} &= \frac{\partial \xi_1}{\partial y}\frac{\partial f}{\partial \xi_1} + \frac{\partial \xi_2}{\partial y}\frac{\partial f}{\partial \xi_2} + \frac{\partial \xi_3}{\partial y}\frac{\partial f}{\partial \xi_3} \\ &= c_1 \frac{\partial f}{\partial \xi_1} + c_2 \frac{\partial f}{\partial \xi_2} + c_3 \frac{\partial f}{\partial \xi_3} \end{aligned}$$

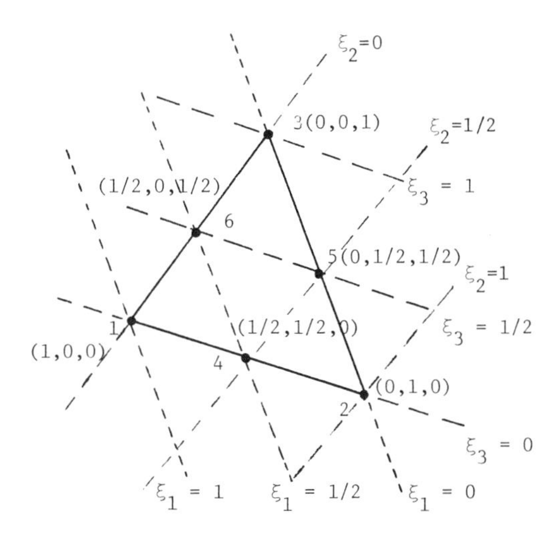

Figure 2.53 Numbering and local coordinates.

where b_α and c_α, $\alpha = 1, 2, 3$, are used in (2.165). Using index notation, we could write

$$\frac{\partial f}{\partial x} = b_i \frac{\partial f}{\partial \xi_i} \qquad \text{and} \qquad \frac{\partial f}{\partial y} = c_i \frac{\partial f}{\partial \xi_i} \tag{2.172}$$

where the range of i is 1, 2, and 3.

For linear interpolation in a three-node triangular element, a function f is approximated by

$$f = f_\alpha \xi_\alpha$$

In this case, we have

$$\frac{\partial f}{\partial x} = b_i \frac{\partial}{\partial \xi_i}(f_\alpha \xi_\alpha) = b_\alpha f_\alpha, \qquad \frac{\partial f}{\partial y} = c_\alpha f_\alpha \tag{2.173}$$

Thus, we obtain the same expression as in section 2.4.

For quadratic interpolation, six nodes (three vertices and three midnodes) are used to "sample" the values of a function. According to the numbering system shown in Figure 2.53, the shape functions N_α, $\alpha = 1, \ldots, 6$, are defined as

$$\begin{aligned} N_1 &= 2(\xi_1 - \tfrac{1}{2})\xi_1 = \xi_1 - \tfrac{1}{2}N_6 - \tfrac{1}{2}N_4 \qquad & N_4 &= 4\xi_1\xi_2 \\ N_2 &= 2(\xi_2 - \tfrac{1}{2})\xi_2 = \xi_2 - \tfrac{1}{2}N_4 - \tfrac{1}{2}N_5 \qquad & N_5 &= 4\xi_2\xi_3 \\ N_3 &= 2(\xi_3 - \tfrac{1}{2})\xi_3 = \xi_3 - \tfrac{1}{2}N_5 - \tfrac{1}{2}N_6 \qquad & N_6 &= 4\xi_3\xi_1 \end{aligned} \tag{2.174}$$

It is clear that

$$N_\alpha(\xi_{1\beta}, \xi_{2\beta}, \xi_{3\beta}) = \delta_{\alpha\beta} \qquad \text{and} \qquad \sum_{\alpha=1}^{6} N_\alpha = 1$$

Then a function T is interpolated as

$$T = T_\alpha N_\alpha(\xi_1, \xi_2, \xi_3) \tag{2.175}$$

in a six-node triangular element, where T_α is the value of the function T at the α node of the finite element.

Table 2.4 *Weight and integration points of the quadrature for a triangular domain from Zienkiewicz (1976)*

Order	Error	Points	Triangular coordinates	Weights
Linear	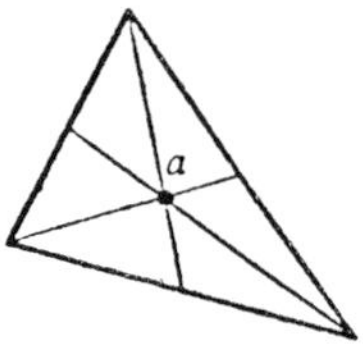 $R = O(h^2)$	a	$\frac{1}{3}, \frac{1}{3}, \frac{1}{3}$	1
Quadratic	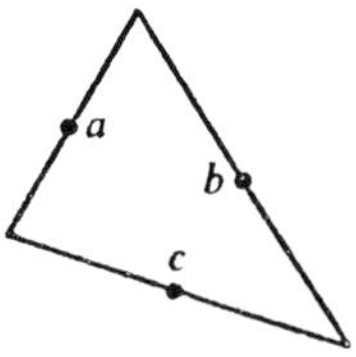 $R = O(h^3)$	a	$\frac{1}{2}, \frac{1}{2}, 0$	$\frac{1}{3}$
		b	$0, \frac{1}{2}, \frac{1}{2}$	$\frac{1}{3}$
		c	$\frac{1}{2}, 0, \frac{1}{2}$	$\frac{1}{3}$
Cubic	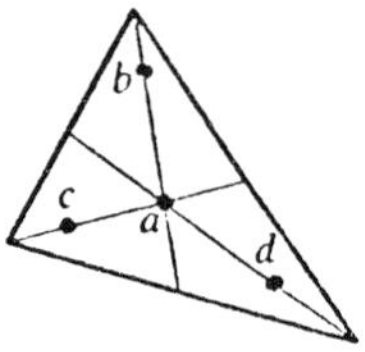 $R = O(h^4)$	a	$\frac{1}{3}, \frac{1}{3}, \frac{1}{3}$	$-\frac{27}{48}$
		b	0.6, 0.2, 0.2	$\frac{25}{48}$
		c	0.2, 0.6, 0.2	
		d	0.2, 0.2, 0.6	
Quintic	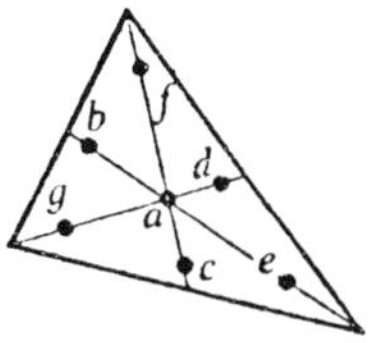 $R = O(h^6)$	a	$\frac{1}{3}, \frac{1}{3}, \frac{1}{3}$	0.2250000000
		b	$\alpha_1, \beta_1, \beta_1$	0.1323941527
		c	$\beta_1, \alpha_1, \beta_1$	
		d	$\beta_1, \beta_1, \alpha_1$	
		e	$\alpha_2, \beta_2, \beta_2$	0.1259391805
		f	$\beta_2, \alpha_2, \beta_2$	
		g	$\beta_2, \beta_2, \alpha_2$	

with
$\alpha_1 = 0.0597158717$
$\beta_1 = 0.4701420641$
$\alpha_2 = 0.7974269853$
$\beta_2 = 0.1012865073$

Recall that the component of an element stiffness matrix is given by

$$K^e_{\alpha\beta} = \int_{\Omega_e} k_{ij} \frac{\partial N_\alpha}{\partial x_i} \frac{\partial N_\beta}{\partial x_j} \, d\Omega \tag{2.176}$$

Because of the relations (2.172) for the derivatives with respect to x and y, the integrand of the stiffness matrix is a function of the area coordinates ξ_1, ξ_2, and ξ_3. The integrals can then be evaluated either analytically, using the formula

$$\int_{\Omega_e} \xi_1^m \xi_2^n \xi_3^p \, d\Omega = 2\Delta \frac{m!n!p!}{(m+n+p+2)!} \tag{2.177}$$

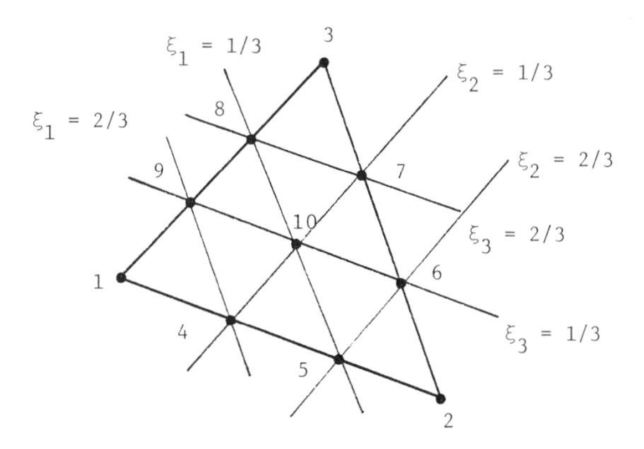

Figure 2.54 Ten-node cubic element.

or numerically, using the quadrature rules*

$$\int_{\Omega_e} f(\xi_1, \xi_2, \xi_3)\, d\Omega \approx \sum_{i=1}^{I} w_i f(\xi_{1i}, \xi_{2i}, \xi_{3i})\Delta \tag{2.178}$$

where

$$2\Delta = (x_1 - x_3)(y_2 - y_3) - (x_2 - x_3)(y_1 - y_3) \tag{2.179}$$

$\{w_i\}$ and $\{(\xi_{1i}, \xi_{2i}, \xi_{3i})\}$, $i = 1, 2, \ldots, I$, are the sets of weights and integration points shown in Table 2.4, and I is the number of integration points. For quadratic interpolation, $I = 3$ is generally assumed for the quadrature rule in the computation of the element stiffness matrix.

For cubic interpolation, 10 nodal points and 10 "sampling" points for a function are used according to the numbering system in a triangle, as shown in Figure 2.54. In this case, the shape functions are defined by

$$\begin{aligned}
N_1 &= \tfrac{9}{2}(\xi_1 - \tfrac{2}{3})(\xi_1 - \tfrac{1}{3})\xi_1, & N_2 &= \tfrac{9}{2}(\xi_2 - \tfrac{2}{3})(\xi_2 - \tfrac{1}{3})\xi_2 \\
N_3 &= \tfrac{9}{2}(\xi_3 - \tfrac{2}{3})(\xi_3 - \tfrac{1}{3})\xi_3, & N_4 &= \tfrac{27}{2}(\xi_1 - \tfrac{1}{3})\xi_1\xi_2 \\
N_5 &= \tfrac{27}{2}(\xi_2 - \tfrac{1}{3})\xi_1\xi_2, & N_6 &= \tfrac{27}{2}(\xi_2 - \tfrac{1}{3})\xi_2\xi_3 \\
N_7 &= \tfrac{27}{2}(\xi_3 - \tfrac{1}{3})\xi_2\xi_3, & N_8 &= \tfrac{27}{2}(\xi_3 - \tfrac{1}{3})\xi_3\xi_1 \\
N_9 &= \tfrac{27}{2}(\xi_1 - \tfrac{1}{3})\xi_3\xi_1, & N_{10} &= 27\xi_1\xi_2\xi_3
\end{aligned} \tag{2.180}$$

The shape functions for other higher-order interpolation can be similarly defined and are left as exercises for the reader.

Let us now define the shape functions for an element that connects a three-node and a six-node triangular element. While there are two possibilities, we shall consider the five-node element shown in Figure 2.55. This case arises when the connecting element must join two six-node elements. In this case, we should have linearity along the edge 1–3. Thus, the shape functions are defined by

$$\begin{aligned}
N_1 &= \xi_1 - \tfrac{1}{2}N_4 & N_4 &= 4\xi_1\xi_2 \\
N_2 &= \xi_2 - \tfrac{1}{2}N_4 - \tfrac{1}{2}N_5 & N_5 &= 4\xi_2\xi_3 \\
N_3 &= \xi_3 - \tfrac{1}{2}N_5
\end{aligned} \tag{2.181}$$

* Quadrature rules on a triangular domain are given in Cowper (1973).

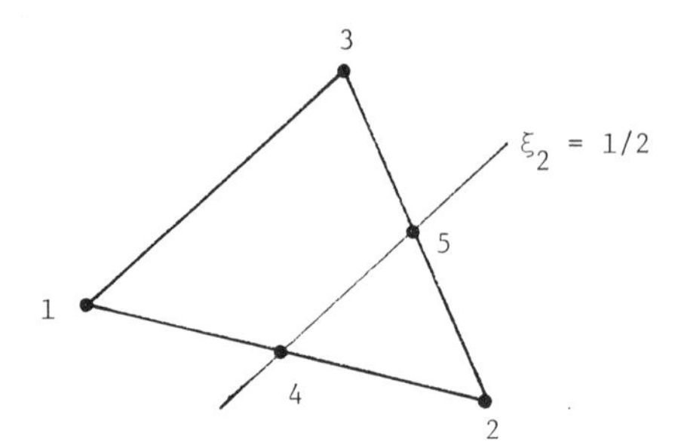

Figure 2.55 Five-node transition triangular element.

Figure 2.56 Four-node transition element.

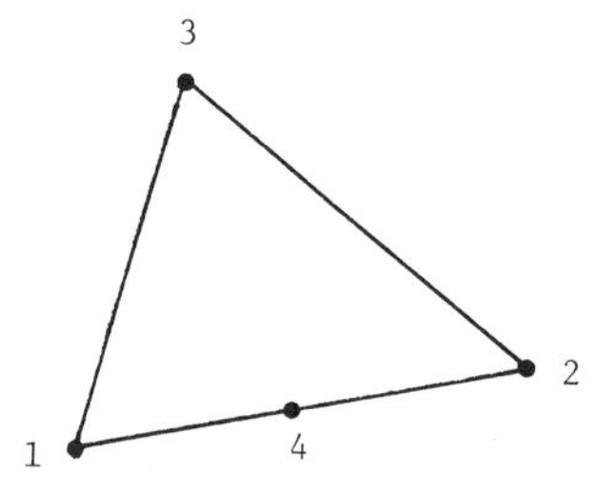

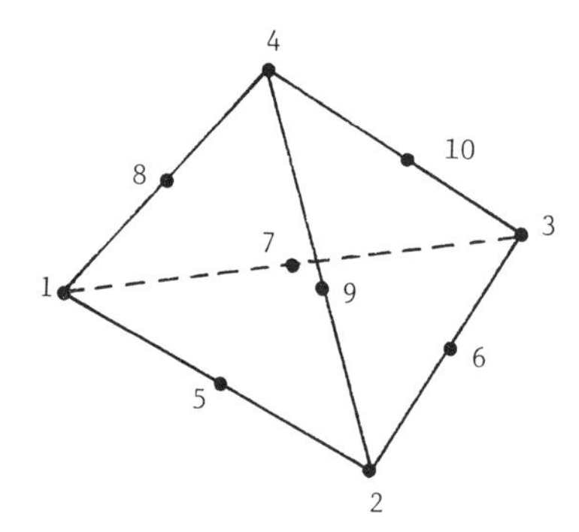

Figure 2.57 Ten-node tetrahedron element.

It is easy to show that

$$\sum_{\alpha=1}^{5} N_\alpha = 1 \qquad \text{and} \qquad N_\alpha(\xi_{1\beta}, \xi_{2\beta}, \xi_{3\beta}) = \delta_{\alpha\beta}$$

Exercise 2.12: Develop a subroutine to compute the element stiffness matrix $[K^e_{\alpha\beta}]$ using quadratic interpolation and the formula (2.177).

Exercise 2.13: Develop a subroutine to compute the element stiffness matrix $[K^e_{\alpha\beta}]$ using cubic interpolation and the quadrature rule (2.178).

Exercise 2.14: Find the shape functions for the four-node triangular element shown in Figure 2.56. The results of this exercise would be needed to connect a single six-node element.

Exercise 2.15: Find the shape functions for the 10-node tetrahedron element shown in Figure 2.57. In this case we shall use the *volume* *coordinates* ξ_1, ξ_2, ξ_3, and ξ_4 such that

$$\xi_i = V_i/V, \qquad \sum_{i=1}^{4} \xi_i = 1 \tag{2.182}$$

and

$$\xi_i = a_i + b_i x + c_i y + d_i z, \quad i = 1, \ldots, 4 \tag{2.183}$$

where a_i, b_i, c_i, d_i are defined in (2.156).

References

Babuska, I., Chandra, J., and Flaherty, J. E., *Adaptive computational methods for partial differential equations,* SIAM, Philadelphia, 1983.

Babuska, I., Oliveira, E. R. A., and Zienkiewicz, O. C., *Accuracy estimates and adaptive refinements in finite element computation,* Technical University of Lisbon, Lisbon, 1984.

Ciarlet, P. G., *The finite element method for elliptic problems,* North-Holland, Amsterdam, 1978.

Cowper, G. R., "Gaussian quadrature formulas for triangles," *International Journal for Numerical Methods in Engineering* 7: 405–8, 1973.

Dowson, D., and Higginson, G. R., *Elasto-hydrodynamic lubrication,* SI edition, Pergamon, Oxford, 1977.

Huebner, K., *The finite element method for engineers,* Wiley, New York, 1975.

Ikegawa, M., and Washizu, K., "Finite element method applied to analysis of flow over a spillway crest," *International Journal for Numerical Methods in Engineering* 6: 179–89, 1973.

Kinderlehrer, D., and Stampaccia, G., *An introduction to variational inequalities and their applications,* Academic, New York, 1980.

Neumann, S. P., and Witherspoon, P. L., "Finite element method of analyzing steady seepage with a free surface," *Water Resources Research* 6(3): 889–97, 1970.

Oden, J. T., *Finite elements IV: mathematical aspects,* Prentice-Hall, Englewood Cliffs, New Jersey, 1983.

Oden, J. T., and Reddy, J. N., *An introduction to the mathematical theory of finite elements,* Wiley-Interscience, New York, 1975.

Pode, L., and Schneider, D. I., *Handbook of BASIC for the IBM PC,* Robert J. Brady Co./Prentice-Hall, Bowie, Maryland, 1983.

Shephard, M. S., and Gallagher, R. H., *Finite element grid optimization,* PVP-38, American Society of Mechanical Engineers, New York, 1979.

Strang, G., and Fix, G. J., *An analysis of the finite element method,* Prentice-Hall, Englewood Cliffs, New Jersey, 1973, chap. 3.

Taylor, R. L., and Brown, C. B., "Darcy flow solutions with a free surface," *Journal of the Hydraulics Division, Proceedings of the ASCE* HY2: 25–33, 1967.

Zienkiewicz, O. C., *The finite element method,* McGraw-Hill, London, 1976.

3
GENERALIZATION OF THE FINITE ELEMENT METHOD FOR HEAT CONDUCTION PROBLEMS

In the previous chapter, we have used the simplest finite element method to solve several problems in applied mechanics and have introduced techniques such as penalization, regularization, and iteration. We return to the problem of heat conduction in this chapter and include the convective terms in the time derivative. To do this, the initial boundary value problem of heat conduction with fluid flow is recalled from Chapter 2:

$$\begin{aligned} \rho c_p\left(\frac{\partial T}{\partial t} + u_i \frac{\partial T}{\partial x_i}\right) &= \frac{\partial}{\partial x_i}\left(k_{ij}\frac{\partial T}{\partial x_j}\right) + f \quad \text{in } \Omega \\ T &= g \quad \text{on } \Gamma_1 \\ k_{ij}\frac{\partial T}{\partial x_j} n_i &= h \quad \text{on } \Gamma_2 \\ k_{ij}\frac{\partial T}{\partial x_j} n_i &= -k_0(T - T_\infty) \quad \text{on } \Gamma_3 \\ T &= T_0 \quad \text{at } t = 0 \end{aligned} \tag{3.1}$$

Here $\mathbf{u} = u_i\mathbf{i}_i$ is the velocity vector of fluid flow and is assumed to be known.* The velocity vector $\mathbf{u}$ might be obtained by solving the equation for potential flows studied in the previous chapter.

3.1 Upwind technique for the convection

We start from a finite element approximation of the convective term $\rho c_p u_i\, \partial T/\partial x_i$. To do this, let us first consider a one-dimensional steady-state problem:

$$\begin{aligned} \rho c_p u \frac{dT}{dx} &= \frac{d}{dx} k \frac{dT}{dx} + f \quad \text{in } \Omega = (0, 1) \\ T(0) &= 0, \qquad T(1) = 1 \end{aligned} \tag{3.2}$$

To derive the finite element approximation studied in Chapter 2, the corresponding weak form must be obtained. Multiplying both sides of (3.2) by an arbitrary virtual

* Actually, the full equations in (3.1) are coupled with those of fluid flow. The assumption that $\mathbf{u}$ is known is equivalent to assuming that the coupling is "weak."

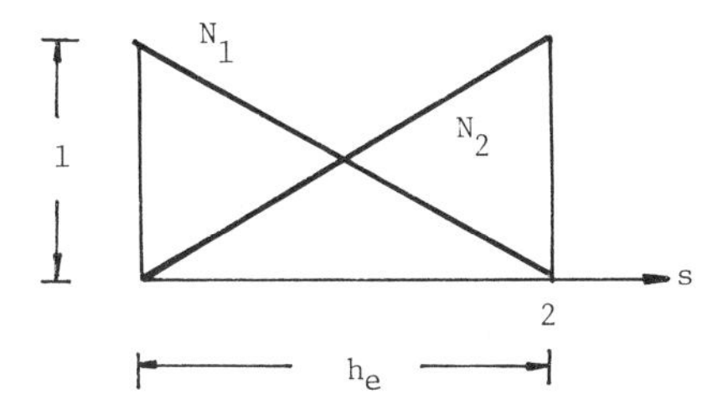

Figure 3.1 Line element Ω_e and its shape functions, $N_1(s) = 1 - s/h_e$ and $N_2(s) = s/h_e$.

temperature $\bar{T}$ and integrating by parts, we have

$$\int_\Omega \left(\rho c_p u \frac{dT}{dx} \bar{T} + k \frac{dT}{dx} \frac{d\bar{T}}{dx} \right) d\Omega = \int_\Omega f \bar{T} \, d\Omega$$
$$\forall \bar{T} \ni \bar{T}(0) = \bar{T}(1) = 0 \tag{3.3}$$

To repeat, the boundary conditions are

$$T(0) = 0, \qquad T(1) = 1 \tag{3.4}$$

Applying the two-node line element for the discretization of the domain $\Omega = (0, 1)$ and the temperatures T and $\bar{T}$ described in Figure 3.1, equation (3.3) is approximated by

$$\sum_{e=1}^{E} \bar{T}_\alpha (K^c_{\alpha\beta} + K^d_{\alpha\beta}) T_\beta = \sum_{e=1}^{E} \bar{T}_\alpha f^e_\alpha \tag{3.5}$$

where

$$K^c_{\alpha\beta} = \int_{\Omega_e} \rho c_p u \frac{dN_\beta}{dx} N_\alpha \, d\Omega, \qquad K^d_{\alpha\beta} = \int_{\Omega_e} k \frac{dN_\beta}{dx} \frac{dN_\alpha}{dx} \, d\Omega \tag{3.6}$$

and

$$f^e_\alpha = \int_{\Omega_e} f N_\alpha \, d\Omega$$

where Ω_e is an arbitrary line element. Note that the stiffness matrix $[K^e_{\alpha\beta}]$ consists of two parts $[K^c_{\alpha\beta}]$ and $[K^d_{\alpha\beta}]$ and that $[K^e_{\alpha\beta}] = [K^c_{\alpha\beta}] + [K^d_{\alpha\beta}]$ is no longer symmetric because $[K^c_{\alpha\beta}] \neq [K^c_{\beta\alpha}]$. Thus, the finite element equation

$$\sum_{e=1}^{E} (K^c_{\alpha\beta} + K^d_{\alpha\beta}) T_\beta = \sum_{e=1}^{E} f^e_\alpha$$
$$T_h(0) = 0, \qquad T_h(1) = 1 \tag{3.7}$$

must be solved by a routine that does not require symmetry.

Example 3.1: Suppose that

$$\rho c_p u = 100 \qquad \text{and} \qquad f = 0$$

in the boundary value problem (3.2), and that the domain $\Omega = (0, 1)$ is uniformly divided into 10 elements. Let us solve the finite element equation (3.7) for $k = 100, 10, 1$. Figures 3.2 and 3.3. display the numerical

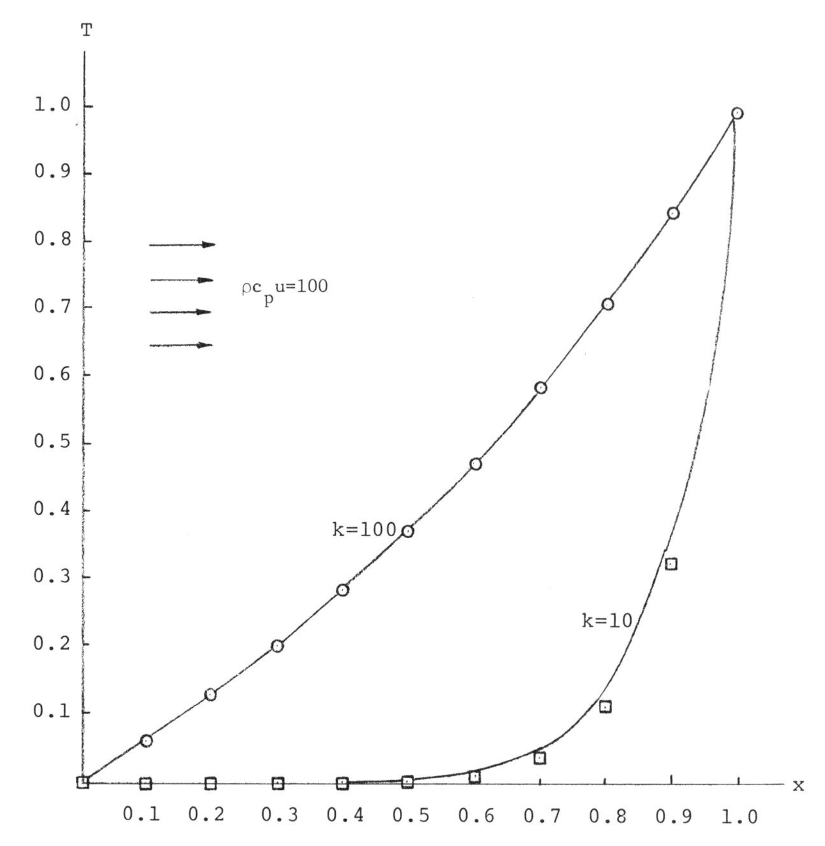

Figure 3.2 Finite element solutions to the one-dimensional convective heat conduction problem, $k = 100, 10$.

results and compare them with the exact solution

$$T(x) = \frac{e^{P_e x} - 1}{e^{P_e} - 1}, \qquad P = \frac{\rho c_p u}{k}$$

Strong oscillations in the numerical solutions are to be noted for the smaller conductivity constant k.

Should we abandon the finite element approximation (3.7) for small conductivity k? Yes, but then what is an alternative to (3.7)? To answer this, we consider the finite element equation (3.7). Suppose that $\rho c_p u$, k, and h_e are constant for $e = 1, \ldots, E$. Then (3.7) can be represented by

$$\frac{\rho c_p u}{2}(T_{\alpha+1} - T_{\alpha-1}) + \frac{k}{h_e}(-T_{\alpha+1} + 2T_\alpha - T_{\alpha-1}) = f_\alpha h_e, \quad \alpha = 2, \ldots, E \quad (3.8)$$

where $T_1 = 0$ and $T_{E+1} = 1$. It is clear that this is the same as the finite difference approximation of equation (3.2) using the central difference scheme

$$\frac{dT}{dx} \approx \frac{T_{\alpha+1} - T_{\alpha-1}}{2h_e}$$

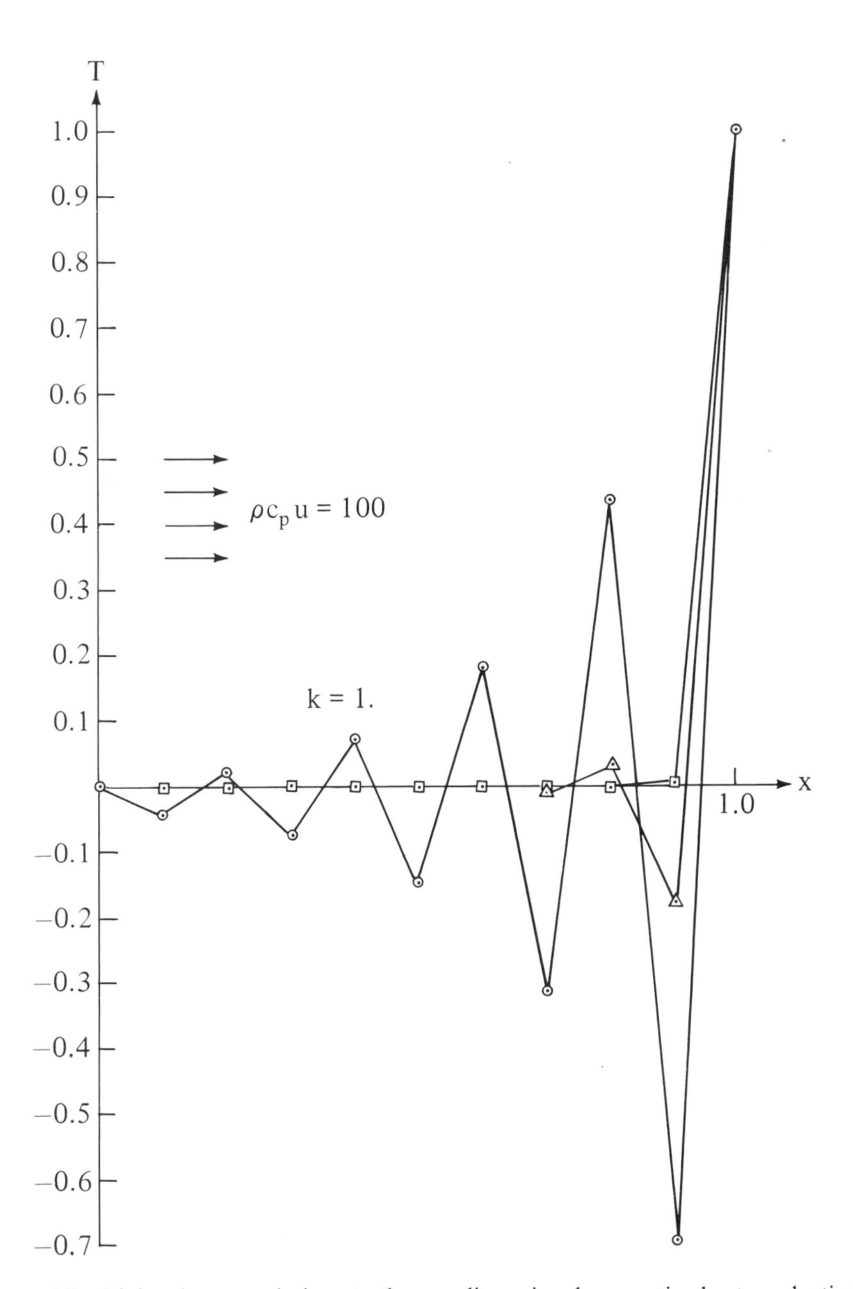

Figure 3.3 Finite element solutions to the one-dimensional convective heat conduction problem, $k = 1$: ⊙, $\alpha_0 = 0$; △, $\alpha_0 = 0.5$; ⊡, $\alpha_0 = 1$.

and the three-point scheme

$$\frac{d^2T}{dx^2} \approx \frac{T_{\alpha+1} - 2T_\alpha + T_{\alpha-1}}{h_e^2}$$

Defining a parameter as the *discrete Peclet number*

$$P_{eh} = \frac{\rho c_p u h_e}{k} \tag{3.9}$$

equation (3.8) can be written as

$$P_{eh}\tfrac{1}{2}(T_{\alpha+1} - T_{\alpha-1}) - T_{\alpha+1} + 2T_\alpha - T_{\alpha-1} = f_\alpha h_e^2 \tag{3.10}$$

This further implies

$$2T_\alpha = (1 - \tfrac{1}{2}P_{eh})T_{\alpha+1} + (1 + \tfrac{1}{2}P_{eh})T_{\alpha-1} + f_\alpha h_e^2$$

For $f_\alpha = 0$, if $P_{eh} > 2$, we have "negative" diffusion from the side of the $(\alpha + 1)$th node. On the other hand, if $P_{eh} < -2$, negative diffusion is obtained from the side of the $(\alpha - 1)$th node. These are physically unrealistic. Thus, condition

$$|P_{eh}| < 2, \qquad \text{i.e.,} \quad h_e \leq 2|k/\rho c_p u| \tag{3.11}$$

is required in order to have physically sound numerical results from the finite element approximation (3.8). Condition (3.11) means that the size of finite elements must be small if k is small or if u is large.

Now, what can we do if the size of finite elements h_e is fixed? One method is the *upwind method,* which has been extensively studied in computational fluid dynamics (Roache, 1972). Noting that

$$\tfrac{1}{2}(T_{\alpha+1} - T_{\alpha-1}) = T_\alpha - T_{\alpha-1} + \tfrac{1}{2}(T_{\alpha+1} - 2T_\alpha + T_{\alpha-1})$$

equation (3.8) written as

$$\frac{\rho c_p u}{2}(T_{\alpha+1} - T_{\alpha-1}) + \left(\frac{\rho c_p u h_e}{2h_e} + \frac{k}{h_e}\right)(-T_{\alpha+1} + 2T_\alpha - T_{\alpha-1}) = f_\alpha h_e, \ \alpha = 2, 3, \ldots, E$$

yields the upwind (backward) difference scheme

$$\frac{dT}{dx} \approx \frac{T_\alpha - T_{\alpha-1}}{h_e}$$

The second term is also similar to the three-point scheme

$$-\frac{\rho c_p u h_e}{2}\frac{d^2T}{dx^2} \approx \frac{\rho c_p u h_e}{2}\,\frac{-T_{\alpha+1} + 2T_\alpha - T_{\alpha-1}}{h_e^2}$$

to the "additional" conduction term by the *artificial* conductivity

$$k_{ad} = \tfrac{1}{2}\rho c_p u h_e \tag{3.12}$$

Thus, we modify the finite element equation (3.7) by adding the term

$$K_{\alpha\beta}^{ad} = \int_{\Omega_e} k_{ad}\frac{dN_\beta}{dx}\frac{dN_\alpha}{dx}\,d\Omega \tag{3.13}$$

that is, the finite element approximation (3.7) is now changed to

$$\sum_{e=1}^{E}(K_{\alpha\beta}^c + \alpha_0 K_{\alpha\beta}^{ad} + K_{\alpha\beta}^d)T_\beta = \sum_{e=1}^{E} f_\alpha^e$$
$$T_h(0) = 0, \qquad T_h(1) = 1 \tag{3.14}$$

in which $0 \leq \alpha_0 \leq 1$.

If $\alpha_0 = 0$, then no artificial conductivity is applied. If $\alpha_0 = 1$, then the approximation scheme (3.14) is equivalent to using upwind differences for convection and three-point differences for conduction.

Table 3.1 *Numerical results by variable meshes*

Node/β	10	15	20	30
1	-4.77×10^{-11}	-1.13×10^{-11}	2.42×10^{-10}	1.96×10^{-7}
2	9.78×10^{-8}	2.33×10^{-8}	-4.99×10^{-7}	-4.08×10^{-4}
3	-5.67×10^{-8}	-9.00×10^{-9}	1.56×10^{-7}	1.05×10^{-4}
4	4.26×10^{-7}	5.94×10^{-8}	-1.02×10^{-6}	-6.98×10^{-4}
5	-2.91×10^{-6}	-1.51×10^{-7}	1.72×10^{-6}	7.76×10^{-4}
6	2.04×10^{-4}	8.46×10^{-7}	-6.56×10^{-6}	-2.36×10^{-3}
7	2.67×10^{-3}	-7.52×10^{-6}	2.66×10^{-5}	5.34×10^{-3}
8	1.77×10^{-2}	1.88×10^{-4}	-1.57×10^{-4}	-1.17×10^{-2}
9	8.31×10^{-2}	8.97×10^{-3}	1.36×10^{-3}	5.50×10^{-2}
10	3.12×10^{-1}	1.21×10^{-1}	2.08×10^{-2}	-2.17×10^{-1}
11	9.99×10^{-1}	9.99×10^{-1}	9.99×10^{-1}	9.99×10^{-1}

Note: $x_i = \left(\frac{i-1}{E}\right)^{p/\beta}$, E = number of elements, and $p = u/k$ ($u = 100$, $k = 1$, $\rho c_p = 1$).

Example 3.2 (continuation): Let us apply the approximate scheme (3.14) to the conduction problem discussed above. We have already observed that (3.7) yields an oscillatory solution for the case $k = 1$ and $\rho c_p u = 100$. Numerical results for this case are shown in Figure 3.3 for $\alpha_0 = 0.5, 1$. It is clear that if $\alpha_0 = 1$ is chosen, the numerical solution does not have any oscillation at all.

It should be noted that the addition of the artificial conductivity is *just a technique* that produces reasonable numerical solutions by finite element or difference methods. If it is feasible to use a sufficiently small mesh size in a given problem, we *can* obtain a solution whose quality is fairly good. However, in practice, it is hard to make a very refined finite element model that provides good numerical solutions. Then some technique must be introduced, and the above method is such a technique.

Another way to obtain a reasonable numerical solution is to use a variable mesh. In the above, a uniform finite element mesh was used for the discretization and led to a solution with strong oscillations. If the mesh is defined by

$$x_i = \left(\frac{i-1}{E}\right)^{p/\beta} \qquad p = \frac{u}{k} \tag{3.15}$$

where x_i is the coordinate of the i node, β is a parameter, and E is the total number of elements, then the numerical solution of the above example is as in Table 3.1. The result for $\beta = 20$ is similar in quality to the case of $\alpha_0 = 1$.

In the above example, we knew the portion of the domain where the temperature is rapidly changed. This implied the mesh defined by (3.15). If there is no information on the solution a priori, the adaptive methods discussed in section 2.10.2 are applicable to define finite element discretization, which dramatically reduces oscillation in a numerical solution without introducing any artificial conductivity k_{ad}. Indeed, Diaz et al. (1983) applied adaptive methods to solve a convection heat conduction problem similar to (3.2).

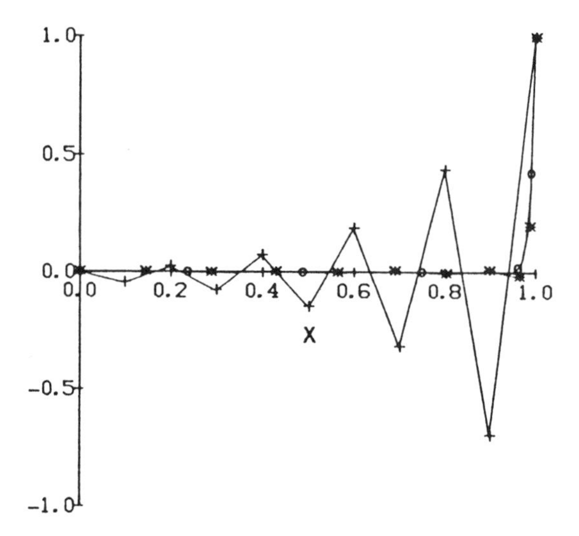

Figure 3.4 Solutions corresponding to uniform grid and improved grid using r-method: $+$, u_h, uniform grid; $*$, u_h, improved grid; ○, u, exact.

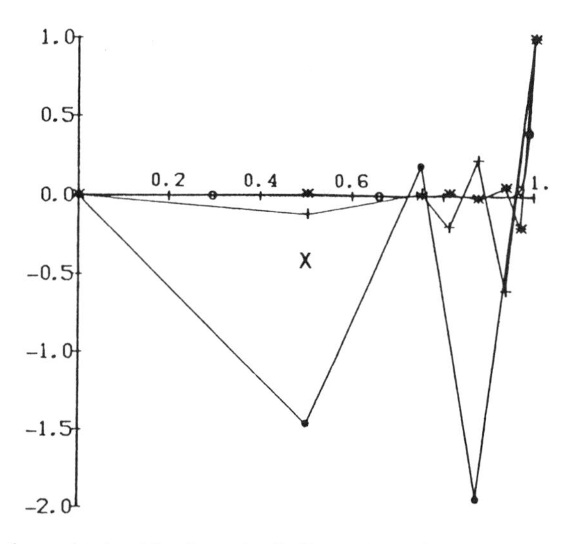

Figure 3.5 Solutions obtained by h-method: ●, u_h, $M = 4$; $+$, u_h, $M = 6$; $*$, u_h, $M = 7$; ○, u (M is the number of elements).

Example 3.3 (continuation): Let us solve the same problem in example 3.1 using both r- and h-adaptive methods. Starting from the uniform mesh by 10 intervals, the results shown in Figures 3.4 and 3.5 are obtained. It is clear that both adaptive methods almost eliminate oscillation on the solutions.

There are certainly other techniques for dealing with oscillation in numerical solutions. However, we shall not explore these in this chapter. We shall mainly apply the technique of artificial conductivity, since it is independent of the dimension of the domain Ω and only one parameter α_0 is involved.

The essential point of the upwind technique is the additional term of artificial conductivity (3.13). The form of this term depends on the given velocity field, conductivity, and mesh size of the finite element model. We shall extend this idea to multidimensional problems. In this case, the velocity is a vector $\mathbf{u} = u_i \mathbf{i}_i$ and the convective term becomes

$$\mathbf{u} \cdot \nabla T = u_i T_{,i}$$
$$\left(= u_x \frac{\partial T}{\partial x} + u_y \frac{\partial T}{\partial y} \quad \text{for two-dimensional problems} \right) \tag{3.16}$$

Since the treatment of three-dimensional problems is quite similar to that used for two-dimensional ones, we shall not specifically study three-dimensional problems in this section.

To extend the idea obtained by studying one-dimensional problems to multidimensional cases, we first point out the fact of physics that temperature is convected by the flow of fluids. Thus, those portions of the flow region where the magnitude of the velocity vector is large must be convection dominated *along the direction of the flow*. This means that the artificial conductivity may be added along the "stream" line of the flow defined by the unit vector $\mathbf{n} = n_i \mathbf{i}_i = n_x \mathbf{i} + n_y \mathbf{j}$ such that

$$n_x = \frac{u_x}{|\mathbf{u}|}, \qquad n_y = \frac{u_y}{|\mathbf{u}|}, \qquad |\mathbf{u}| = (u_i u_i)^{1/2} \tag{3.17}$$

By analogy with its one-dimensional form (3.13), we shall define the additional artificial conductivity $\mathbf{k}^{ac}$ by

$$k_{ij}^{ac} = \alpha_0 \tfrac{1}{3}(\rho c_p |\mathbf{u}| h) n_i n_j \tag{3.18}$$

where α_0 is a magnifying parameter, $|\mathbf{u}|$ is the speed of the flow, and h is a representative mesh size, which may be defined by $h = \max|\hat{\mathbf{x}} - \mathbf{x}|$, where $\hat{\mathbf{x}}, \mathbf{x} \in \Omega_e$.

Therefore, when there are portions where the flow velocity is much larger than the conductivity, we may modify the steady heat conduction problem (3.1) to read

$$\rho c_p u_i T_{,i} = \{(k_{ij} + k_{ij}^{ac}) T_{,j}\}_{,i} + f \quad \text{in } \Omega \tag{3.19}$$

We call this the artificial conductivity or upwind technique for problems that are convection dominated. Thus, the element stiffness matrix $[K_{\alpha\beta}^e]$ for the heat conduction problem (3.1), with convection terms, is given by

$$[K_{\alpha\beta}^e] = [K_{\alpha\beta}^{c,e}] + [K_{\alpha\beta}^{ac,e}] + [K_{\alpha\beta}^{d,e}] \tag{3.20}$$

where

$$K_{\alpha\beta}^{c,e} = \int_{\Omega_e} \rho c_p u_i \frac{\partial N_\beta}{\partial x_i} N_\alpha \, d\Omega \qquad \text{(convection)} \tag{3.21}$$

$$K_{\alpha\beta}^{ac,e} = \int_{\Omega_e} k_{ij}^{ac} \frac{\partial N_\beta}{\partial x_j} \frac{\partial N_\alpha}{\partial x_i} \, d\Omega \qquad \text{(artificial conductivity)} \tag{3.22}$$

$$K_{\alpha\beta}^{d,e} = \int_{\Omega_e} k_{ij} \frac{\partial N_\beta}{\partial x_j} \frac{\partial N_\alpha}{\partial x_i} \, d\Omega \qquad \text{(diffusion)} \tag{3.23}$$

The other terms in the equations of (3.1) can be discretized in Chapter 2, that is, for the case of steady state $\partial T/\partial t = 0$,

$$T_\beta = g_\beta \quad \text{on } \Gamma_1$$

$$\sum_{e=1}^{E} K^e_{\alpha\beta} T_\beta + \sum_{e=1}^{E_3} K^{e,3}_{\alpha\beta} T_\beta = \sum_{e=1}^{E} f^e_\alpha + \sum_{e=1}^{E_2} f^{e,2}_\alpha + \sum_{e=1}^{E_3} f^{e,3}_\alpha \tag{3.24}$$

It is noted that the stiffness matrix $[K^e_{\alpha\beta}]$ is no longer symmetric because of the convective term (3.21). Thus, the finite element equation after assembling

$$K_{\alpha\beta} T_\beta = f_\alpha \tag{3.25}$$

cannot be solved by the symmetric solver used in the program FEM1. It is necessary to introduce, for example, a program of unsymmetric band solver for the heat conduction problem with the convective term.

Exercise 3.1: In (3.19), we have modified only the conduction term by adding artificial conductivity. Solve the convective problem by modifying the heat source term in (3.19) as follows:

$$\int_\Omega f\bar{T}\, d\Omega \to \int_\Omega f\left(\bar{T} + k_{ad}\frac{u_j}{|\mathbf{u}|^2}\bar{T}_{,j}\right) d\Omega \tag{3.26}$$

for some positive constant α_0 and

$$k_{ad} = \tfrac{1}{3}\alpha_0(\rho c_p|\mathbf{u}|h)$$

The constant α_0 is a number close to 1 and is obtained by numerical experiments.

More details on the upwind method can be found in Hughes (1979), Brooks and Hughes (1982), Kelly et al. (1980), and elsewhere.

3.2 θ-Method for time integration

So far, we have neglected the term $\partial T/\partial t$ for simplicity; that is, we have ensured that the temperature field T be independent of time. We now discuss the case of $T = T(\mathbf{x}, t)$ in this section by using the *θ-method* for the time integration.

Noting that the spatial discretization of the domain need not be changed as t varies, we use the finite element discretization of the domain at time $t = 0$ for all time t and the discretization of functions T and $\bar{T}$ such that

$$T_h(\mathbf{x}, t) = T_\alpha(t)N_\alpha(\mathbf{x}), \qquad \bar{T}_h(\mathbf{x}, t) = \bar{T}_\alpha(t)N_\alpha(\mathbf{x}) \tag{3.27}$$

on each finite element Ω_e. That is, the shape functions $\{N_\alpha(\mathbf{x})\}$, $\alpha = 1, 2, 3$, are independent of time and are defined at the initial stage $t = 0$. If the relations (3.27) are substituted into the weak form of the initial boundary value problem (3.1),

$$\int_\Omega \left(\rho c_p \frac{\partial T}{\partial t}\bar{T} + \rho c_p u_i \frac{\partial T}{\partial x_i}\bar{T} + k_{ij}\frac{\partial T}{\partial x_j}\frac{\partial \bar{T}}{\partial x_i}\right) d\Omega + \int_{\Gamma_3} k_0 T\bar{T}\, d\Gamma$$

$$= \int_\Omega f\bar{T}\, d\Omega + \int_{\Gamma_2} h\bar{T}\, d\Gamma + \int_{\Gamma_3} k_0 T_\infty \bar{T}\, d\Gamma$$

$$\forall \bar{T} \ni \bar{T} = 0 \quad \text{on } \Gamma_1 \tag{3.28}$$

$$T = g \quad \text{on } \Gamma_1, \qquad T = T_0 \quad \text{at } t = 0$$

we have the additional term

$$\sum_{e=1}^{E} M_{\alpha\beta}^{e} \frac{dT_\beta}{dt} \quad \text{where } M_{\alpha\beta}^{e} = \int_\Omega \rho c_p N_\alpha N_\beta \, d\Omega \tag{3.29}$$

on the left side of the finite element equation (3.24) as well as the initial condition

$$T_\beta = T_0(\mathbf{x}_\beta) \quad \text{at } t = 0$$

That is, we have

$$\sum_{e=1}^{E} \left(M_{\alpha\beta}^{e} \frac{dT_\beta}{dt} + K_{\alpha\beta}^{e} T_\beta \right) + \sum_{e=1}^{E_3} T_\beta = \sum_{e=1}^{E} f_\alpha^e + \sum_{e=1}^{E_2} f_\alpha^{2,e} + \sum_{e=1}^{E_3} f_\alpha^{3,e} \tag{3.30}$$

the essential boundary condition

$$T_\alpha = g_\alpha \quad \text{on } \Gamma_1 \tag{3.31}$$

and the initial condition

$$T_\alpha = T_{0\alpha} \quad \text{at } t = 0 \tag{3.32}$$

where $T_{0\alpha} = T_0(\mathbf{x}_\alpha)$. Thus, the ordinary differential equation (3.30) is obtained by the finite element approximation in the domain (or space) Ω.

The next step is the time integration of this differential equation. To see how this can be done, let us first consider the simpler problem

$$\frac{dx}{dt} = f(x), \quad x = x_0 \quad \text{at } t = 0 \tag{3.33}$$

For a given time increment Δt, we use the notation $x^n = x(n\,\Delta t)$ and then equation (3.33) is approximated by

$$\frac{x^n - x^{n-1}}{\Delta t} = \theta f(x^n) + (1 - \theta) f(x^{n-1}) \tag{3.34}$$

where θ is a parameter such that $0 \leq \theta \leq 1$. The use of the extreme values $\theta = 0$ and $\theta = 1$ in (3.34) corresponds to the forward and backward difference schemes, respectively. As an example, consider

$$f(x) = -kx + f_0 \qquad \text{and} \qquad x(0) = 1 \tag{3.35}$$

We shall investigate the quality of the approximation (3.34):

$$\frac{x^n - x^{n-1}}{\Delta t} = \theta(-kx^n + f_0) + (1 - \theta)(-kx^{n-1} + f_0)$$
$$= -\theta k x^n - (1 - \theta) k x^{n-1} + f_0$$

that is,

$$\left(\frac{1}{\Delta t} + \theta k \right) x^n = \left[\frac{1}{\Delta t} - (1 - \theta) k \right] x^{n-1} + f_0 \tag{3.36}$$

Numerical results when $k = 1$, $f_0 = 0$, and $\Delta t = 0.1$ are shown in Table 3.2. In this case, it is clear that the choice of $\theta = \frac{1}{2}$ is the best of three choices. The forward difference scheme ($\theta = 0$) damps the result, and the backward difference scheme

Table 3.2 *Numerical results by the θ-method*

T	$\theta = 0$	$\theta = \frac{1}{2}$	$\theta = 1$	Exact
0.	1.	1.	1.	1.
0.1	0.9	0.904761905	0.909090909	0.904837418
0.2	0.81	0.818594104	0.826446281	0.818730753
0.3	0.729	0.740632761	0.7513414801	0.740818221
0.4	0.6561	0.670096308	0.683013456	0.670320046
0.5	0.59049	0.606277612	0.620921323	0.60653066
0.6	0.531441	0.548536887	0.56447393	0.548811636
0.7	0.4782969	0.496295278	0.513158119	0.496585304
0.8	0.43046721		0.46650738	0.449328964
0.9	0.387420489	0.406264389	0.414097619	0.40656966
1.0	0.34867844	0.367572542	0.38554329	0.367879441

Note: $k = 1$, $f_0 = 0$, and $\Delta t = 0.1$.

($\theta = 1$) provides a slightly slower decay of the solution. If we keep $k = 1$ but increase the time increment Δt, the choices $\theta = 0$ and $\theta = \frac{1}{2}$ give numerical solutions with oscillations. Further, these in general diverge as $t \to +\infty$. However, if the backward scheme ($\theta = 1$) is applied, we avoid the oscillations and divergence in time, but the decay is much slower than that of the exact solution. The results using the θ-method are very similar to those of the artificial conductivity (or upwind) method for the convection problem. In that case, the parameter α_0 corresponds to $2\theta - 1$. That is:

	Forward	Central	Backward
θ	0	$\frac{1}{2}$	1
α_0	-1	0	1

It is clear that if Δt is sufficiently small, numerical solutions are quite good, as is the case in convection problems when a reasonably fine mesh is used, that is, when h is small enough.

The critical value of Δt at which numerical solutions with oscillations begin to appear can be found by checking the sign of the coefficient of the term x^{n-1} in (3.36). If Δt is chosen so that

$$\frac{1}{\Delta t} - (1 - \theta)k < 0, \qquad \text{i.e.,} \quad \Delta t > \frac{1}{(1 - \theta)k} \tag{3.37}$$

numerical solutions, in general, contain oscillations. The value of Δt at which divergent results arise can be found by checking the ratio of the coefficients of x^n and x^{n-1}. That is, if

$$\left| \frac{1/\Delta t - (1 - \theta)k}{1/\Delta t + \theta k} \right| > 1, \qquad \text{i.e.,} \quad \frac{1}{\Delta t} < (1 - 2\theta)k \tag{3.38}$$

then the numerical solution of (3.36) in general diverges as $t \to +\infty$.

In view of the above considerations, let us now describe the θ-method for the ordinary differential equation (3.30) obtained by the finite element approximation in space. Using the notation $T^n_\beta = T_\beta(n\,\Delta t)$, (3.30) is approximated by

$$\sum_{e=1}^{E}\left(\frac{1}{\Delta t}M^e_{\alpha\beta} + \theta K^e_{\alpha\beta}\right)T^n_\beta + \sum_{e=1}^{E_3}\theta K^{3,e}_{\alpha\beta}T^n_\beta$$
$$= \sum_{e=1}^{E}\left[\frac{1}{\Delta t}M^e_{\alpha\beta} - (1-\theta)K^e_{\alpha\beta}\right]T^{n-1}_\beta - \sum_{e=1}^{E_3}(1-\theta)K^{3,e}_{\alpha\beta}T^{n-1}_\beta$$
$$+ \sum_{e=1}^{E} f^e_\alpha + \sum_{e=1}^{E_2} f^{2,e}_\alpha + \sum_{e=1}^{E_3} f^{3,e}_\alpha \tag{3.39}$$

The boundary condition (3.31) and the initial condition (3.32) become

$$T_\alpha = g_\alpha \quad \text{on } \Gamma_1 \qquad \text{and} \qquad T^0_\alpha = T_{0\alpha} \tag{3.40}$$

or $n = 1, 2, \ldots, N$. For the computation of T^n_β at the nth time step corresponding to time $n\,\Delta t$, the value T^{n-1}_β is already known. Thus, the system of linear equations (3.39) in T^n_β can be solved. Since the element mass matrix $[M^e_{\alpha\beta}]$ is symmetric, if $K^e_{\alpha\beta} = K^e_{\beta\alpha}$, the symmetric solver is applicable.

The size of the time increment Δt for $\theta < 1$ may be decided by checking the sign of the diagonal term of the matrix

$$\frac{1}{\Delta t}[M^e_{\alpha\beta}] - (1-\theta)[K^e_{\alpha\beta}]$$

by analogy with what was done to get (3.37). It is desirable to keep these diagonal terms positive.

A systematic way to determine the size of the increment Δt is the von Neumann stability analysis. Following Roache (1972), we shall briefly describe this using the one-dimensional problem that yields the finite element equation

$$\frac{\rho c_p h_e}{6\,\Delta t}(T^{n+1}_{\alpha+1} + 4T^{n+1}_\alpha + T^{n+1}_{\alpha-1}) + \frac{\theta k}{h_e}(-T^{n+1}_{\alpha+1} + 2T^{n+1}_\alpha - T^{n+1}_{\alpha-1})$$
$$= \frac{\rho c_p h_e}{6\,\Delta t}(T^n_{\alpha+1} + 4T^n_\alpha + T^n_{\alpha-1}) - \frac{(1-\theta)k}{h_e}(-T^n_{\alpha+1} + 2T^n_\alpha - T^n_{\alpha-1}) \tag{3.41}$$

for $f = 0$, where $T^m_\beta = T(\beta h_e, m\,\Delta t)$ and h_e is the size of finite elements. Substituting

$$T^n_\beta = v^n e^{i\beta A} \tag{3.42}$$

into (3.41), we have

$$\left\{\frac{\rho c_p h_e}{6\,\Delta t}(e^{iA} + 4 + e^{-iA}) + \frac{\theta k}{h_e}(-e^{iA} + 2 - e^{-iA})\right\}v^{n+1}e^{i\alpha A}$$
$$= \left\{\frac{\rho c_p h_e}{6\,\Delta t}(e^{iA} + 4 + e^{-iA}) - \frac{(1-\theta)k}{h_e}(-e^{iA} + 2 - e^{-iA})\right\}v^n e^{i\alpha A}$$

Noting that

$$e^{iA} + e^{-iA} = 2\cos A$$

we have

$$\{\tfrac{1}{3}(2+\cos A) + 2Q\theta(1-\cos A)\}v^{n+1} = \{\tfrac{1}{3}(2+\cos A) - 2Q(1-\theta)(1-\cos A)\}v^n$$

where

$$Q = k\,\Delta t/\rho c_p h_e^2 \tag{3.43}$$

Since a necessary condition to reach the nondivergent steady state is given by

$$|v^{n+1}/v^n| \le 1 \tag{3.44}$$

Δt and h_e must satisfy the condition

$$\left|\frac{(2+\cos A)/3 - 2Q(1-\theta)(1-\cos A)}{(2+\cos A)/3 + 2Q\theta(1-\cos A)}\right| \le 1 \tag{3.45}$$

that is

$$Q(1-\cos A)(1-2\theta) \le \tfrac{1}{3}(2+\cos A) \tag{3.46}$$

Thus, if the parameter θ is chosen as

$$\theta > \tfrac{1}{2} \tag{3.47}$$

the stability condition (3.46) is always satisfied. If $\theta < \frac{1}{2}$, then

$$Q \le \frac{2+\cos A}{3(1-\cos A)(1-2\theta)}$$

Since the right side becomes the minimum if $\cos A = -1$, Q must satisfy

$$Q \le \frac{1}{6(1-2\theta)}$$

That is, the size of the time increment Δt must be small enough,

$$\Delta t \le \frac{\rho c_p h_e^2}{6(1-2\theta)k} \tag{3.48}$$

in order to have a stable solution that is not divergent.

Another important comment regarding the system of linear equations (3.39) is that if ρc_p, $\mathbf{u}$, k, and k_0 do not depend on time, the matrices $[M_{\alpha\beta}^e]$, $[K_{\alpha\beta}^e]$, and $[K_{\alpha\beta}^{3,e}]$ are constant in time. Thus, if a fixed time increment Δt is used, forward elimination (or LDU decomposition) of the (global) stiffness matrix needs to be taken only at the first step in order to solve the system of equations (3.39). Since reduction of the load vector and back substitution require much less computational time than forward elimination, the choice of $\theta > 0$ may be comparable to the choice $\theta = 0$.

It is certain that the time increment Δt need not be constant for every step of time integration. There are many other methods for solving the ordinary differential equation (3.30), especially when the coefficient matrices $[M_{\alpha\beta}^e]$, $[K_{\alpha\beta}^e]$, and $[K_{\alpha\beta}^{3,e}]$ are also functions of the temperature $\{T_\beta\}$. We shall not discuss these advanced techniques here. Let us leave them for advanced courses in finite element methods.

3.3 Other methods of time integration

There are many other methods for solving time-dependent problems, although the θ-method has been widely used in finite element methods. In this section, we shall briefly describe several other methods of time integration that are commonly used.

As described earlier, the explicity method in the θ-method (i.e., Euler's explicity method) requires a very small time increment for reasonable accuracy. The *Runge–Kutta methods* are explicity methods that involve evaluating the function at selected times on each time interval and are known to be quite accurate. To explain the Runge–Kutta methods, let us consider the initial value problem

$$\frac{d\mathbf{u}}{dt} = \mathbf{f}(t, \mathbf{u}), \quad \mathbf{u} = \mathbf{u}_0 \quad \text{at } t = 0 \tag{3.49}$$

Suppose that

$$\mathbf{u}^{n+1} = \mathbf{u}^n + a\mathbf{v} + b\mathbf{w} \tag{3.50}$$

where $\mathbf{u}^n = \mathbf{u}(n\,\Delta t)$, Δt is the time increment,

$$\mathbf{v} = \Delta t\, \mathbf{f}(t^n, \mathbf{u}^n) \qquad \text{and} \qquad \mathbf{w} = \Delta t\, \mathbf{f}(t^n + \alpha\,\Delta t, \mathbf{u}^n + \beta\mathbf{v}) \tag{3.51}$$

Here $t^n = n\,\Delta t$, and a, b, α, and β are constants to be determined so that (3.50) yields good agreement with the Taylor formula:

$$\mathbf{u}^{n+1} = \mathbf{u}^n + \Delta t \frac{d\mathbf{u}}{dt} t^n + \frac{1}{2}(\Delta t)^2 \frac{d^2\mathbf{u}}{dt^2} t^n + \frac{1}{6}(\Delta t)^3 \frac{d^3\mathbf{u}}{dt^3} t^n + \cdots$$

From (3.49),

$$\frac{d^2\mathbf{u}}{dt^2} = \frac{\partial \mathbf{f}}{\partial t} + \frac{\partial \mathbf{f}}{\partial \mathbf{u}} \frac{d\mathbf{u}}{dt}$$

where the notation used in the second term is a convenient alternative to

$$\frac{\partial \mathbf{f}}{\partial \mathbf{u}^n} \frac{d\mathbf{u}^n}{dt}$$

Substituting these into the Taylor formula, we get

$$\begin{aligned}\mathbf{u}^{n+1} &= \mathbf{u}^n + \Delta t\, \mathbf{f}(t^n, \mathbf{u}^n) + \frac{1}{2}(\Delta t)^2 \left(\frac{\partial \mathbf{f}}{\partial t} + \frac{\partial \mathbf{f}}{\partial \mathbf{u}} \mathbf{f}\right)(t^n, \mathbf{u}^n) \\ &\quad + \frac{1}{6}(\Delta t)^3 \left(\frac{\partial^2 \mathbf{f}}{\partial t^2} + 2\frac{\partial^2 \mathbf{f}}{\partial t\, \partial \mathbf{u}} \mathbf{f} + \frac{\partial^2 \mathbf{f}}{\partial \mathbf{u}^2} \mathbf{f}\mathbf{f} + \frac{\partial \mathbf{f}}{\partial \mathbf{u}} \frac{\partial \mathbf{f}}{\partial t} + \frac{\partial \mathbf{f}}{\partial \mathbf{u}} \frac{\partial \mathbf{f}}{\partial \mathbf{u}} \mathbf{f}\right)(t^n, \mathbf{u}^n) \\ &\quad + (\text{higher-order terms})\end{aligned} \tag{3.52}$$

On the other hand, we have

$$\begin{aligned}\frac{\mathbf{w}}{\Delta t} &= \mathbf{f}(t^n + \alpha\,\Delta t, \mathbf{u}^n + \beta\mathbf{v}) = \mathbf{f}(t^n, \mathbf{u}^n) + \alpha\,\Delta t \frac{\partial \mathbf{f}}{\partial t} \\ &\quad + \beta \frac{\partial \mathbf{f}}{\partial \mathbf{u}}(t^n, \mathbf{u}^n)\mathbf{v} + \frac{1}{2}\alpha^2\,\Delta t^2 \frac{\partial^2 \mathbf{f}}{\partial t^2}(t^n, \mathbf{u}^n) + \alpha\beta\,\Delta t \frac{\partial^2 \mathbf{f}}{\partial t\, \partial \mathbf{u}}(t^n, \mathbf{u}^n)\mathbf{v} \\ &\quad + \frac{1}{2}\beta^2 \frac{\partial^2 \mathbf{f}}{\partial \mathbf{u}^2}(t^n, \mathbf{u}^n)\mathbf{v}\mathbf{v} + (\text{higher-order terms})\end{aligned} \tag{3.53}$$

Substituting (3.53) into the form (3.50) yields

$$\mathbf{u}^{n+1} = \mathbf{u}^n + (a+b)\,\Delta t\,\mathbf{f}(t^n, \mathbf{u}^n) + b\,\Delta t^2\left(\alpha\frac{\partial \mathbf{f}}{\partial t} + \beta\frac{\partial \mathbf{f}}{\partial \mathbf{u}}\mathbf{f}\right)(t^n, \mathbf{u}^n)$$
$$+\, b\,\Delta t^3\left(\frac{1}{2}\alpha^2\frac{\partial^2 \mathbf{f}}{\partial t^2} + \alpha\beta\frac{\partial^2 \mathbf{f}}{\partial t\,\partial \mathbf{u}}\mathbf{f} + \frac{1}{2}\beta^2\frac{\partial^2 \mathbf{f}}{\partial \mathbf{u}^2}\mathbf{f}\mathbf{f}\right)(t^n, \mathbf{u}^n)$$
$$+\ (\text{higher-order terms}) \tag{3.54}$$

Comparison of (3.54) and (3.52) implies

$$a + b = 1, \qquad b\alpha = b\beta = \tfrac{1}{2}$$

Thus, we may take

$$a = b = \tfrac{1}{2} \quad \text{and} \quad \alpha = \beta = 1 \tag{3.55}$$

Then we have the second-order Runge–Kutta method:

$$\begin{aligned}\mathbf{u}^{n+1} &= \mathbf{u}^n + \tfrac{1}{2}(\mathbf{v} + \mathbf{w})\\ &= \mathbf{u}^n + \tfrac{1}{2}[\mathbf{f}(t^n, \mathbf{u}^n) + \mathbf{f}(t^{n+1}, \mathbf{u}^n + \Delta t\,\mathbf{f}(t^n, \mathbf{u}^n))]\end{aligned} \tag{3.56}$$

Similarly, the fourth-order Runge–Kutta method is given as

$$\mathbf{u}^{n+1} = \mathbf{u}^n + \tfrac{1}{6}(\mathbf{v}_1 + 2\mathbf{v}_2 + 2\mathbf{v}_3 + \mathbf{v}_4) \tag{3.57}$$

where

$$\begin{aligned}\mathbf{v}_1 &= \Delta t\,\mathbf{f}(t^n, \mathbf{u}^n), & \mathbf{v}_2 &= \Delta t\,\mathbf{f}(t^n + \tfrac{1}{2}\Delta t, \mathbf{u}^n + \tfrac{1}{2}\mathbf{v}_1)\\ \mathbf{v}_3 &= \Delta t\,\mathbf{f}(t^n + \tfrac{1}{2}\Delta t, \mathbf{u}^n + \tfrac{1}{2}\mathbf{v}_2), & \mathbf{v}_4 &= \Delta t\,\mathbf{f}(t^{n+1}, \mathbf{u}^n + \mathbf{v}_3)\end{aligned} \tag{3.58}$$

Another type of integration formula can be obtained by quadrature rules. More precisely, if the initial value problem (3.49) is integrated over the interval (t^n, t^{n+1}), we have

$$\mathbf{u}^{n+1} - \mathbf{u}^n = \int_{t^n}^{t^{n+1}} \frac{d\mathbf{u}}{dt}\,dt = \int_{t^n}^{t^{n+1}} \mathbf{f}(t, \mathbf{u}(t))\,dt \tag{3.59}$$

If the trapezoid formula is applied,

$$\mathbf{u}^{n+1} = \mathbf{u}^n + \tfrac{1}{2}\Delta t\,[\mathbf{f}(t^n, \mathbf{u}^n) + \mathbf{f}(t^{n+1}, \mathbf{u}^{n+1})] \tag{3.60}$$

Note that formula (3.60) involves the unknown quantity $\mathbf{u}^{n+1}$ on the right side. This is resolved by the iteration scheme described below. As a first approximation, we pick the solution by the Euler explicit formula, that is,

$$\mathbf{u}_0^{n+1} = \mathbf{u}^n + \Delta t\,\mathbf{f}(t^n, \mathbf{u}^n) \tag{3.61}$$

Then the iteration

$$\mathbf{u}_{k+1}^{n+1} = \mathbf{u}^n + \tfrac{1}{2}\Delta t\,[\mathbf{f}(t^n, \mathbf{u}^n) + \mathbf{f}(t^{n+1}, \mathbf{u}_k^{n+1})] \tag{3.62}$$

is performed until the difference between $\mathbf{u}_{k+1}^{n+1}$ and $\mathbf{u}_k^{n+1}$ is small enough. This method is called a *second-order predictor–corrector method,* since the prediction by the Euler explicity method is corrected by the iteration scheme (3.62).

Exercise 3.2 (Stefan one-dimensional problem): Let us consider a transition of the physical state of a body from a liquid phase to a solid phase by changing the temperature through the point of fusion. As the phase transition is taking place, the temperature remains constant, and latent heat of fusion is liberated. The additional condition that must be fulfilled at the surface of separation of the solid and liquid phases defined by

$$x = \xi(t) \tag{3.63}$$

is obtained as

$$k_1 \left.\frac{\partial T_1}{\partial x}\right|_{x=\xi} - k_2 \left.\frac{\partial T_2}{\partial x}\right|_{x=\xi} = \lambda\rho \frac{\partial \xi}{\partial t} \tag{3.64}$$

using the balance of heat in the domain, where k_1 and k_2 are the coefficients of heat conduction of the first and second phases, and λ is the latent heat of fusion. This relation holds for both solidification and fusion processes.

Now, let us consider the freezing of water, where the temperature of the phase transition equals zero. Let an interval $(0, L)$ be initially occupied with water, the temperature of which is given by $T = T_0(x) > 0$. If a constant temperature $T = T_L < 0$ is maintained at the surface $x = 0$, the freezing surface $x = \xi$ propagates into the liquid. If the other end $x = L$ maintains a constant temperature $T = T_R > 0$, the speed of propagation of the freezing surface becomes small as $t \to \infty$. The problem of the temperature distribution inside the ice and water is then described by the following initial boundary value problem:

$$\begin{aligned}
&(\rho c_p)_1 \frac{\partial T_1}{\partial t} = \frac{\partial}{\partial x}\left(k_1 \frac{\partial T_1}{\partial x}\right) \quad \text{in } x \in (0, \xi) \\
&(\rho c_p)_2 \frac{\partial T_2}{\partial t} = \frac{\partial}{\partial x}\left(k_2 \frac{\partial T_2}{\partial x}\right) \quad \text{in } x \in (\xi, L) \\
&T_1(0, t) = T_L < 0, \qquad T_2(L, t) = T_R > 0, \quad \forall t \\
&T_1(x, 0) = T_0(x) > 0, \quad \forall x
\end{aligned} \tag{3.65}$$

and

$$\begin{aligned}
&T_1(\xi, t) = T_2(\xi, t) = 0 \\
&k_1 \left.\frac{\partial T_1}{\partial x}\right|_{x=\xi} - k_2 \left.\frac{\partial T_2}{\partial x}\right|_{x=\xi} = \lambda\rho \frac{\partial \xi}{\partial t}
\end{aligned} \tag{3.66}$$

Find the temperature distribution and the location of freezing surface $x = \xi$ using the finite element methods and time integration schemes described above. A method to solve this problem is to use "moving" finite elements. Suppose that E_1 and E_2 number of elements are assumed in the frozen and liquid portions, respectively. Coordinates of nodes are then defined by

$$x_i = \frac{i-1}{E_1}\xi, \quad i = 1, 2, \ldots, E_1 + 1 \tag{3.67}$$

and

$$x_{E_1+i} = \xi + \frac{i-1}{E_2}(L-\xi), \quad i = 1, \ldots, E_2+1 \tag{3.68}$$

if ξ is given, $\xi > 0$. Since the location of the freezing surface $x = \xi$ is unknown, let us assume its location by

$$\xi_1^{n+1} = \xi^n + \frac{\Delta t}{\lambda\rho}\left(k_1 \frac{\partial T_1}{\partial x}\bigg|_{\substack{x=\xi^n \\ t=n\Delta t}} - k_2 \frac{\partial T_2}{\partial x}\bigg|_{\substack{x=\xi^n \\ t=n\Delta t}}\right) \tag{3.69}$$

The location may be corrected by the iterative method at each incremental step:

$$\xi_{i+1}^{n+1} = \xi^n + \frac{\Delta t}{\lambda\rho}\left(k_1 \frac{\partial T_1}{\partial x}\bigg|_{\substack{x=\xi_i^{n+1} \\ t=(n+1)\Delta t}} - k_2 \frac{\partial T_2}{\partial x}\bigg|_{\substack{x=\xi_i^{n+1} \\ t=(n+1)\Delta t}}\right) \tag{3.70}$$

for $i = 1, 2, \ldots$.

Extension to two-dimensional problems can be found in Samarskii and Moiseyenko (1965), which provides details of computational methods to solve multidimensional Stefan problems using finite difference schemes. Further details of physics of the Stefan problem can be found in Okendon and Hodgkins (1975).

References

Brooks, A., and Hughes, T. J. R., "Streamline upwind/Petrov–Galerkin formulation for convection dominated flows with particular emphasis on the incompressible Navier–Stokes equations," *Computer Methods in Applied Mechanics and Engineering* 32: 199–259, 1982.

Diaz, A. R., Kikuchi, N., Taylor, J. E., "A method of grid optimization for finite element methods," *Computer Methods in Applied Engineering* 41: 29–45, 1983.

Hughes, T. J. R. (ed.), *Finite element methods for convection dominated flows,* American Society of Mechanical Engineers, New York, 1979.

Kelly, D. W., Nakzawa, S., Zienkiewicz, O. C., and Heinrich, J. C., "A note on upwinding and anisotropic balancing dissipation in finite element approximations to convective diffusion problems," *International Journal for Numerical Methods in Engineering* 15: 1705–11, 1980.

Okendon, J. R., and Hodgkins, W. R. (eds.), *Moving boundary problems in heat flow and diffusion,* Clarendon, Oxford, 1975.

Roache, P. J., *Computational fluid dynamics,* Hermosa Publishers, Albuquerque, New Mexico, 1972.

Samarskii, A. A., and Moiseyenko, B. D., "An economic continuous calculation scheme for the Stefan multidimensional problem," *Zhurnal Vychislitel'noi Matematiki i Matematicheskoi Fiziki* 5: 816–27, 1965.

4
SIMPLE ELASTIC STRUCTURES AND THEIR FREE VIBRATION PROBLEMS

We studied a finite element approximation of heat conduction problems in which the unknown is a scalar function, and we used a family of triangular elements in the process. One of the characteristics of the study is that each node in a finite element model has only one degree of freedom, namely, the value of the temperature at that node. In the following chapters, we shall study finite element approximations of problems in structural, solid, and fluid mechanics such that two or more degrees of freedom occur at each node of a finite element model.

As a preliminary study, let us first consider finite element approximations of simple elastic structures that do not require the detailed theory of continuum mechanics for solids and fluids. Since dynamics of structures is an important area of structural mechanics, we shall briefly cover the eigenvalue approach to the free vibrations of structures and some of the direct integration methods used to solve vibration problems.

4.1 Elastic bar element

The simplest structural element is the bar element, which can transmit only an axial force. Also, the following linear stress–strain relation is assumed:

$$\sigma = E(\varepsilon - \varepsilon_0) - \alpha E(T - T_0) \tag{4.1}$$

where σ is the axial stress, ε the axial strain, E Young's modulus, ε_0 the initial strain, α the thermal expansion coefficient, T the temperature field, and T_0 the reference temperature. If the balance of forces in the differential element of the bar is considered as shown in Figure 4.1, we have

$$-N(s - \tfrac{1}{2}\,\Delta s) - \rho_A \ddot{w}\,\Delta s + N(s + \tfrac{1}{2}\,\Delta s) + f\,\Delta s = 0$$

and then

$$\rho_A \ddot{w}\,\Delta s = \frac{dN}{ds}\,\Delta s + O(\Delta s^2) + f\,\Delta s$$

where ρ_A is the mass of the bar per unit length, w the displacement of the bar along the axial direction, f the applied axial force per unit length, and N the internal force satisfying $N = \sigma A$, where A is the cross-sectional area of the bar. Passing to the limit $\Delta s \to 0$, we have the equilibrium equation

$$\rho_A \ddot{w} = \frac{d}{ds}(A\sigma) + f \tag{4.2}$$

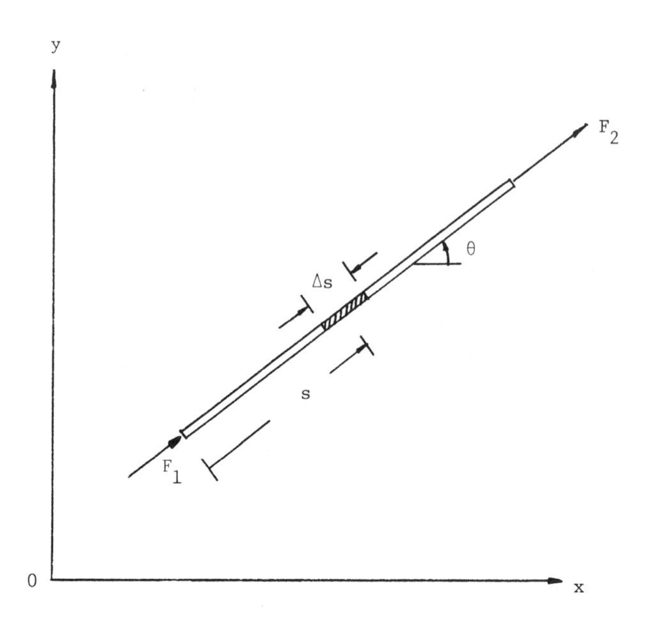

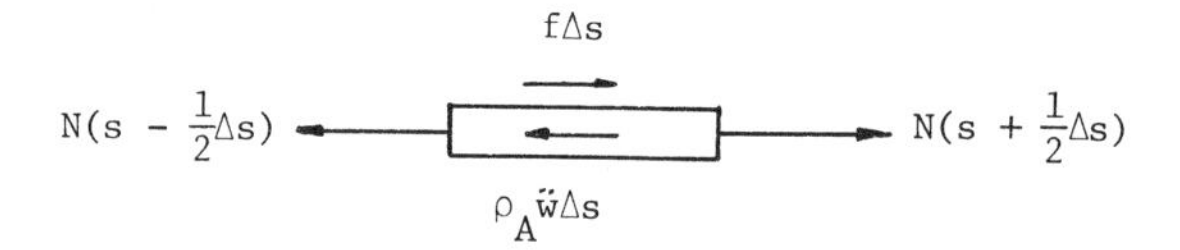

Figure 4.1 Balance of forces in the bar element.

Under the loading conditions at the ends of the bar

$$N_1 = -F_1 \qquad \text{and} \qquad N_2 = F_2 \tag{4.3}$$

we have the weak form of the equilibrium equation (4.2) with the boundary conditions (4.3):

$$\int_0^L \rho_A \ddot{w}\bar{w}\, ds = \int_0^L \frac{d}{ds}(A\sigma)\bar{w}\, ds + \int_0^L f\bar{w}\, ds$$

that is,

$$\int_0^L \rho_A \ddot{w}\bar{w}\, ds + \int_0^L A\sigma \frac{d\bar{w}}{ds}\, ds = F_2\bar{w}_2 + F_1\bar{w}_1 + \int_0^L f\bar{w}\, ds$$

where $\bar{w}$ is the virtual displacement and $\{\bar{w}_1, \bar{w}_2\}$ is the set of its values at the end points. Applying the strain–displacement relation

$$\varepsilon = \frac{dw}{ds} \tag{4.4}$$

and the stress–strain relation (4.1), we have the *weak form* of the bar element

$$\int_0^L \rho_A \ddot{w}\bar{w}\, ds + \int_0^L AE \frac{dw}{ds}\frac{d\bar{w}}{ds}\, ds$$
$$= \{F_1, F_2\}\begin{Bmatrix} \bar{w}_1 \\ \bar{w}_2 \end{Bmatrix} + \int_0^L f\bar{w}\, ds + \int_0^L \{AE\varepsilon_0 + \alpha AE(T - T_0)\}\frac{d\bar{w}}{ds}\, ds, \quad \forall \bar{w} \quad (4.5)$$

Assuming that the actual displacement field w and the virtual displacement field $\bar{w}$ are linear in the bar element, that is,

$$\begin{aligned} w(s) &= c_0 + c_1 s = w_1\left(1 - \frac{s}{L}\right) + w_2 \frac{s}{L} \\ \bar{w}(s) &= \bar{c}_0 + \bar{c}_1 s = \bar{w}_1\left(1 - \frac{s}{L}\right) + \bar{w}_2 \frac{s}{L} \end{aligned} \quad (4.6)$$

we can discretize the weak form (4.5) as

$$\{\bar{w}_1, \bar{w}_2\}\left(\frac{\rho_A L}{6}\begin{bmatrix} 2 & 1 \\ 1 & 2 \end{bmatrix}\begin{Bmatrix} \ddot{w}_1 \\ \ddot{w}_2 \end{Bmatrix} + \frac{AE}{L}\begin{bmatrix} 1 & -1 \\ -1 & 1 \end{bmatrix}\begin{Bmatrix} w_1 \\ w_2 \end{Bmatrix}\right)$$
$$= \{\bar{w}_1, \bar{w}_2\}\left(\begin{Bmatrix} F_1 \\ F_2 \end{Bmatrix} + \frac{fL}{2}\begin{Bmatrix} 1 \\ 1 \end{Bmatrix} + [AE\varepsilon_0 + \alpha AE(T - T_0)_e]\begin{Bmatrix} -1 \\ 1 \end{Bmatrix}\right)$$

under the assumption that the distributed axial force f is constant in the bar element and that $(T - T_0)_e$ is the value of $T - T_0$ at the center of the bar, namely, $\frac{1}{2}[(T - T_0)_1 + (T - T_0)_2]$. Note that the strain ε and the initial strain ε_0 must be constant in the bar element since the displacement w is assumed to be linear. Thus, because of the arbitrary nature of the virtual displacement $\bar{w}$, we have the following matrix equation in a bar element:

$$M^e_{\alpha\beta}\ddot{w}_\beta + K^e_{\alpha\beta}w_\beta = f^e_\alpha \quad (4.7)$$

where

$$\begin{aligned} [M^e_{\alpha\beta}] &= \frac{\rho_A L}{6}\begin{bmatrix} 2 & 1 \\ 1 & 2 \end{bmatrix}, \qquad [K^e_{\alpha\beta}] = \frac{AE}{L}\begin{bmatrix} 1 & -1 \\ -1 & 1 \end{bmatrix} \\ \{f^e_\alpha\} &= \begin{Bmatrix} F_1 \\ F_2 \end{Bmatrix} + \frac{fL}{2}\begin{Bmatrix} 1 \\ 1 \end{Bmatrix} + [AE\varepsilon_0 + \alpha AE(T - T_0)_e]\begin{Bmatrix} -1 \\ 1 \end{Bmatrix} \end{aligned} \quad (4.8)$$

Note that expression (4.7) is based on the coordinate s measured along the axis of the bar element. We shall now rewrite equation (4.7) in terms of an xy coordinate system in which the data can be formed easily. To do this, let us use the following relations.

> The displacement w along the s coordinate is equivalent to the displacements u and v in the x and y directions:
>
> $$w = u\cos\theta + v\sin\theta \quad (4.9)$$
>
> where θ is the angle between the s and x axis. An axial force F can be decomposed into its x and y components:
>
> $$F_x = F\cos\theta, \qquad F_y = F\sin\theta \quad (4.10)$$

Relations (4.9) and (4.10) yield

$$\begin{Bmatrix} w_1 \\ w_2 \end{Bmatrix} = \underbrace{\begin{bmatrix} \cos\theta & \sin\theta & 0 & 0 \\ 0 & 0 & \cos\theta & \sin\theta \end{bmatrix}}_{[T_{\alpha\beta}]} \begin{Bmatrix} u_1 \\ v_1 \\ u_2 \\ v_2 \end{Bmatrix} \tag{4.11}$$

and

$$\begin{Bmatrix} f^e_{1x} \\ f^e_{1y} \\ f^e_{2x} \\ f^e_{2y} \end{Bmatrix} = \begin{bmatrix} \cos\theta & 0 \\ \sin\theta & 0 \\ 0 & \cos\theta \\ 0 & \sin\theta \end{bmatrix} \begin{Bmatrix} f^e_1 \\ f^e_2 \end{Bmatrix} \tag{4.12}$$

where $[T_{\alpha\beta}]$ is called the transformation matrix. Applying these relations to equation (4.7) yields

$$\begin{Bmatrix} f^e_{1x} \\ f^e_{1y} \\ f^e_{2x} \\ f^e_{2y} \end{Bmatrix} = \begin{bmatrix} \cos\theta & 0 \\ \sin\theta & 0 \\ 0 & \cos\theta \\ 0 & \sin\theta \end{bmatrix} \frac{\rho_A L}{6} \begin{bmatrix} 2 & 1 \\ 1 & 2 \end{bmatrix} \begin{bmatrix} \cos\theta & \sin\theta & 0 & 0 \\ 0 & 0 & \cos\theta & \sin\theta \end{bmatrix} \begin{Bmatrix} \ddot{u}_1 \\ \ddot{v}_1 \\ \ddot{u}_2 \\ \ddot{v}_2 \end{Bmatrix}$$

$$+ \begin{bmatrix} \cos\theta & 0 \\ \sin\theta & 0 \\ 0 & \cos\theta \\ 0 & \sin\theta \end{bmatrix} \frac{AE}{L} \begin{bmatrix} 1 & -1 \\ -1 & 1 \end{bmatrix} \begin{bmatrix} \cos\theta & \sin\theta & 0 & 0 \\ 0 & 0 & \cos\theta & \sin\theta \end{bmatrix} \begin{Bmatrix} u_1 \\ v_1 \\ u_2 \\ v_2 \end{Bmatrix}$$

$$= \underbrace{\frac{\rho_A L}{6} \begin{bmatrix} 2\cos^2\theta & 2\cos\theta\sin\theta & \cos^2\theta & \cos\theta\sin\theta \\ & 2\sin^2\theta & \sin\theta\cos\theta & \sin^2\theta \\ & & 2\cos^2\theta & 2\cos\theta\sin\theta \\ \text{SYM} & & & 2\sin^2\theta \end{bmatrix}}_{[\bar{M}_{pq}]} \begin{Bmatrix} \ddot{u}_1 \\ \ddot{v}_1 \\ \ddot{u}_2 \\ \ddot{v}_2 \end{Bmatrix}$$

$$+ \underbrace{\frac{AE}{L} \begin{bmatrix} \cos^2\theta & \cos\theta\sin\theta & -\cos^2\theta & -\cos\theta\sin\theta \\ & \sin^2\theta & -\sin\theta\cos\theta & -\sin^2\theta \\ & & \cos^2\theta & \cos\theta\sin\theta \\ & \text{SYM} & & \sin^2\theta \end{bmatrix}}_{[\bar{K}_{pq}]} \begin{Bmatrix} u_1 \\ v_1 \\ u_2 \\ v_1 \end{Bmatrix} \tag{4.13}$$

Note that the last two terms of the load vector $\{f^e_\alpha\}$ in (4.8) cannot be decomposed into x and y directions, but we still have the general expression

$$\begin{Bmatrix} f^e_{1x} \\ f^e_{1y} \\ f^e_{2x} \\ f^e_{2y} \end{Bmatrix} = \begin{Bmatrix} F_{1x} \\ F_{1y} \\ F_{2x} \\ F_{2y} \end{Bmatrix} + \begin{bmatrix} \cos\theta & 0 \\ \sin\theta & 0 \\ 0 & \cos\theta \\ 0 & \sin\theta \end{bmatrix} \left[\frac{fL}{2} \begin{Bmatrix} 1 \\ 1 \end{Bmatrix} + [AE\varepsilon_0 + \alpha AE(T - T_0)_e] \begin{Bmatrix} -1 \\ 1 \end{Bmatrix} \right] \tag{4.14}$$

for the applied load.

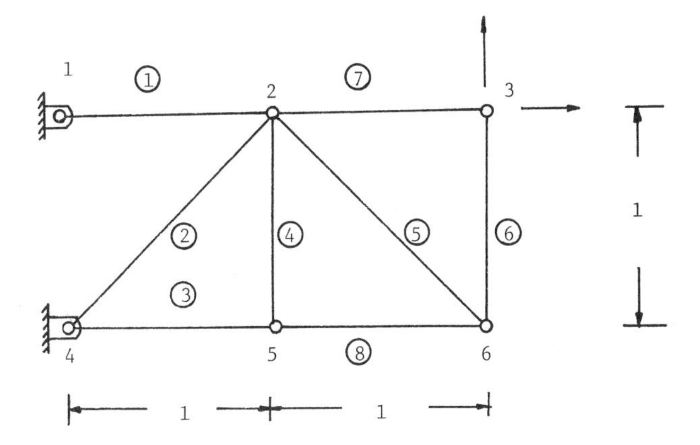

Figure 4.2 Truss structure: circled numbers are for elements, uncircled for nodes.

Once the displacement $\{u_1, v_1, u_2, v_2\}$ is obtained, the axial force in the bar element is computed by

$$N = A\sigma = AE\varepsilon = AE\frac{dw}{ds} = AE\{-1, 1\}\begin{Bmatrix} w_1 \\ w_2 \end{Bmatrix}$$

$$= AE\{-\cos\theta \; -\sin\theta \; \cos\theta \; \sin\theta\}\begin{Bmatrix} u_1 \\ v_1 \\ u_2 \\ v_2 \end{Bmatrix} \tag{4.15}$$

The equation of motion (4.13) holds for a single bar element in a structure. Thus, if a structure consists of several pin-connected bars as shown in Figure 4.2, we *have to assemble* the finite structure as was done in the problem of heat conduction in Chapter 2. As we did for heat conduction problems, we shall introduce element connectivities to assemble element stiffness and mass matrices to the global ones. The only difference between the present case and that of heat conduction is that each node has two degrees of freedom, its x and y components of displacement. For the example problem shown in Figure 4.2, the total number of bar elements is 8, the total number of nodes (or joints) is 6, and the element connectivities IJK are given in the following table:

NEL	IJK(1,NEL)	IJK(2,NEL)	NEL	IJK(1,NEL)	IJK(2,NEL)
1	1	2	5	2	6
2	4	2	6	6	3
3	4	5	7	2	3
4	5	2	8	5	6

Once the element connectivity is defined for each member of the structure, assembling of the mass matrix $[\bar{M}^e_{2(\alpha-1)+i,\, 2(\beta-1)+j}]$, the stiffness matrix $[\bar{K}^e_{2(\alpha-1)+i,\, 2(\beta-1)+j}]$, and the load vector $\{f^e_{2(\alpha-1)+i}\}$ can be performed by an algorithm similar to the one used in Chapter 2. Here α and β indicate the local node numbers 1 and 2, and i and j indicate the two degrees of freedom in the x and y directions, that is, 1 and 2 as in the index notation. If we introduce the following correspondences:

In the notes	In the FORTRAN code
$\bar{M}^e_{2(\alpha-1)+i,\,2(\beta-1)+j}$	SME(2 * (IA − 1) + I,2 * (JB − 1) + J)
$\bar{K}^e_{2(\alpha-1)+i,\,2(\beta-1)+j}$	SKE(2 * (IA − 1) + I,2 * (JB − 1) + J)
$f^e_{2(\alpha-1)+i}$	FE(2 * (IA − 1) + I)

then the assembling routine becomes

```
      DO 100 NEL = 1,NELX
      ----------
      DO 200 IA = 1,2
      IJKIA = IJK(IA,NEL)
      DO 200 I = 1,2
      IJKIAI = 2 * (IJKIA - 1) + I
      IAI = 2 * (IA - 1) + I
      F(IJKIAI) = F(IJKIAI) + FE(IAI)
      DO 200 JB = 1,2
      IJKJB = IJK(JB,NEL)                                        (4.16)
      DO 200 J = 1,2
      IJKJBJ = 2 * (IJKJB - 1) + J
      JBJ = 2 * (JB - 1) + J
      IAJB1 = IJKJBJ - IJKIAI + 1
      IF(IAJB1.LE.0) GO TO 200
      SM(IJKIAI,IAJB1) = SM(IJKIAI,IAJB1) + SME(IAI,JBJ)
      SK(IJKIAI,IAJB1) = SK(IJKIAI,IAJB1) + SKE(IAI,JBJ)
  200 CONTINUE
```

Here SM and SK are the global, *banded,* mass and stiffness matrices, respectively, and IJKIA is the node number in the structure that corresponds to the IAth local node number. After this assembling, we formally have the global finite element equation

$$\mathbf{M}\ddot{\mathbf{U}} + \mathbf{K}\mathbf{U} = \mathbf{F} \qquad \text{or} \qquad M_{IJ}\ddot{U}_J + K_{IJ}U_J = F_I \tag{4.17}$$

where the vector $\mathbf{U}$ is defined as $\{u_1, v_1, u_2, v_2, u_3, v_3, \ldots, u_N, v_N\}^{\mathrm{T}}$, and N is the total number of nodes in the structure. The arrays M_{IJ} and K_{IJ} are stored as rectangular matrices. It is also noted that if $f = 0$, $\varepsilon_0 = 0$, $\alpha = 0$, and no external forces are applied at a node, say I, then, after assembling the element load vectors, $F_{2(I-1)+1}$ and F_{2I} become zero since the summation of $\{F_{1x}, F_{1y}, F_{2x}, F_{2y}\}$ of all the bars for the Ith node yields zero for equilibrium.

Example 4.1: As the first example, let us consider the deformation of a truss structure neglecting inertia effects but involving a nonlinear stress–strain relation instead of the linear form (4.1). Indeed, the stress–strain relation is given as

$$\sigma = E(\sigma)\varepsilon, \qquad E(\sigma) = \frac{E_0}{1 + (\sigma/\sigma_0)^2}$$

where E_0 is Young's modulus of the material for small stress levels. Furthermore, let the cross-sectional area of the bar also depend on the strain (i.e., equivalently the stress) as

$$A(\varepsilon) = A_0 + A_1\varepsilon$$

for a given coefficient A_1.

Although we have been studying only linear problems so far, it is possible to deal with nonlinear problems without much additional effort. Indeed, if a stepwise linear approximation of the nonlinear stress–strain relation is taken as

$$\Delta\sigma^n = E(\sigma^{n-1})\,\Delta\varepsilon^n$$

where $\Delta\sigma^n$ and $\Delta\varepsilon^n$ are the increments of stress and strain, respectively, during the nth incremental step, and σ^{n-1} is the stress accumulated in reaching the $(n-1)$th step, then we can solve the nonlinear problem using the program developed for linear problems. The material constant E is simply updated, and the computation for linear problems is repeated until the desired load level is achieved by accumulating each increment of applied load at an incremental step. This type of stepwise method, which uses the program for linear problems, is in general called the *incremental method.* We shall elaborate on this method in the present example problem by delineating seven steps:

1. Make an incremental path of the applied force. The simplest way is to define the increment of the applied load by $\Delta\mathbf{f} = \mathbf{f}/\mathrm{NS}$, where NS is the number of steps in the incremental procedure.
2. At the initial strain level ε_0, before applying the load, define the initial Young's modulus E and the area A of the cross section.
3. Solve the static problem (4.13) and (4.14) by assuming $\ddot{\mathbf{u}}$ and $\ddot{\mathbf{v}}$ are zero for the load increment $\Delta\mathbf{f}$. Let the displacement obtained be denoted by $\Delta\mathbf{u}$ and $\Delta\mathbf{v}$ in each direction.
4. Keep the results of the displacement as $\mathbf{u} = \mathbf{u} + \Delta\mathbf{u}$ and $\mathbf{v} = \mathbf{v} + \Delta\mathbf{v}$, where $\mathbf{u}$ and $\mathbf{v}$ are initially zero. Then compute the incremental strain $\Delta\varepsilon$, stress $\Delta\sigma$, and the member force ΔP using the results of $\Delta\mathbf{u}$ and $\Delta\mathbf{v}$. Update the strain $\varepsilon = \varepsilon + \Delta\varepsilon$, the stress $\sigma = \sigma + \Delta\sigma$, and the member force $P = P + \Delta P$. Here $\varepsilon = \varepsilon_0$ initially.
5. Update the configuration of the structure by $\mathbf{x} = \mathbf{x} + \Delta\mathbf{u}$ and $\mathbf{y} = \mathbf{y} + \Delta\mathbf{v}$, that is, the location of the joints (or nodes) is modified by the increments of the displacement.
6. Reevaluate Young's modulus E and the area A of the cross section at the stress level σ (or equivalently at the strain level ε) by using the stress–strain relation obtained, for example, by the simple tension test.
7. Repeat steps 3–6 until all the incremental steps have been performed. According to step 1, NS iterations of steps 3–6 will be necessary.

Let us obtain the deformation of the truss shown in Figure 4.3 under the assumption that $E_0 = 100$ kN/cm^2, $\sigma_0 = 10$ kN/cm^2, $A_0 = 1$ cm^2, $A_1 = 0.01$ cm^2, and the applied load $\mathbf{P} = 0.45$ kN $\mathbf{i} - 1.8$ kN $\mathbf{j}$ at the

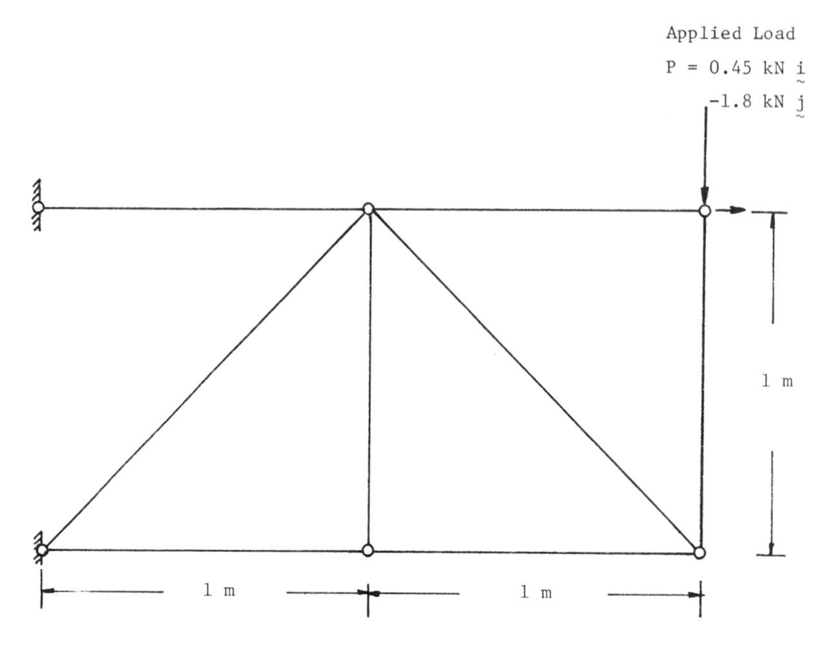

Figure 4.3 Truss structure for example 4.1.

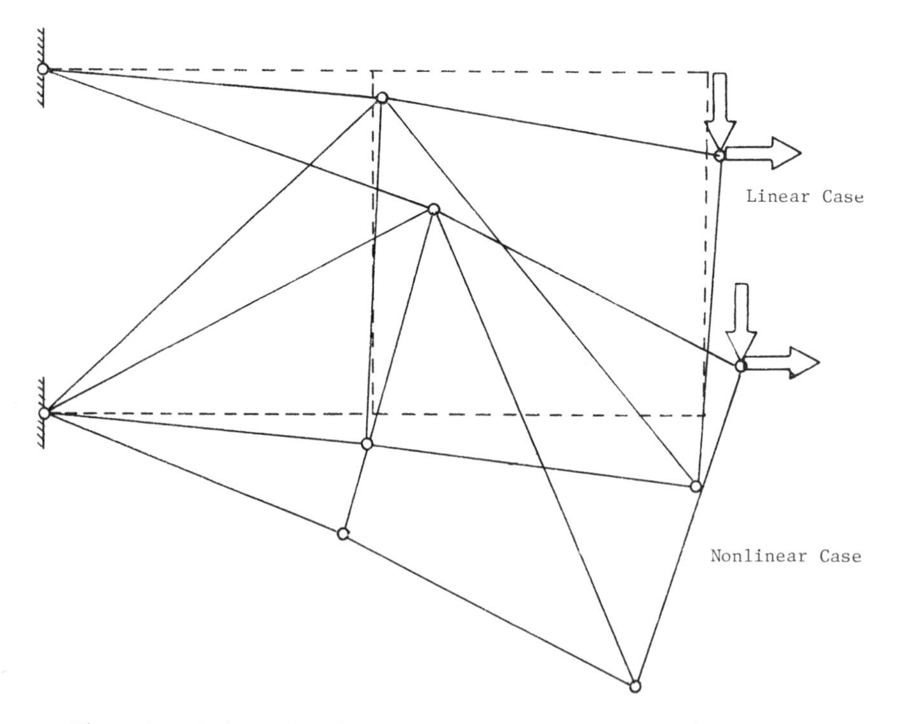

Figure 4.4 Deformed configurations, with deflections magnified 100 times.

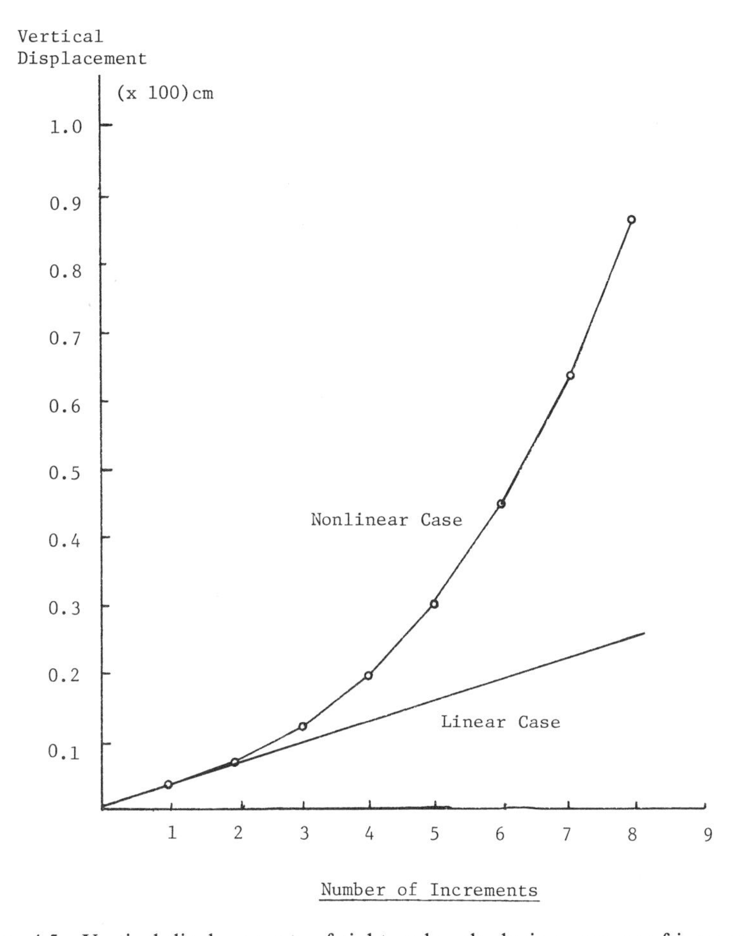

Figure 4.5 Vertical displacements of right end node during process of increment.

upper right edge. Each member has the same material constants and initial cross-sectional area.

Numerical results are shown in Figures 4.4 and 4.5. Nine incremental steps were taken in the present example.

Example 4.2: Let us consider a minimum-weight problem for a plane truss structure as an application of finite element methods. Suppose that a 10-member cantilever truss, shown in Figure 4.6, is considered with design data

$$E = 10^4 \text{ ksi}, \qquad \rho = 0.1 \text{ lb/in.}^3, \qquad l = 360 \text{ in.}$$

The lower limit on the area of the cross section is $\underline{A} = 0.1$ in.^{2} under the applied force $P = 100$ kip at the third and fifth nodes. The design problem is to find the area of the cross section of each member to minimize the weight of the truss under the design restriction that

$$|\sigma_e| \leq \bar{\sigma}, \quad e = 1, 2, \ldots, 10$$

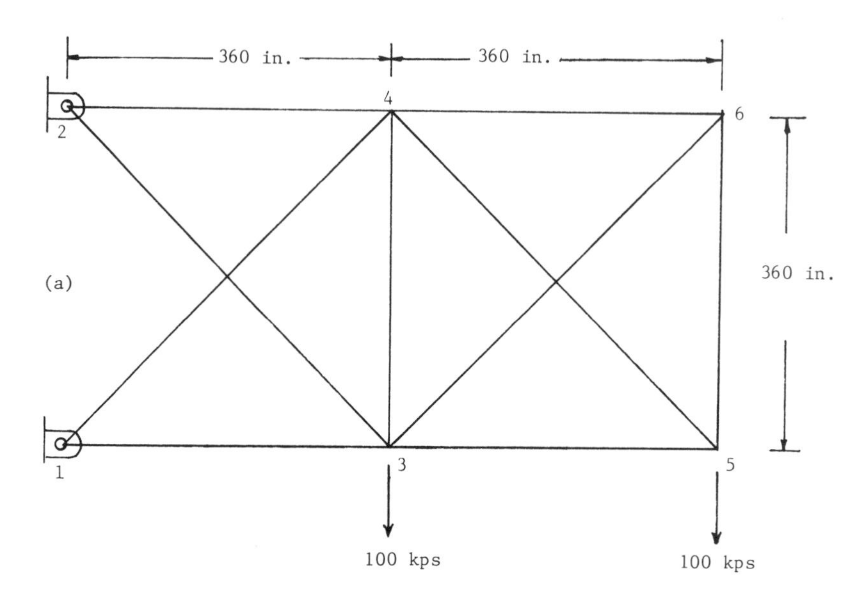

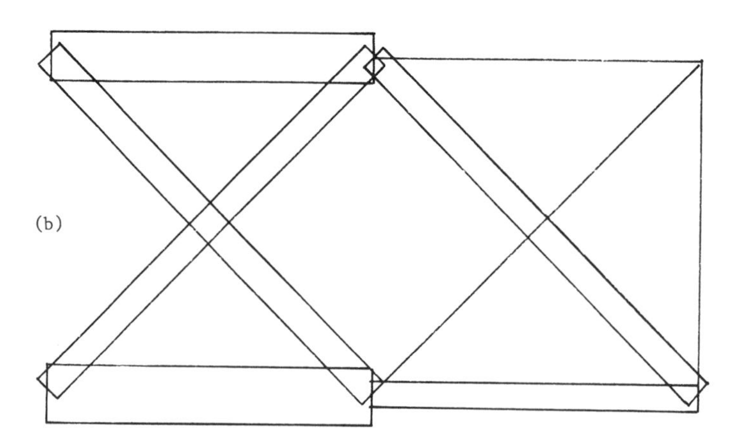

Figure 4.6 Initial design of truss structure and its minimum-weight design configuration: (a) truss, (b) after optimization.

where σ_e is the axial stress in the eth member and $\bar{\sigma}$ is the upper bound on the stress, taken to be $\bar{\sigma} = 25$ ksi. Note that the weight of the truss is given by

$$\sum_{e=1}^{10} \rho\, l_e A_e$$

where l_e and A_e are the length and area of the cross section of the eth member; the minimum-weight design problem can be represented by the following minimization problem:

$$\min_{A_e} \sum_{e=1}^{10} \rho l_e A_e$$

subject to

$$\underline{A} \le A_e, \qquad -\bar{\sigma} \le \sigma_e \le \bar{\sigma}, \quad e = 1, 2, \ldots, 10$$

Applying Lagrange multiplier methods to reduce the constraints, we define the Lagrangian

$$\psi(A_e, \mu_e, \lambda_{1e}, \lambda_{2e}) = \sum_{e=1}^{10} [\rho l_e A_e - \mu_e(\underline{A} - A_e) - \lambda_{1e}(-\bar{\sigma} - \sigma_e) - \lambda_{2e}(\sigma_e - \bar{\sigma})]$$

where $\lambda_{1e} \le 0$, $\lambda_{2e} \le 0$, and $\mu_e \le 0$, $e = 1, \ldots, 10$. Noting that $\sigma_e = P_e/A_e$, where P_e is the axial force in the eth member, the stationary conditions for ψ are

$$\rho l_e + \mu_e - \frac{\lambda_{1e}\sigma_e}{A_e} + \frac{\lambda_{2e}\sigma_e}{A_e} = 0$$

$$\begin{array}{llll} \mu_e \le 0, & \underline{A} - A_e \le 0, & \mu_e(\underline{A} - A_e) & = 0 \\ \lambda_{1e} \le 0, & -\bar{\sigma} - \sigma_e \le 0, & -\lambda_{1e}(\bar{\sigma} + \sigma_e) & = 0 \\ \lambda_{2e} \le 0, & \sigma_e - \bar{\sigma} \le 0, & \lambda_{2e}(\sigma_e - \bar{\sigma}) & = 0 \end{array}$$

for $e = 1, 2, \ldots, 10$.

Now let us consider the meaning of the above stationary conditions. To do this, suppose that $\lambda_{1e} = 0$, that is, $-\bar{\sigma} < \sigma_e$, $e = 1, \ldots, 10$. Then if $\underline{A} < A_e$, $e = 1, \ldots, 10$, the first equation yields

$$\rho l_e + \frac{\lambda_{2e}\sigma_e}{A_e} = 0$$

since $\mu_e = 0$ has to be satisfied. This means $\lambda_{2e} \ne 0$, in general. Therefore, $\sigma_e = \bar{\sigma}$ must be satisfied. Similarly, if $\sigma_e < \bar{\sigma}$ and $\underline{A} < A_e$ are assumed, then we have $\sigma_e = -\bar{\sigma}$. Thus, we have the condition

$$\text{if } \underline{A} < A_e, \quad \text{then } |\sigma_e| = \bar{\sigma}$$

For the case $\underline{A} = A_e$, we expect $|\sigma_e| \le \bar{\sigma}$.

How can we use the above discussion to find the area of the cross section of each member that yields the minimum weight? One suggestion may be the following four-step iterative method:

1. Assume A_e^0, $e = 1, \ldots, 10$, as an initial approximation.
2. For $i = 0, 1, \ldots$, solve the equilibrium equation by the finite element method for the prescribed areas A_e^i, $e = 1, \ldots, 10$. Then compute σ_e in each member.
3. Find a new approximation of the cross-sectional areas by
$$A_e^{i+1/2} = A_e^i + C(|\sigma_e| - \bar{\sigma}), \qquad A_e^{i+1} = \max(\underline{A}, A_e^{i+1/2})$$
for a proper positive "constant" C related to $\underline{A}$ and $\bar{\sigma}$.
4. Compute the quantity
$$\sqrt{\sum_{e=1}^{10} (A_e^{i+1} - A_e^i)^2} \Bigg/ \sqrt{\sum_{e=1}^{10} (A_e^{i+1})^2}$$

Table 4.1 *Stresses and areas of the optimized truss*

NEL	Axial stress	Area of cross section
1	25.1138252	7.89920624
2	15.9164693	0.1
3	−24.6743834	8.16286588
4	−25.0501744	3.93045679
5	−1.43411096	0.1
6	15.9164693	0.1
7	24.6508119	5.82255824
8	−25.1481465	5.53188435
9	25.1836133	5.53357405
10	−22.5092867	0.1

If this is small enough, then we suppose A_e^{i+1} is the solution. If otherwise, go back to step 2.

Starting from the truss consisting of members whose area of the cross section is 10 in.^{2}, the choice of design magnification factor $C = C_0 A_0/\bar{\sigma}$ ($C_0 = 0.2$, $A_0 = 5$ in.^{2}, and $\bar{\sigma} = 25$ ksi) yields convergence at the 34th iteration with the tolerance 0.89×10^{-2}. The weight of the truss becomes 1595 lb while the initial weight is 3836 lb. Stresses and areas of truss members are given in Table 4.1 and Figure 4.6. It is clear that members 2, 5, 6, and 10 are not essential to support the applied forces at the third and fifth nodes.

Exercise 4.1: Solve the same design problem in example 4.2 under an additional constraint

$$|w_e| \leq \bar{w}, \quad e = 1, \ldots, 10$$

where w_e is the axial displacement of the eth member, and $\bar{w} = 2$ in. is the upper bound of the axial displacement. To solve this, note that the first variation of the relation

$$\frac{EA_e}{l_e}\begin{bmatrix} 1 & -1 \\ -1 & 1 \end{bmatrix}\begin{Bmatrix} w_{e1} \\ w_{e2} \end{Bmatrix} = \begin{Bmatrix} f_{e1} \\ f_{e2} \end{Bmatrix}$$

could be applied to find the variation of $w_e = w_{e1} - w_{e2}$ with respect to the design change:

$$\delta\left(\frac{EA_e}{l_e} w_e\right) = \delta f_e, \qquad \text{i.e.,} \quad A_e\, \delta w_e + w_e\, \delta A_e = 0$$

Exercise 4.2: In the above, the pin-jointed truss is considered to be planar. Let us extend the idea to cover the case of a space truss.

Suppose that the direction cosines of the bar axis (i.e., s axis) relative to a fixed Cartesian coordinate system (x, y, z) are given by (l, m, n) (see Figure 4.7). If the displacement of an arbitrary point P in the bar is

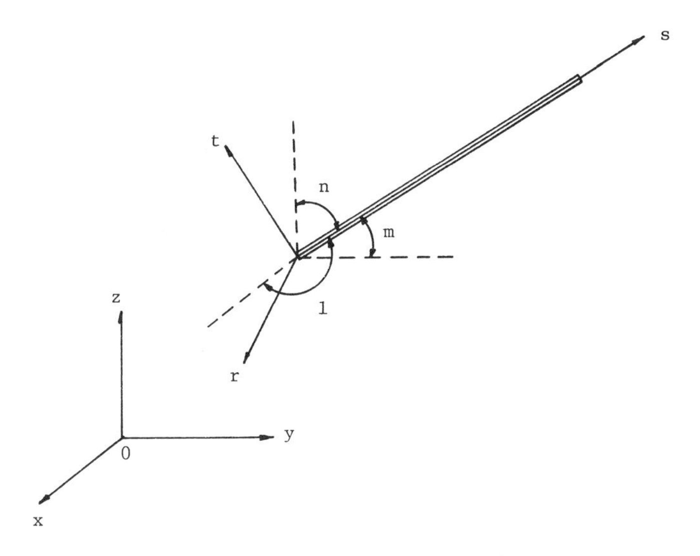

Figure 4.7 Bar element in three-dimensional space.

given as $\mathbf{u} = u_x\mathbf{i} + u_y\mathbf{j} + u_z\mathbf{k}$ using the components in the fixed xyz coordinate system, then the displacement along the bar axis is obtained by

$$u_{ps} = lu_{px} + mu_{py} + nu_{pz} \tag{4.9b}$$

Identifying P first as one of the ends, and then the other, the displacements u_{1s} and u_{2s} of these points along the bar axis are represented, in terms of the displacements $\mathbf{u}_1$ and $\mathbf{u}_2$, by

$$\begin{Bmatrix} u_{1s} \\ u_{2s} \end{Bmatrix} = \underbrace{\begin{bmatrix} l & m & n & 0 & 0 & 0 \\ 0 & 0 & 0 & l & m & n \end{bmatrix}}_{\substack{[T_{\alpha p}] \\ \text{Transformation matrix}}} \begin{Bmatrix} u_{1x} \\ u_{1y} \\ u_{1z} \\ u_{2x} \\ u_{2y} \\ u_{2z} \end{Bmatrix} \tag{4.11b}$$

Applying the relation (4.11b), the element stiffness matrix is obtained as

$$\bar{\mathbf{K}}^e = \mathbf{T}^T\mathbf{K}^e\mathbf{T} \qquad (\text{i.e.,} \quad \bar{K}^e_{pq} = T_{\alpha p}K^e_{\alpha\beta}T_{\beta q}) \tag{4.13b}$$

Obtain the relation similar to (4.14) for the generalized load vector and modify (4.16) for a three-dimensional truss.

Furthermore, using the coordinates (x_1, y_1, z_1) and (x_2, y_2, z_2) of the end points of the bar in the (x, y, z) coordinate system, obtain the direction cosines (l, n, m) of the bar axis.

Now we return to the dynamics of a truss structure. First, the free vibration problem for equation (4.17) will be considered. To do this, let the displacement U_J be assumed to depend on the time according to

$$U_J = e^{i\omega t}\hat{U}_J \tag{4.18}$$

where $i = \sqrt{-1}$ and ω is the frequency of the free vibration obtained by solving the *generalized* eigenvalue problem

$$(-\omega^2 M_{IJ}\hat{U}_J + K_{IJ}\hat{U}_J)e^{i\omega t} = 0$$

that is,

$$K_{IJ}\hat{U}_J - \omega^2 M_{IJ}\hat{U}_J = 0 \tag{4.19}$$

Equation (4.19) is obtained from (4.17) by substituting (4.18). Since the mass matrix $[M_{IJ}]$ is, in general, invertible, the form (4.19) can also be written as

$$(M_{IK})^{-1}K_{KJ}\hat{U}_J = \omega^2\hat{U}_I \tag{4.20}$$

or

$$\hat{\mathbf{K}}\hat{\mathbf{U}} = \lambda\mathbf{U} \tag{4.21}$$

where $\lambda = \omega^2$, and $\hat{K}_{IJ} = (M_{IK})^{-1}K_{KJ}$.

Even if the mass matrix $[M_{IK}]$ is *lumped* before its inverse is taken, that is, if the mass matrix is modified to be $[\hat{M}_{IK}]$ where

$$\hat{M}_{II} = \sum_{K=1}^{N} M_{IK} \quad \text{(no sum on } I\text{)}, \qquad \hat{M}_{IJ} = 0 \quad \text{if } I \neq J$$

for $I = 1, \ldots, N$, the modified stiffness matrix $\hat{K}_{IJ} = (\hat{M}_{IK})^{-1}K_{KJ}$ is not, in general, symmetric. Thus (4.21) cannot be solved by the Jacobi method studied in section 1.3. Therefore, we have to study methods for solving a generalized eigenvalue problem such as (4.19).

4.2 Review of eigenvalue problems

Let us briefly review linear algebra for eigenvalue problems defined by the standard form

$$(A_{ij} - \lambda\delta_{ij})u_j = 0 \tag{4.22a}$$

and by the generalized form

$$(A_{ij} - \lambda B_{ij})u_j = 0 \tag{4.22b}$$

Clearly, $u_j = 0, j = 1, \ldots, n$, is a solution to problem (4.22). It is called the trivial solution. Our interest is to find a λ that provides a nontrivial solution $\mathbf{u} = u_j\mathbf{i}_j$. In general, if $[A_{ij}]$ and $[B_{ij}]$ are $n \times n$ square matrices, there are n values of λ and n $\mathbf{u}$'s, that is, there are n pairs $\{\lambda_1, \mathbf{u}_1\}, \{\lambda_2, \mathbf{u}_2\}, \ldots, \{\lambda_n, \mathbf{u}_n\}$ that satisfy equation (4.22). If the matrices

$$\begin{matrix} V_{ji} = u_{ij}, & \mathbf{u}_i = u_{ij}\mathbf{i}_j \\ \Lambda_{ii} = \lambda_i, & \Lambda_{ij} = 0 \quad \text{if } i \neq j \end{matrix} \qquad \text{(no sum on } i\text{)} \tag{4.23}$$

are defined, equation (4.22) yields the form

$$A_{ik}V_{kj} = B_{ik}V_{km}\Lambda_{mj} \tag{4.24a}$$

or

$$\mathbf{AV} = \mathbf{BV\Lambda} \tag{4.24b}$$

The matrix **V** is called the *modal matrix*. Multiplying both sides of (4.24) by $\mathbf{V}^T$ on the left, we have

$$V_{in}A_{ik}V_{kj} = V_{in}B_{ik}V_{km}\Lambda_{mj}$$

If the eigenvectors $\mathbf{u}_i$, $i = 1, \ldots, n$, are normalized so that

$$V_{in}B_{ik}V_{km} = \delta_{nm} \qquad \text{or} \qquad \mathbf{V}^T\mathbf{B}\mathbf{V} = \mathbf{I} \tag{4.25}$$

(*caution* – many codes do not have this normalization),

$$V_{in}A_{ik}V_{kj} = \Lambda_{nj} \qquad \text{or} \qquad \mathbf{V}^T\mathbf{A}\mathbf{V} = \mathbf{\Lambda} \tag{4.26}$$

In finite element analyses for mathematically well-posed problems, both matrices **A** and **B** are real, symmetric, and nonnegative. Furthermore, we need not find all of the eigenvalues* λ_i and eigenvectors $\mathbf{u}_i$ in most of the applications. Thus, the Jacobi method introduced in Chapter 1 becomes unrealistic for solving eigenvalue problems obtained by finite element approximations if the total number of degrees of freedom is very large. We shall study more about eigensolvers in this chapter. To do this, we first note some results obtained in linear algebra:

(a) If the matrices **A** and **B** are real and symmetric, and if **B** is nonsingular, then the eigenvalues and eigenvectors are real.
(b) If **A** and **B** are real and symmetric matrices, and if **B** is positive definite, then there exist n independent eigenvalues and eigenvectors. Moreover,

$$V_{mi}B_{mn}V_{nj} = 0 \quad \text{if } i \neq j$$

and it is possible to normalize so that

$$V_{mi}B_{mn}V_{nj} = \delta_{ij}$$

Note that if **B** is not positive definite, then the n eigenvectors are not necessarily independent.

There are many numerical methods that compute eigenvalues and eigenvectors. For example, we have the following standard methods:

1. *Power iteration method* to obtain the maximum eigenvalue: Both forms (4.22a) and (4.22b) can be solved, and symmetry of the matrices **A** and **B** need not be assumed if there exist n independent eigenvectors.
2. *Inverse iteration method* to obtain the minimum eigenvalue: Both forms (4.22a) and (4.22b) can be solved, and symmetry of **A** and **B** is not necessary if there exist n independent eigenvectors.
3. *Subspace iteration method* to obtain several eigenvalues and eigenvectors at the same time: Both forms (4.22a) and (4.22b) can be solved under the assumption that the matrices **A** and **B** are real and symmetric.
4. *Jacobi method* to compute all eigenvalues and eigenvectors at the same time: This uses the similarity transformation for both forms (4.22a) and (4.22b) eigenvalue problems; however, the matrices **A** and **B** must be real and symmetric.
5. *Householder method* for the form (4.22a) and a symmetric matrix **A**.

* Note that indeed the underlying "mechanics" may not be valid at the higher frequencies.

6. *QR method* for the form (4.22a) under the assumption that there exist n independent eigenvectors.
7. *Determinant search method* for an arbitrary eigenvalue problem.

The first and second methods are feasible for middle-sized eigenvalue problems, and the third and fifth methods are better suited to large-size problems, whereas the rest of the methods are adequate for small-size problems.

One of the general references on the eigenvalue problem is by Wilkinson (1965). Here we shall study only Jacobi's method and the inverse iteration method, since rather small size problems are considered in this chapter. Let us leave the subspace iteration method for more advanced studies.

4.2.1 The generalized Jacobi method

We have derived the numerical scheme of Jacobi's method for finding eigenvalues and eigenvectors in Chapter 1. Jacobi's method is a method that uses the concept of *similarity transformation* of a given matrix $\mathbf{A} = A_{ij}\mathbf{i}_i\mathbf{i}_j$. Let $\mathbf{P} = P_{ij}\mathbf{i}_i\mathbf{i}_j$ be a regular $n \times n$ square matrix. Then the matrix $\hat{\mathbf{A}} = \hat{A}_{ij}\mathbf{i}_i\mathbf{i}_j$ is said to be the matrix obtained by the similarity transformation of $\mathbf{A}$ if

$$\hat{A}_{ij} = P_{ik}^{-1}A_{km}P_{mj} \tag{4.27}$$

where $\mathbf{P}^{-1}$ is the inverse of the matrix $\mathbf{P}$. Note that if the matrix $\mathbf{P}$ is also *orthogonal,* that is, if

$$\det \mathbf{P} = 1$$

then the inverse $\mathbf{P}^{-1}$ is the same as the *transpose* $\mathbf{P}^{\mathrm{T}}$ of $\mathbf{P}$; that is, $P_{ij}^{-1} = P_{ji}$. Then, (4.27) can be written as

$$\hat{A}_{ij} = P_{ki}A_{km}P_{mj} \tag{4.28}$$

for an orthogonal matrix $\mathbf{P}$.

Suppose that λ and $\mathbf{u}$ are, respectively, an eigenvalue and an eigenvector of the standard form (4.22a). Then multiplying (4.22a) on the left by $\mathbf{P}^{-1}$ yields

$$P_{ik}^{-1}A_{km}u_m = \lambda P_{im}^{-1}u_m$$

Identifying

$$u_m = P_{mj}w_j, \qquad \text{i.e.,} \quad w_j = P_{jm}^{-1}u_m \tag{4.29}$$

we have

$$P_{ik}^{-1}A_{km}P_{mj}w_j = \lambda P_{im}^{-1}P_{mj}w_j = \lambda w_i$$

since $P_{im}^{-1}P_{mj} = \delta_{ij}$. Thus,

$$\hat{A}_{ij}w_j = \lambda w_i \tag{4.30}$$

This means that the eigenvalue λ is invariant under the similarity transformation (4.27) or (4.28) whereas the eigenvector is transformed to w as in (4.29). Thus, we have the following property.

(c) Similar matrices have the same eigenvalues.

Using the above property, it is possible to find an orthogonal matrix **P** that yields a transformed matrix whose eigenvalues can be easily obtained. For example, if we transform it to a diagonal matrix, then the diagonal elements are the eigenvalues. This idea connects Jocobi's method with the other property:

(d) If (4.22) has n independent eigenvectors, then **A** can be reduced to a diagonal matrix.

The Jacobi method described in Chapter 1 is the sequential procedure that uses similarity transformation to force the maximum off-diagonal element $A_{ij} = A_{ji}$ to zero at each step. We repeat the similarity transformations until the maximum off-diagonal term becomes $O(\varepsilon)$ for some specified ε. How many transformations are necessary? We do not know the exact answer. But if we repeat infinitely many times, a diagonal matrix should be obtained if **A** satisfies all necessary conditions. How do we know this?

Recall the transformed matrix $\hat{S}_{ij}$ by the formula (1.54) in section 1.3. Computation of the sum of squares of the off-diagonal terms yields

$$\sum_{m<n} \hat{S}_{mn}^2 = \sum_{m<n} S_{mn}^2 - S_{ij}^2 \tag{4.31}$$

That is, the summation of squares of the off-diagonal terms after the transformation is smaller than that before the transformation. This indicates the convergence to a diagonal matrix.

We shall consider an extension of Jacobi's method to the generalized form (4.22b) of the eigenvalue problem:

$$A_{ij}u_j - \lambda B_{ij}u_j = 0 \tag{4.32}$$

To do this, let us introduce a rotation matrix $\mathbf{P}^{(ij)}$ such that

$$\mathbf{P}^{(ij)} = \begin{array}{c} \\ i \\ \\ j \\ \\ \end{array}\left[\begin{array}{ccccc} \ddots & \vdots & & \vdots & \\ \cdots & 1 & \cdots & \alpha & \cdots \\ & \vdots & \ddots & \vdots & \\ \cdots & \beta & \cdots & 1 & \cdots \\ & \vdots & & \vdots & \ddots \end{array}\right] \tag{4.33}$$

(columns i and j marked above the matrix)

where α and β are defined so that the terms $\hat{A}_{ij}$ and $\hat{B}_{ij}$ are zero (i, j are specified and the index notation does not apply). That is,

$$\begin{aligned} \hat{A}_{ij} &= \alpha A_{ii} + (1 + \alpha\beta)A_{ij} + \beta A_{jj} = 0 \\ \hat{B}_{ij} &= \alpha B_{ii} + (1 + \alpha\beta)B_{ij} + \beta B_{jj} = 0 \end{aligned} \tag{4.34}$$

If we solve (4.34), the constants α and β are obtained as

$$\alpha = \frac{\bar{A}_{jj}}{\gamma} \quad \text{and} \quad \beta = \frac{-\bar{A}_{ii}}{\gamma} \tag{4.35}$$

where, without summing in repeated indices,

$$\begin{aligned}
\bar{A}_{ii} &= A_{ii}B_{ij} - A_{ij}B_{ii} \\
\bar{A}_{jj} &= A_{jj}B_{ij} - A_{ij}B_{jj} \\
\gamma &= 0.5s + (\text{sgn } s)\sqrt{0.25s^2 + \bar{A}_{ii}\bar{A}_{jj}} \\
s &= A_{ii}B_{jj} - A_{jj}B_{ii}
\end{aligned} \tag{4.36}$$

Once the rotation matrix $\mathbf{P}^{(ij)}$ is specified with the form (4.33) through (4.35) and (4.36), the similarity transformation is obtained by the matrix multiplication (4.28) since $\mathbf{P}$ is orthogonal.

The program listed below is BASIC GENERALIZED JACOBI developed on an IBM PC-XT personal computer. Input and output format is very similar to the program BASIC JACOBI in Chapter 1. We shall not describe the details of the program here.

```
1000 CLS
1010 PRINT "************************************"
1020 PRINT "   GENERALIZED JACOBI'S METHOD"
1030 PRINT "************************************"
1040 PRINT : PRINT : PRINT
1050 PRINT "SET UP THE MASS AND STIFFNESS MATRICES"
1060 PRINT "--------------------------------------------------------------------------"
1070 PRINT
1080 INPUT "Number of equations   NEQ = ";NEQ
1090 PRINT : PRINT " A = STIFFNESS MATRIX " : PRINT " B = MASS MATRIX" : PRINT
1100 DIM A(NEQ,NEQ),B(NEQ,NEQ),T(NEQ,NEQ),EV(NEQ)
1110 FOR I=1 TO NEQ : FOR J=I TO NEQ
1120 PRINT I;TAB(5);J;TAB(10);" A, B = ";: INPUT A(I,J),B(I,J)
1130 A(J,I)=A(I,J) : B(J,I)=B(I,J)
1140 NEXT J : NEXT I
1150 GOSUB 1170
1160 GOTO 1880
1170 REM    Subroutine  Generalized Jacobi's Method
1180 REM  ----------------------------------------------------------------------
1190 FOR I=1 TO NEQ : FOR J=1 TO NEQ : T(I,J)=0 : NEXT J : T(I,I)=1 : NEXT I
1200 CLS : PRINT
1210 PRINT "Generalized Jacobi's Method"
1220 PRINT "--------------------------------------------------------------------------"
1230 PRINT : INPUT "Maximum number of iterations  IX = ";IX
1240 INPUT "Tolerance   TR = ";TR
1250 PRINT :
     INPUT "Number of degrees of freedom for adjustment of tolerance  NB = ";NB
1260 TR=TR*ABS(A(NB,NB))
1270 IT=1 : MT=1
1280 PRINT :
     PRINT "Iteration Process ..............................................." :
     PRINT
1290 AM=0
1300 FOR I=1 TO NEQ-1 : FOR J=I+1 TO NEQ
1310 AK=ABS(A(I,J)) : IF AK<=AM  THEN GOTO 1330
1320 AM=AK : IR=I : JR=J
1330 NEXT J : NEXT I
1340 IF AM<TR THEN GOTO 1710
1350 A1=A(IR,IR) : A2=A(JR,JR) : A3=A(IR,JR)
1360 B1=B(IR,IR) : B2=B(JR,JR) : B3=B(IR,JR)
1370 C1=A1*B3-A3*B1 : C2=A2*B3-A3*B2 : C3=A1*B2-A2*B1
1380 C4=(.5*C3)*(.5*C3)+C1*C2 : C4=SQR(C4)
1390 IF C3>=0 THEN C5=.5*C3+C4
1400 IF C3<0 THEN C5=.5*C3-C4
```

```
1410 IF C5=0 THEN GOTO 1430
1420 AL=C2/C5 : BT=-C1/C5 : GOTO 1470
1430 IF A1+A3=0 THEN GOTO 1450
1440 AL=-(A3+A2)/(A1+A3) : BT=1 : GOTO 1470
1450 AL=-(A3-A2)/(A1-A3) : BT=-1
1460 REM < MODIFY THE MATRIX A & B >
1470 FOR K=1 TO NEQ : IF K=IR OR K=JR THEN GOTO 1520
1480 A4=A(IR,K) : A5=A(JR,K) : B4=B(IR,K) : B5=B(JR,K)
1490 A6=A4+BT*A5 : A7=AL*A4+A5 : B6=B4+BT*B5 : B7=AL*B4+B5
1500 A(IR,K)=A6 : A(K,IR)=A6 : A(JR,K)=A7 : A(K,JR)=A7
1510 B(IR,K)=B6 : B(K,IR)=B6 : B(JR,K)=B7 : B(K,JR)=B7
1520 NEXT K
1530 A(IR,IR)=A1+(BT*BT)*A2+2*BT*A3
1540 B(IR,IR)=B1+(BT*BT)*B2+2*BT*B3
1550 A(JR,JR)=(AL*AL)*A1+A2+2*AL*A3
1560 B(JR,JR)=(AL*AL)*B1+B2+2*AL*B3
1570 A(IR,JR)=0 : A(JR,IR)=0 : B(IR,JR)=0 : B(JR,IR)=0
1580 REM  < COMPUTE MODAL MATRIX >
1590 FOR K=1 TO NEQ
1600 T1=T(K,IR) : T2=T(K,JR)
1610 T(K,IR)=T1+BT*T2 : T(K,JR)=AL*T1+T2
1620 NEXT K
1630 IF IT/MT < 1 THEN GOTO 1660
1640 PRINT IT;TAB(10);"Maximum off diagonal term   AM = ";AM
1650 MT=MT+10
1660 IF IT<IX THEN GOTO 1690
1670 INPUT "Do you want to iterate more ?   (YES/NO) = ";L$
1680 IF L$="YES" THEN INPUT " Additional number of iteration IB = ";IB : IX=IX+IB
1690 IT=IT+1
1700 GOTO 1290
1710 PRINT :
     PRINT "Convergence : Generalized Jacobi's Method ..........................."
1720 PRINT : PRINT "  NUMBER OF ITERATION = ";IT : PRINT "   TOLERANCE = ";AM
1730 REM  < EIGENVALUES >
1740 FOR I=1 TO NEQ : BB=B(I,I) : IF BB>=1E-10 THEN GOTO 1760
1750 EV(I)=1E+10 : FOR J=1 TO NEQ : T(J,I)=0 : NEXT J : T(I,I)=1 : GOTO 1770
1760 B(I,I)=BB : EV(I)=A(I,I)/BB : BB=SQR(B(I,I)) :
     FOR J=1 TO NEQ : T9J,I)=T(J,I)/BB : NEXT J
1770 NEXT I
1780 REM  < OUTPUT THE RESULTS >
1790 PRINT : PRINT : PRINT "EIGENVALUES AND EIGENVECTORS" :
     PRINT "-----------------------------------------------------------------------"
1800 PRINT : PRINT
1810 INPUT "Which eigenvalue and eigenvector will be output ?   LC = ";LC
1820 IF LC<=0 THEN GOTO 1870
1830 PRINT : PRINT "EIGENVALUE = ";EV(LC)
1840 PRINT : PRINT "EIGENVECTOR" : PRINT
1850 FOR J=1 TO NEQ : PRINT J;TAB(10);INT(100000!*T(J,LC))/100000! : NEXT J
1860 GOTO 1800
1870 RETURN
1880 END
```

Example 4.3: Using the program BASIC GENERALIZED JACOBI, we shall solve the eigenvalue problem

$$\begin{bmatrix} 2 & -1 & 0 & 0 \\ -1 & 2 & -1 & 0 \\ 0 & -1 & 2 & -1 \\ 0 & 0 & -1 & 1 \end{bmatrix} \begin{Bmatrix} u_1 \\ u_2 \\ u_3 \\ u_4 \end{Bmatrix} = \lambda \begin{bmatrix} 0 & 0 & 0 & 0 \\ 0 & 2 & 0 & 0 \\ 0 & 0 & 0 & 0 \\ 0 & 0 & 0 & 1 \end{bmatrix} \begin{Bmatrix} u_1 \\ u_2 \\ u_3 \\ u_4 \end{Bmatrix}$$

After five iterations, the maximum of the off-diagonal terms becomes zero, and the solution is as in Table 4.2. As shown, the program can be used even if some of the diagonal terms are zero.

Table 4.2

	1	2	3	4
λ	2×10^{10}	0.853553391	2×10^{10}	0.146446609
u_1	100000	0.25	0	0.25
u_2	0	0.5	0	0.5
u_3	0	−0.10355	100000	0.60355
u_4	0	−0.70711	0	0.70711

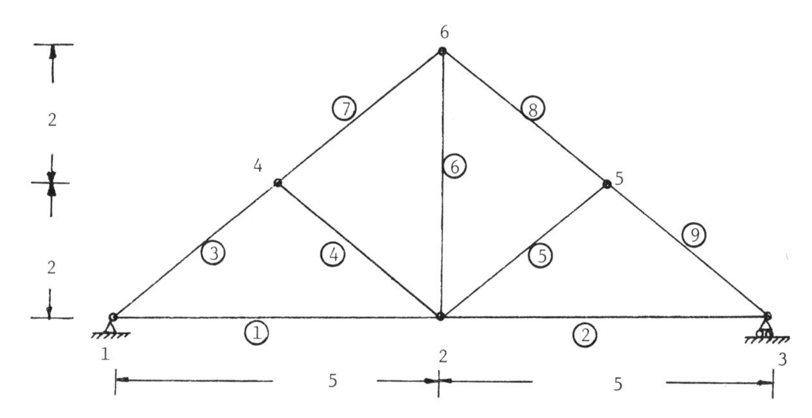

Figure 4.8 Truss structure for example 4.4.

Example 4.4: Now we use the program BASIC GENERALIZED JACOBI to solve an eigenvalue problem of the truss structure shown in Figure 4.8. For simplicity, the stiffness and mass matrices are normalized by taking the values $E = 1$, $A = 1$, and $\rho_A = 1$ for the material constants and cross-sectional areas of the bars. Physical dimensions for the nine-bar truss are given in Figure 4.8. In this case, 12 eigenvalues exist. Convergence of the iteration is achieved by 138 iterations with the maximum off-diagonal terms being 9.13×10^{-4}. The first 6 eigenvalues are as follows:

$$\lambda_1 = 0.0065255$$
$$\lambda_2 = 0.021722$$
$$\lambda_3 = 0.057997$$
$$\lambda_4 = 0.124427$$
$$\lambda_5 = 0.219540$$
$$\lambda_6 = 0.460945$$

The first four eigenmodes corresponding to the eigenvalues are shown in Figures 4.9 and 4.10.

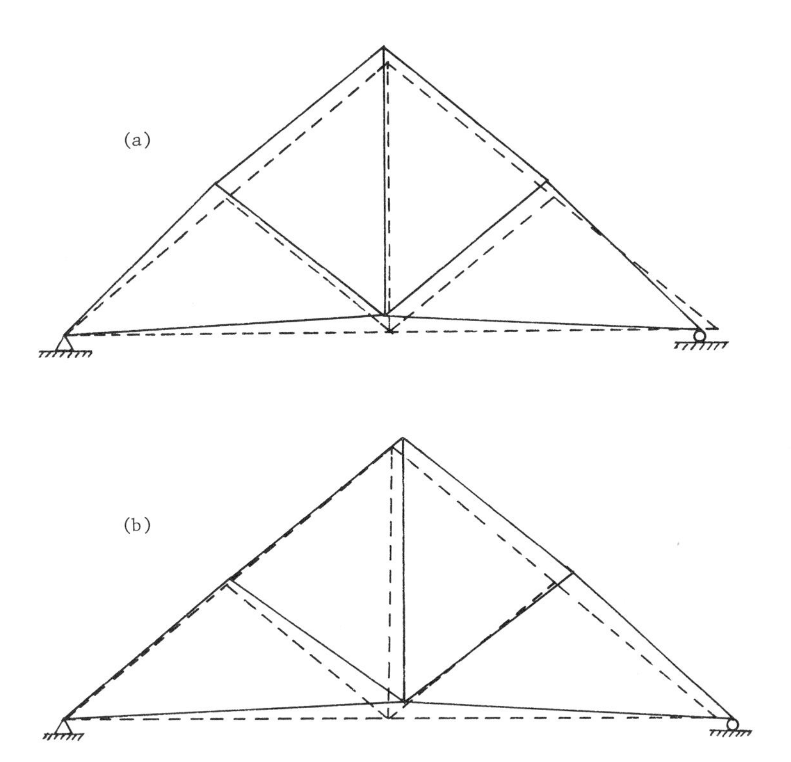

Figure 4.9 First and second modes of free vibration: (a) $\lambda_1 = 0.0065255$, (b) $\lambda_2 = 0.021722$.

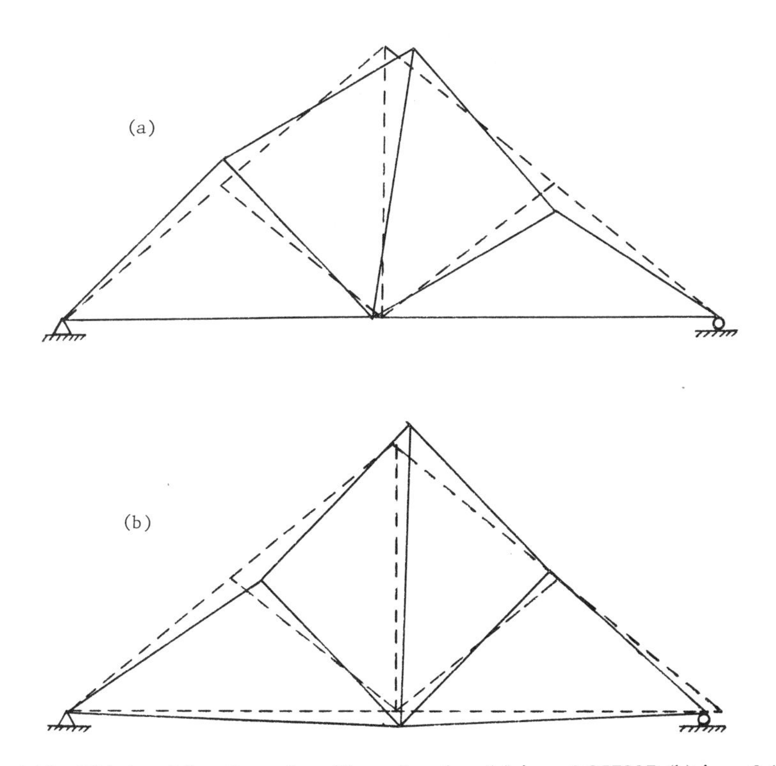

Figure 4.10 Third and fourth modes of free vibration: (a) $\lambda_3 = 0.057997$, (b) $\lambda_4 = 0.124427$.

4.2.2 Inverse iteration methods

Another commonly used method for finding eigenvalues and eigenvectors is the *inverse iteration method,* which uses a type of iterative method. This method, in general, yields the minimum eigenvalue and its corresponding eigenvector and is applicable to middle- and even large-size problems.

We shall first consider the standard form (4.22a). To do this, let us assume that there exist n independent eigenvectors of the matrix $\mathbf{A}$ and that

$$|\lambda_1| < |\lambda_2| \leq |\lambda_3| \leq \cdots \leq |\lambda_n| \tag{4.37}$$

for problem

$$A_{ij}v_j - \lambda v_i = 0 \tag{4.38}$$

For a nonzero vector $\mathbf{x}^0$ initially chosen, let us define the sequence of vectors $\{\mathbf{x}^k\}$ obtained as the solutions of the problems

$$A_{ij}x_j^k = x_i^{k-1} \tag{4.39}$$

for $k = 1, 2, \ldots$. Since any vector can be represented as a linear combination of eigenvectors, we put

$$\mathbf{x}^0 = c_m \mathbf{v}_m \tag{4.40}$$

for a set of proper numbers c_m, $m = 1, \ldots, n$. Substitution of (4.40) into (4.39) implies

$$A_{ij}x_j^1 = c_m v_{mi}$$

for $k = 1$. Since

$$A_{ij}v_{mj} = \lambda_m v_{mi} \quad \text{(no sum on } m\text{)}$$

that is,

$$\frac{v_{mj}}{\lambda_m} = A_{ji}^{-1} v_{mi}$$

for $m = 1, 2, \ldots, n$, we have

$$x_j^1 = \sum_{m=1}^{n} \frac{c_m}{\lambda_m} v_{mj}$$

Repeating this process up to k, we have

$$x_j^k = \sum_{m=1}^{n} \frac{c_m}{\lambda_m^k} v_{mj} = \frac{c_1}{\lambda_1^k} \left[v_{1j} + \sum_{m=2}^{n} \left(\frac{\lambda_1}{\lambda_m} \right)^k \frac{c_m}{c_1} v_{mj} \right] \tag{4.41}$$

from the iteration scheme (4.39). Because of assumption (4.37), (4.41) yields

$$x_j^k \to \frac{c_1}{\lambda_1^k} v_{1j} \quad \text{as } k \to \infty \tag{4.42}$$

Since the magnitude of an eigenvector is not essential, we can say that the vector $\mathbf{x}^k$ obtained by (4.39) converges to the eigenvector $\mathbf{v}_1$ as $k \to \infty$.

The eigenvalue λ_1^k corresponding to $\mathbf{x}^k$ by (4.39) is then computed by the Rayleigh quotient

$$\lambda_1^k = \frac{x_i^k A_{ij} x_j^k}{x_i^k x_i^k} \quad \text{(no sum on } k) \tag{4.43}$$

because of the real symmetric matrix $\mathbf{A}$.

It is easily seen that the speed of convergence is decided by the ratio

$$|\lambda_1|/|\lambda_2| \tag{4.44}$$

as shown in (4.41). Thus, if the ratio (4.44) is small enough, the speed of convergence is fast.

A pertinent comment on the iterative method of (4.39) and (4.43) is that we have to "shift the matrix $\mathbf{A}$" if $\mathbf{A}$ is not invertible (or equivalently if $\mathbf{A}$ has a zero eigenvalue). As seen in (4.39), invertibility of the matrix $\mathbf{A}$ is necessary to obtain the kth approximation from the vector $\mathbf{x}^{k-1}$.

$$\bar{A}_{ij} = A_{ij} + \mu\delta_{ij} \tag{4.45}$$

for a given $\mu \in \mathbb{R}$ makes the matrix $\bar{\mathbf{A}}$ invertible. Then we shall consider an eigenvector problem

$$\bar{A}_{ij} v_j - \bar{\lambda} v_i = 0 \tag{4.46}$$

Because of (4.45), we have

$$A_{ij} v_j - (\bar{\lambda} - \mu) v_i = 0$$

so that

$$\lambda = \bar{\lambda} - \mu \tag{4.47}$$

is an eigenvalue of the original if $\bar{\lambda}$ is an eigenvalue of (4.46).

Extension to the generalized form (4.22b) is straightforward. Indeed, for a given nonzero $\mathbf{x}^0$, we have

$$A_{ij} x_j^k = B_{ij} x_j^{k-1} \tag{4.48}$$

and

$$\lambda_1^k = \frac{x_i^k A_{ij} x_j^k}{x_i^k B_{ij} x_j^k} \quad \text{(no sum on } k) \tag{4.49}$$

with a normalization of the approximate eigenvector

$$\mathbf{v}_1^k = \mathbf{x}^k / \sqrt{x_i^k B_{ij} x_j^k} \quad \text{(no sum on } k) \tag{4.50}$$

If $\lambda_2, \lambda_3, \ldots, \lambda_n$ are also desired, we simply shift the matrix $\mathbf{A}$ by $\lambda_1, \lambda_2, \ldots, \lambda_{n-1}$. If orthogonal eigenvectors are desired, we can apply the Schmidt process of orthogonalization. As is easily observed from the iterative procedure (4.39) or (4.48), the inverse iteration method can be implemented simply by using solvers of a system of linear equations. At the first step, the triangulation of the stiffness matrix $\mathbf{A}$ (or equivalently the LU decomposition of $\mathbf{A}$) is required. However, after the first step, we need only compute the updated load vector $\mathbf{Bx}^{k-1}$, its reduction according to the triangulation, and the back substitution to obtain the next approximation $\mathbf{x}^k$ of the "eigenvector."

The program listed below, BASIC INVERSE ITERATION, was developed for an IBM PC-XT personal computer with 256K memory to solve eigenvalue problems of the generalized form (4.22b). In the program, the matrices **A** and **B** are stored as banded matrices SK and SM, the **K** and the **M** being suggestive of the stiffness and the mass matrices in finite element approximations. The vector **x** is denoted by U, the updated load vector **Bx** is stored in W, the eigenvector **v** is described by V, and the eigenvalue λ is R in the program. As a solver of a system of linear equations, we use a band solver for a symmetric coefficient matrix SK in the present program.

Necessary input data are as follows:

1. The matrices SK and SM, after specifying the number of equations and the bandwidth.
2. The number of eigenvalues, NG, that we want to obtain using the inverse iteration method: If NG = 3 is given, the first three eigenvalues and eigenvectors are computed.
3. Because of the nature of an iterative method, it is best to specify the maximum number of iterations in order to monitor whether or not the computation is going on well.
4. The tolerance, TR, on the difference $|\lambda^{k+1} - \lambda^k|$ between two successive estimates of an eigenvalue during the iteration process must be given.
5. As the last input item, the amount of shift, SH, must be specified. If no shifting is necessary, input SH = 0.

```
1000 PRINT "**********************************"
1010 PRINT "    INVERSE ITERATION METHOD"
1020 PRINT "**********************************"
1030 PRINT "                  Version 1.0 / Fall 1985"
1040 PRINT : PRINT : PRINT
1050 INPUT "Size of the matrices, that is, number of equations   NEQ = ";NEQ
1060 INPUT "Half band width of the matrices   MB = ";MB
1070 DIM SK(NEQ,MB),SM(NEQ,MB),U(NEQ),V(NEQ),W(NEQ)
1080 PRINT : PRINT "SET UP THE MATRICES  [A] AND [B] ........................."
1090 PRINT
1100 FOR I=1 TO NEQ : FOR J=1 TO MB : IF I+J-1>NEQ THEN GOTO 1120
1110 PRINT " A(";I;",";I+J-1;") & B(";I;",";I+J-1;") = "; : INPUT SK(I,J),SM(I,J)
1120 NEXT J : NEXT I
1130 GOSUB 1150
1140 GOTO 2720
1150 REM  Subroutine Inverse Iteration Method
1160 REM ---------------------------------------------------------------------
1170 REM  < SET THE DATA : INPUT >
1180 PRINT : PRINT "INVERSE ITERATION METHOD : GENERALIZED EIGENVALUE PROBLEM"
1190 PRINT "---------------------------------------------------------------------"
1200 PRINT : INPUT "How many eigenvalues will be obtained ?   NG = ";NG
1210 INPUT "Maximum number of iteration    IX =";IX
1220 INPUT "Tolerance   TR = ";TR
1230 INPUT "Amount of Initial Shift  SO = ";SO
1240 SH=SO : TR=TR*TR
1250 REM  < SAVE THE STIFFNESS MATRIX >
1260 INPUT "File Name which stores the stiffness matrix  F$ = ";F$
1270 OPEN F$ FOR OUTPUT AS #1
1280 FOR I=1 TO NEQ : FOR J=1 TO MB : PRINT#1, SK(I,J) : NEXT J : NEXT I
1290 CLOSE #1
1300 NI=1
1310 NJ=NI-1
1320 PRINT : PRINT NI;" -TH EIGENVALUE AND EIGENVECTOR" :
     PRINT "--------------------------------------------------------" : PRINT
1330 LT=1 : IT=1
```

```
1340 PRINT : PRINT "ITERATION PROCESS ..........................................."
1350 REM  < INITIALIZATION >
1360 FOR I=1 TO NEQ : U(I)=1 : W(I)=1 : NEXT I
1370 REM  < RETRIEVE THE STIFFNESS MATRIX >
1380 OPEN F$ FOR INPUT AS #1
1390 FOR I=1 TO NEQ : FOR J=1 TO MB : INPUT#1, SK(I,J) : NEXT J : NEXT I
1400 CLOSE#1
1410 REM  < SHIFT THE MATRIX >
1420 FOR I=1 TO NEQ : FOR J=1 TO MB : SK(I,J)=SK(I,J)-SH*SM(I,J) : NEXT J : NEXT I
1430 REM  < INITIAL SET : V=SM*U >
1440 FOR K=1 TO NEQ : SU=0 : IM=K-1
1450 IF IM>MB-1 THEN IM=MB-1
1460 IF IM<=0 THEN GOTO 1480
1470 FOR I=1 TO IM : KI=K-I : SU=SU+SM(KI,I+1)*U(KI) : NEXT I
1480 IM=NEQ-K+1 : IF IM>MB THEN IM=MB
1490 FOR I=1 TO IM : SU=SU+SM(K,I)*U(K+I-1) : NEXT I
1500 W(K)=SU
1510 NEXT K
1520 REM  < SOLVE  SK*U=W  IN  U  >
1530 IT=1 : RM=0
1540 FOR K=1 TO NEQ : U(K)=W(K) : NEXT K
1550 KK=2 : IF IT=1 THEN KK=0
1560 GOSUB 2350
1570 REM  < ORTHOGONALIZATION >
1580 IF NJ<=0 THEN GOTO 1740
1590 FOR N=1 TO NJ : AL=0
1600 FOR K=1 TO NEQ : SU=0 : IM=K-1
1610 IF IM>MB-1 THEN IM=MB-1
1620 IF IM<=0 THEN GOTO 1640
1630 FOR I=1 TO IM : KI=K-I : SU=SU+SM(KI,I+1)*U(KI) : NEXT I
1640 IM=NEQ-K+1 : IF IM>MB THEN IM=MB
1650 FOR I=1 TO IM : SU=SU+SM(K,I)*U(K+I-1) : NEXT I
1660 AL=AL+EV(K,N)*SU
1670 NEXT K
1680 AF(N)=AL
1690 NEXT N
1700 FOR K=1 TO NEQ : SU=0
1710 FOR N=1 TO NJ : SU=SU+AF(N)*EV(K,N) : NEXT N
1720 U(K)=U(K)-SU
1730 NEXT K
1740 REM  < COMPUTE NUMERATOR : RN >
1750 RN=0
1760 FOR K=1 TO NEQ : RN=RN+W(K)*U(K) : NEXT K
1770 REM  < V=SM*U & COMPUTE DENOMINATOR : RD >
1780 RD=0
1790 FOR K=1 TO NEQ : SU=0 : IM=K-1
1800 IF IM>MB-1 THEN IM=MB-1
1810 IF IM<=0 THEN GOTO 1830
1820 FOR I=1 TO IM : KI=K-I : SU=SU+SM(KI,I+1)*U(KI) : NEXT I
1830 IM=NEQ-K+1 : IF IM>MB THEN IM=MB
1840 FOR I=1 TO IM : SU=SU+SM(K,I)*U(K+I-1) : NEXT I
1850 RD=RD+U(K)*SU : V(K)=SU
1860 NEXT K
1870 REM  < NORMALIZATION : W*SM*W=1 >
1880 IF ABS(RD)<1E-10 THEN RD=1E-10
1890 SA=0 : SB=0 : SR=SQR(RD)
1900 FOR K=1 TO NEQ : UK=V(K)/SR : WK=W(K)
1910 SA=SA+(UK-WK)*(UK-WK) : SB=SB+UK*UK
1920 W(K)=UK
1930 NEXT K
1940 REM  < COMPUTE EIGENVALUE >
1950 RA=RN/RD : IF ABS(RA)<1E-10 THEN RA=1E-10
1960 SC=ABS(RA-RM)/ABS(RA)
1970 RM=RA
1980 REM  < CHECK CONVERGENCE >
1990 IF ABS(SB)<1E-10 THEN SB=1E-10
2000 SB=SQR(SA/SB)
2010 IF SB>SC THEN SC=SB
2020 IF SC<=TR THEN GOTO 2120
2030 IF IT<=IX THEN GOTO 2070
```

```
2040 IF IT>IX THEN INPUT "Do you need more iterations? If no, input zero. IB = ";IB
2050 IF IB<=0 THEN GOTO 2120
2060 IX=IX+IB
2070 IF IT<>LT THEN GOTO 2100
2080 LT=LT+10
2090 PRINT IT;TAB(10);"RM = ";RM;TAB(40);"TOL = ";SC
2100 IT=IT+1
2110 GOTO 1540
2120 REM  < OUTPUT >
2130 PRINT : PRINT "RESULTS OF THE ";NI;" EIGENVALUE ......................"
2140 PRINT : PRINT " EIGENVALUE = ";RM;TAB(40);"TOLERANCE = ";SC
2150 PRINT : PRINT " EIGENVECTOR" : PRINT
2160 FOR K=1 TO NEQ : VK=U(K)/SR : V(K)=VK : EV(K,NI)=VK : NEXT K
2170 FOR K=1 TO NEQ : PRINT K;TAB(10);V(K)  : NEXT K
2180 REM  < STORE EIGENVALUES >
2190 R(NI)=RM+SH
2200 SH=SO+.999*R(NI)
2210 IF SH<=.00001 THEN SH=-.1
2220 NI=NI+1
2230 IF NI<=NG THEN GOTO 1310
2240 REM  < OUTPUT : RESULTS >
2250 PRINT : PRINT :
     PRINT "RESULTS : EIGENVALUES AND EIGENVECTORS ........................." : PRINT
2260 INPUT "Which eigenvalue and eigenvector will be output ?   (1....NG) = ";LC
2270 IF LC<=0 GOTO 2340
2280 IF LC>NG THEN LC=NG
2290 PRINT
2300 PRINT LC;TAB(9);"EIGENVALUE = ";R(LC)  : PRINT : PRINT "EIGENVECTOR" : PRINT
2310 FOR I=1 TO NEQ : PRINT I;TAB(8);EV(I,LC)  : NEXT I
2320 PRINT
2330 GOTO 2260
2340 RETURN
2350 REM  Subroutine Band Method / Symmetric
2360 REM  -----------------------------------------------------------------
2370 REM  < FORWARD ELIMINATION >
2380 N1=NEQ-1
2390 IF KK=2 THEN GOTO 2530
2400 FOR N=1 TO N1
2410 M = N - 1:MR = MB:NM = NEQ-M: IF MR>NM THEN MR=NM
2420 PI = SK(N,1)
2430 FOR L=2 TO MR
2440 CP=SK(N,L)/PI  : IF CP=0 THEN GOTO 2490
2450 I = M + L:J = 0
2460 FOR K=L TO MR : J=J+1
2470 SK(I,J) = SK(I,J) - CP * SK(N,K)
2480 NEXT K
2490 SK(N,L) = CP
2500 NEXT L
2510 NEXT N
2520 IF KK=1 THEN GOTO 2710
2530 REM  < REDUCTION OF THE RIGHT HAND SIDE >
2540 FOR N=1 TO N1
2550 M = N - 1:MR = MB:NM = NEQ-M
2560 IF MR>NM THEN MR=NM
2570 CP=U(N)  : IF CP=0 THEN GOTO 2620
2580 U(N) = CP / SK(N,1)
2590 FOR L=2 TO MR : I=M+L
2600 U(I) = U(I) - SK(N,L) * CP
2610 NEXT L
2620 NEXT N
2630 REM  < BACK SUBSTITUTION >
2640 U(NEQ)= U(NEQ)/ SK(NEQ,1)
2650 FOR I=1 TO N1
2660 N = NEQ- I:M = N - 1:MR = MB:NM = NEQ- M : IF MR>NM THEN MR=NM
2670 FOR K=2 TO MR : L=M+K
2680 U(N) = U(N) - SK(N,K) * U(L)
2690 NEXT K
2700 NEXT I
2710 RETURN
2720 END
```

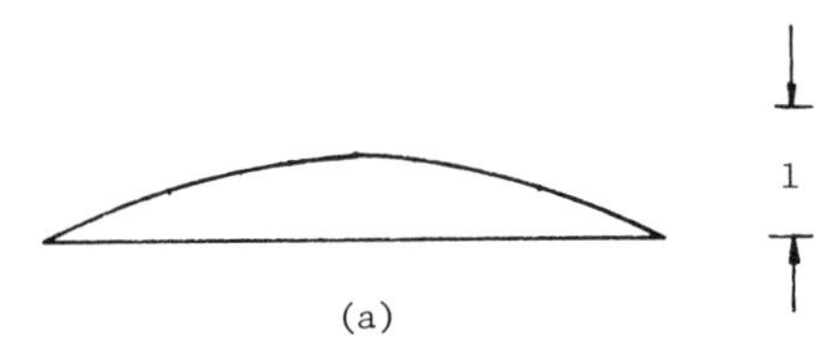

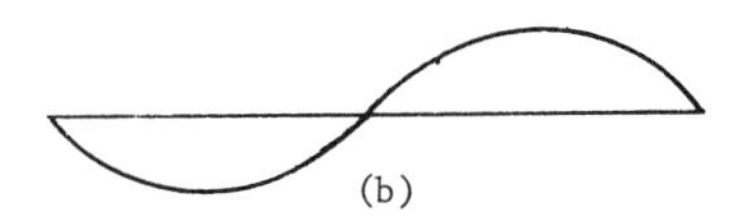

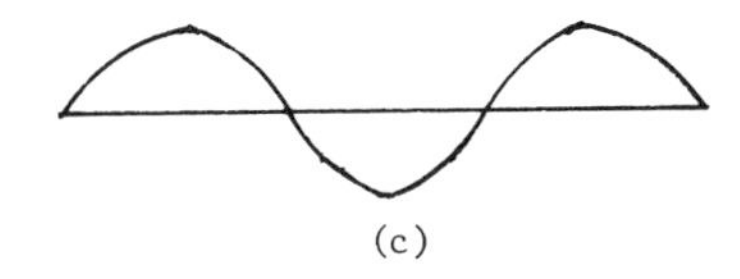

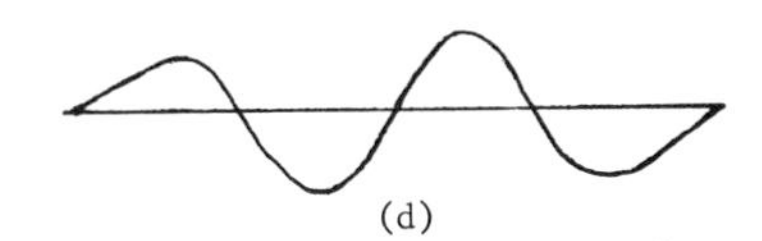

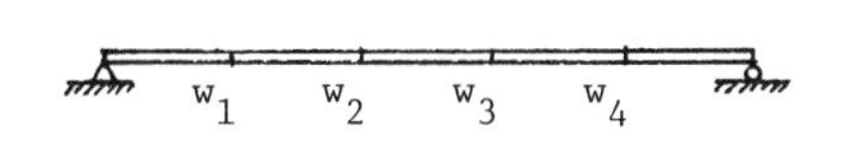

Figure 4.11 Eigenvectors of matrix in example 4.5: (a) $\lambda_1 = 0.1459$, (b) $\lambda_2 = 1.9098$, (c) $\lambda_3 = 6.8541$, (d) $\lambda_4 = 13.0901$.

Example 4.5: We shall apply the program BASIC INVERSE ITERATION to solve the eigenvalue problem

$$\begin{bmatrix} 5 & -4 & 1 & 0 \\ -4 & 6 & -4 & 1 \\ 1 & -4 & 6 & -4 \\ 0 & 1 & -4 & 5 \end{bmatrix} \begin{Bmatrix} w_1 \\ w_2 \\ w_3 \\ w_4 \end{Bmatrix} = \lambda \begin{bmatrix} 1 & 0 & 0 & 0 \\ 0 & 1 & 0 & 0 \\ 0 & 0 & 1 & 0 \\ 0 & 0 & 0 & 1 \end{bmatrix} \begin{Bmatrix} w_1 \\ w_2 \\ w_3 \\ w_4 \end{Bmatrix}$$

Since the stiffness matrix is not singular, the shifting procedure need not be introduced. For the tolerance $TR = 10^{-8}$, the eigenvalues are obtained as

$$\lambda_1 = 0.14589034, \qquad \lambda_2 = 1.90983006$$
$$\lambda_3 = 6.85410197, \qquad \lambda_4 = 13.0901699$$

The corresponding eigenvectors are shown in Figure 4.11.

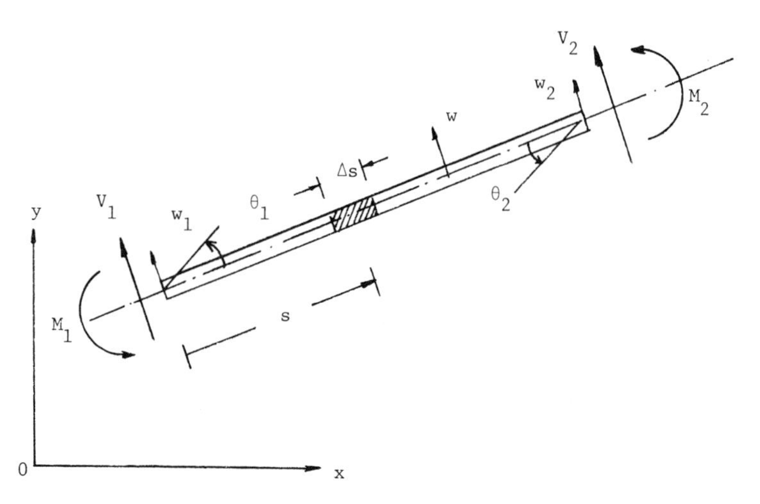

Figure 4.12 Beam-bending element: V is the shear force, M is the bending moment, θ is the gradient of w, and w is the deflection of the axis; subscripts 1 and 2 refer to the "left" and "right" ends, respectively. Not shown is $f(s)$, the distributed normal load.

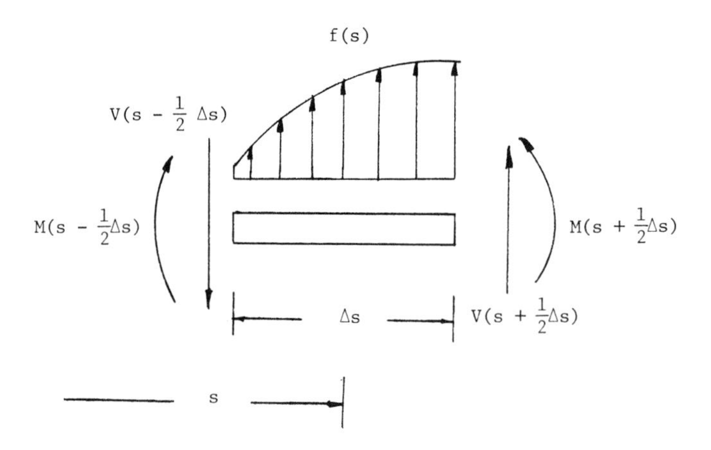

Figure 4.13 Equilibrium of a differential element.

4.3 Beam-bending element

The second simplest structural element is the planar beam-bending element. Let the s axis coincide with the axis of a symmetric beam and let w be the deflection of the beam's axis in the transverse symmetry axis, as shown in Figure 4.12. Bending moments are applied about the axes perpendicular to the plane of bending. Let us now isolate a differential element Δs of the beam and consider its equilibrium as shown in Figure 4.13. Considering equilibrium yields

$$V(s + \tfrac{1}{2}\,\Delta s) - V(s - \tfrac{1}{2}\,\Delta s) + f(\bar{s})\,\Delta s = 0$$

and

$$\begin{aligned} M(s + \tfrac{1}{2}\,\Delta s) - M(s - \tfrac{1}{2}\,\Delta s) + \tfrac{1}{2}\,\Delta s\; V(s - \tfrac{1}{2}\,\Delta s) + \tfrac{1}{2}\,\Delta s\; V(s + \tfrac{1}{2}\,\Delta s) \\ - [f(\bar{s})\,\Delta s](s - \bar{s}) = 0 \end{aligned}$$

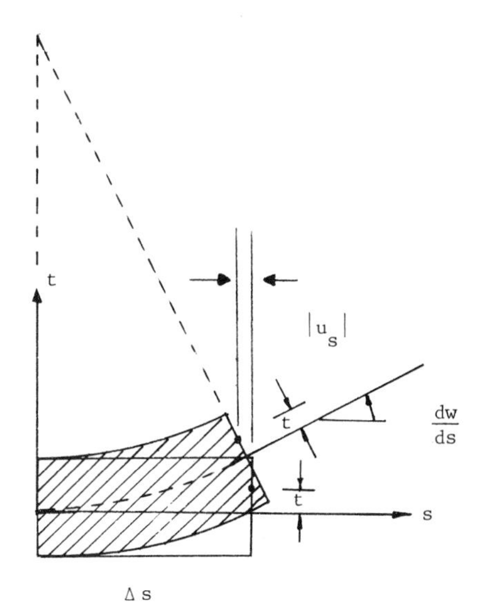

Figure 4.14 Beam-bending element.

where

$$s - \tfrac{1}{2}\Delta s < \bar{s} < s + \tfrac{1}{2}\Delta s$$

Divide by Δs and let $\Delta s \to 0$ to get

$$\frac{dV}{ds} + f = 0 \qquad \text{and} \qquad \frac{dM}{ds} + V = 0 \tag{4.51}$$

Combining these two equations, we have the equilibrium equation

$$\frac{d^2M}{ds^2} = f \tag{4.52}$$

for the bending moment.

The bending moment M in (4.52) is the resultant of the distribution of the stress σ_s on the cross section of the beam:

$$M = -\int_A \sigma_s t \, dA \tag{4.53}$$

where t is perpendicular to the beam axis s and the deflection of the beam occurs along the t axis. Here A is the cross section of the beam.

Suppose that the displacement u_s along the beam is *approximated* by

$$u_s = -\frac{dw}{ds} t \tag{4.54}$$

as shown in Figure 4.14. Thus, there is no deformation on the beam axis, and planes normal to the beam axis before deformation remain normal after deformation. Then the stress σ_s on the cross section is given by

$$\sigma_s = E\varepsilon_s = E\frac{du_s}{ds} = -E\frac{d^2w}{ds^2} t \tag{4.55}$$

Here the effect of shear has been neglected, and linearity of the material is assumed. Substituting (4.55) into (4.53) yields

$$M = \left(\int_A Et^2 \, dA \right) \frac{d^2 w}{ds^2}$$

If Young's modulus E is constant across the cross section, we have

$$M = E \frac{d^2 w}{ds^2} \int_A t^2 \, dA$$

Introducing the moment of inertia,

$$I = \int_A t^2 \, dA \tag{4.56}$$

we have the moment-curvature equation for the beam,

$$M = EI \frac{d^2 w}{ds^2} \tag{4.57}$$

Combining (4.52) and (4.57), we obtain the beam equation

$$\frac{d^2}{ds^2} EI \frac{d^2 w}{ds^2} = f \tag{4.58}$$

in terms of transverse deflection w of the beam's axis.

We have not as yet considered inertial effects. However, these can be included in the applied force term f. That is, we can make use of d'Alembert's principle and replace f by*

$$f - \rho_s \ddot{w} \tag{4.59}$$

where ρ_s is the weight of the beam per unit length, that is,

$$\rho_s = \int_A \rho \, dA \tag{4.60}$$

Thus, (4.58) is modified to

$$\rho_s \ddot{w} + \frac{\partial^2}{\partial s^2} EI \frac{\partial^2 w}{\partial s^2} = f \tag{4.61}$$

Note that d/ds must be changed to $\partial/\partial s$ because w now depends on the time $\hat{t}$ in additional to its dependence on s.

Now the weak form of the beam element shown in Figure 4.12 is obtained from (4.61). Multiplying by an arbitrary virtual deflection $\bar{w}$ on both sides of (4.61) and integrating over the beam element, we have

$$\int_0^L \left(\rho_s \ddot{w} + \frac{\partial^2}{\partial s^2} EI \frac{\partial^2 w}{\partial s^2} \right) \bar{w} \, ds = \int_0^L f \bar{w} \, ds$$

* Inertial effects due to the rotation of the beam elements (rotary inertia) are neglected here.

where L is the length of the beam. Two integrations by parts gives

$$\int_0^L \rho_s \ddot{w}\bar{w}\,ds + \int_0^L EI\frac{\partial^2 w}{\partial s^2}\frac{\partial^2 \bar{w}}{\partial s^2}\,ds$$
$$+\left[\frac{\partial}{\partial s}EI\frac{\partial^2 w}{\partial s^2}\bar{w}\right]_0^L - \left[EI\frac{\partial^2 w}{\partial s^2}\frac{\partial \bar{w}}{\partial s}\right]_0^L = \int_0^L f\bar{w}\,ds$$

In the beam element of Figure 4.11, we have

$$-V_1 = -\frac{\partial}{\partial s}EI\frac{\partial^2 w}{\partial s^2}; \qquad -M_1 = EI\frac{\partial^2 w}{\partial s^2}$$
$$V_2 = -\frac{\partial}{\partial s}EI\frac{\partial^2 w}{\partial s^2}; \qquad M_2 = EI\frac{\partial^2 w}{\partial s^2}$$

Using these, at $s = 0$ and $s = L$, yields

$$\int_0^L \rho_s \ddot{w}\bar{w}\,ds + \int_0^L EI\frac{\partial^2 \bar{w}}{\partial s^2}\frac{\partial^2 w}{\partial s^2}\,ds$$
$$= V_1\bar{w}_1 + V_2\bar{w}_2 + M_1\bar{\theta}_1 + M_2\bar{\theta}_2 + \int_0^L f\bar{w}\,ds \tag{4.62}$$

for every virtual deflection $\bar{w}$. Suppose that the actual deflection w and the virtual deflection $\bar{w}$ are approximated by the cubic polynomials:

$$\begin{aligned} w(s, \hat{t}) &= a_0 + a_1 s + a_2 s^2 + a_3 s^3 \\ \bar{w}(s, \hat{t}) &= \bar{a}_0 + \bar{a}_1 s + \bar{a}_3 s^2 + \bar{a}_3 s^3 \end{aligned} \tag{4.63}$$

where a_i and $\bar{a}_i$, $i = 0, 1, 2, 3$, are functions of time $\hat{t}$. These polynomial forms for w and $\bar{w}$ can be expressed in terms of values of w, $\bar{w}$ and their gradients, that is, $\theta = w'$, $\bar{\theta} = \bar{w}'$, at the end points $s = 0, L$ by

$$\begin{aligned} w(s, \hat{t}) &= \{N_1(\xi), N_2(\xi), N_3(\xi), N_4(\xi)\}\begin{Bmatrix} w_1 \\ \theta_1 \\ w_2 \\ \theta_2 \end{Bmatrix} \\ \bar{w}(s, \hat{t}) &= \{N_1(\xi), N_2(\xi), N_3(\xi), N_4(\xi)\}\begin{Bmatrix} \bar{w}_1 \\ \bar{\theta}_1 \\ w_2 \\ \bar{\theta}_2 \end{Bmatrix} \end{aligned} \tag{4.64}$$

in which $w_1 = w_1(\hat{t})$ means $w(0, \hat{t}), \ldots,$ and

$$\begin{aligned} N_1(\xi) &= 1 - 3\xi^2 + 2\xi^3 \\ N_2(\xi) &= L\xi(\xi - 1)^2 \\ N_3(\xi) &= 3\xi^2 - 2\xi^3 \\ N_4(\xi) &= L\xi(\xi^2 - \xi) \end{aligned} \tag{4.65}$$

with $\xi = s/L$. If we assume that ρ_s and EI are constants, then substitution of (4.64) into (4.62) yields the equation

$$\{\bar{w}_1, \bar{\theta}_1, \bar{w}_2, \bar{\theta}_2\}\left[\frac{\rho_s L}{420}\begin{bmatrix} 156 & 22L & 54 & -13L \\ & 4L^2 & 13L & -3L^2 \\ & & 156 & -22L \\ \text{SYM} & & & 4L^2 \end{bmatrix}\begin{Bmatrix} \ddot{w}_1 \\ \ddot{\theta}_1 \\ \ddot{w}_2 \\ \ddot{\theta}_2 \end{Bmatrix}\right.$$

$$\left. + \frac{EI}{L^3}\begin{bmatrix} 12 & 6L & -12 & 6L \\ & 4L^2 & -6L & 2L^2 \\ & & 12 & -6L \\ \text{SYM} & & & 4L^2 \end{bmatrix}\begin{Bmatrix} w_1 \\ \theta_1 \\ w_2 \\ \theta_2 \end{Bmatrix}\right]$$

$$= \{\bar{w}_1, \bar{\theta}_1, \bar{w}_2, \bar{\theta}_2\}\left[\begin{Bmatrix} V_1 \\ M_1 \\ V_2 \\ M_2 \end{Bmatrix} + \begin{Bmatrix} f_1 \\ f_2 \\ f_3 \\ f_4 \end{Bmatrix}\right] \tag{4.66}$$

where

$$f_i = \int_0^L f N_i(s)\, ds, \quad i = 1, 2, 3, 4 \tag{4.67}$$

Since the virtual deflection $\bar{w}$ is arbitrary, it follows from (4.66) that

$$\frac{\rho_s L}{420}\begin{bmatrix} 156 & 22L & 54 & -13L \\ & 4L^2 & 13L & -3L^2 \\ & & 156 & -22L \\ \text{SYM} & & & 4L^2 \end{bmatrix}\begin{Bmatrix} \ddot{w}_1 \\ \ddot{\theta}_1 \\ \ddot{w}_2 \\ \ddot{\theta}_2 \end{Bmatrix}$$

$$+ \frac{EI}{L^3}\begin{bmatrix} 12 & 6L & -12 & 6L \\ & 4L^2 & -6L & 2L^2 \\ & & 12 & -6L \\ \text{SYM} & & & 4L^2 \end{bmatrix}\begin{Bmatrix} w_1 \\ \theta_1 \\ w_2 \\ \theta_2 \end{Bmatrix} = \begin{Bmatrix} V_1 \\ M_1 \\ V_2 \\ M_2 \end{Bmatrix}^* + \begin{Bmatrix} f_1 \\ f_2 \\ f_3 \\ f_4 \end{Bmatrix} \tag{4.68}$$

We shall symbolically denote this finite element equation for a beam element as

$$M^e_{ab}\ddot{W}_b + K^e_{ab}W_b = f^e_a \tag{4.69}$$

The matrices $[M^e_{ab}]$ and $[K^e_{ab}]$ are, respectively, the element mass and stiffness matrices, and $\{f^e_a\}$ is the element generalized load vector. If a structure consists of several beam-type members, assembling of the element mass and stiffness matrices is required as was the case for a truss structure.

It is noted that the deflection w is in the thickness direction. Thus, in order to consider a planar structure, the components u and v of the deflection with respect to a fixed coordinate system (x, y) might be convenient. To do this, we shall use the transformation rules

$$\begin{aligned} w &= -\sin\theta u + \cos\theta v \\ V_x &= -V\sin\theta, \qquad V_y = V\cos\theta \end{aligned} \tag{4.70}$$

* This vector is, in general, canceled with that of the adjacent beam element after assembling of all element stiffness matrices and generalized load vectors, unless externally applied point forces or moments exist at end points of beam elements.

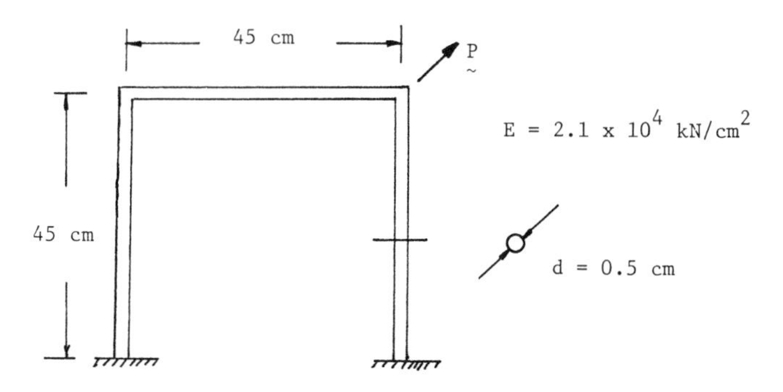

Figure 4.15 Frame structure.

Then the 4 × 4 matrices $[M^e_{ab}]$ and $[K^e_{ab}]$ will be transformed to 6 × 6 matrices $[\bar{M}^e_{pq}]$ and $[\bar{K}^e_{pq}]$ by the transformation matrix $[T_{ap}]$ defined by

$$[T_{ap}] = \begin{bmatrix} -\sin\theta & \cos\theta & 0 & 0 & 0 & 0 \\ 0 & 0 & 1 & 0 & 0 & 0 \\ 0 & 0 & 0 & -\sin\theta & \cos\theta & 0 \\ 0 & 0 & 0 & 0 & 0 & 1 \end{bmatrix} \tag{4.71}$$

Indeed, we have

$$\bar{M}^e_{pq} = T_{ap} M^e_{ab} T_{bq} \quad \text{and} \quad \bar{K}^e_{pq} = T_{ap} K^e_{ab} T_{bq} \tag{4.72}$$

Similarly, the load vector $\{f^e_a\}$ is transformed to $\{\bar{f}^e_p\}$ by

$$\bar{f}^e_p = T_{ap} f^e_a \tag{4.73}$$

Therefore, there are three degrees of freedom (u, v, θ) at each nodal point (i.e., at the end of each beam element). Assembling the individual element equations:

$$\bar{M}^e_{pq}\ddot{U}_q + \bar{K}^e_{pq}U_q = \bar{f}^e_p \tag{4.74}$$

to the global one, we have to consider the fact that each node has three degrees of freedom (u, v, θ).

Exercise 4.3: Obtain the element stiffness and mass matrices when both axial forces and bending moments are present by superimposing the matrices obtained in (4.13) and (4.72). In this case, the matrices become 6 × 6 as in the bending problem. Using these, find the deformation of the structure shown in Figure 4.15 for the case $\mathbf{P} = 10\text{ kN }\mathbf{i} + 400\text{ kN }\mathbf{j}$.

A final remark in this section concerns the boundary conditions. While the essential (or first type of) boundary condition was $T = g$ for heat conduction problems, this type consists of w and its derivative $\theta = w'$ for the beam-bending problem. To see the reason for this, let us compare the temperature in the heat conduction problem with the displacement in the elasticity problem. Since in beam theory a prescribed slope leads to an axial displacement [see equation (4.54)], for the beam element, the essential boundary condition requires the prescription of both deflection and slopes.

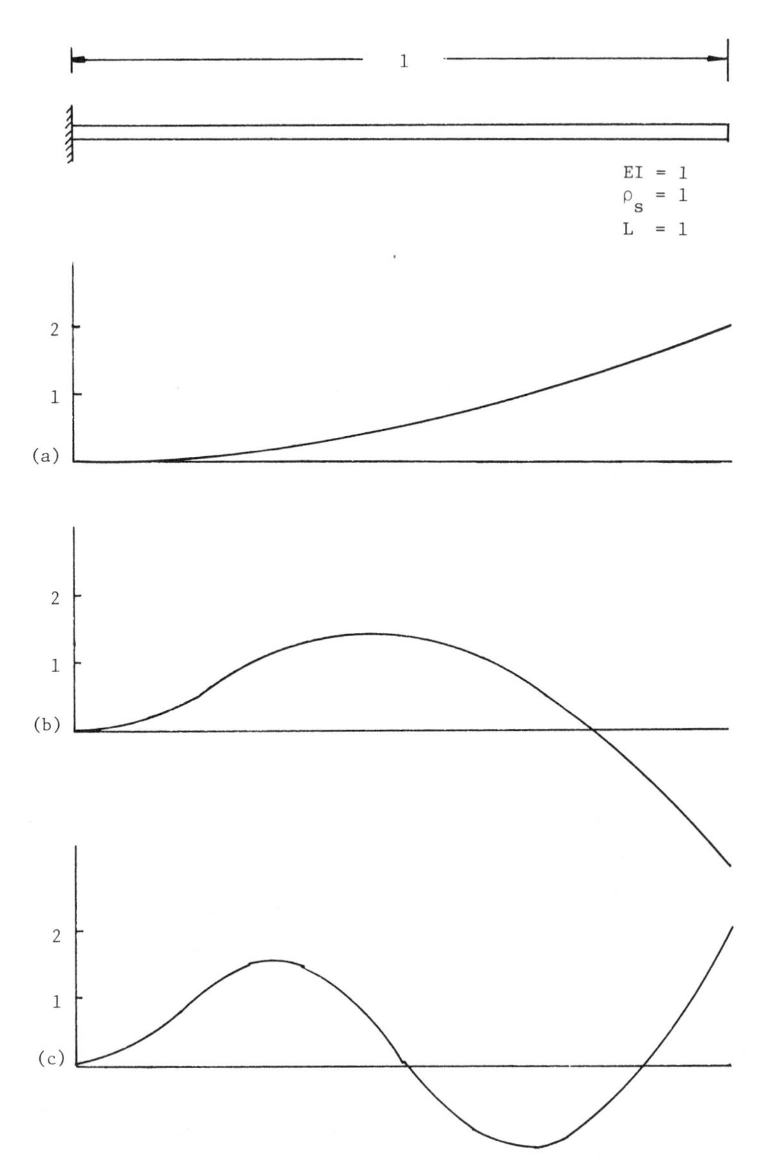

Figure 4.16 First three eigenvectors: (a) $\lambda_1 = (1.875)^4$, (b) $\lambda_2 = (4.697)^4$, (c) $\lambda_3 = (7.885)^4$.

Example 4.6: Let us consider a free vibration problem for the cantilever beam shown in Figure 4.16 under the condition, numerically, $EI = \rho_s L^4$, where L is the length of the cantilever. For simplicity, suppose that $EI = 1$ without specifying the units of the quantities. In this case, the first three eigenvalues are analytically obtained as

$$\lambda_1 = (1.875)^4, \qquad \lambda_2 = (4.694)^4, \qquad \text{and} \qquad \lambda_3 = (7.855)^4$$

If the program BASIC INVERSE ITERATION is applied to solve the eigenvalue problem deduced from the free vibration of the cantilever,

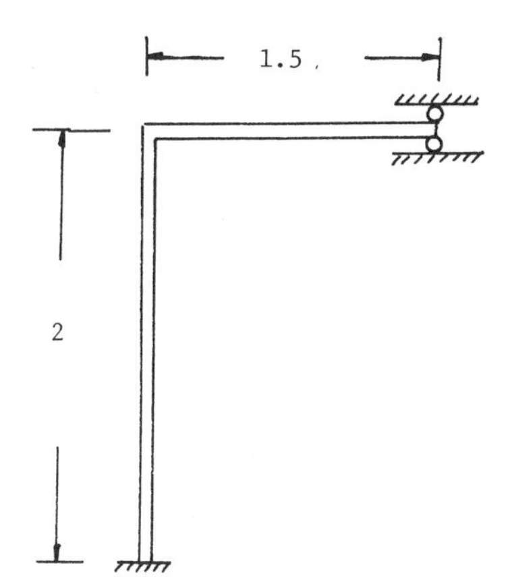

Figure 4.17 Frame structure for free vibration problem.

the first three eigenvalues are computed as

$$\lambda_1 = (1.875)^4, \qquad \lambda_2 = (4.697)^4, \qquad \text{and} \qquad \lambda_3 = (7.885)^4$$

and plots of the corresponding eigenfunctions are as shown in Figure 4.16.

Exercise 4.4: Consider the free vibration problem for the frame structure shown in Figure 4.17. For simplicity, assume that the numerical values of EI, ρ_s, and EA are all unity.

Example 4.7: Let us design the thickness $h(x)$ of a simply supported beam maintaining constant width so that the minimum eigenvalue(s) of the free vibration problem could be the maximum among all possible designs (i.e., thickness function h) under the volume constraint

$$V_0 = \int_0^L bh(x)\, dx$$

For simplicity, we shall assume that E and ρ are constants. Note that the minimum eigenvalue(s) can be obtained by the inverse iteration method and satisfies the relation

$$\lambda = \int_0^L EI(w'')^2\, dx \bigg/ \int_0^L \rho A w^2\, dx$$

where w is the corresponding eigenvector. Using this, the problem can be formally formulated by the maximization problem

$$\max_h \lambda \qquad \text{subject to} \quad V_0 = \int_0^L bh\, dx$$

Applying Lagrange multiplier methods, we have the extremum problem for the functional

$$\psi(h, \mu) = \int_0^L \tfrac{1}{12} Ebh^3 w''^2\, dx \bigg/ \int_0^L \rho bhw^2\, dx - \mu\left(\int_0^L bh\, dx - V_0\right)$$

In other words, we must obtain a stationary point (h, μ) of the functional ψ that is a function of the design variable h and the Lagrange multiplier

$\mu \in \mathbb{R}$ to the volume constraint. As in section 1.4, the stationary condition is obtained by taking the "first variation" of the functional ψ. Defining two constants a and B by

$$a = \int_0^L \tfrac{1}{12} Ebh^3 w''^2 dx \qquad \text{and} \qquad B = \int_0^L \rho bhw^2\, dx$$

the stationary condition is written by

$$\int_0^L \left(\frac{1}{4B} Ebh^2 w''^2 - \frac{a}{B^2} \rho bw^2 - \mu b \right) \bar{h}\, dx = 0, \quad \forall \bar{h}$$

$$\bar{\mu} \left(\int_0^L bh\, dx - V_0 \right) = 0, \quad \forall \bar{\mu}$$

that is,

$$\frac{1}{4B} Eh^2 w''^2 - \frac{a}{B^2} \rho w^2 = \mu, \qquad \int_0^L bh\, dx = V_0$$

If the eigenvector w is normalized by

$$\int_0^L \rho bhw^2\, dx = 1$$

then the stationary condition becomes

$$\tfrac{1}{4} Eh^2 w''^2 - a\rho w^2 = \mu, \qquad \int_0^L bh\, dx = V_0$$

This means that the optimal design h has to provide the eigenvector, which yields the quantity

$$Q = \tfrac{1}{4} Eh^2 w''^2 - a\rho w^2$$

being constant at every point of the beam. We may use this condition to determine the optimum thickness h_e.

Now we shall approximate the beam with variable thickness by a stepped beam, the thickness of which is piecewise constant. Suppose that N_e steps are assumed in this example. If we identify each step of the beam as a beam element, the original beam is modeled by N_e beam elements studied above. A similar approximation will be extended to the stationary condition as follows:

$$Q_e = \tfrac{1}{4} E_e h_e^2 (w''^2)_e - \lambda \rho_e (w^2)_e = \mu$$

$$\sum_{e=1}^{N_e} h_e = \frac{V_0}{b}$$

where λ is the minimum eigenvalue(s) in the above approximation, and $(\phi)_e$ means the quantity ϕ evaluated at the centroid of the eth beam element. In order to find an h_e, $e = 1, \ldots, N_e$, that satisfies the stationary condition, the following five-step iteration procedure is considered:

1. Choose h_e^0, $e = 1, \ldots, N_e$, so that

$$\sum_{e=1}^{N_e} h_e^0 = \frac{V_0}{b}$$

is satisfied.

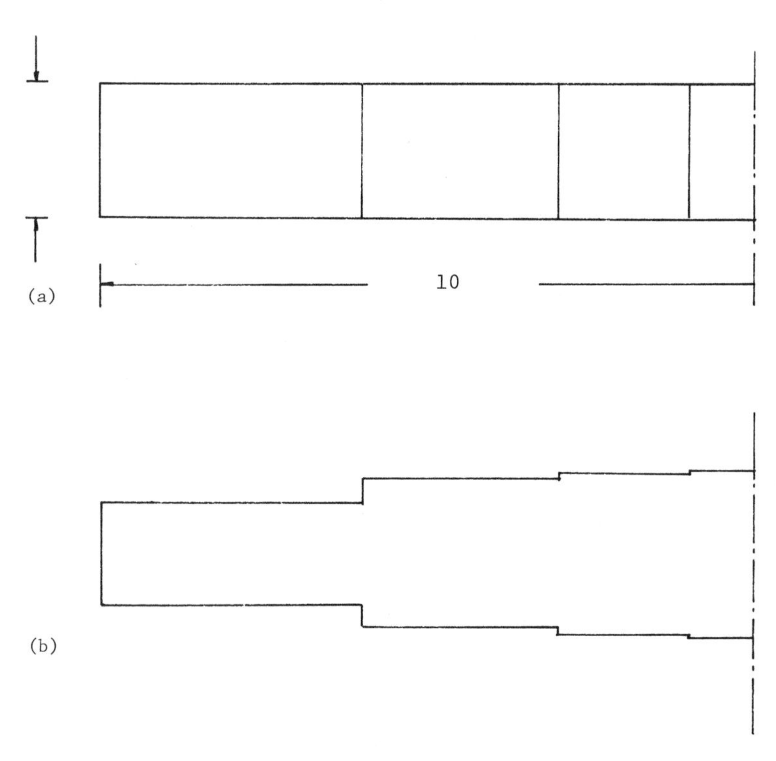

Figure 4.18 (a) Initial and (b) optimal design of a simply supported beam.

2. For $i = 1, 2, \ldots,$ solve the eigenvalue problem for h_e^{i-1}, $e = 1, \ldots, N_e$. That is, prescribing the height of the stepped beam by h_e^{i-1}, obtain the minimum eigenvalue(s) λ^i and eigenvector(s) w^i.
3. For each e, $e = 1, \ldots, N_e$, compute

$$Q_e^i = \tfrac{1}{4}E_e(h_e^{i-1})^2(w^{i''^2})_e - \lambda^i \rho_e (w^{i^2})_e$$

and taking the average

$$\bar{Q}^i = \sum_{e=1}^{N_e} \frac{Q_e^i}{N_e}$$

4. Modify the height h_e, $e = 1, \ldots, N_e$, using Q_e^i:

$$h_e^i = h_e^{i-1} - C(Q_e^i - \bar{Q}^i)$$

where C is a magnification factor chosen by numerical experiments.
5. Check the maximum deviation in Q_e^i, $e = 1, \ldots, N_e$, defined by

$$T_0 = \max_e |Q_e^i - \bar{Q}^i|$$

If T_0 is small enough, then we assume stationarity is obtained. If otherwise, repeat steps 2–5 until T_0 becomes small enough.

The result shown in Figure 4.18 is the final design of the thickness of the simply supported beam after nine iterations from the initial uniform thickness. In this example, we have assumed

$$E = \rho = b = 1, \qquad L = 10, \qquad \text{and} \qquad V_0 = 10$$

Table 4.3 *First two eigenvalues at each iteration step*

Iteration	Tolerance	$\lambda_1\ (\times 10^{-4})$	λ_2
1	0.84413	8.1178	0.0130083
2	0.71410	8.8148	0.0129822
3	0.54629	9.0633	0.01255
4	0.36915	9.2301	0.012159
5	0.21663	9.2414	0.011732
6	0.11411	9.2302	0.011455
7	0.05534	9.2088	0.011288
8	0.02586	9.1978	0.011206
9	0.01176	9.1919	0.011166

Table 4.3 shows the change in the first two eigenvalues as the design changes in the iterative scheme described above. The iteration factor C is chosen as $C = 0.1h_0/\bar{Q}$, where h_0 is the initial thickness of the beam and $\bar{Q}$ is the average value of Q_e defined in each element. As the design reaches the optimum, the first eigenvalue λ_1 increases as much as 13% from the initial and the second eigenvalue λ_2 gradually decreases.

For cantilevered beams and beams with fixed ends, it might be necessary to add a constraint on the design

$$\underline{h} \leq h(x) \leq \bar{h}$$

where $\underline{h}$ and $\bar{h}$ are the lower and upper bound of the design allowable, respectively. Without this, we may end up with very unrealistic results (and possible violation of the underlying mechanics).

Example 4.8: If the effect of the axial force P is taken into account in the beam-bending problem, equation (4.51) has to be changed by

$$\frac{dV}{ds} + \frac{d}{ds} T\theta + f = 0, \qquad \frac{dM}{ds} + V = 0$$

where T is the axial force and θ is the gradient of the deflection w. Positive sign of T means tension. Substitution of

$$M = EI\frac{d^2w}{ds^2} \quad \text{and} \quad \theta = \frac{dw}{ds}$$

yields the equilibrium equation for stationary problems,

$$\frac{d^2}{ds^2} EI \frac{d^2w}{ds^2} - \frac{d}{ds} T \frac{dw}{ds} = f$$

If there are no distributed, transverse loads acting on the beam, except the constant axial force T at the beam ends, instability problems of the beam column can be considered using the above equilibrium equation.

Indeed,

$$\frac{d^2}{ds^2} EI \frac{d^2 w}{ds^2} = T \frac{d^2 w}{ds^2}$$

forms on eigenvalue problem, and the eigenvalue $\lambda = -T$ becomes a critical load to the beam whose stiffness is EI. As in example 4.7, eigenvalues are characterized by the Rayleigh quotient

$$\lambda = \int_0^L EI(w'')^2 \, dx \Big/ \int_0^L (w')^2 \, dx$$

where $\phi' = d\phi/dx$. Note that if the cross section of the beam is rectangular, the moment of inertia, I, is written by $I = bh^3/12$, where b is the width and h is the height of the beam.

Under the *isoperimetric* volume constraint

$$\int_0^L bh(x) \, dx \leq V_0$$

an optimal design problem could be defined by the maximization problem

$$\max_h \int_0^L \frac{Ebh^3}{12} (w'')^2 \, dx \Big/ \int_0^L (w')^2 \, dx$$

Introducing a Lagrange multiplier, $\mu \in \mathbb{R}$, the Lagrangian $\psi(h, \mu)$ is defined as

$$\psi(h, \mu) = \int_0^L \frac{Ebh^3}{12} (w'')^2 \, dx \Big/ \int_0^L (w')^2 \, dx + \mu\left(\int_0^L bh \, dx - V_0\right)$$

where $\mu \leq 0$ and w is the eigenvector for the minimum eigenvalue(s). Taking the first variation for ψ with respect to h and μ, we have the following stationary condition:

$$\int_0^L \left[\frac{Ebh^2}{4} (w'')^2 + \mu b\right] \bar{h} \, dx = 0, \quad \forall \bar{h}$$

$$\mu \leq 0, \quad \mu\left(\int_0^L bh \, dx - V_0\right) = 0, \quad \int_0^L bh \, dx \leq V_0$$

that is,

$$\tfrac{1}{4} Ebh^2 (w'')^2 = -\mu = \text{const}$$

$$\int_0^L bh \, dx = V_0, \quad \mu < 0$$

Using a similar iteration algorithm given in example 4.7, an optimal height $h(x)$ could be obtained. As an example, we consider a step beam with the fixed condition of both ends such that $E = 1$, $b = 1$, and $V_0 = 1$. Assuming 20 steps for the beam, the design shown in Figure 4.19 is obtained from the uniform beam. Improvement by this design change is 32% in the first eigenvalue.

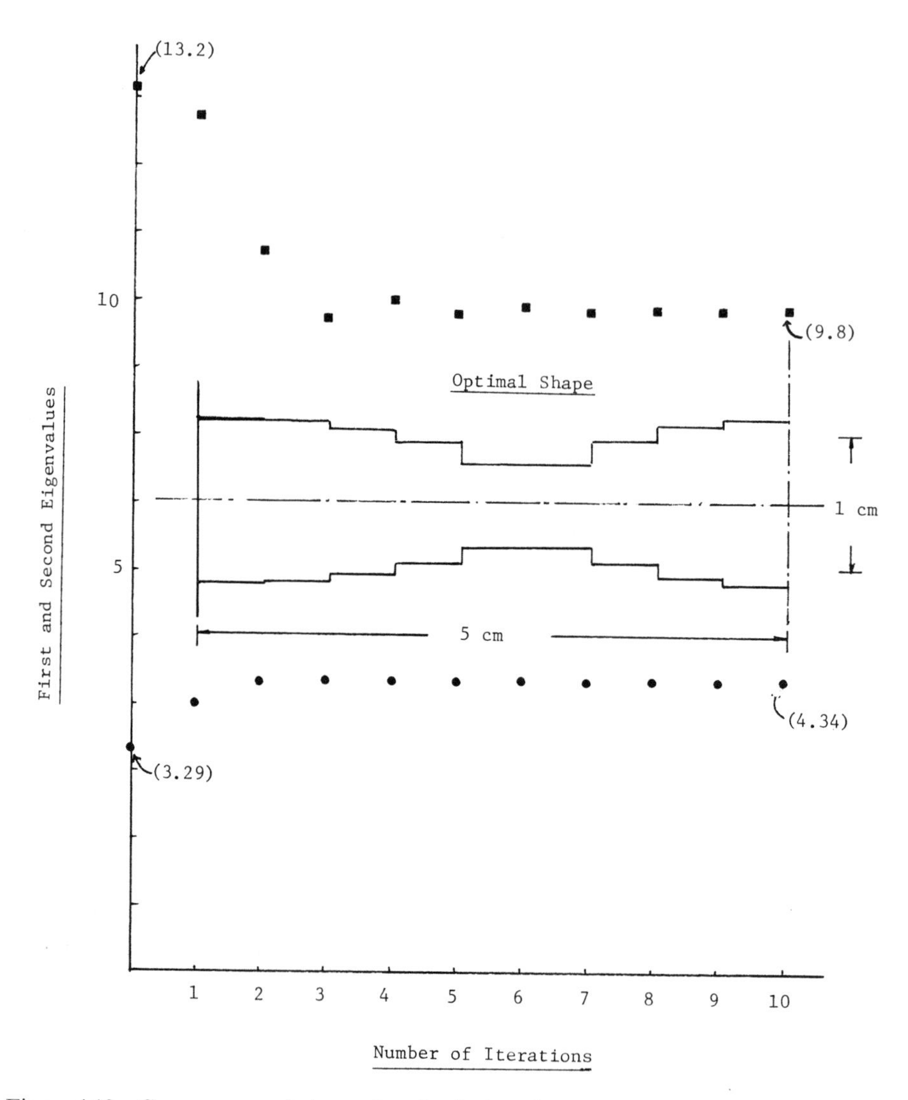

Figure 4.19 Convergence of eigenvalues by design change and the optimum height: ●, first eigenvalue; ■, second eigenvalue.

Exercise 4.5: For the same design problem in example 4.8, obtain a method to consider the restriction on the design of the radius r of the beam cross section by the consideration of axial and shear stresses. More precisely, find the optimal design under the following additional constraints:

$$\sigma_A = \lambda/A \leq \sigma_Y, \qquad \tau = T/J \leq \tau_Y$$

where λ is the minimum eigenvalue (i.e., critical load), A the area of the cross section, T the applied torque (a known quantity), J the polar moment of inertia, σ_Y the yield stress in compression, and τ_Y the yield stress in shear.

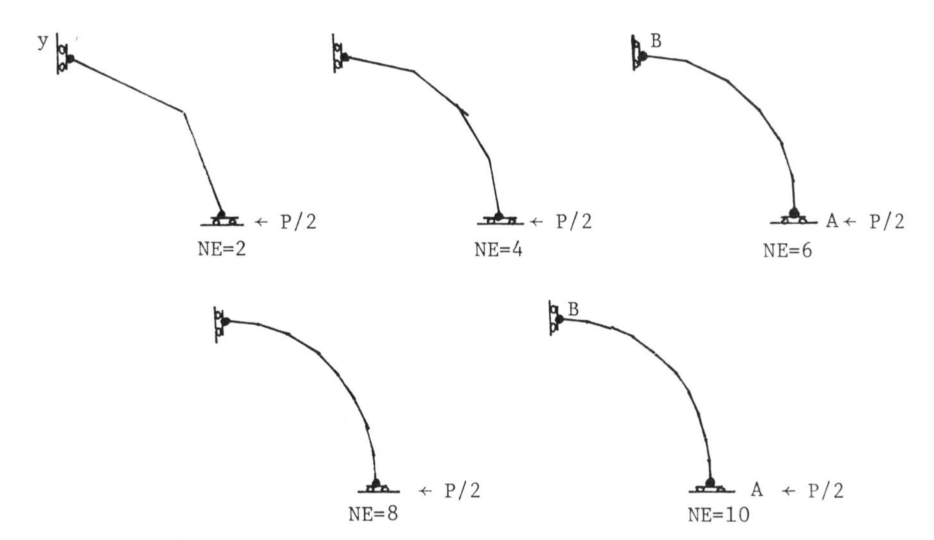

Figure 4.20 Finite element models for a ring problem, $R = 10$ cm (NE is number of elements).

Exercise 4.6: If a beam column is standing vertically in the gravity field, the effect of weight must be taken into account in the consideration of buckling. Furthermore, suppose that an axial force T is applied at the top end of the beam column; the buckling problem is governed by the differential equation

$$\frac{d^2}{ds^2} EI \frac{d^2w}{ds^2} - \frac{d}{ds}\left\{[T - w_0(L - s)] \frac{dw}{ds}\right\} = 0$$

in $(0, L)$, where w_0 is the weight of the beam column per unit length; and the origin of the s coordinate is the bottom end of the beam column, which is fixed on the ground. If w_0 is small enough, the buckling problem can be defined as finding $T < 0$ and w such that

$$\frac{d^2}{ds^2} EI \frac{d^2w}{ds^2} + \frac{d}{ds}\left[w_0(L - s)\frac{dw}{ds}\right] = T\frac{d^2w}{ds^2}$$

in $(0, L)$. Solve this problem in a manner similar to example 4.8.

Example 4.9: We shall check the convergence property of plane frame elements involving bending and axial deformation using a circular ring loaded by a point force P (shown in Figure 4.20). It has a radius R, Young's modulus E, and a constant moment of inertia I of the cross section.

Using symmetry of the problem, we need analyze only a quarter part of a circular ring. For a convergence test, we shall solve the problem of 2, 4, 6, 8, and 10 elements. The result shown in Table 4.4 is the deflection of the ring at points A and B for $E = 2.1 \times 10^4$ kN/cm^2, $I = \pi/64$ cm^4, $R = 10$ cm, and $P = 10$ kN. The rate of convergence is shown

Table 4.4 *Comparison of deflections*

Number of elements	d_A	d_B	Error
2	−0.6543	0.5953	0.1048
4	−0.7087	0.6452	0.0304
6	−0.7208	0.7563	0.0138
8	−0.7252	0.6603	0.0078
10	−0.7272	0.6622	0.0050
Exact	−0.7309	0.6655	—

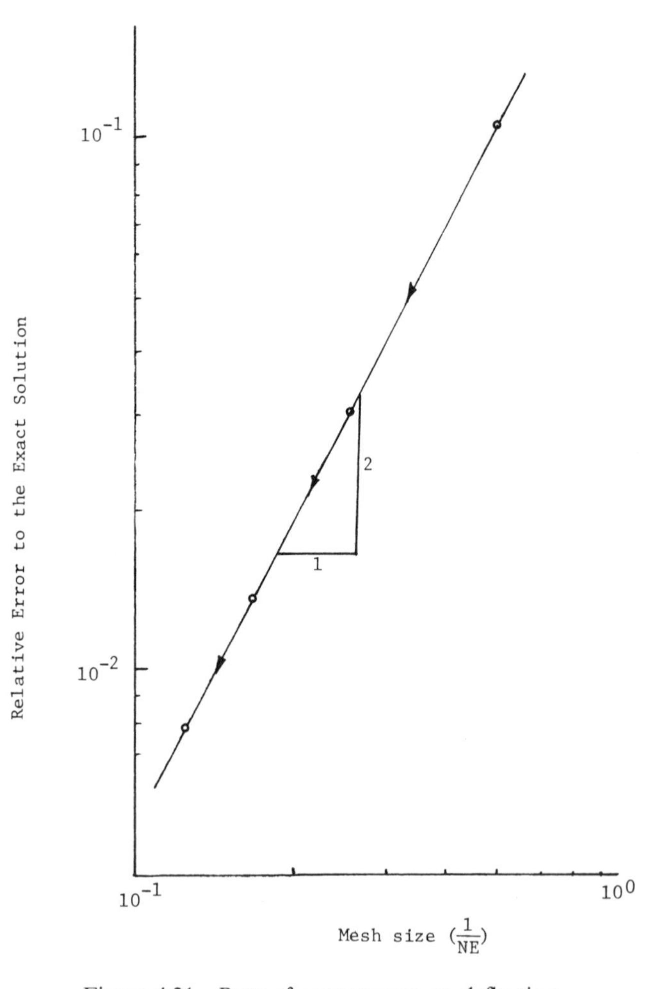

Figure 4.21 Rate of convergence on deflection.

in Figure 4.21 for the deflection. Since the present problem involves bending and axial deformations, the rate of convergence is dominated by the piecewise linear approximation of the axial deformation instead of the piecewise cubic approximation of the transverse deflection of a beam element. Thus, the rate $O(h^2)$ obtained by numerical experiments coincides to the analytical one obtained by error analysis of finite element methods (see, e.g., Ciarlet, 1977).

4.4 Direct integration methods for vibration problems

In this section, we shall briefly review some of the direct integration methods used to solve vibration problems that are governed by the equation

$$\mathbf{M}\ddot{\mathbf{u}} + \mathbf{C}\dot{\mathbf{u}} + \mathbf{K}\mathbf{u} = \mathbf{f} \tag{4.75}$$

in which **M**, **C**, and **K** are, respectively, the mass, viscous damping, and stiffness matrices and **f** is the generalized load vector.

One of the most popular methods is the *Newmark β method,* which is based on the approximation

$$\begin{aligned} \dot{\mathbf{u}}_{k+1} &= \dot{\mathbf{u}}_k + \Delta t\,[(1-\theta)\ddot{\mathbf{u}}_k + \theta\ddot{\mathbf{u}}_{k+1}] \\ \mathbf{u}_{k+1} &= \mathbf{u}_k + \Delta t\,\dot{\mathbf{u}}_k + \tfrac{1}{2}\,\Delta t^2\,\ddot{\mathbf{u}}_k + \beta\,\Delta t^2\,(\ddot{\mathbf{u}}_{k+1} - \ddot{\mathbf{u}}_k) \end{aligned} \tag{4.76}$$

where $\mathbf{u}_k$ means the "value" of the function $\mathbf{u}$ at the kth step in time. If θ and β are taken to be zero in (4.76), then we have an *explicit scheme* for time integration. Substitution of (4.76) into the equation of motion (4.75) at the $k+1$ step yields

$$\begin{aligned} &(\mathbf{M} + \theta\,\Delta t\,\mathbf{C} + \beta\,\Delta t^2\,\mathbf{K})\ddot{\mathbf{u}}_{k+1} \\ &\qquad = \mathbf{f} - \mathbf{K}\mathbf{u}_k - (\mathbf{C} + \Delta t\,\mathbf{K})\dot{\mathbf{u}}_k - [(1-\theta)\,\Delta t\,\mathbf{C} + (\tfrac{1}{2} - \beta)\,\Delta t^2\,\mathbf{K}]\ddot{\mathbf{u}}_k \end{aligned}$$

Putting

$$\begin{aligned} \mathbf{A} &= \mathbf{M} + \theta\,\Delta t\,\mathbf{C} + \beta\,\Delta t^2\,\mathbf{K} \\ \mathbf{b} &= \mathbf{f} - \mathbf{K}\mathbf{u}_k - (\mathbf{C} + \Delta t\,\mathbf{K})\dot{\mathbf{u}}_k - [(1-\theta)\,\Delta t\,\mathbf{C} + (\tfrac{1}{2} - \beta)\Delta t^2\,\mathbf{K}]\ddot{\mathbf{u}}_k \end{aligned} \tag{4.77}$$

we have

$$\mathbf{A}\ddot{\mathbf{u}}_{k+1} = \mathbf{b} \tag{4.78}$$

Solving this yields $\ddot{\mathbf{u}}_{k+1}$, and then (4.76) gives $\dot{\mathbf{u}}_{k+1}$ and $\mathbf{u}_{k+1}$. As a special case of (4.77), we assume $\theta = \beta = 0$. Then (4.78) becomes

$$\mathbf{M}\ddot{\mathbf{u}}_{k+1} = \mathbf{f} - \mathbf{K}\mathbf{u}_k - (\mathbf{C} + \Delta t\,\mathbf{K})\dot{\mathbf{u}}_k - (\Delta t\,\mathbf{C} + \tfrac{1}{2}\,\Delta t^2\,\mathbf{K})\ddot{\mathbf{u}}_k \tag{4.79}$$

If $\theta = \frac{1}{2}$ and $\beta = \frac{1}{4}$ are assumed, the scheme becomes *unconditionally stable.* In the *Wilson θ-method,* we assume that approximation

$$\begin{aligned} \dot{\mathbf{u}}_{k+\theta} &= \dot{\mathbf{u}}_k + \tfrac{1}{2}\theta\,\Delta t\,(\ddot{\mathbf{u}}_k + \ddot{\mathbf{u}}_{k+\theta}) \\ \mathbf{u}_{k+\theta} &= \mathbf{u}_k + \theta\,\Delta t\,\dot{\mathbf{u}}_k + \tfrac{1}{6}\theta^2\,\Delta t^2\,(2\ddot{\mathbf{u}}_k + \ddot{\mathbf{u}}_{k+\theta}) \end{aligned} \tag{4.80}$$

where $\theta \geq 1$, then the equation of motion becomes

$$\mathbf{A}\ddot{\mathbf{u}}_{k+\theta} = \mathbf{b} \tag{4.81}$$

where

$$\begin{aligned} \mathbf{A} &= \mathbf{M} + \tfrac{1}{2}\theta\,\Delta t\,\mathbf{C} + \tfrac{1}{6}\theta^2\,\Delta t^2\,\mathbf{K} \\ \mathbf{b} &= \mathbf{f} - \mathbf{K}\mathbf{u}_k - (\mathbf{C} + \theta\,\Delta \mathbf{t}\,\mathbf{K})\dot{\mathbf{u}}_k - (\tfrac{1}{2}\theta\,\Delta t\,\mathbf{C} + \tfrac{1}{3}\theta^2\,\Delta t^2\,\mathbf{K})\ddot{\mathbf{u}}_k \end{aligned} \tag{4.82}$$

After obtaining the quantities $\ddot{\mathbf{u}}_{k+\theta}$, we compute

$$\begin{aligned} \ddot{\mathbf{u}}_{k+1} &= \ddot{\mathbf{u}}_k + \frac{1}{\theta}(\ddot{\mathbf{u}}_{k+\theta} - \ddot{\mathbf{u}}_k) \\ \dot{\mathbf{u}}_{k+1} &= \dot{\mathbf{u}}_k + \Delta t\,\ddot{\mathbf{u}}_k + \frac{\Delta t}{2\theta}(\ddot{\mathbf{u}}_{k+\theta} - \ddot{\mathbf{u}}_k) \\ \mathbf{u}_{k+1} &= \mathbf{u}_k + \Delta t\,\dot{\mathbf{u}}_k + \tfrac{1}{2}\Delta t^2\,\ddot{\mathbf{u}}_k + \frac{\Delta t^2}{6\theta}(\ddot{\mathbf{u}}_{k+\theta} - \ddot{\mathbf{u}}_k) \end{aligned} \tag{4.83}$$

To obtain an unconditionally stable scheme, $\theta \geq 1.37$ has to be assumed in (4.80).

Other methods are applicable if (4.75) is first rewritten as the following system of two first-order equations:

$$\begin{bmatrix} \mathbf{M} & \mathbf{0} \\ \mathbf{0} & -\mathbf{K} \end{bmatrix} \begin{Bmatrix} \dot{\mathbf{v}} \\ \dot{\mathbf{u}} \end{Bmatrix} + \begin{bmatrix} \mathbf{C} & \mathbf{K} \\ \mathbf{K} & \mathbf{0} \end{bmatrix} \begin{Bmatrix} \mathbf{v} \\ \mathbf{u} \end{Bmatrix} = \begin{Bmatrix} \mathbf{f} \\ \mathbf{0} \end{Bmatrix} \tag{4.84}$$

in which $\mathbf{K}$ must satisfy the condition: $\mathbf{K}\mathbf{x} = \mathbf{0} \Rightarrow \mathbf{x} = \mathbf{0}$. Symbolically, let us describe (4.84) by

$$\hat{\mathbf{C}}\dot{\mathbf{w}} + \hat{\mathbf{K}}\mathbf{w} = \hat{\mathbf{f}}, \qquad \mathbf{w}^{\mathrm{T}} = \{\mathbf{v}, \mathbf{u}\} \tag{4.85}$$

In direct integration methods for structural dynamics, A-stability of the integration schemes becomes important since the ratio of the largest and smallest diagonal terms of the stiffness matrix becomes very large. If the time step Δt is chosen according to the largest diagonal term of the stiffness matrix, Δt must be quite small and is sometimes not realistic. In this case, in order to use a time step Δt that corresponds to a rather small diagonal term of the stiffness matrix, the integration schemes must satisfy the *A-stability* condition, which means that the numerical solution $\mathbf{u}_k$ for the differential equation

$$\dot{\mathbf{u}}_k = \mathbf{B}\mathbf{u}_k$$

converge to zero as $k \to +\infty$ for any size of time integration step Δt. However, the requirement of A-stability is too restrictive for many problems. To avoid this, Gear introduced the concept of *stiff stability*, which is weaker than A-stability. Then the *second-order Gear method* can be derived from the approximation.

$$\mathbf{w}_{k+1} = \mathbf{w}_k + \tfrac{1}{3}(\mathbf{w}_k - \mathbf{w}_{k-1}) + \tfrac{2}{3}\Delta t\,\dot{\mathbf{w}}_{k+1} \tag{4.86}$$

Substitution of (4.86) into (4.85) yields

$$\mathbf{A}\mathbf{w}_{k+1} = \mathbf{b} \tag{4.87}$$

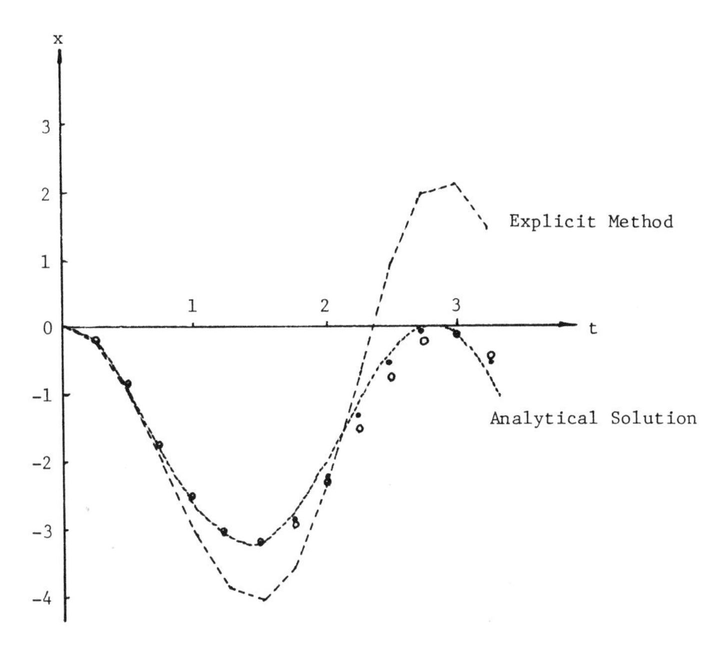

Figure 4.22 Comparison of integration schemes for a single-degree-of-freedom system: ○, Wilson's θ-method; ●, Newmark β-method.

where

$$\mathbf{A} = \hat{\mathbf{C}} + \tfrac{2}{3}\,\Delta t\,\hat{\mathbf{K}}, \qquad \mathbf{b} = \hat{\mathbf{f}} - \hat{\mathbf{K}}\mathbf{w}_k - \tfrac{1}{3}\hat{\mathbf{K}}(\mathbf{w}_k - \mathbf{w}_{k-1}) \tag{4.88}$$

The integration scheme based on (4.86) is stiffly stable. Details of stiffly stable schemes can be found in Gear (1967).

Example 4.10: We shall apply the explicit Newmark β- and Wilson θ-methods to solve a problem of a single degree of freedom:

$$\ddot{x} + 5x = -10\sqrt{\tfrac{2}{3}}$$

the exact solution of which is given by

$$x = 2\sqrt{\tfrac{2}{3}}(-1 + \cos\sqrt{5}t)$$

For the choice of $\Delta t = 0.25$, numerical results are shown in Figure 4.22. Here $\theta = 0.5$ and $\beta = 0.25$ are assumed for Newmark's method, and $\theta = 1.4$ is taken in Wilson's method. It is clear that although a large time increment Δt is assumed, both the Newmark β- and Wilson θ-methods provide fairly good agreement to the exact solution. Indeed, the errors at $t = 3$ are 2.8% for Newmark's method and 1.4% for Wilson's method. However, the explicit scheme yields a divergent result, as shown in Figure 4.22.

Example 4.11: Let us consider the dynamic response of a uniform simply supported beam subjected to a central step function loading. Suppose that ρ_s and EI are constant and that the magnitude of the load is P_0. Applying the mode superposition analysis (Clough and Penzien,

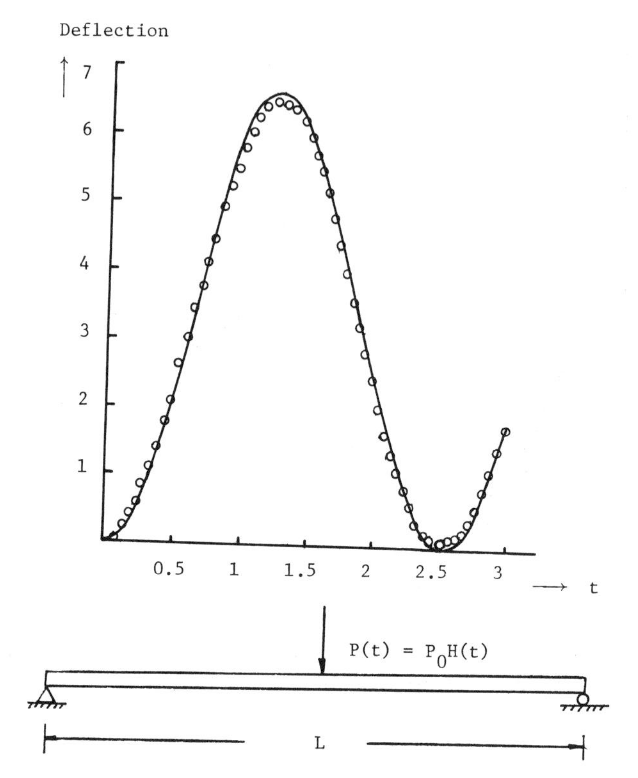

Figure 4.23 Numerical result of beam vibration by Newmark β-method: ○, β-method; ———, analytical solution.

1975), the deflection of the beam can be obtained as

$$w(x, t) = \sum_{n=1}^{\infty} \frac{2P_0\alpha_n}{\rho_s L\omega_n^2}(1 - \cos \omega_n t) \sin \frac{n\pi x}{L}$$

where

$$\alpha_n = \begin{cases} 1 & \text{if } n = 1, 5, 9, \ldots \\ -1 & \text{if } n = 3, 7, 11, \ldots, \\ 0 & \text{if even} \end{cases} \qquad \omega_n = (n\pi)^2, \; n = 1, 2, 3, \ldots$$

If the deflection of the beam is computed at $x = L/2$, we have

$$W(\tfrac{1}{2}L, t) = \frac{2P_0L^3}{\pi^4 EI}\left[(1 - \cos \omega_1 t) - \frac{1 - \cos \omega_3 t}{81} + \frac{1 - \cos \omega_5 t}{625} + \ldots\right]$$

For the case that $EI = 1$, $\rho_s = 1$, $L = 2$, and $P_0 = 20$, we solve the problem using the Newmark β-method with $\beta = \frac{1}{4}$ and using four equal-size beam elements. Numerical results are shown in Figure 4.23 for the deflection of the beam at the center together with the analytical solution. It is clear that the results are very close to the exact solution even if a rather large time increment $\Delta t = 0.05$ is assumed.

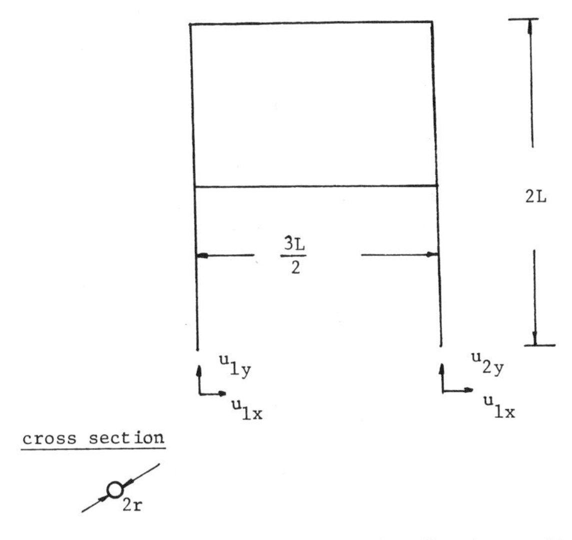

Figure 4.24 Frame structure in plane for vibration problem.

Exercise 4.7: Solve the dynamic response of the plane frame structure involving the bending and axial deformation shown in Figure 4.24. As the initial condition, we assume $\mathbf{u} = \mathbf{0}$, $\dot{\mathbf{u}} = \mathbf{0}$, $\ddot{\mathbf{u}} = \mathbf{0}$, and $\theta = 0$ at $t = 0$. Using the data

$$E = 2.1 \times 10^4 \text{ kN/cm}^2, \qquad r = 1 \text{ cm}, \qquad L = 100 \text{ cm}$$

solve the problem under the condition

$$u_{1x}(t) = u_{1y}(t) = 0$$
$$u_{2x}(t) = \sin 10t, \qquad u_{2y}(t) = 5 \cos 20t$$

4.5 Saint-Venant torsional element

Another typical structural element is the *Saint-Venant torsional element,* whose displacement field is assumed to be

$$\begin{aligned} u_r(r, s, t) &= -t\theta(s) \\ u_t(r, s, t) &= r\theta(s) \\ u_s(r, s, t) &= \psi(r, t)\theta'(s) \end{aligned} \tag{4.89}$$

where the coordinate system (r, t) has an origin located at the centroid of the cross section, $\theta(s)$ is the rotation of the bar's cross section at s around the s axis (i.e., the bar axis), and ψ is a warping function such that*

$$\int_A \psi \, dA = \int_A \psi r \, dA = \int_A \psi t \, dA = 0 \tag{4.90}$$

where A is the cross section of the bar.

* These conditions ensure that the cross section transmits only a torque.

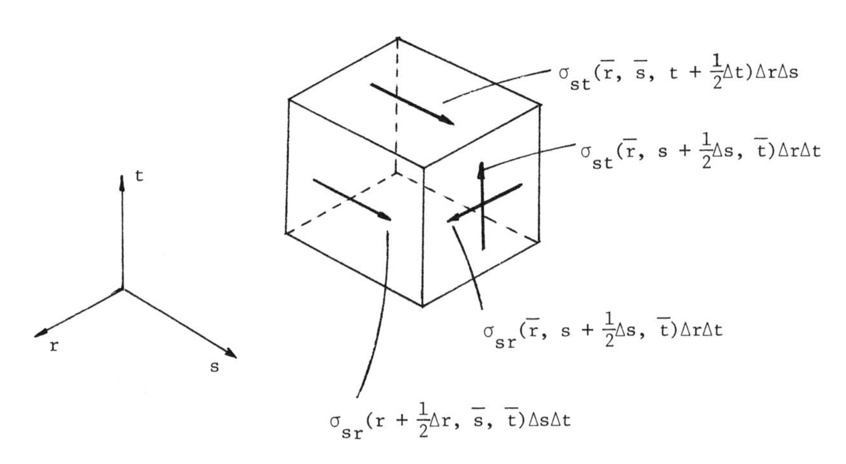

Figure 4.25 Forces in the s direction.

If there is a uniform torsion, that is, Saint-Venant torsion is considered, we assume that

$$\frac{d^2\theta}{ds^2} = 0 \tag{4.91}$$

Thus, the strain due to the torsion is given by

$$\begin{aligned}
&\varepsilon_{rr} = \frac{\partial u_r}{\partial r} = 0, \qquad \varepsilon_{ss} = \frac{\partial u_s}{\partial s} = 0, \qquad \varepsilon_{tt} = \frac{\partial u_t}{\partial t} = 0 \\
&\gamma_{rs} = \frac{\partial u_r}{\partial s} + \frac{\partial u_s}{\partial r} = \left(\frac{\partial \psi}{\partial r} - t\right)\frac{d\theta}{ds} \\
&\gamma_{rt} = \frac{\partial u_r}{\partial t} + \frac{\partial u_t}{\partial r} = 0 \\
&\gamma_{ts} = \frac{\partial^u t}{\partial s} + \frac{\partial^u s}{\partial t} = \left(\frac{\partial \psi}{\partial t} + r\right)\frac{d\theta}{ds}
\end{aligned} \tag{4.92}$$

The only nonzero components of the strain are γ_{rs} and γ_{ts}. If linear isotropic elasticity is assumed, the stress–strain relation yields the nonzero components of the stress tensor:

$$\begin{aligned}
\sigma_{rs} &= G\gamma_{rs} = G\left(\frac{\partial \psi}{\partial r} - t\right)\frac{d\theta}{ds} \\
\sigma_{ts} &= G\gamma_{ts} = G\left(\frac{\partial \psi}{\partial t} + r\right)\frac{d\theta}{ds}
\end{aligned} \tag{4.93}$$

Noting that $\sigma_{ss} = 0$, equilibrium of the resultant forces on a differential element in the s axis (see Figure 4.25) yields

$$\begin{aligned}
&(\sigma_{rs} + \tfrac{1}{2}\,\Delta\sigma_{rs})\,\Delta t\,\Delta s - (\sigma_{rs} - \tfrac{1}{2}\,\Delta\sigma_{rs})\,\Delta t\,\Delta s \\
&\quad + (\sigma_{ts} + \tfrac{1}{2}\,\Delta\sigma_{ts})\,\Delta r\,\Delta s - (\sigma_{ts} - \tfrac{1}{2}\,\Delta\sigma_{ts})\,\Delta r\,\Delta s = 0
\end{aligned}$$

that is,

$$\Delta\sigma_{rs}\,\Delta t\,\Delta s + \Delta\sigma_{ts}\,\Delta r\,\Delta s = 0 \tag{4.94}$$

Dividing by $\Delta r\,\Delta s\,\Delta t$, and passing to the limit $\Delta r \to 0$, $\Delta s \to 0$, and $\Delta t \to 0$, we have

$$\frac{\partial \sigma_{rs}}{\partial r} + \frac{\partial \sigma_{ts}}{\partial t} = 0 \tag{4.95}$$

Similarly, equilibrium in the r and t directions yields

$$\frac{\partial \sigma_{rs}}{\partial s} = 0 \qquad \text{and} \qquad \frac{\partial \sigma_{ts}}{\partial s} = 0 \tag{4.96}$$

respectively. From these three equilibrium equations, we now derive a weak form. To do this, multiply (4.95) and (4.96), respectively, by the virtual displacements $\bar{u}_s$, $\bar{u}_r$, and $\bar{u}_t$ and integrate over the bar. The result, after some integrations by parts, is

$$\int_0^L \int_A \left[\sigma_{rs}\left(\frac{\partial \bar{u}_s}{\partial r} + \frac{\partial \bar{u}_r}{\partial s}\right) + \sigma_{ts}\left(\frac{\partial \bar{u}_s}{\partial t} + \frac{\partial \bar{u}_t}{\partial s}\right)\right] dA\,ds$$
$$- \int_\Gamma [(\sigma_{rs}n_r + \sigma_{ts}n_t)\bar{u}_s + \sigma_{rs}n_s\bar{u}_r + \sigma_{ts}n_s\bar{u}_t]\,d\Gamma = 0$$

where (n_r, n_s, n_t) are the components of the unit outward normal to the boundary Γ of the bar. Since no tractions are applied on this lateral surface of the bar, and since $n_r = n_t = 0$ on both ends, we obtain

$$\int_0^L \int_A \left[\sigma_{rs}\left(\frac{\partial \bar{u}_s}{\partial r} + \frac{\partial \bar{u}_r}{\partial s}\right) + \sigma_{ts}\left(\frac{\partial \bar{u}_s}{\partial t} + \frac{\partial \bar{u}_t}{\partial s}\right)\right] dA\,ds$$
$$= \int_0^L \int_A (\sigma_{rs}\bar{\gamma}_{rs} + \sigma_{ts}\bar{\gamma}_{ts})\,dA\,ds = T\bar{\theta}\Big|_{s=0} + T\bar{\theta}\Big|_{s=L}$$

where $\bar{\gamma}_{rs}$ and $\bar{\gamma}_{ts}$ are the virtual strains corresponding to $\bar{u}_r$, $\bar{u}_s$, and $\bar{u}_t$, and

$$T = \int_A (-t\sigma_{rs} + r\sigma_{ts})\,dA \tag{4.97}$$

defines the torque that each cross section must transmit. Applying the relation (4.89), we have the weak form

$$\int_0^L GK\frac{d\theta}{ds}\frac{d\bar{\theta}}{ds}\,ds = T\bar{\theta}\Big|_{s=0} + T\bar{\theta}\Big|_{s=L} \tag{4.98}$$

where

$$K = \int_A \left[\left(\frac{\partial \psi}{\partial r} - t\right)^2 + \left(\frac{\partial \psi}{\partial t} + r\right)^2\right] dA \tag{4.99}$$

is the *torsional rigidity* defined for a given cross section in terms of the associated warping function ψ, and $\bar{\theta}$ is an arbitrary virtual angle of rotation of the bar's cross sections.

Now, using linear interpolation for the angle of rotation,

$$\theta(s) = \theta_1 N_1(s) + \theta_2 N_2(s) \tag{4.100}$$

where

$$N_1(s) = 1 - s/L \qquad \text{and} \qquad N_2(s) = s/L \tag{4.101}$$

the weak form (4.98) is discretized to

$$\{\bar{\theta}_1, \bar{\theta}_2\} \frac{GK}{L} \begin{bmatrix} 1 & -1 \\ -1 & 1 \end{bmatrix} \begin{Bmatrix} \theta_1 \\ \theta_2 \end{Bmatrix} = \{\bar{\theta}_1, \bar{\theta}_2\} \begin{Bmatrix} T_1 \\ T_2 \end{Bmatrix} \tag{4.102}$$

Thus, the element stiffness matrix is given by

$$[K^e_{\alpha\beta}] = \frac{GK}{L} \begin{bmatrix} 1 & -1 \\ -1 & 1 \end{bmatrix} \tag{4.103}$$

and the element generalized load vector is

$$\begin{Bmatrix} f^e_1 \\ f^e_2 \end{Bmatrix} = \begin{Bmatrix} T_1 \\ T_2 \end{Bmatrix} \tag{4.104}$$

If the dynamical problem of torsional wave motion is to be considered, we have to add the term

$$\int_0^L \int_A \rho_0 \{\ddot{u}_r \bar{u}_r + \ddot{u}_s \bar{u}_s + \ddot{u}_t \bar{u}_t\} \, dA \, ds \tag{4.105}$$

to the left side of the weak form (4.98). Applying the relations (4.89), this term is written as

$$\int_0^L J \ddot{\theta} \bar{\theta} \, ds + \int_0^L \hat{J} \frac{\partial \ddot{\theta}}{\partial s} \frac{\partial \bar{\theta}}{\partial s} \, ds \tag{4.106}$$

where

$$J = \int_A \rho_0 (r^2 + t^2) \, dA \qquad \text{and} \qquad \hat{J} = \int_A \rho_0 \psi^2 \, dA \tag{4.107}$$

Here ρ_0 is the density of the material of the bar. Discretization of (4.106) yields the element mass matrix

$$[M^e_{\alpha\beta}] = \frac{JL}{6} \begin{bmatrix} 2 & 1 \\ 1 & 2 \end{bmatrix} + \frac{\hat{J}}{L} \begin{bmatrix} 1 & -1 \\ -1 & 1 \end{bmatrix} \tag{4.108}$$

If the mass matrix is lumped, we get

$$[\bar{M}^e_{\alpha\beta}] = \frac{JL}{2} \begin{bmatrix} 1 & 0 \\ 0 & 1 \end{bmatrix} \tag{4.109}$$

Then the discretized equation for governing the dynamical torsion problem becomes

$$\bar{M}^e_{\alpha\beta} \ddot{\theta}_\beta + K^e_{\alpha\beta} \theta_\beta = f^e_\alpha \tag{4.110}$$

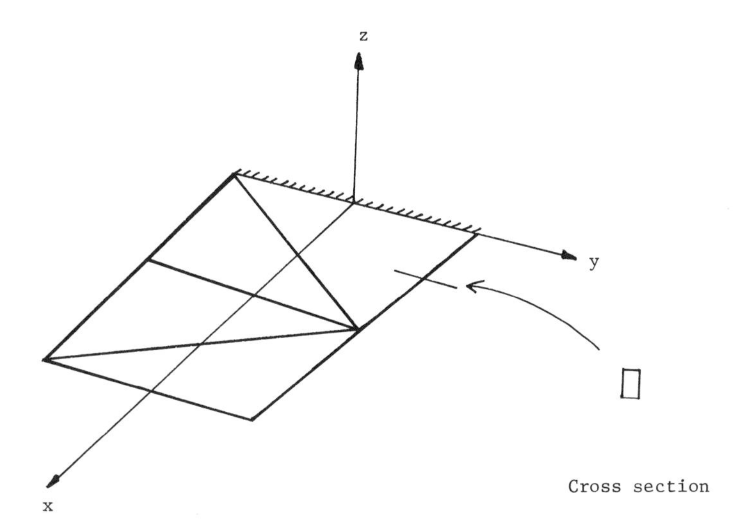

Figure 4.26 Model of airplane wing.

For a structure consisting of several torsional bar elements, the global equation becomes

$$\sum_{e=1}^{E} (\bar{M}^e_{\alpha\beta}\ddot{\theta}_\beta + K^e_{\alpha\beta}\theta_\beta) = \sum_{e=1}^{E} f^e_\alpha \tag{4.111}$$

after assembling discrete equation (4.110) for a bar element, where E is the total number of elements.

If natural (torsional) frequencies are of interest, we put

$$f^e_\alpha = 0 \qquad \text{and} \qquad \theta_\beta = e^{-i\omega\hat{t}}\hat{\theta}_\beta$$

and obtain the eigenvalue problem

$$\sum_{e=1}^{E} K^e_{\alpha\beta}\hat{\theta}_\beta = \omega^2 \sum_{e=1}^{E} \bar{M}^e_{\alpha\beta}\hat{\theta}_\beta \tag{4.112}$$

Example 4.12: Combining two fundamental structural elements of bending and torsion, let us develop the element stiffness and mass matrices to find the eigenvalues and eigenfunctions for the free transverse vibration of the frame structure shown in Figure 4.26. The degrees of freedom of the beam in this example problem are the transverse deflection w, the rotation θ_x about the x axis, and the rotation θ_y about the y axis. Indeed, each beam resists both bending and torsion.

Note that the angle between the beam and the x axis is given by α, and the rotation θ_r by pure bending and the angle of twist θ_s are related to (θ_x, θ_y) by

$$\theta_r = \sin\alpha\theta_x - \cos\alpha\theta_y, \qquad \theta_s = \cos\alpha\theta_x + \sin\alpha\theta_y \tag{4.113}$$

Using these two relations, 6×6 element stiffness and lumped mass matrices can be obtained for the frame structure involving torsion and bending.

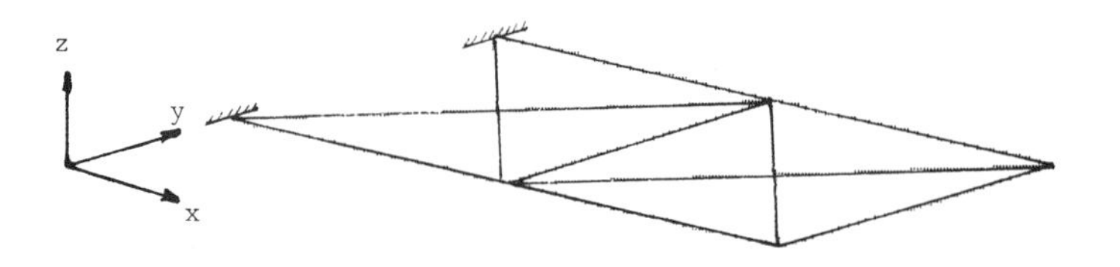

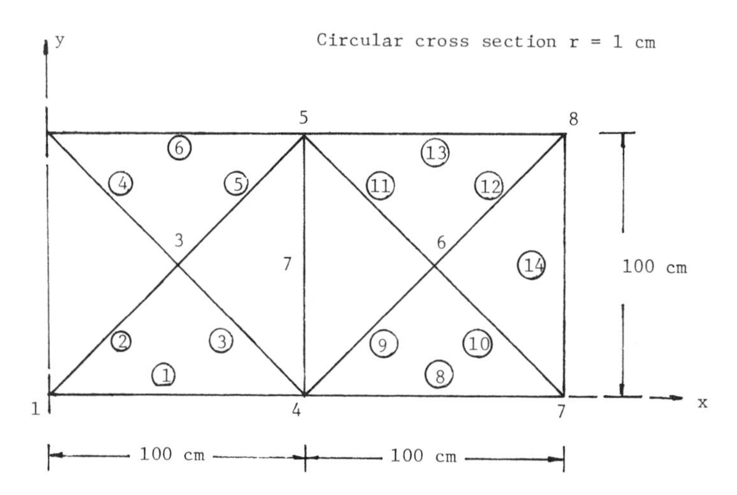

Figure 4.27 Free vibration problem of plane frame (transverse deflection with torsion): $E = 2.1 \times 10^4$ kN/cm^2, $I = \frac{1}{4}r^4 = 0.785$ cm^4, $J = \frac{1}{2}r^4 = 1.57$ cm^4, $G = 0.81 \times 10^4$ kN/cm^2, $\rho = 0.0079$ kg/cm^3, $K = J$.

We shall consider a free vibration problem to the structure described in Figure 4.27 using 14 beam elements that can take both bending and torsion. Using the inverse iteration method developed above, the first 10 eigenvalues and eigenvectors are obtained as shown in Figures 4.28(a)–(d). As shown clearly in the second and fourth eigenvectors, torsion effect dominates the vibration mode of the frame structure.

4.6 Three-dimensional frame element

By combining the developments for simple structural elements such as bars (for axial forces), beams (for bending moments), and shafts (for torques), it is possible to construct (by assembling) the element stiffness matrix and the (lumped) element mass matrix for a rod element subjected to all three types of loads. Alternatively, the same results can be obtained by starting with the assumed displacement field:

$$\begin{aligned} u_r &= u_0(s) - (t - t_0)\theta(s) \\ u_s &= w_0(s) - ru_0'(s) - tv_0'(s) + \psi(r, t)\theta'(s) \\ u_t &= v_0(s) + (r - r_0)\theta(s) \end{aligned} \tag{4.114}$$

where u_0 and v_0 are the displacements of the shear center (r_0, t_0) in the r and t directions (see Figure 4.29), respectively, and w_0 is the average displacement of the cross

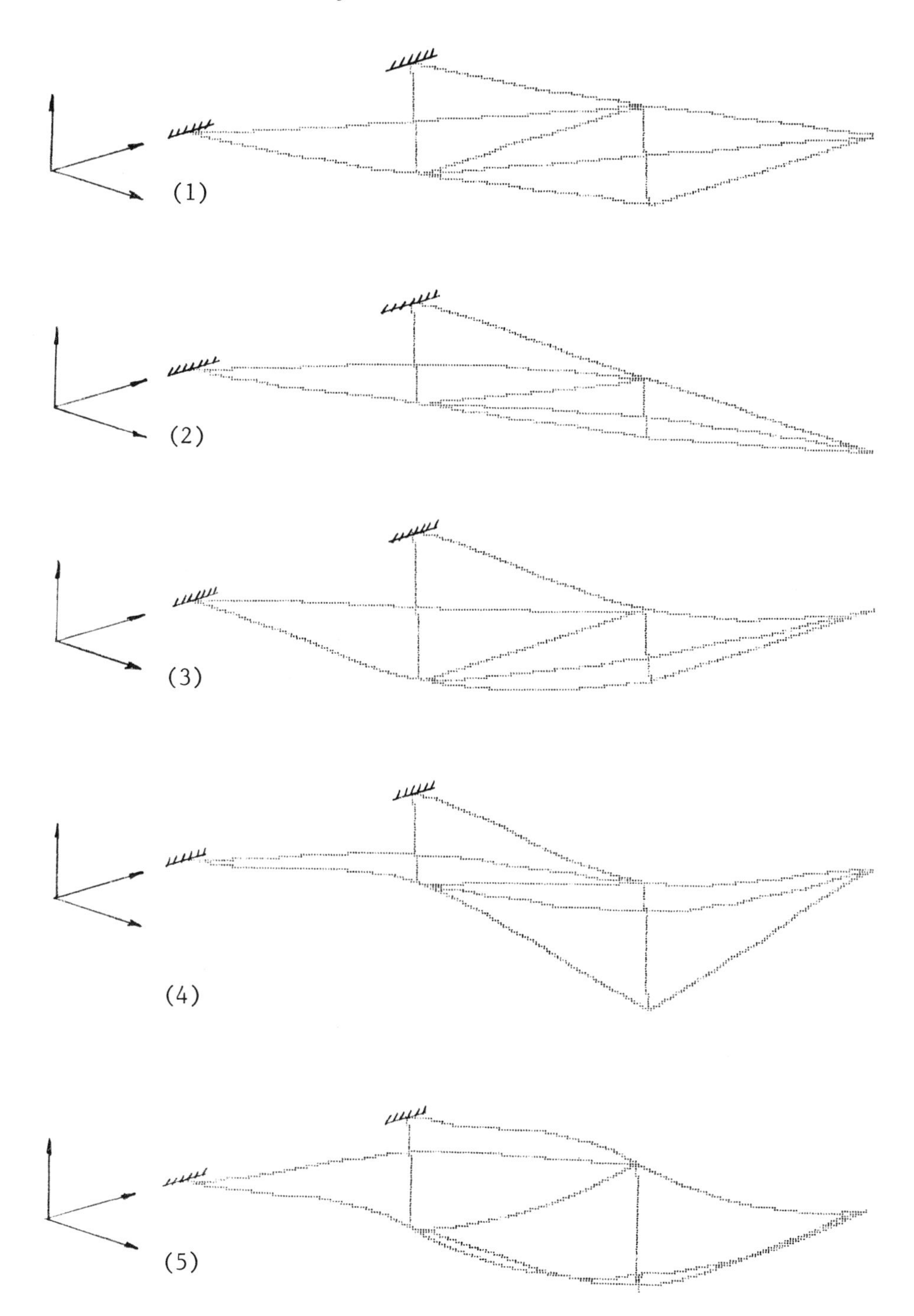

Figure 4.28 First 10 eigenvectors.

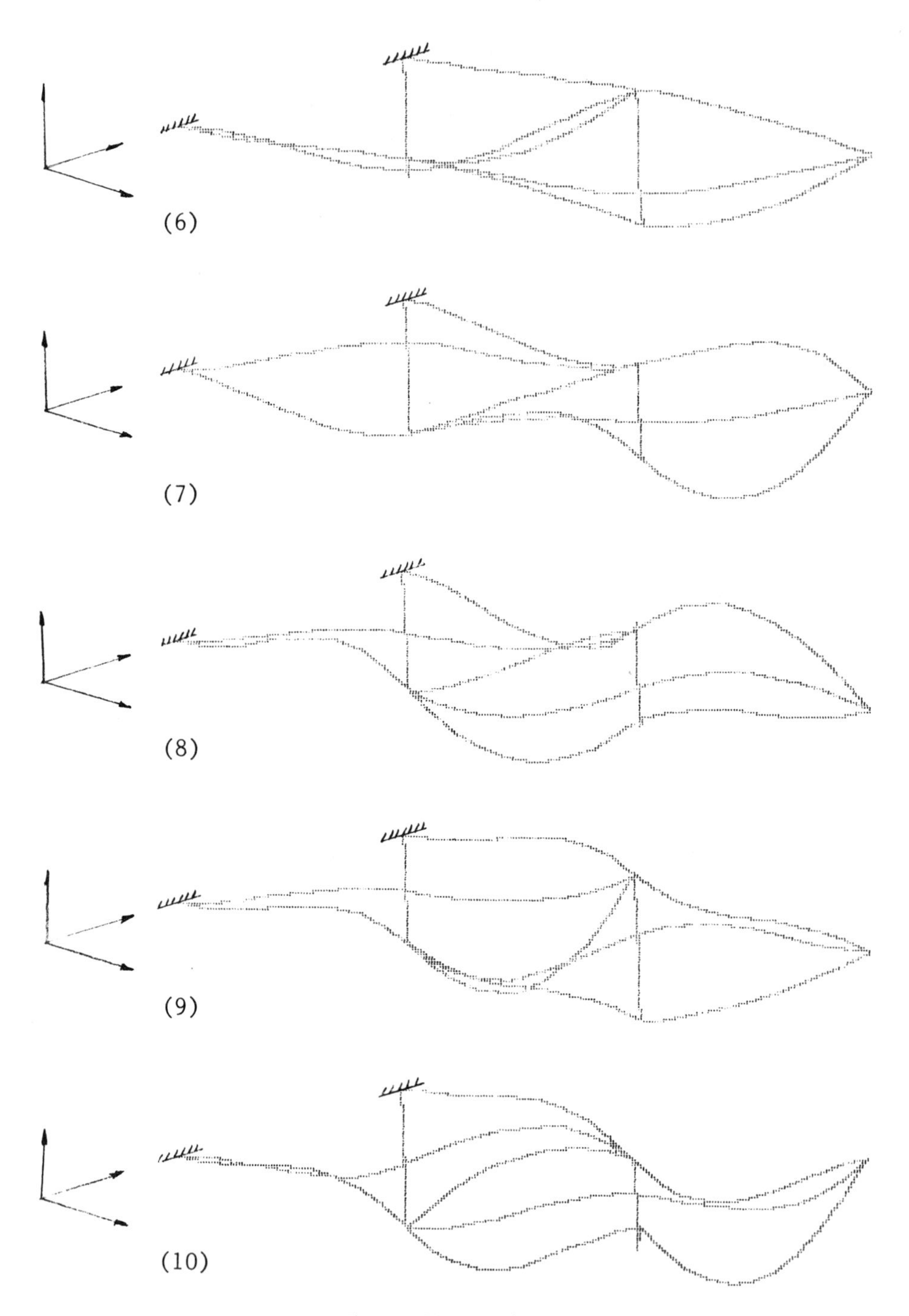

Figure 4.28 (continued)

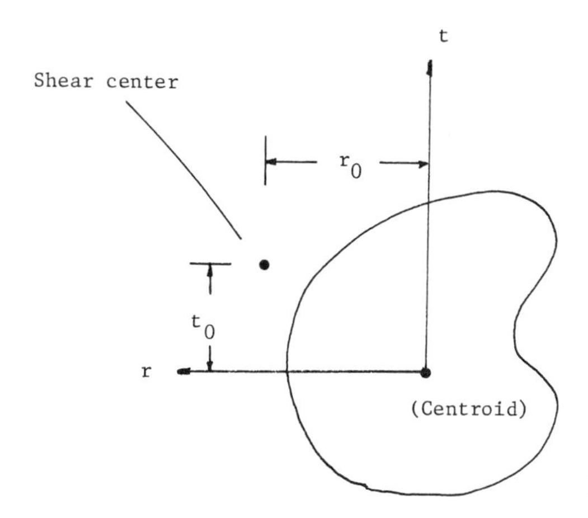

Figure 4.29 Shear center (r_0, t_0).

section of the bar.* In arriving at (4.114), we have applied the *Bernoulli–Euler assumption* and the torsion theory of Saint-Venant.

For the applied shear forces and moments at both end points shown in Figure 4.30, we have the finite element equations (4.115) for the static problem:

$$
\begin{bmatrix}
\frac{EA}{L} & 0 & 0 & 0 & 0 & 0 & -\frac{EA}{L} & 0 & 0 & 0 & 0 & 0 \\
0 & \frac{12EI}{L^3} & 0 & 0 & 0 & \frac{6EI}{L^2} & 0 & -\frac{12EI}{L^3} & 0 & 0 & 0 & \frac{6EI}{L^2} \\
0 & 0 & \frac{12E\hat{I}}{L^3} & 0 & -\frac{6E\hat{I}}{L^2} & & 0 & 0 & -\frac{12E\hat{I}}{L^3} & 0 & -\frac{6EI^4}{L^2} & 0 \\
0 & 0 & 0 & \frac{GK}{L} & 0 & 0 & 0 & 0 & 0 & -\frac{GK}{L} & 0 & 0 \\
0 & 0 & -\frac{6E\hat{I}}{L^2} & 0 & \frac{4E\hat{I}}{L} & 0 & 0 & 0 & \frac{6E\hat{I}}{L^2} & 0 & \frac{2E\hat{I}}{L} & 0 \\
0 & \frac{6EI}{L^2} & 0 & 0 & 0 & \frac{4EI}{L} & 0 & -\frac{6EI}{L^2} & 0 & 0 & 0 & \frac{2EI}{L} \\
-\frac{EA}{L} & 0 & 0 & 0 & 0 & 0 & \frac{EA}{L} & 0 & 0 & 0 & 0 & 0 \\
0 & -\frac{12EI}{L^3} & 0 & 0 & 0 & -\frac{6EI}{L^2} & 0 & \frac{12EI}{L^3} & 0 & 0 & 0 & -\frac{6EI}{L^2} \\
0 & 0 & \frac{12E\hat{I}}{L^2} & 0 & \frac{6E\hat{I}}{L^2} & 0 & 0 & 0 & \frac{12E\hat{I}}{L^3} & 0 & \frac{6E\hat{I}}{L^2} & 0 \\
0 & 0 & 0 & -\frac{GK}{L} & 0 & 0 & 0 & 0 & 0 & \frac{GK}{L} & 0 & 0 \\
0 & 0 & -\frac{6E\hat{I}}{L^2} & 0 & \frac{2E\hat{I}}{L} & 0 & 0 & 0 & \frac{6E\hat{I}}{L^2} & 0 & \frac{4E\hat{I}}{L} & 0 \\
0 & \frac{6EI}{L^2} & 0 & 0 & 0 & \frac{2EI}{L} & 0 & -\frac{6EI}{L^2} & 0 & 0 & 0 & \frac{4EI}{L}
\end{bmatrix}
\begin{Bmatrix}
u_{1s} \\ u_{1t} \\ u_{1r} \\ \theta_{1s} \\ -\theta_{1t} \\ \theta_{1r} \\ u_{2s} \\ u_{2t} \\ u_{2r} \\ \theta_{2s} \\ -\theta_{2t} \\ \theta_{2r}
\end{Bmatrix}
$$

$$
= \{P_{1s}, P_{1t}, P_{1r}, M_{1s}, M_{1t}, M_{1r}, P_{2s}, P_{2t}, P_{2r}, M_{2s}, M_{2t}, M_{2r}\}^{\mathrm{T}} \tag{4.115}
$$

* The coordinates of the *shear center* are r_0, t_0. In general, these are very difficult to obtain, but they can be readily obtained for thin-walled structures (see Oden and Ripperger, 1981).

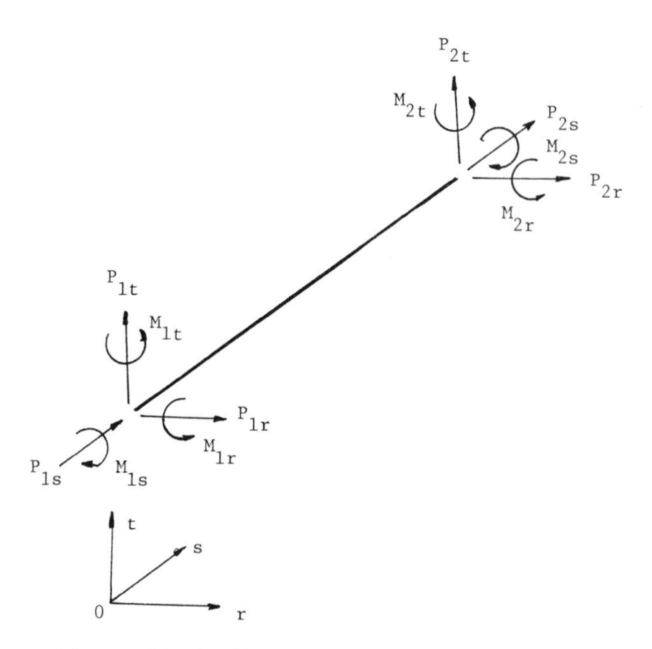

Figure 4.30 Applied loads and moments to a frame.

Exercise 4.8: Find the transformation matrix $[T_{p\alpha}]$ that transforms the stiffness matrix expressed in *rst* coordinates to one in *xyz* coordinates, where $p, \alpha = 1, 2, \ldots, 12$.

Exercise 4.9: Construct the mass matrix for a three-dimensional frame element by assembling the element mass matrices for truss, beam, and torsion elements.

Example 4.13: Let us solve a problem of the three-dimensional frame structure shown in Figure 4.31. The cross section of the frame is uniform and rectangular. If a point force P_x is applied, the total (complementary) strain energy of the frame is given by

$$U = \int_0^L \left(\frac{P_x^2 s^2}{2EI} + \frac{1.2P^2}{2GA} \right) ds + \int_0^L \left(\frac{P_x^2 L^2}{2EI} + \frac{P_x^2}{2EA} \right) ds + \int_0^L \left(\frac{P_x^2 s^2}{2EI} + \frac{1.2P_x^2}{2GA} + \frac{P_x^2 L^2}{2GJ} \right) ds$$

Applying the Castigliano theorem, the deflection in the x direction at the point P becomes

$$\Delta_x = \frac{\partial U}{\partial P_x} = \frac{5}{3} \frac{P_x L^3}{EI} + \frac{2.4 P_x L}{GA} + \frac{P_x L}{EA} + \frac{P_x L^3}{GJ}$$

substituting $E = 1000$, $G = 400$, $b = h = 1$, $I = \frac{1}{2}$, $J = 0.141$, $A = 1$,

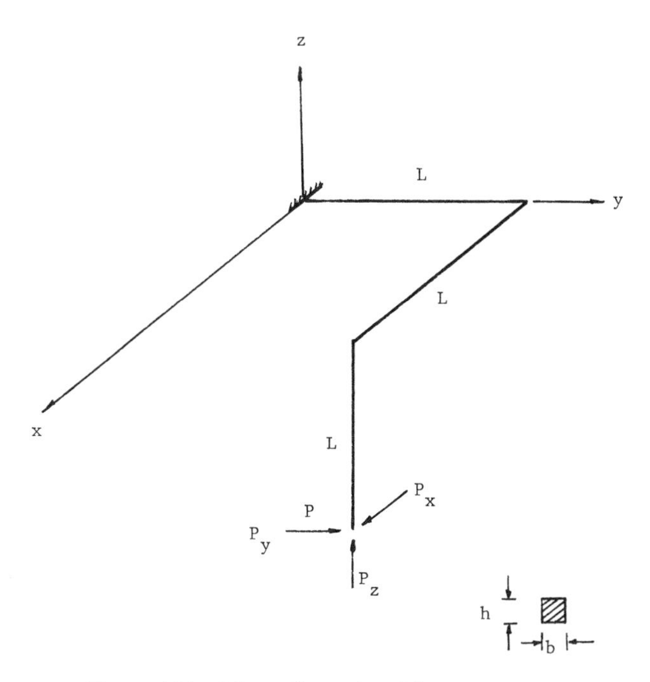

Figure 4.31 Three-dimensional frame structure.

$L = 10$, and $P_x = 1$, we have

$$\Delta_x = 37.8$$

Using the stiffness matrix in (4.115) to a three-dimensional frame, we can obtain the deflection $\Delta_x = 37.8$ at P, which coincides with the exact solution. Similarly, if $P_y = 1$ is applied at P, the deflection Δy is 49.8, which is again the same as the exact solution. For $P_z = 1$, $\Delta_z = 25.8$ is obtained.

As the last example of this chapter, let us check convergence of the finite element approximation of a frame structure. To do this, a semicircular ring is considered, one end of which is fixed. Applying a point force P vertically at the middle of the ring and assuming that the other end is free, the Castigliano theorem yields the vertical deflection at the free end,

$$\Delta_z = \frac{PR^3}{2}\left(\frac{1}{EI} + \frac{(\pi - 1)}{GJ}\right)$$

Here R is the radius of the ring, and its cross section is circular, the diameter of which is $D = 2$. Assuming $I = \pi D^4/64 = 0.785$, $J = \pi D^4/32 = 0.392$, $A = \pi D^2/4 = 3.1416$, $R = 5$, $P = 1$, $E = 1$, and $G = 0.4$ we have $\Delta_z = 933.25$.

Using 2, 4, 8, and 16 straight frame elements, the semicircular ring is approximated. The results are shown in Figure 4.32. It is clear that the finite element solutions converge to the analytical one as the number of finite elements is increased.

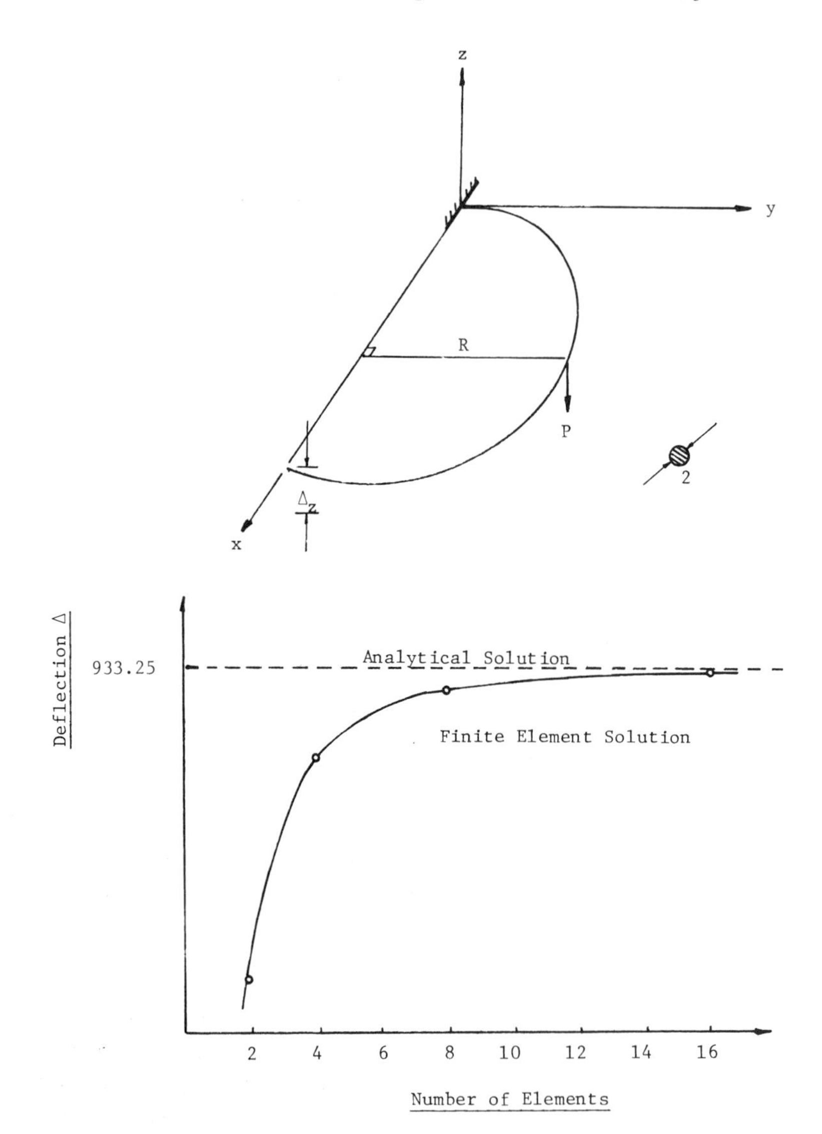

Figure 4.32 Convergence test of three-dimensional frame element: $I = 0.785$, $J = 0.392$, $A = 3.142$, $R = 5$, $P = 1$, $E = 1$, $G = 0.4$.

References

Ciarlet, P. G., *The finite element method for elliptic problems,* North-Holland, Amsterdam, 1977.

Clough, R. W., and Penzien, J., *Dynamics of structure,* McGraw-Hill, New York, 1975, pp. 331–6.

Gear, C. W., "The numerical integration of ordinary differential equations," *Mathematics of Computation* 21(2): 146–56, 1967.

Oden, J. T., and Ripperger, E. A., *Mechanics of elastic structures,* McGraw-Hill, New York, 1981.

Wilkinson, J. H., *The algebraic eigenvalue problem,* Clarendon, Oxford, 1965.

5

FINITE ELEMENT APPROXIMATIONS FOR PROBLEMS IN LINEAR ELASTICITY

A very popular application of finite element methods is the stress analysis of elastic structures under the assumption of infinitesimal deformation. In this chapter, we shall study a finite element approximation of plane elasticity problems using mainly four-node quadrilateral isoparametric elements together with the technique of Gaussian numerical integration. The study can be easily extended to other types of two-dimensional elements and to finite element approximations for three-dimensional solids.

5.1 Deformation, strain, and stress

Let a deformable body (or structure) Ω be referred to three-dimensional Euclidean space $\mathbb{R}^3$, and let it be subjected to an applied load. Each material point P of the body is identified with the coordinates (a_1, a_2, a_3) of the position of the point P relative to some fixed Cartesian coordinate system in $\mathbb{R}^3$. After applying the load, let the material point P change its position to (x_1, x_2, x_3). Then, the position vectors of the material point P are given by

$$\mathbf{r}_0 = a_\alpha \mathbf{i}_\alpha \qquad \text{and} \qquad \mathbf{r} = x_i \mathbf{i}_i \tag{5.1}$$

in the initial and deformed configurations, respectively, as shown in Figure 5.1. The displacement is then defined as the difference of these two position vectors:

$$\mathbf{u} = \mathbf{r} - \mathbf{r}_0 \qquad (\text{or} \quad u_\alpha = x_\alpha - a_\alpha) \tag{5.2}$$

Suppose that P', a point near the material point P, has coordinates $(a_1 + da_1, a_2 + da_2, a_3 + da_3)$ in the initial configuration. Let us now consider the deformation of the differential line segment $d\mathbf{r}_0 = \mathbf{r}'_0 - \mathbf{r}_0 = da_\alpha \mathbf{i}_\alpha$ by the applied load. Let the position of the material point P' be given by $\mathbf{r}' = \mathbf{r} + d\mathbf{r}$, $d\mathbf{r} = dx_i\, \mathbf{i}_i$, in the deformed configuration. Then it is easily seen that the difference in the squares of the lengths of these two line segments can be written as

$$|d\mathbf{r}|^2 - |d\mathbf{r}_0|^2 = dx_i\, dx_i - da_\alpha\, da_\alpha$$

If the differential rule

$$da_\alpha = \frac{\partial a_\alpha}{\partial x_i}\, dx_i \tag{5.3}$$

is applied,

$$dx_i\, dx_i - da_\alpha\, da_\alpha = \left(\delta_{ij} - \frac{\partial a_\alpha}{\partial x_i} \frac{\partial a_\alpha}{\partial x_j} \right) dx_i\, dx_j$$

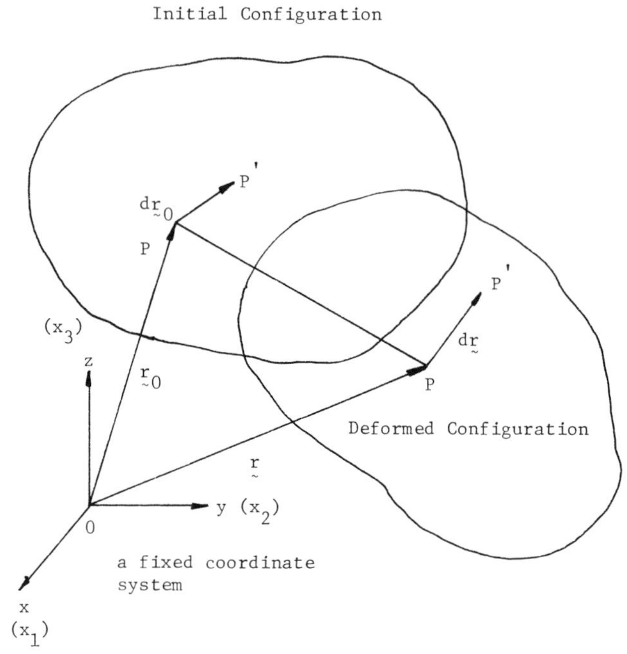

Figure 5.1 Deformation of a body.

If the strain tensor $\mathbf{e} = e_{ij}\mathbf{i}_i\mathbf{i}_j$ is defined as

$$e_{ij} = \frac{1}{2}\left(\delta_{ij} - \frac{\partial a_\alpha}{\partial x_i}\frac{\partial a_\alpha}{\partial x_j}\right) \tag{5.4}$$

then

$$dx_i\, dx_i - da_\alpha\, da_\alpha = 2e_{ij}\, dx_i\, dx_j \tag{5.5}$$

The strain tensor $\mathbf{e}$ was introduced by Cauchy for infinitesimal deformations and by Almansi and Hamel when the deformation is finite. A characteristic of $\mathbf{e}$ is that the strain is measured with respect to the deformed configuration. Note that (5.2) yields

$$\frac{\partial u_\alpha}{\partial x_i} = \delta_{\alpha i} - \frac{\partial a_\alpha}{\partial x_i} \tag{5.6}$$

Substitution of (5.6) into (5.4) gives the displacement–strain relations

$$e_{ij} = \frac{1}{2}\left(\frac{\partial u_i}{\partial x_j} + \frac{\partial u_j}{\partial x_i} - \frac{\partial u_\alpha}{\partial x_i}\frac{\partial u_\alpha}{\partial x_j}\right) \tag{5.7}$$

If $|\partial u_i/\partial x_j| \ll 1$ so that the nonlinear terms can be neglected, the tensor $\mathbf{e}$ becomes the Cauchy *infinitesimal strain tensor* $\boldsymbol{\varepsilon} = \varepsilon_{ij}\mathbf{i}_i\mathbf{i}_j$, where

$$\varepsilon_{ij} = \frac{1}{2}\left(\frac{\partial u_i}{\partial x_j} + \frac{\partial u_j}{\partial x_i}\right) \tag{5.8}$$

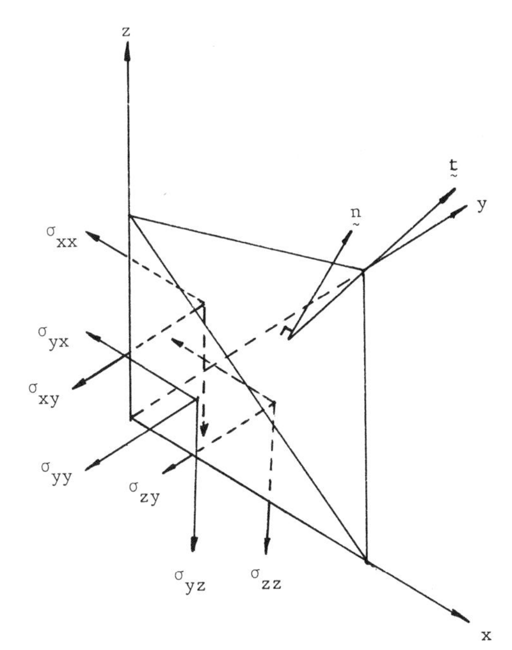

Figure 5.2 Stress components acting on a tetrahedral element.

or, without using the index notation,

$$\begin{aligned} \varepsilon_{xx} &= \frac{\partial u}{\partial x}, & \varepsilon_{xy} &= \varepsilon_{yx} = \frac{1}{2}\left(\frac{\partial u}{\partial y} + \frac{\partial v}{\partial x}\right) \\ \varepsilon_{yy} &= \frac{\partial v}{\partial y}, & \varepsilon_{yz} &= \varepsilon_{zy} = \frac{1}{2}\left(\frac{\partial v}{\partial z} + \frac{\partial w}{\partial y}\right) \\ \varepsilon_{zz} &= \frac{\partial w}{\partial z}, & \varepsilon_{zx} &= \varepsilon_{zx} = \frac{1}{2}\left(\frac{\partial w}{\partial x} + \frac{\partial u}{\partial z}\right) \end{aligned} \tag{5.9}$$

The next step in the development is to introduce the stress tensor $\boldsymbol{\sigma} = \sigma_{ij}\mathbf{i}_j\mathbf{i}_j$ in the deformed configuration. To do this, let us define the traction (or stress) vector $\mathbf{t}$ on a surface characterized by a unit vector normal to it. Let $\Delta\mathbf{F}$ be the net force applied on a differential area ΔA of the surface.* We assume that the limit of $\Delta\mathbf{F}/\Delta A$ as $\Delta A \to 0$ exists and define this limit to be the *traction vector* $\mathbf{t}$ acting on the surface characterized by the normal unit vector $\mathbf{n}$. Since $\mathbf{t}$ is a vector, it can be represented by $\mathbf{t} = t_i\mathbf{i}_i$ in the Cartesian coordinate system. Applying the Cauchy principle, that is, considering the equilibrium of forces acting on the tetrahedron shown in Figure 5.2, we can define the components of the *stress tensor* σ_{ji} by

$$t_i = \sigma_{ji} n_j \tag{5.10}$$

Indeed, in the x direction, we have

$$\sigma_{xx}\,\Delta A_x + \sigma_{yx}\,\Delta A_y + \sigma_{zx}\,\Delta A_z - \rho f_x\,\Delta V = t_x\,\Delta A$$

* In general a couple $\Delta\mathbf{C}$ would also be involved. We assume, however, that $\Delta\mathbf{C}/\Delta A \to 0$ as $\Delta A \to 0$.

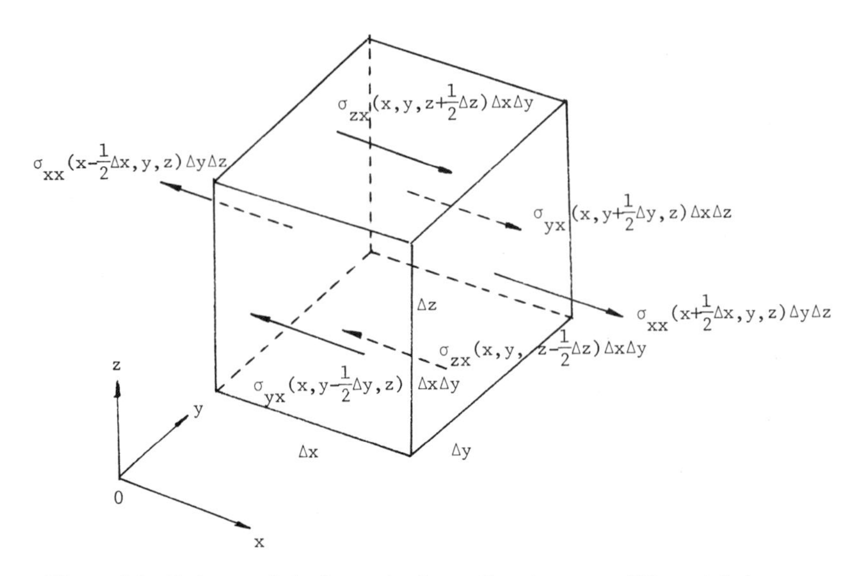

Figure 5.3 Balance of the forces in the x direction on a differential element.

Applying the relations

$$\Delta A_x = n_x \, \Delta A, \qquad \Delta A_y = n_y \, \Delta A, \qquad \Delta A_z = n_z \, \Delta A$$

yields

$$\sigma_{j1} n_j \, \Delta A - \rho f_1 \, \Delta V = t_1 \, \Delta A$$

using index notation. Since $\Delta V / \Delta A \to 0$ as the size of the tetrahedron becomes smaller and smaller, the Cauchy principle (5.10) is obtained.

Let us now consider the equilibrium of forces in the differential volume element inside of the deformed body shown in Figure 5.3. Equilibrium in the x direction becomes

$$\begin{aligned}
&\{-\sigma_{xx}(x - \tfrac{1}{2}\,\Delta x, y, z) + \sigma_{xx}(x + \tfrac{1}{2}\,\Delta x, y, z)\}\,\Delta y\,\Delta z \\
&\quad + \{-\sigma_{yx}(x, y - \tfrac{1}{2}\,\Delta y, z) + \sigma_{yx}(x, y + \tfrac{1}{2}\,\Delta y, z)\}\,\Delta z\,\Delta x \\
&\quad + \{-\sigma_{zx}(x, y, z - \tfrac{1}{2}\,\Delta z) + \sigma_{zx}(x, y, z + \tfrac{1}{2}\,\Delta z)\}\,\Delta x\,\Delta y \\
&\quad + \rho f_x\,\Delta x\,\Delta y\,\Delta z = 0
\end{aligned}$$

Divide by $\Delta x \, \Delta y \, \Delta z$ and pass to the limits Δx, Δy, and $\Delta z \to 0$ to obtain the equation

$$\frac{\partial \sigma_{xx}}{\partial x} + \frac{\partial \sigma_{yx}}{\partial y} + \frac{\partial \sigma_{zx}}{\partial z} + \rho f_x = 0 \tag{5.11}$$

Similar equations are obtained for the other two directions. In index notation, these can be expressed as

$$\frac{\partial \sigma_{ji}}{\partial x_j} + \rho f_i = 0 \tag{5.12}$$

The result of imposing moment equilibrium is the symmetry of the stress tensor.* That is,

$$\sigma_{ij} = \sigma_{ji} \qquad (\text{or} \quad \sigma_{xy} = \sigma_{yx}, \sigma_{yz} = \sigma_{zy}, \sigma_{zx} = \sigma_{xz}) \tag{5.13}$$

If inertia effects are included, the equilibrium equation (5.12) must be changed to

$$\rho \ddot{u}_i = \frac{\partial \sigma_{ji}}{\partial x_j} + \rho f_i \tag{5.14}$$

Note that "$\cdot$" is the material derivative. That is, the convective terms are included because we are describing the deformation and motion with respect to the deformed configuration. However, if only infinitesimal deformation is considered, it is assumed that there is no distinction between the undeformed and deformed configurations *as far as geometrical quantities are concerned*. We shall restrict our interests to such infinitesimal deformation in this chapter.

Exercise 5.1 (cylindrical coordinate system): If the cylindrical system (r, θ, z) is applied instead of the Cartesian coordinate system (x, y, z) to describe deformation of a body, the displacement vector $\mathbf{u}$ is represented by

$$\mathbf{u} = u_r \mathbf{e}_r + u_\theta \mathbf{e}_\theta + u_z \mathbf{e}_z \tag{5.15}$$

where $\mathbf{e}_r$, $\mathbf{e}_\theta$, and $\mathbf{e}_z$ are unit vectors along the r, θ, and z axes, respectively. Noting that $\mathbf{e}_r$ and $\mathbf{e}_\theta$ change their directions in θ and their derivatives in θ are given by

$$\frac{d\mathbf{e}_r}{d\theta} = \mathbf{e}_\theta \qquad \text{and} \qquad \frac{d\mathbf{e}_\theta}{d\theta} = -\mathbf{e}_r$$

the (linearized) strain tensor $\boldsymbol{\varepsilon}$ is obtained by the definition (5.8) after neglecting higher-order terms. Indeed, the square of the distance of two very close points P and P' in the undeformed configuration is given by

$$\Delta S_0^2 = (\mathbf{r}_{P'} - \mathbf{r}_P) \cdot (\mathbf{r}_{P'} - \mathbf{r}_p) = \Delta r^2 + (r\,\Delta\theta)^2 + \Delta\, z^2 \tag{5.16}$$

where the position vectors of P and P' are defined by

$$\mathbf{r}_P = r\mathbf{e}_r(\theta) + z\mathbf{e}_z$$

and

$$\mathbf{r}_{P'} = (r + \Delta r)\mathbf{e}_r(\theta + \Delta\theta) + (z + \Delta z)\mathbf{e}_z$$

respectively. If P and P' are displaced by

$$\mathbf{u}_P = u_r \mathbf{e}_r + u_\theta \mathbf{e}_\theta + u_z \mathbf{e}_z$$

and

$$\mathbf{u}_{P'} = (u_r + \Delta u_r)\mathbf{e}_r(\theta + \Delta\theta) + (u_\theta + \Delta u_\theta)\mathbf{e}_\theta(\theta + \Delta\theta) + (u_z + \Delta u_z)\mathbf{e}_z$$

* Had we *not* assumed $\lim(\Delta\mathbf{C}/\Delta A) = \mathbf{0}$, the stress tensor would not be symmetric. For more on this, consult Fung (1965).

respectively, after deformation, the square of the distance of P and P' in the deformed configuration is approximately obtained as follows by neglecting the higher-order terms of u_θ^2, $(\partial u_r/\partial r)^2$, and so forth:

$$\Delta S^2 = \Delta S_0^2 + 2\left[\frac{\partial u_r}{\partial r}\Delta r^2 + r\left(\frac{\partial u_\theta}{\partial \theta} + u_r\right)\Delta\theta^2 + \frac{\partial u_z}{\partial z}\Delta z^2 + \left(\frac{\partial u_r}{\partial \theta} + r\frac{\partial u_\theta}{\partial r} - u_\theta\right)\Delta r\,\Delta\theta + \left(r\frac{\partial u_\theta}{\partial z} + \frac{\partial u_z}{\partial \theta}\right)\Delta\theta\,\Delta z + \left(\frac{\partial u_z}{\partial r} + \frac{\partial u_r}{\partial z}\right)\Delta z\,\Delta r\right] \tag{5.17}$$

Here the relation

$$\Delta u_r = \frac{\partial u_r}{\partial r}\Delta r + \frac{\partial u_r}{\partial \theta}\Delta\theta + \frac{\partial u_r}{\partial z}\Delta z$$

and other similar relations on Δu_θ and Δu_z have been applied. Thus, the definition of the strain tensor (5.8) implies the physical components of the strain ε in the cylindrical coordinate system (r, θ, z):

$$\varepsilon_{rr} = \frac{\partial u_r}{\partial r}, \qquad \varepsilon_{\theta\theta} = \frac{1}{r}\left(\frac{\partial u_\theta}{\partial \theta} + u_r\right), \qquad \varepsilon_{zz} = \frac{\partial u_z}{\partial z}$$
$$\varepsilon_{r\theta} = \frac{1}{2}\left(\frac{1}{r}\frac{\partial u_r}{\partial \theta} + \frac{\partial u_\theta}{\partial r} - \frac{u_\theta}{r}\right), \qquad \varepsilon_{\theta z} = \frac{1}{2}\left(\frac{\partial u_\theta}{\partial z} + \frac{1}{r}\frac{\partial u_z}{\partial \theta}\right) \tag{5.18}$$
$$\varepsilon_{zr} = \frac{1}{2}\left(\frac{\partial u_z}{\partial r} + \frac{\partial u_r}{\partial z}\right)$$

Derive the physical components of the strain ε without neglecting the higher-order terms, that is, for large deformation.

Now, considering the balance of forces in an infinitesimal volume element in a cylindrical coordinate system, derive the equilibrium equation

$$-\frac{1}{r}\frac{\partial}{\partial r}(r\sigma_{rr}) - \frac{1}{r}\frac{\partial}{\partial \theta}\sigma_{r\theta} - \frac{\partial}{\partial z}\sigma_{rz} + \frac{\sigma_{\theta\theta}}{r} = \rho f_r$$
$$-\frac{1}{r^2}\frac{\partial}{\partial r}(r^2\sigma_{\theta r}) - \frac{1}{r}\frac{\partial}{\partial \theta}\sigma_{\theta\theta} - \frac{\partial}{\partial z}\sigma_{\theta z} = \rho f_\theta \tag{5.19}$$
$$-\frac{1}{r}\frac{\partial}{\partial r}(r\sigma_{zr}) - \frac{1}{r}\frac{\partial}{\partial \theta}\sigma_{z\theta} - \frac{\partial}{\partial z}\sigma_{zz} = \rho f_z$$

where σ_{rr}, $\sigma_{r\theta}$, and others are the physical components of the stress tensor $\boldsymbol{\sigma}$, and $\mathbf{f}$ is the body force per unit mass.

Assuming axisymmetry about the z-axis such that all the functions are independent of the coordinate θ, obtain the following strain–displacement relation and equilibrium equation for the axisymmetric problem:

$$\varepsilon_{rr} = \frac{\partial u_r}{\partial r}, \qquad \varepsilon_{\theta\theta} = \frac{u_r}{r}, \qquad \varepsilon_{zz} = \frac{\partial u_z}{\partial z}$$
$$\varepsilon_{r\theta} = \frac{1}{2}\left(\frac{\partial u_\theta}{\partial r} - \frac{u_\theta}{r}\right), \qquad \varepsilon_{\theta z} = \frac{1}{2}\frac{\partial u_\theta}{\partial z}, \qquad \varepsilon_{zr} = \frac{1}{2}\left(\frac{\partial u_z}{\partial r} + \frac{\partial u_r}{\partial z}\right) \tag{5.20}$$

and

$$-\frac{1}{r}\frac{\partial}{\partial r}(r\sigma_{rr}) - \frac{\partial}{\partial z}\sigma_{rz} + \frac{1}{r}\sigma_{\theta\theta} = \rho f_r$$
$$-\frac{1}{r^2}\frac{\partial}{\partial r}(r^2\sigma_{\theta r}) - \frac{\partial}{\partial z}\sigma_{\theta z} = \rho f_\theta \tag{5.21}$$
$$-\frac{1}{r}\frac{\partial}{\partial r}(r\sigma_{zr}) - \frac{\partial}{\partial z}\sigma_{zz} = \rho f_z$$

5.2 Linear elastic materials

The displacement–strain relations and the equilibrium equations have been obtained so far. The remaining mechanical issue is to introduce the material characteristics for the solid body. In this section, we shall restrict our attention to linear materials described by the constitutive equations

$$\sigma_{ij} = E_{ijkl}\varepsilon_{kl} \tag{5.22}$$

by using the *elasticity tensor* **E**. Since the stress and strain tensors are symmetric, the components of **E** must satisfy

$$E_{ijkl} = E_{jikl} = E_{ijlk} \tag{5.23a}$$

If, furthermore, the existence of a strain energy density is assumed, the **E**'s must satisfy the additional symmetry conditions

$$E_{ijkl} = E_{klij} \tag{5.23b}$$

Under the conditions (5.23), the strain energy density U_0 is defined as

$$U_0 = \tfrac{1}{2}E_{ijkl}\varepsilon_{kl}\varepsilon_{ij} \tag{5.24}$$

if **E** does not depend on the strain and stress. Moreover, the relation

$$\sigma_{ij} = \frac{\partial U_0}{\partial \varepsilon_{ij}} \tag{5.25}$$

is satisfied. If the material is nonlinear, and if the stress–strain relation (5.25) holds for a strain energy density U_0, such materials are called *hyperelastic*.

The tensor **E** assumes its simplest form for the *isotropic* materials, and we have

$$E_{ijkl} = \lambda\delta_{ij}\delta_{kl} + \mu(\delta_{ik}\delta_{jl} + \delta_{il}\delta_{jk}) \tag{5.26}$$

where λ and μ are the Lamé constants related to Young's modulus E and Poisson's ratio ν by

$$\lambda = \frac{E\nu}{(1-2\nu)(1+\nu)}, \qquad \mu = \frac{E}{2(1+\nu)} \tag{5.27}$$

In this case, the constitutive equation (5.22) becomes

$$\sigma_{ij} = \lambda\varepsilon_{kk}\delta_{ij} + 2\mu\varepsilon_{ij} \tag{5.28}$$

If usual notations are used, these are equivalent to

$$\begin{aligned}
\varepsilon_{xx} &= \frac{1}{E}(\sigma_{xx} - \nu\sigma_{yy} - \nu\sigma_{zz}), & \varepsilon_{xy} &= \frac{1}{2\mu}\sigma_{xy} \\
\varepsilon_{yy} &= \frac{1}{E}(-\nu\sigma_{xx} + \sigma_{yy} - \nu\sigma_{zz}), & \varepsilon_{yz} &= \frac{1}{2\mu}\sigma_{yz} \\
\varepsilon_{zz} &= \frac{1}{E}(-\nu\sigma_{xx} - \nu\sigma_{yy} + \sigma_{zz}), & \varepsilon_{zx} &= \frac{1}{2\mu}\sigma_{zx}
\end{aligned} \tag{5.29}$$

Let us now decompose the strain $\boldsymbol{\varepsilon}$ into the sum of a volumetric and a pure shear contribution:

$$\varepsilon_{ij} = \tfrac{1}{3}\varepsilon_{kk}\delta_{ij} + \varepsilon_{ij}^{D} \tag{5.30}$$

where ε_{ij}^{D} is called the strain deviator defined by

$$\varepsilon_{ij}^{D} = \varepsilon_{ij} - \tfrac{1}{3}\varepsilon_{kk}\delta_{ij} \tag{5.31}$$

When the decomposition (5.30) is applied, the constitutive equation (5.29) becomes

$$\sigma_{ij} = (\lambda + \tfrac{2}{3}\mu)\varepsilon_{kk}\delta_{ij} + 2\mu\varepsilon_{ij}^{D} \tag{5.32}$$

where the coefficient $\lambda + \frac{2}{3}\mu$, which is often denoted by κ, that is,

$$\kappa = \lambda + \tfrac{2}{3}\mu \tag{5.33}$$

is called the *bulk modulus*. The *mean stress* is defined by

$$\sigma_m = \tfrac{1}{3}\sigma_{ii} = \kappa\varepsilon_{ii} \tag{5.34}$$

In the above, we have used the index (or tensor) notation to describe the material characteristics. Although this notation is widely used, another description is also frequently used in the literature. We shall call the following the *contracted* notation. Let

$$\boldsymbol{\sigma} = \begin{Bmatrix} \sigma_1 \\ \sigma_2 \\ \sigma_3 \\ \sigma_4 \\ \sigma_5 \\ \sigma_6 \end{Bmatrix} = \begin{Bmatrix} \sigma_{11} \\ \sigma_{12} \\ \sigma_{33} \\ \sigma_{23} \\ \sigma_{31} \\ \sigma_{12} \end{Bmatrix}, \qquad \boldsymbol{\varepsilon} = \begin{Bmatrix} \varepsilon_1 \\ \varepsilon_2 \\ \varepsilon_3 \\ \varepsilon_4 \\ \varepsilon_5 \\ \varepsilon_6 \end{Bmatrix} = \begin{Bmatrix} \varepsilon_{11} \\ \varepsilon_{22} \\ \varepsilon_{33} \\ 2\varepsilon_{23} = \gamma_{23} \\ 2\varepsilon_{31} = \gamma_{31} \\ 2\varepsilon_{23} = \gamma_{23} \end{Bmatrix} \tag{5.35}$$

and let $\mathbf{C}$ be a contracted version of $\mathbf{E}$ (to be defined below), which is symmetric $(C_{ij} = C_{ji})$. Corresponding to the constitutive equation (5.22) we have

$$\sigma_i = C_{ij}\varepsilon_j \tag{5.36}$$

The strain energy density U_0 and the stress–strain relation given by (5.24) and (5.25), respectively, become

$$U_0 = \tfrac{1}{2}C_{ij}\varepsilon_j\varepsilon_i \tag{5.37a}$$

$$\sigma_i = \frac{\partial U_0}{\partial \varepsilon_i}, \quad i = 1, \ldots, 6 \tag{5.37b}$$

The correspondence between **E** and **C** is given as

$$[C_{ij}] = \begin{bmatrix} E_{1111} & E_{1122} & E_{1133} & E_{1123} & E_{1131} & E_{1112} \\ & E_{2222} & E_{2233} & E_{2223} & E_{2231} & E_{2212} \\ & & E_{3333} & E_{3323} & E_{3331} & E_{3312} \\ & & & E_{2323} & E_{2331} & E_{2312} \\ & & & & E_{3131} & E_{3112} \\ \text{SYM} & & & & & E_{1212} \end{bmatrix} \tag{5.38}$$

In general, we have 21 independent constants to describe the material characteristics.

If the mechanical material behavior is symmetric about the plane $x_3 = 0$, then the number of independent constants is reduced to 13 and is shown as follows:

$$[C_{ij}] = \begin{bmatrix} E_{1111} & E_{1122} & E_{1133} & 0 & 0 & E_{1112} \\ & E_{2222} & E_{2223} & 0 & 0 & E_{2212} \\ & & E_{3333} & 0 & 0 & E_{3312} \\ & & & E_{2323} & E_{2331} & 0 \\ & & & & E_{3131} & 0 \\ \text{SYM} & & & & & E_{1212} \end{bmatrix} \tag{5.39}$$

Such materials are called *monoclinic*. If material symmetry exists for two orthogonal planes, the material is said to be *orthotropic*, and we have

$$[C_{ij}] = \begin{bmatrix} E_{1111} & E_{1122} & E_{1133} & 0 & 0 & 0 \\ & E_{2222} & E_{2233} & 0 & 0 & 0 \\ & & E_{3333} & 0 & 0 & 0 \\ & & & E_{2323} & 0 & 0 \\ & & & & E_{3131} & 0 \\ \text{SYM} & & & & & E_{1212} \end{bmatrix} \tag{5.40}$$

If the material behavior is independent of direction in the plane $x_3 = 0$, the material is said to be *transversely isotropic* and is characterized by five constants as

$$[C_{ij}] = \begin{bmatrix} E_{1111} & E_{1122} & E_{1133} & 0 & 0 & 0 \\ & E_{1111} & E_{1133} & 0 & 0 & 0 \\ & & E_{3333} & 0 & 0 & 0 \\ & & & E_{2323} & 0 & 0 \\ & & & & E_{2323} & 0 \\ \text{SYM} & & & & & \tfrac{1}{2}(E_{1111} - E_{1122}) \end{bmatrix} \tag{5.41}$$

If there are no directional references at all, the material is said to be *isotropic* and is characterized by (5.32); that is,

$$[C_{ij}] = \begin{bmatrix} E_{1111} & E_{1122} & E_{1122} & 0 & 0 & 0 \\ & E_{1111} & E_{1122} & 0 & 0 & 0 \\ & & E_{1111} & 0 & 0 & 0 \\ & & & \frac{1}{2}(E_{1111}-E_{1122}) & 0 & 0 \\ & & & & \frac{1}{2}(E_{1111}-E_{1122}) & 0 \\ \text{SYM} & & & & & \frac{1}{2}(E_{1111}-E_{1122}) \end{bmatrix} \tag{5.42a}$$

or

$$[C_{ij}] = \begin{bmatrix} \lambda+2\mu & \lambda & \lambda & 0 & 0 & 0 \\ & \lambda+2\mu & \lambda & 0 & 0 & 0 \\ & & \lambda+2\mu & 0 & 0 & 0 \\ & & & \mu & 0 & 0 \\ & & & & \mu & 0 \\ \text{SYM} & & & & & \mu \end{bmatrix} \tag{5.42b}$$

5.2.1 Deformation with temperature change

Let us now consider deformation accompanied by a change in the temperature of the body. Suppose that the initial (i.e., undeformed) configuration has no external forces and is at a given temperature T_0. If the temperature changes inside the body, either as a result of the deformation process itself or from external causes, the first two laws of thermodynamics and the law of thermal expansion provide the constitutive equation:

$$\sigma_{ij} = -\beta_{ij}(T - T_0) + E_{ijkl}\varepsilon_{kl} \tag{5.43}$$

where $\boldsymbol{\beta} = \beta_{ij}\mathbf{i}_i\mathbf{i}_j$ is the material tensor related to thermal expansion.* Since the stress is symmetric, $\boldsymbol{\beta}$ must satisfy

$$\beta_{ij} = \beta_{ji} \tag{5.44}$$

If the material is isotropic, we may put

$$\beta_{ij} = \kappa\alpha\delta_{ij} \tag{5.45}$$

which implies the relation

$$\varepsilon_{kk} = \alpha(T - T_0) \tag{5.46}$$

if $\sigma_{ii} = 0$. [Recall that for an isotropic material, $\mathbf{E}$ is given by (5.26).]

If isothermal deformation takes place, we must put $T = T_0$, and then the coefficients κ and μ are called isothermal moduli. In an adiabatic deformation, the change of temperature $T - T_0$ due to the deformation can be computed. This results in a change in the constitutive equation. For isotropic materials, we have

$$\sigma_{ij} = \kappa_{ad}\varepsilon_{kk}\delta_{ij} + 2\mu\varepsilon_{ij}^D \tag{5.47}$$

* Changes in E_{ijkl} with T are ignored, which is reasonable for temperature changes of a few hundred degrees Fahrenheit.

where κ_{ad} is the adiabatic bulk modulus related to the isothermal bulk modulus κ by

$$\frac{1}{\kappa_{ad}} = \frac{1}{\kappa} - \frac{T_\alpha^2}{c_p} \tag{5.48}$$

Here c_p is the specific heat per unit volume at constant pressure.

If the deformation is infinitesimal, and if "decoupling" of the deformation process from heat conduction is assumed, the temperature distribution T can be computed by solving the heat conduction equation discussed in Chapters 2 and 3. Then we can determine the deformation of the body by substituting the computed temperature T into (5.43). The term $-\beta_{ij}(T - T_0)$ is considered as an "initial" stress in the body created by the temperature distribution.

In summary, we have the following relations for infinitesimal deformation, uncoupled from thermal effects:

Equations of motion

$$\rho \ddot{u}_i - \frac{\partial}{\partial x_j}\sigma_{ji} = \rho f_i \quad \text{in } \Omega$$

Displacement–strain relation

$$\varepsilon_{ij} = \frac{1}{2}\left(\frac{\partial u_i}{\partial x_j} + \frac{\partial u_j}{\partial x_i}\right) \quad \text{in } \Omega \tag{5.49}$$

Stress–strain relation

$$\sigma_{ij} = -\beta_{ij}(T - T_0) + E_{ijkl}\varepsilon_{kl} \quad \text{in } \Omega$$

Heat conduction

$$\rho c_p \dot{T} - \frac{\partial}{\partial x_i}\left(k_{ij}\frac{\partial T}{\partial x_j}\right) = f \quad \text{in } \Omega$$

In the following discussion, we shall treat only the static problem so that

$$\ddot{\mathbf{u}} = \mathbf{0} \tag{5.50}$$

and shall assume that the temperature T is known in the body as a result of solving the steady-state heat conduction equation with proper boundary conditions.

5.3 Finite element approximations

Let us now consider a stationary boundary value problem for a linearly elastic body:

$$\left.\begin{aligned} -\frac{\partial}{\partial x_j}\sigma_{ji} &= \rho f_i \\ \varepsilon_{ij} &= \frac{1}{2}\left(\frac{\partial u_i}{\partial x_j} + \frac{\partial u_j}{\partial x_i}\right) \\ \sigma_{ij} &= -\beta_{ij}(T - T_0) + E_{ijkl}\varepsilon_{kl} \end{aligned}\right\} \quad \text{in } \Omega \tag{5.51}$$

and the boundary conditions

$$\begin{aligned} u_i &= g_i \quad \text{on } \Gamma_{i1} \\ \sigma_{ji} n_j &= h_i \quad \text{on } \Gamma_{i2} \\ \sigma_{ji} n_j &= -\hat{k}_{ij}(u_j - u_{\infty j}) \quad \text{on } \Gamma_{i3} \end{aligned} \tag{5.52}$$

where $\Gamma = \Gamma_{i1} \cup \Gamma_{i2} \cup \Gamma_{i3}$ for each i, $i = 1, 2, 3$. Note that we specify the i component of the displacement vector $\mathbf{u}$ on the boundary Γ_{i1}. Similarly, the i component of the traction is specified on Γ_{i2}. It is certain that we have assumed Γ_{i1}, Γ_{i2}, and Γ_{i3} are mutually disjointed. The first boundary condition is the specification of the boundary displacement of the body. For example, if a structure is supported rigidly over Γ_{i1}, then g_i becomes zero. The second boundary condition is nothing but the traction condition. If the boundary is free, then $h_i = 0$. The third boundary condition describes a generalization of a spring support on the boundary.

The first step in a finite element approximation is to derive a weak form of the boundary value problem (5.51) and (5.52). Since the displacement is a vector-valued function, let us multiply the first of (5.51) by a virtual displacement $\bar{\mathbf{u}}$ and integrate over Ω. Integration by parts then gives

$$\int_\Omega \sigma_{ji}(\mathbf{u}) \bar{u}_{i,j} \, d\Omega - \int_\Gamma \sigma_{ji}(\mathbf{u}) n_j \bar{u}_i \, d\Gamma = \int_\Omega \rho f_i \bar{u}_i \, d\Omega$$

where $\sigma_{ji}(\mathbf{u})$ indicates that the stress component is computed from the actual displacement field $\mathbf{u}$. Similarly, let $\varepsilon_{ij}(\bar{\mathbf{u}})$ denote the strain component computed from the virtual displacement $\bar{\mathbf{u}}$. Applying the symmetry of the stress,

$$\begin{aligned} \sigma_{ji}(\mathbf{u}) \bar{u}_{i,j} &= \tfrac{1}{2}[\sigma_{ij}(\mathbf{u}) + \sigma_{ji}(\mathbf{u})] \bar{u}_{i,j} \\ &= \tfrac{1}{2}\sigma_{ij}(\mathbf{u}) \bar{u}_{i,j} + \tfrac{1}{2}\sigma_{ij}(\mathbf{u}) \bar{u}_{j,i} \\ &= \sigma_{ij}(\mathbf{u}) \varepsilon_{ij}(\bar{\mathbf{u}}) \end{aligned}$$

and the boundary conditions (5.52), we obtain the weak form

$$\begin{aligned} & u_i = g_i \quad \text{on } \Gamma_{i1} \\ & \int_\Omega \sigma_{ij}(\mathbf{u}) \varepsilon_{ij}(\bar{\mathbf{u}}) \, d\Omega + \int_{\Gamma_{i3}} \hat{k}_{ij} u_j \bar{u}_i \, d\Gamma \\ & \quad = \int_\Omega \rho f_i \bar{u}_i \, d\Omega + \int_{\Gamma_{i2}} h_i \bar{u}_i \, d\Gamma + \int_{\Gamma_{i3}} \hat{k}_{ij} u_{\infty j} \bar{u}_i \, d\Gamma, \\ & \forall \bar{\mathbf{u}} \ni \bar{u}_i = 0 \text{ on } \Gamma_{i1} \end{aligned} \tag{5.53}$$

where we assume the following conventions:

$$\int_{\Gamma_{i3}} \hat{k}_{ij} u_j \bar{u}_i \, d\Gamma = \sum_{i=1}^{3} \int_{\Gamma_{i3}} \hat{k}_{ij} u_j \bar{u}_i \, d\Gamma$$

$$\int_{\Gamma_{i2}} h_i \bar{u}_i \, d\Gamma = \sum_{i=1}^{3} \int_{\Gamma_{i2}} h_i \bar{u}_i \, d\Gamma$$

and

$$\int_{\Gamma_{i3}} \hat{k}_{ij} u_{\infty j} \bar{u}_i \, d\Gamma = \sum_{i=1}^{3} \int_{\Gamma_{i3}} \hat{k}_{ij} u_{\infty j} \bar{u}_i \, d\Gamma$$

In the subsequent study, we shall apply these conventions. The first term $\int_\Omega \sigma_{ij}(\mathbf{u})\varepsilon_{ij}(\bar{\mathbf{u}})\, d\Omega$ is the *internal work* done due to the virtual displacement $\bar{\mathbf{u}}$ from the equilibrium configuration defined by the displacement field $\mathbf{u}$. Similarly, the second term is the virtual work of the "spring" along the boundary Γ_{i3}. The right side becomes the *virtual work* by the body force $\rho\mathbf{f}$ and others. Thus, the weak form is nothing but the *principle of virtual work* in mechanics!

Since (5.23b) and (5.44) are assumed here, if the spring constants $\hat{k}_{ij}$ are also symmetric (i.e., if $\hat{k}_{ij} = \hat{k}_{ji}$), the weak form (5.53) could be derived from the *principle of minimum potential energy*. Indeed, if a functional F is defined as

$$F(\mathbf{v}) = \tfrac{1}{2}\int_\Omega \sigma_{ij}(\mathbf{v})\varepsilon_{ij}(\mathbf{v})\, d\Omega + \tfrac{1}{2}\int_{\Gamma_{i3}} \hat{k}_{ij} v_j v_i\, d\Gamma$$

$$- \int_\Omega \rho f_i v_i\, d\Omega - \int_{\Gamma_{i2}} h_i v_i\, d\Gamma - \int_{\Gamma_{i3}} \hat{k}_{ij} u_{\infty j} v_i\, d\Gamma \tag{5.54}$$

for an arbitrary displacement field $\mathbf{v}$ such that $v_i = g_i$ on Γ_{i1}, then F becomes the total potential energy of the body Ω consisting of the total strain energy, the strain energy of the springs attached to the boundary Γ_{i3}, and the potential of the applied forces ($\mathbf{f}$, $\mathbf{h}$, $\hat{\mathbf{k}}\mathbf{u}_\infty$) in the body and on its boundary. The principle of minimum potential energy then says that the displacement field $\mathbf{u}$ corresponding to the equilibrium configuration minimizes the total potential energy F among all possible displacements that satisfy the kinematic condition $v_i = g_i$ on the boundary Γ_{i1}, that is,

$$u_i = g_i \quad \text{on } \Gamma_{i1} \qquad \text{and} \qquad F(\mathbf{u}) \le F(\mathbf{v}), \quad \forall \mathbf{v} \ni v_i = g_i \text{ on } \Gamma_{i1} \tag{5.55}$$

If, in (5.55), we take $\mathbf{v} = \mathbf{u} \pm \theta\bar{\mathbf{u}}$ with $\theta \in (0, 1)$ and $\bar{\mathbf{u}}$ an arbitrary displacement field such that $\bar{u}_i = 0$ on Γ_{i1}, we have

$$\pm\theta\left\{\int_\Omega \sigma_{ij}(\mathbf{u})\varepsilon_{ij}(\bar{\mathbf{u}})\, d\Omega + \int_{\Gamma_{i3}} \hat{k}_{ij} u_j \bar{u}_i\, d\Gamma\right\}$$

$$\mp\theta\left\{\int_\Omega \rho f_i \bar{u}_i\, d\Omega + \int_{\Gamma_{i2}} h_i \bar{u}_i\, d\Gamma + \int_{\Gamma_{i3}} \hat{k}_{ij} u_{\infty j} \bar{u}_i\, d\Gamma\right\}$$

$$+ \tfrac{1}{2}\theta^2\left\{\int_\Omega \sigma_{ij}(\bar{\mathbf{u}})\varepsilon_{ij}(\bar{\mathbf{u}})\, d\Omega + \int_{\Gamma_{i3}} \hat{k}_{ij} \bar{u}_j \bar{u}_i\, d\Gamma\right\} \ge 0$$

Dividing by $\pm\theta$ and letting $\theta \to 0$, we obtain the form

$$\int_\Omega \sigma_{ij}(\mathbf{u})\varepsilon_{ij}(\bar{\mathbf{u}})\, d\Omega + \int_{\Gamma_{i3}} \hat{k}_{ij} u_j \bar{u}_i\, d\Gamma$$

$$= \int_\Omega \rho f_i \bar{u}_i\, d\Omega + \int_{\Gamma_{i2}} h_i \bar{u}_i\, d\Gamma + \int_{\Gamma_{i3}} \hat{k}_{ij} u_{\infty j} \bar{u}_i\, d\Gamma,$$

$$\forall \bar{\mathbf{u}} \ni \bar{u}_i = 0 \text{ on } \Gamma_{i1}$$

This is same as (5.53). Because of the choice $\mathbf{v} = \mathbf{u} + \theta\bar{\mathbf{u}}$, it can be understood that $\bar{\mathbf{u}}$ is an arbitrary variation from the true displacement field $\mathbf{u}$ and θ is a magnification factor. In the literature on the calculus of variations, $\theta\bar{\mathbf{u}}$ is denoted by $\delta\mathbf{u}$ in order to emphasize "*variation* δ." Thus, the minimum problem represented by (5.54) and (5.55) could also be a basis of finite element approximations.

The next step is to discretize the domain Ω into a set of finite elements and to define the shape functions $\{N_\alpha(\mathbf{x})\}$ on each finite element Ω_e. Then the displacement $\mathbf{u}$ and the virtual displacement $\bar{\mathbf{u}}$ are approximated by linear combinations of their nodal values and their shape functions:

$$u_j = u_{j\beta} N_\beta(\mathbf{x}), \qquad \bar{u}_i = \bar{u}_{i\alpha} N_\alpha(\mathbf{x}) \tag{5.56}$$

in each finite element Ω_e.

The discretization of the first term of the weak form (5.53) follows from (5.43) and (5.56):

$$\begin{aligned} \int_\Omega \sigma_{ij}(\mathbf{u})\varepsilon_{ij}(\bar{\mathbf{u}})\, d\Omega &= -\int_\Omega \beta_{ij}(T - T_0)\bar{u}_{i,j}\, d\Omega + \int_\Omega E_{ijkl} u_{k,l} \bar{u}_{i,j}\, d\Omega \\ &= -\sum_{e=1}^{E} \bar{u}_{i\alpha} f_{i\alpha}^{0,e} + \sum_{e=1}^{E} \bar{u}_{i\alpha} E_{i\alpha j\beta}^{e} u_{i\beta} \end{aligned} \tag{5.57}$$

where

$$f_{i\alpha}^{0,e} = \int_{\Omega_e} (T - T_0)\beta_{ij} N_{\alpha,j}\, d\Omega \tag{5.58}$$

and

$$E_{i\alpha j\beta}^{e} = \int_{\Omega_e} E_{ikjl} N_{\alpha,k} N_{\beta,l}\, d\Omega \tag{5.59}$$

If we define

$$\begin{aligned} E_{i\alpha j\beta}^{3,e} &= \int_{\Gamma_{i3}^e} \hat{k}_{ij} N_\alpha N_\beta\, d\Gamma \\ f_{i\alpha}^{1,e} &= \int_{\Omega_e} \rho f_i N_\alpha\, d\Omega \qquad \text{(no sum on } i\text{)} \\ f_{i\alpha}^{2,e} &= \int_{\Gamma_{i2}^e} h_i N_\alpha\, d\Gamma \\ f_{i\alpha}^{3,e} &= \int_{\Gamma_{i3}^e} k_{ij} u_{\infty j} N_\alpha\, d\Gamma \end{aligned} \tag{5.60}$$

then the weak form (5.53) becomes

$$\begin{aligned} & u_{i\alpha} = g_{i\alpha} \quad \text{on } \Gamma_{i1} \\ & \sum_{e=1}^{E} \bar{u}_{i\alpha} E_{i\alpha j\beta}^{e} u_{j\beta} + \sum_{e=1}^{E_3} \bar{u}_{i\alpha} E_{i\alpha i\beta}^{3,e}\, u_{j\beta} \\ & \quad = \sum_{e=1}^{E} \bar{u}_{i\alpha}(f_{i\alpha}^{0,e} + f_{i\alpha}^{1,e}) + \sum_{e=1}^{E_2} \bar{u}_{i\alpha} f_{i\alpha}^{2,e} + \sum_{e=1}^{E_3} \bar{u}_{i\alpha} f_{i\alpha}^{3,e} \\ & \qquad \forall \bar{u}_{i\alpha} \ni \bar{u}_{i\alpha} = 0 \text{ on } \Gamma_{i1} \end{aligned} \tag{5.61}$$

Since the virtual displacement is arbitrary, (5.61) yields the *finite element equation*

$$\begin{aligned} & u_{i\alpha} = g_{i\alpha} \quad \text{on } \Gamma_{i1} \\ & \sum_{e=1}^{E} E_{i\alpha j\beta}^{e} u_{j\beta} + \sum_{e=1}^{E_3} E_{i\alpha j\beta}^{3,e} \\ & \quad = \sum_{e=1}^{E} (f_{i\alpha}^{0,e} + f_{i\alpha}^{1,e}) + \sum_{e=1}^{E_2} f_{i\alpha}^{2,e} + \sum_{e=1}^{E_2} f_{i\alpha}^{3,e} \end{aligned} \tag{5.62}$$

The next step is to define the shape functions $\{N_\alpha(\mathbf{x})\}$ for a particular choice of finite elements. If three-node triangular elements are used for plane problems, we know the shape functions from Chapter 2. Then the forming of the element stiffness matrix and the generalized load vector can be performed as was done in the case of heat conduction. In the following section, we shall introduce the simplest element from the family of quadrilateral isoparametric elements for plane problems, which enables us to extend the scope of finite element methods. The isoparametric elements have characteristics that are very different from those of the three-node triangular element used so far.

5.4 Four-node quadrilateral element

Let us now restrict our attention to the plane problems in linear elasticity. In this case, we have two different categories: *plane strain* and *generalized plane stress* (GPS). These are mathematically identical, but physically quite different. If plane strain problems are considered for isotropic materials, we have

$$\sigma_{ij} = \lambda \varepsilon_{kk} \delta_{ij} + 2\mu \varepsilon_{ij}, \quad i, j = 1, 2 \tag{5.63}$$

where λ and μ are defined by (5.27). For GPS problems (for isotropic materials), we have

$$\sigma_{ij} = \hat{\lambda} \varepsilon_{kk} \delta_{ij} + 2\mu \varepsilon_{ij}, \quad i, j = 1, 2 \tag{5.64}$$

where μ is given by (5.27) and $\hat{\lambda}$ is now

$$\hat{\lambda} = \frac{E\nu}{1 - \nu^2} \tag{5.65}$$

If the deformation is restricted so that the components u and v of the displacement vector $\mathbf{u} = u\mathbf{i} + v\mathbf{j} + w\mathbf{k}$ are independent of z and $w \equiv 0$, then we have plane strain (parallel to the xy plane). Since the displacement components determine everything else, it is to be noted that in plane strain the only nonzero strain components are ε_{xx}, ε_{yy}, and ε_{xy}.

In order to maintain some generality in our treatment of finite element approximations, we shall assume that the constitutive relations for plane problems have the form

$$\sigma_{ij} = -\hat{\beta}_{ij}(T - T_0) + \hat{E}_{ijkl} \varepsilon_{kl} \tag{5.66}$$

where $\hat{\beta}_{ij} = \hat{\beta}_{ji}$, $\hat{E}_{ijkl} = \hat{E}_{jikl}$, and $i, j, k, l = 1, 2$. The form (5.66) is not limited to isotropic materials.

Now let us define the four-node quadrilateral isoparametric element referred to as the $Q4$ element. Suppose that an element Ω_e used in the discretization of the body is obtained by a mapping of the *master element* defined in the st coordinate system as shown in Figure 5.4 through the rule

$$x = x_\alpha \hat{N}_\alpha(s, t), \qquad y = y_\alpha \hat{N}_\alpha(s, t) \tag{5.67}$$

in which (x_α, y_α) are the coordinates of the four corner nodes in the global coordinate system and $\hat{N}_\alpha(s, t)$ are the shape functions corresponding to the four corner nodes of the master element in the st coordinate system. Since the master

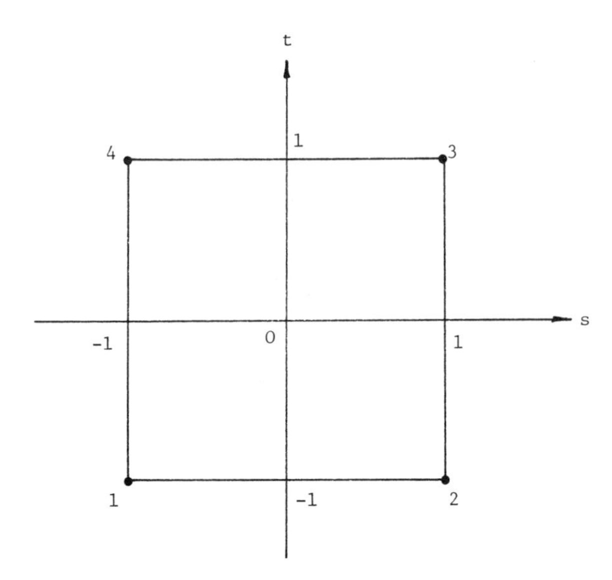

Figure 5.4 Master element for four-node isoparametric element.

element is a square defined by $(-1, 1) \times (-1, 1)$, the shape functions $\{\hat{N}_\alpha(s, t)\}$ are defined as

$$\begin{aligned} \hat{N}_1(s, t) &= \tfrac{1}{4}(1 - s)(1 - t), \qquad & \hat{N}_2(s, t) &= \tfrac{1}{4}(1 + s)(1 - t) \\ \hat{N}_3(s, t) &= \tfrac{1}{4}(1 + s)(1 + t), \qquad & \hat{N}_4(s, t) &= \tfrac{1}{4}(1 - s)(1 + t) \end{aligned} \tag{5.68}$$

that is,

$$\hat{N}_\alpha(s, t) = \tfrac{1}{4}(1 + s_\alpha s)(1 + t_\alpha t)$$

where $\{(s_\alpha, t_\alpha)\}$ is the set of coordinates of nodes in the st system. Note that

$$\hat{N}_\alpha(s_\beta, t_\beta) = \delta_{\alpha\beta}, \qquad \sum_{\alpha=1}^{4} \hat{N}_\alpha(s, t) \equiv 1 \tag{5.69}$$

In terms of the shape functions (5.68), any function can be approximated by a bilinear polynomial. That is,

$$f(s, t) = f_\alpha \hat{N}_\alpha(s, t) = c_0 + c_1 s + c_2 t + c_3 st$$

If the actual displacement field $\mathbf{u}$ and the virtual displacement $\bar{u}$ are approximated similarly to (5.67),

$$u_j = u_{j\beta} \hat{N}_\beta(s, t), \qquad \bar{u}_i = \bar{u}_{i\alpha} \hat{N}_\alpha(s, t) \tag{5.70}$$

we call the four-node element an *isoparametric* element.

A difficulty with the approximations (5.70) is the shape functions are functions of the coordinates s and t in the master element. However, in (5.61), we have approximated $\mathbf{u}$ and $\bar{\mathbf{u}}$ by

$$u_j = u_{j\beta} N_\beta(x, y), \qquad \bar{u}_i = \bar{u}_{i\alpha} N_\alpha(x, y) \tag{5.71}$$

in which the N's are the shape functions for the element Ω_e itself. Because of the mapping (5.67), it is clear that*

$$\hat{N}_\alpha(s,t) = N_\alpha(\overbrace{x_\beta \hat{N}_\beta(s,t)}^{x}, \overbrace{y_\beta \hat{N}_\beta(s,t)}^{y}) \tag{5.72}$$

Thus, for a given element, the shape function $N_\alpha(x, y)$ is defined by (5.72) through specifying s and t in the master element. Applying the chain rule to (5.72), we obtain the first derivatives of the shape functions:

$$\begin{aligned}\frac{\partial \hat{N}_\alpha}{\partial s} &= \frac{\partial N_\alpha}{\partial x}\frac{\partial x}{\partial s} + \frac{\partial N_\alpha}{\partial y}\frac{\partial y}{\partial s}\\ \frac{\partial \hat{N}_\alpha}{\partial t} &= \frac{\partial N_\alpha}{\partial x}\frac{\partial x}{\partial t} + \frac{\partial N_\alpha}{\partial y}\frac{\partial y}{\partial t}\end{aligned} \tag{5.73}$$

that is,

$$\begin{Bmatrix} \dfrac{\partial \hat{N}_\alpha}{\partial s} \\ \dfrac{\partial \hat{N}_\alpha}{\partial t} \end{Bmatrix} = \begin{bmatrix} \dfrac{\partial x}{\partial s} & \dfrac{\partial y}{\partial s} \\ \dfrac{\partial x}{\partial t} & \dfrac{\partial y}{\partial t} \end{bmatrix} \begin{Bmatrix} \dfrac{\partial N_\alpha}{\partial x} \\ \dfrac{\partial N_\alpha}{\partial y} \end{Bmatrix} \tag{5.74}$$

Because of (5.67), each term of the matrix can be evaluated as long as the coordinates $\{(x_\alpha, y_\alpha)\}$ are known. Then by taking the inverse of the relation (5.74), we obtain the first derivatives of the shape functions with respect to the global coordinates:

$$\begin{Bmatrix} \dfrac{\partial N_\alpha}{\partial x} \\ \dfrac{\partial N_\alpha}{\partial y} \end{Bmatrix} = \frac{1}{J} \begin{bmatrix} \dfrac{\partial y}{\partial t} & -\dfrac{\partial y}{\partial s} \\ -\dfrac{\partial x}{\partial t} & \dfrac{\partial x}{\partial s} \end{bmatrix} \begin{Bmatrix} \dfrac{\partial \hat{N}_\alpha}{\partial s} \\ \dfrac{\partial \hat{N}_\alpha}{\partial t} \end{Bmatrix} \tag{5.75}$$

where

$$J = \frac{\partial x}{\partial s}\frac{\partial y}{\partial t} - \frac{\partial y}{\partial s}\frac{\partial x}{\partial t} \tag{5.76}$$

The remaining question is then the evaluation of the "area" element $d\Omega$ for the domain in the global system. To do this, we note that a line element $d\hat{\mathbf{r}}\ (= \mathbf{i}\,ds + \mathbf{j}\,dt)$, defined in the master element, is transformed to its image $d\mathbf{r}\,(= \mathbf{i}\,dx + \mathbf{j}\,dy)$ in Ω_e. The two vectors are related by (5.67) in that

$$dx = \frac{\partial x}{\partial s}\,ds + \frac{\partial x}{\partial t}\,dt \qquad \text{and} \qquad dy = \frac{\partial y}{\partial s}\,ds + \frac{\partial y}{\partial t}\,dt \tag{5.77}$$

Thus, the vectors $ds\,\mathbf{i}$ and $dt\,\mathbf{j}$ are transformed to

$$\mathbf{g}_s = \frac{\partial x}{\partial s}\,ds\,\mathbf{i} + \frac{\partial y}{\partial s}\,ds\,\mathbf{j} \qquad \text{and} \qquad \mathbf{g}_t = \frac{\partial x}{\partial t}\,dt\,\mathbf{i} + \frac{\partial y}{\partial t}\,dt\,\mathbf{j} \tag{5.78}$$

* Note that the inverse mapping, which is nonlinear in general, is not needed for the subsequent calculations.

respectively. Then

$$d\Omega\,\mathbf{k} = \mathbf{g}_s \times \mathbf{g}_t = J\,ds\,dt\,\mathbf{k} = J\,d\hat{\Omega}\,\mathbf{k}$$

that is,

$$d\Omega = J\,ds\,dt \qquad \text{or} \qquad d\Omega = J\,d\hat{\Omega} \tag{5.79}$$

Applying (5.75) and (5.79), the terms in the finite element equation (5.62) can be computed using the coordinates of the four corner nodes. We shall discuss details of the procedure to evaluate the element stiffness matrix $[E^e_{i\alpha j\beta}]$. For the isoparametric elements, the evaluation of the integral cannot be obtained in closed form in general. Thus, a commonly used technique is the numerical integration that is represented by

$$\int_{\Omega_e} E_{ikjl} N_{\alpha,k} N_{\beta,l}\, d\Omega = \int_{-1}^{1}\int_{-1}^{1} E_{ijkl}\hat{N}_{\alpha,k}\hat{N}_{\beta,l} J\, d\hat{\Omega}$$

$$= \sum_{IX=1}^{ITX}\sum_{IY=1}^{ITY} (E_{ikjl}\hat{N}_{\alpha,k}\hat{N}_{\beta,1} J)\Big|_{\substack{s=s_{IX}\\ t=s_{IY}}} * W_{IX} * W_{IY} \tag{5.80}$$

where (s_{IX}, s_{IY}) is the integration point, and (W_{IX}, W_{IY}) is the corresponding weight. A typical integration rule for four-node elements is the 2×2 Gaussian rule. (Details of quadrature rules used in finite element methods can be found in section 5.9.) The expression 2×2 means that there are two integration points in both s and t directions. If 2×3 is used, this means two integration points in the s direction and three points in the t direction, respectively. Let us apply this formula for obtaining the stiffness matrix. At first, define integration points and their weights:

```
ITX = 2,ITY = 2
GX(1) = -0.577350269189626,GX(2) = -GX(1)
GW(1) = 1,GW(2) = 1
```

Then we start with the loops for numerical integration:

```
DO 200 IX = 1,ITX
W1 = GW(IX)
S = GX(IX)
DO 200 IY = 1,ITY
W2 = GW(IY)
T = GX(IY)
```

Let us define DN(IA,I), IA = 1, . . . , 4; I = 1, 2, to be the first derivatives of the shape functions $N_{IA}(s, t)$ in the s and t direction at a given integration point. That is,

```
DN(1,1) = -0.25 * (1. - T), DN(1,2) = -0.25 * (1. - S)
DN(2,1) =  0.25 * (1. - T), DN(2,2) = -0.25 * (1. + S)
DN(3,1) =  0.25 * (1. + T), DN(3,2) =  0.25 * (1. + S)
DN(4,1) = -0.25 * (1. + T), DN(4,2) =  0.25 * (1. - S)
```

The determinant J = DET and the inverse of the Jacobian matrix must be computed by using the coordinates of the four corner nodes stored in the array XE(IA,I), IA = 1, . . . , 4, I = 1,2. Note that these values are measured in the global coordinate system x and y but are not the coordinates of the four corner nodes of the master element.

```
      DO 100 I = 1,2
      DO 100 J = 1,2
      DJIJ = 0.
      DO 102 IA = 1,4
  102 DJIJ = DJIJ + XE(IA,J) * DN(IA,I)
  100 DJ(I,J) = DJIJ
      DET = DJ(1,1) * DJ(2,2) - DJ(1,2) * DJ(2,1)
      DJ11 = DJ(1,1)
      DJ22 = DJ(2,2)
      DJ(1,1) = DJ22/DET
      DJ(1,2) = -DJ(1,2)/DET
      DJ(2,1) = -DJ(2,1)/DET
      DJ(2,2) = DJ11/DET
```

Once we know the inverse of the Jacobian matrix, the first derivatives of the shape functions in the global coordinate system stored in the array GDN(IA,I), IA = 1, . . . , 4, I = 1,2, can be computed by

```
      DO 104 IA = 1,4
      GDN(IA,1) = DJ(1,1) * DN(IA,1) + DJ(1,2) * DN(IA,2)
  104 GDN(IA,2) = DJ(2,1) * DN(IA,1) + DJ(2,2) * DN(IA,2)
```

If the material constants E_{ikjl} are stored in the array E(I,K,J,L), the element stiffness matrix $[E^e_{i\alpha j\beta}]$ stored in SKE(2 * (IA − 1) + I, 2 * (JB − 1) + J) is evaluated as

```
      DO 106 IA = 1,4
      DO 106 I = 1,2
      IAI = 2 * (IA - 1) + I
      DO 106 JB = 1,4
      DO 106 J = 1,2
      JBJ = 2 * (JB - 1) + J
      EIAJB = 0
      DO 108 K = 1,2
      DO 108 L = 1,2
  108 EIAJB = EIAJB + E(I,K,J,L) * GDN(IA,K) * GDN(JB,L)
  106 SKE(IAI,JBJ) = SKE(IAI,JBJ) + EIAJB * W1 * W2 * DET
```

If a temperature distribution is involved, the load vector $\{f^{0,e}_{i\alpha}\}$, stored in FE(2* (IA − 1) + I), is computed by

```
      DO 110 IA = 1,4
      DO 110 I = 1,2
      IAI = 2 * (IA - 1) + I
      FEIA = 0.
      DO 112 J = 1,2
  112 FEIA = FEIA + TEP * B(I,J) * GDN(IA,J)
  110 FE(IAI) = FE(IAI) + FEIA * W1 * W2 * DET
```

where TEP is the temperature difference $T - T_0$ at the integration point. If the temperature difference $T = T_0$ is given at the four corner nodes of the element Ω_e and is stored in TE(IA), the quantity TEP is obtained by

```
      TEP = 0.
      DO 109 IA = 1,4
  109 TEP = TEP + TE(IA) * SH(IA)
```

where SH(IA) are the shape functions for the master element, that is, $\hat{N}_\alpha$ is stored in SH(IA). The last step is the closing of the loop for the numerical integration:

```
  200 CONTINUE
```

The body force term $f_{i\alpha}^{1,e}$, see (5.60), is similarly evaluated. Then the rest of the terms are just for the boundary. Since four-node elements are used for the inside of the domain, the boundary elements are the two-node line elements studied in Chapter 2. We shall not discuss how to evaluate the terms $E_{i\alpha j\beta}^{3,e}$, $f_{i\alpha}^{2,e}$, and $f_{i\alpha}^{3,e}$ in this chapter, since their evaluation is similar to the evaluations in Chapter 2.

In the above, the element stiffness matrices are obtained using index notation. Another way to derive them involves the use of the contracted notation to describe the stress, strain, and elastic constants. Indeed, noting that the first term of the weak form (5.53) can be written as

$$\int_\Omega \sigma_{ij}(\mathbf{u})\varepsilon_{ij}(\bar{\mathbf{u}})\,d\Omega = \int_\Omega \boldsymbol{\varepsilon}(\bar{\mathbf{u}}) \cdot \mathbf{C}\boldsymbol{\varepsilon}(\mathbf{u})\,d\Omega$$
$$= \int_\Omega \varepsilon_I(\bar{\mathbf{u}}) C_{IJ} \varepsilon_J(\mathbf{u})\,d\Omega$$

Introducing the matrix representation of the strain–displacement relation,

$$\begin{Bmatrix} \varepsilon_1 \\ \varepsilon_2 \\ \varepsilon_3 \\ \varepsilon_4 \\ \varepsilon_5 \\ \varepsilon_6 \end{Bmatrix} = \begin{bmatrix} \dfrac{\partial}{\partial x} & 0 & 0 \\ 0 & \dfrac{\partial}{\partial y} & 0 \\ 0 & 0 & \dfrac{\partial}{\partial z} \\ 0 & \dfrac{\partial}{\partial z} & \dfrac{\partial}{\partial y} \\ \dfrac{\partial}{\partial z} & 0 & \dfrac{\partial}{\partial x} \\ \dfrac{\partial}{\partial y} & \dfrac{\partial}{\partial x} & 0 \end{bmatrix} \begin{Bmatrix} u_x \\ u_y \\ u_z \end{Bmatrix} \tag{5.81}$$

where $\mathbf{u} = u_x\mathbf{i} + u_y\mathbf{j} + u_z\mathbf{k}$, we have

$$\int_{\Omega_e} \sigma_{ij}(\mathbf{u})\varepsilon_{ij}(\bar{\mathbf{u}})\,d\Omega = \bar{u}_p \int_{\Omega_e} B_{Ip} C_{IJ} B_{Jq}\,d\Omega\, u_q$$

Here $p = 3(\alpha - 1) + i$, $q = 3(\beta - 1) + j$, and

$$[B_{Ip}] = \begin{bmatrix} \frac{\partial N_1}{\partial x} & 0 & 0 & \cdots & \frac{\partial N_a}{\partial x} & 0 & 0 \\ 0 & \frac{\partial N_1}{\partial y} & 0 & \cdots & 0 & \frac{\partial N_a}{\partial y} & 0 \\ 0 & 0 & \frac{\partial N_1}{\partial z} & \cdots & 0 & 0 & \frac{\partial N_a}{\partial z} \\ 0 & \frac{\partial N_1}{\partial z} & \frac{\partial N_1}{\partial y} & \cdots & 0 & \frac{\partial N_a}{\partial z} & \frac{\partial N_a}{\partial y} \\ \frac{\partial N_1}{\partial z} & 0 & \frac{\partial N_1}{\partial x} & \cdots & \frac{\partial N_a}{\partial z} & 0 & \frac{\partial N_a}{\partial x} \\ \frac{\partial N_1}{\partial y} & \frac{\partial N_1}{\partial x} & 0 & \cdots & \frac{\partial N_a}{\partial y} & \frac{\partial N_a}{\partial x} & 0 \end{bmatrix} \tag{5.82}$$

$$\{u_q\}^{\mathrm{T}} = \{u_{x1}u_{y1}u_{z1}u_{x2}u_{y2}u_{z2} \cdots u_{z\alpha}\} \tag{5.83}$$

where $u_{x\alpha}$ is the x component of $\mathbf{u}$ at the αth node and α is the total number of nodes in an element. The matrix $[B_{Ip}]$ is called the *B matrix* in the literature of finite element methods. Thus, the stiffness matrix is obtained by

$$\hat{\mathbf{K}}^e = \int_{\Omega_e} \mathbf{B}^{\mathrm{T}}\mathbf{C}\mathbf{B}\, d\Omega \tag{5.84}$$

that is, in component form,

$$\hat{K}^e_{pq} = \int_{\Omega_e} B_{Ip}C_{IJ}B_{Jq}\, d\Omega \tag{5.85}$$

For plane problems, p and q are related to α, β, i, and j in index notation by $p = 2(\alpha - 1) + i$ and $q = 2(\beta - 1) + j$.

Exercise 5.2: Develop a FORTRAN program to compute the stiffness matrix $[\hat{K}^e_{pq}]$ using the B matrix and the contracted elasticity matrix **C**. Use the 2×2 Gaussian quadrature to evaluate the stiffness matrix.

Exercise 5.3 (axisymmetric problem): Based on the equilibrium equation (5.21), we shall obtain the weak form (i.e., the virtual work principle) for axisymmetric problems. Adding three equations obtained by integrating three equilibrium equations over an axisymmetric body multiplied by arbitrary components $\bar{u}_r$, $\bar{u}_\theta$, and $\bar{u}_z$ of a virtual displacement **u**, respectively, we have

$$\begin{aligned} &2\pi \int_\Omega \left[-\frac{\partial}{\partial r}(r\sigma_{rr}) - r\frac{\partial}{\partial z}(\sigma_{rz}) + \sigma_{\theta\theta} \right] \bar{u}_r\, dr\, dz \\ &+ 2\pi \int_\Omega \left[-\frac{1}{r}\frac{\partial}{\partial r}(r^2\sigma_{\theta r}) - r\frac{\partial}{\partial z}(\sigma_{\theta z}) \right] \bar{u}_\theta\, dr\, dz \\ &+ 2\pi \int_\Omega \left[-\frac{\partial}{\partial r}(r\sigma_{zr}) - r\frac{\partial}{\partial z}(\sigma_{zz}) \right] \bar{u}_z\, dr\, dz \\ &= 2\pi \int_\Omega (\rho f_r \bar{u}_r + \rho f_\theta \bar{u}_\theta + \rho f_z \bar{u}_z)\, d\Omega \end{aligned}$$

where Ω is the cross section of the axisymmetric body at $\theta = 0$, and the differential area element $d\Omega$ is given by $d\Omega = r\,dr\,dz$. Applying the divergence theorem yields

$$\int_\Omega \left[\sigma_{rr} \frac{\partial \bar{u}_r}{\partial r} + \sigma_{rz} \frac{\partial \bar{u}_r}{\partial z} + \sigma_{\theta\theta} \frac{\bar{u}_r}{r} + r\sigma_{\theta r} \frac{\partial}{\partial r} \frac{\bar{u}_\theta}{r} + \sigma_{\theta z} \frac{\partial \bar{u}_\theta}{\partial z} + \sigma_{zr} \frac{\partial \bar{u}_z}{\partial r} + \sigma_{zz} \frac{\partial \bar{u}_z}{\partial z} \right] d\Omega$$

$$- \int_\Gamma \left[(\sigma_{rr} n_r + \sigma_{rz} n_z)\bar{u}_r + (\sigma_{\theta r} n_r + \sigma_{\theta z} n_z)\bar{u}_\theta + (\sigma_{zr} n_r + \sigma_{zz} n_z)\bar{u}_z \right] d\Gamma$$

$$= \int_\Omega (f_r \bar{u}_r + f_\theta \bar{u}_\theta + f_z \bar{u}_z)\rho\, d\Omega, \quad \forall \bar{\mathbf{u}}$$

where Γ is the boundary of the cross section Ω. Using the following notations,

$$\bar{\varepsilon}_{rr} = \frac{\partial \bar{u}_r}{\partial r}, \qquad \bar{\varepsilon}_{\theta\theta} = \frac{\bar{u}_r}{r}, \qquad \bar{\varepsilon}_{zz} = \frac{\partial \bar{u}_z}{\partial z}$$

$$\bar{\gamma}_{zr} = 2\bar{\varepsilon}_{zr} = \frac{\partial \bar{u}_r}{\partial z} + \frac{\partial \bar{u}_z}{\partial r}, \qquad \bar{\gamma}_{\theta z} = 2\bar{\varepsilon}_{\theta z} = \frac{\partial \bar{u}_\theta}{\partial z}$$

$$\bar{\gamma}_{r\theta} = 2\bar{\varepsilon}_{r\theta} = \frac{\partial \bar{u}_\theta}{\partial r} - \frac{\bar{u}_\theta}{r}$$

$$t_r = \sigma_{rr} n_r + \sigma_{rz} n_z, \qquad t_\theta = \sigma_{\theta r} n_r + \sigma_{\theta z} n_z, \qquad t_z = \sigma_{zr} n_r + \sigma_{zz} n_z$$

we have the weak form

$$\int_\Omega \{\bar{\varepsilon}_{rr}, \bar{\varepsilon}_{\theta\theta}, \bar{\varepsilon}_{zz}, \bar{\gamma}_{\theta z}, \bar{\gamma}_{zr}, \bar{\gamma}_{r\theta}\} \begin{Bmatrix} \sigma_{rr} \\ \sigma_{\theta\theta} \\ \sigma_{zz} \\ \sigma_{\theta z} \\ \sigma_{zr} \\ \sigma_{r\theta} \end{Bmatrix} d\Omega$$

$$= \int_\Omega \{\bar{u}_r, \bar{u}_\theta, \bar{u}_z\} \begin{Bmatrix} f_r \\ f_\theta \\ f_z \end{Bmatrix} \rho\, d\Omega + \int_\Gamma \{\bar{u}_r, \bar{u}_\theta, \bar{u}_z\} \begin{Bmatrix} t_r \\ t_\theta \\ t_z \end{Bmatrix} d\Gamma \tag{5.86}$$

for every $\bar{\mathbf{u}}$.

The last term of the weak form involves boundary conditions. For example, if the displacement $\mathbf{u}$ is specified on a part Γ_1 of Γ, and if the traction $\mathbf{t}$ is given on $\Gamma_2 = \Gamma - \Gamma_1$, then an arbitrary virtual displacement $\bar{\mathbf{u}}$ must be zero on Γ_1 and the last term of (5.86) becomes

$$\int_{\Gamma_2} \{\bar{u}_r, \bar{u}_\theta, \bar{u}_z\} \begin{Bmatrix} t_r \\ t_\theta \\ t_z \end{Bmatrix} d\Gamma$$

If the stress–strain relation is assumed to be the same as (5.36), the left side of (5.86) becomes

$$\int_\Omega \bar{\boldsymbol{\varepsilon}}^{\mathrm{T}} \mathbf{C} \boldsymbol{\varepsilon} \, d\Omega$$

using the contracted notation. Thus, the weak form (5.86) can be written as

$$\int_\Omega \bar{\boldsymbol{\varepsilon}}^{\mathrm{T}} \mathbf{C} \boldsymbol{\varepsilon} \, d\Omega = \int_\Omega \bar{\mathbf{u}}^{\mathrm{T}} \mathbf{f} \rho \, d\Omega + \int_{\Gamma_2} \bar{\mathbf{u}}^{\mathrm{T}} \mathbf{t} \, d\Gamma$$

$$\forall \bar{\mathbf{u}} \ni \bar{\mathbf{u}} = \mathbf{0} \text{ on } \Gamma_1 \tag{5.87}$$

where $\bar{\mathbf{u}}^{\mathrm{T}} = \{\bar{u}_r, \bar{u}_\theta, \bar{u}_z\}$

Noting that the strain–displacement relation for axisymmetric problems is given by the form

$$\begin{Bmatrix} \varepsilon_1 \\ \varepsilon_2 \\ \varepsilon_3 \\ \varepsilon_4 \\ \varepsilon_5 \\ \varepsilon_6 \end{Bmatrix} = \begin{Bmatrix} \varepsilon_{rr} \\ \varepsilon_{\theta\theta} \\ \varepsilon_{zz} \\ \gamma_{\theta z} \\ \gamma_{zr} \\ \gamma_{r\theta} \end{Bmatrix} = \begin{bmatrix} \dfrac{\partial}{\partial r} & 0 & 0 \\ \dfrac{1}{r} & 0 & 0 \\ 0 & 0 & \dfrac{\partial}{\partial z} \\ 0 & \dfrac{\partial}{\partial z} & 0 \\ \dfrac{\partial}{\partial z} & 0 & \dfrac{\partial}{\partial r} \\ 0 & -\dfrac{1}{r} + \dfrac{\partial}{\partial r} & 0 \end{bmatrix} \begin{Bmatrix} u_r \\ u_\theta \\ u_z \end{Bmatrix} \tag{5.88}$$

obtain the B matrix for axisymmetric problems, where u_r, u_θ, and u_z are approximated by

$$u_r(r, z) = \sum_{\alpha=1}^{N_e} u_{r\alpha} N_\alpha(r, z)$$

$$u_\theta(r, z) = \sum_{\alpha=1}^{N_e} u_{\theta\alpha} N_\alpha(r, z) \tag{5.89}$$

$$u_z(r, z) = \sum_{\alpha=1}^{N_e} u_{z\alpha} N_\alpha(r, z)$$

respectively.

5.5 Assembling and a skyline solver

The next procedure after obtaining the element stiffness matrix and the load vector is assembling to the global stiffness matrix and the global load vector. Assembling cannot be independent of the mode of the solver of the system of linear equations. We have used the band solver for the heat conduction problem. Although

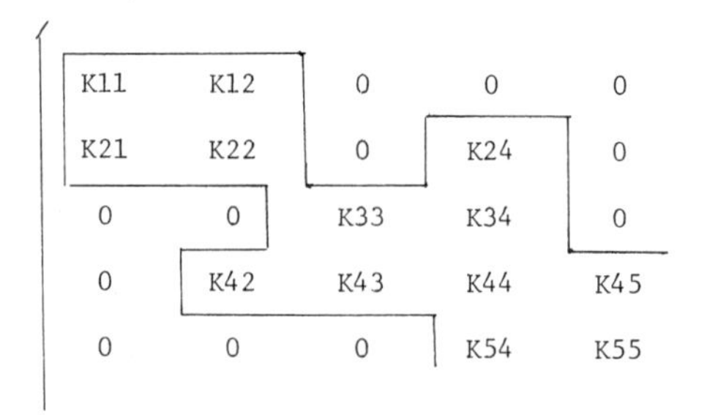

Matrix SK(I,j) : Original Form

Diagonal Pointer JDIAG

I	1	2	3	4	5
JDIAG(I)	1	3	4	7	9

JDIAG(I) indicates the location of SK(I,I) in the one-dimensional arrays A and C .

Lower Triangular Matrix of SK : A

K	1	2	3	4	5	6	7	8	9
A(K)	K11	K21	K22	K33	K42	K43	K44	K54	K55

Upper Triangular Matrix of SK : C

K	1	2	3	4	5	6	7	8	9
C(K)	K11	K12	K22	K33	K24	K34	K44	K45	K55

Figure 5.5 Skyline of the matrix.

it is possible to use it here as well, we shall develop a different solver using the *skyline method*. For simplicity, symmetry of the stiffness matrix is assumed.

Suppose that the stiffness matrix has the form shown in Figure 5.5 so that the profile (skyline) of the nonzero terms can be made. Then, we store only the entries under the skyline, as in Figure 5.5. The array of JDIAG(I) is introduced for the diagonal pointer. Using this, the entry K_{ij} such that $i \leq j$ is stored at SK(JDIAG(J) – (J – I)). If $i > j$, then K_{ij} is transferred to SK(JDIAG(I) – (I – J)). For the skyline solver, Gaussian elimination is inadequate, and the related *Crout method* is used. The main purpose of the Crout method is to find the LU decomposition of the matrix $\mathbf{A}$:

$$A_{ij} = L_{ik} U_{kj} \tag{5.90}$$

where

$$\mathbf{L} = \begin{bmatrix} 1 & & & \\ L_{21} & 1 & & \\ \vdots & & & \\ L_{n1} & L_{n2} & \cdots & 1 \end{bmatrix}, \qquad \mathbf{U} = \begin{bmatrix} U_{11} & U_{12} & \cdots & U_{1n} \\ & U_{22} & \cdots & U_{2n} \\ & & & \vdots \\ & & & U_{nn} \end{bmatrix} \tag{5.91}$$

The decomposition is performed by the routine

$j = 1$: $\quad L_{11} = 1, \quad U_{11} = A_{11}$

$\rightarrow j = j + 1$

for $i = 1, \ldots, j-1$

$$L_{ji} = \left(A_{ji} - \sum_{k=1}^{i-1} L_{jk} U_{ki} \right) \Big/ U_{ii} \quad \text{(no sum on } i\text{)} \tag{5.92}$$

$$U_{ij} = A_{ij} - \sum_{k=1}^{i-1} L_{ik} U_{kj} \tag{5.93}$$

and then

$$L_{jj} = 1, \qquad U_{jj} = A_{jj} - \sum_{k=1}^{j-1} L_{jk} U_{kj} \quad \text{(no sum on } j\text{)} \tag{5.94}$$

$j < n$ (return to $j = j + 1$)

$j \geq n$

Stop

If the matrix $\mathbf{A}$ is symmetric, we have

$$U_{ij} = L_{ji} U_{ii} \tag{5.95}$$

Thus equations (5.92) and (5.93) become

$$\left. \begin{aligned} L_{ji} &= \left(A_{ji} - \sum_{k=1}^{i-1} L_{jk} L_{ik} U_{kk} \right) \Big/ U_{ii} \\ U_{jj} &= A_{jj} - \sum_{k=1}^{j-1} L_{jk} L_{jk} U_{kk} \end{aligned} \right\} \quad (i, j\text{: no sum}) \tag{5.96}$$

respectively.

Once the LU decomposition is obtained, the system of linear equations

$$A_{ij} x_j = b_i \tag{5.97}$$

can be solved by the two-step algorithm

$$L_{ik} y_k = b_i \qquad \text{and} \qquad U_{kj} x_j = y_k \tag{5.98}$$

That is, the reduction of the load vector $\mathbf{b}$ is

$$y_k = b_k - \sum_{j=1}^{k-1} L_{kj} y_j \tag{5.99}$$

where $y_1 = b_1$. After reducing the load vector, the solution $\mathbf{x}$ is obtained by the back substitution

$$x_k = \left(y_k - \sum_{j=k+1}^{n} U_{kj} y_j \right) \Big/ U_{kk} \quad (k\text{: no sum}) \tag{5.100}$$

where $x_n = y_n / U_{nn}$. If the matrix **A** is symmetric, (5.100) is changed to

$$x_k = \frac{y_k}{U_{kk}} - \sum_{j=k+1}^{n} L_{jk} y_j \tag{5.101}$$

A characteristic of the Crout method is that the reduction of the entries L_{ji} and U_{ij} is performed only one time by using the entries of the ith and jth columns and rows. This makes possible a modification of the routine by the skyline method, since the skyline structure is not disturbed by the decomposition of the matrix using Crout's method. If the original Gauss elimination method is applied, we cannot preserve the skyline, although the band structure is maintained. The skyline method in common use involves a modification of the Crout method, which takes advantage of the skyline structure.

Exercise 5.4: The first of the following two programs is the BASIC version of the Crout method. The second is a modification of the LU decomposition portion of first using the skyline method. Develop the remaining two parts: the reduction of the load vector and the back substitution. Note that in the program JDIAG is abbreviated to JD. The upper triangular part **U** of the matrix **A** is stored in the string array C(I), and the lower triangular part **L** is in the string array A(I). The right side **B** is stored in B(I) in the programs.

BASIC Crout method

```
1000   REM
1010   REM ——— SUBROUTINE CROUT'S METHOD ———
1020   REM
1030   REM ⟨LU-DECOMPOSITION⟩
1040   FOR J = 2 TO NQ
1050 J1 = J − 1
1060   FOR I = 1 TO J1
1070 I1 = I − 1
1080 S1 = 0:S2 = 0
1090   IF I1 = < 0 THEN GOTO 1140
1100   FOR K = 1 to I1
1110 S1 = S1 + A(J,K) * A(K,I)
1120 S2 = S2 + A(I,K) * A(K,J)
1130   NEXT K
1140 A(J,I) = (A(J,I) − S1) / A(I,I)
1150 A(I,J) = A(I,J) − S2
1160   NEXT I
1170 S3 = 0
1180   FOR K = 1 TO J1
1190 S3 = S3 + A(J,K) * A(K,J)
1200   NEXT K
1210 A(J,J) = A(J,J) − S3
```

```
1220    NEXT J
1230    REM <REDUCTION OF THE LOAD>
1240    FOR J = 2 TO NQ
1250 S1 = 0
1260    FOR K = 1 TO J - 1
1270 S1 = S1 + A(J,K) * B(K)
1280    NEXT K
1290 B(J) = B(J) - S1
1300    NEXT J
1310    REM <BACK SUBSTITUTION>
1320 J = NQ
1330 B(J) = B(J) / A(J,J)
1340 J = J - 1
1350    IF J = < 0 THEN RETURN
1360 S1 = 0
1370    FOR K = J + 1 TO NQ
1380 S1 = S1 + A(J,K) * B(K)
1390    NEXT K
1400 B(J) = (B(J) - S1) / A(J,J)
1410    GOTO 1340
1420    REM <END>
```

Modification of the LU decomposition by the skyline method

```
1000    REM LU-DECOMPOSITION
1010    FOR J = 2 TO NQ
1020 JJ = JD(J - 1)
1030 JL = JD(J)
1040 IM = JL - JJ - 1
1050    FOR I = 1 TO IM
1060 JI = JD(I - 1)
1070 S1 = 0:S2 = 0
1080 I1 = I - 1
1090    IF I1 = < 0 THEN GOTO 1160
1100    FOR K = 1 TO I1
1110 JK = JJ + K
1120 IK = JI + K
1130 S1 = S1 - A(JK) * C(IK)
1140 S2 = S2 + A(IK) * C(JK)
1150    NEXT K
1160 JM = JJ + I
1170 A(JM) = (A(JM) - S1) / C(JD(I))
1180 C(JM) = C(JM) - S2
1190    NEXT I
1200 S1 = 0
1210    FOR K = 1 TO J - 1
1220 JK = JJ + K
1230 S1 = S1 + A(JK) * C(JK)
1240    NEXT K
1250 C(JL) = C(JL) - S1
1260    NEXT J
```

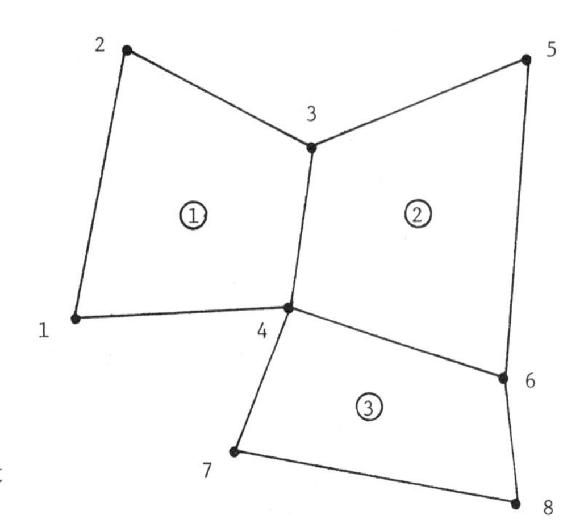

Figure 5.6 Example of finite element model using four-node elements.

For the assembly of element stiffness matrices we must define the skyline and its diagonal pointer. This part of the preparation for the method of skyline is called the *preskyline* procedure. To study this, let us assume that an element has NODE number of nodes at which NDF degrees of freedom are assigned. For the four-node quadrilateral elements for plane problems, NODE = 4 and NDF = 2. Then the skyline is defined by the following routine:

```
      DO 100 NEL = 1,NELX
      DO 100 IA = 1,NODE
      IJKIA = IJK(IA,NEL)
      DO 100 I = 1,NDF
      IJKIAI = NDF * (IJKIA - 1) + I
      DO 100 JB = 1,NODE
      IJKJB = IJK (JB,NEL)                                    (5.102)
      DO 100 J = 1,NDF
      IJKJBJ = NDF * (IJKJB - 1) + J
      IAIJBJ = IJKIAI - IJKJBJ + 1
      IF(IAIJBJ.LT.1) GOTO 100
      IF (IAIJBJ.GT.JDIAG(IJKIAI))JDIAG(IJKIAI) = IAIJBJ
  100 CONTINUE
```

The diagonal pointer is then defined by

```
      DO 102 I = 2,NEQ
  102 JDIAG(I) = JDIAG(I - 1) + JDIAG(I)                      (5.103)
```

where NEQ is the total number of degrees of freedom in a finite element model.

Exercise 5.5: Define the skyline and the diagonal pointer for the finite element model shown in Figure 5.6.

Assembly of element stiffness matrices is performed using the diagonal pointer JDIAG:

```
      DO 100 NEL = 1, NELX
      ——— Form SKE and FE ———
      DO 110 IA = 1, NODE
      IJKIA = IJK(IA,NEL)
      DO 110 I = 1, NDF
      IJKIAI = NDF * (IJKIA − 1) + I
      IAI = NDF * (IA − 1) + I
      F(IJKIAI) = F(IJKIAI) + FE(IAI)
      DO 110 J = 1, NDF
      IJKJB = IJK(JB, NEL)                                (5.104)
      DO 110 J = 1, NDF
      IJKJBJ = NDF * (IJKJB − 1) + J
      JBJ = NDF * (JB − 1) + J
      JBJIAI = IJKJBJ − IJKIAI
      IF (JBJIAI. LT. 0) GOTO 110
      ND = JDIAG (IJKJBJ) − JBJIAI
      SKU (ND) = SKU (ND) + SKE (IAI, JBJ)
  110 CONTINUE
```

where SKU stores the upper triangular matrix of the stiffness matrix and was denoted by A in Figure 5.5.

Exercise 5.6: Suppose that every node has different degrees of freedom, and that NODNDF(I) stores the number of degrees of freedom of the Ith node. Modify the preskyline procedure (5.102) and the assembly routine (5.104).

5.6 Remarks on the four-node quadrilateral isoparametric element

Let us study some additional details associated with the four-node element by studying the eigenvalues and eigenvectors of the stiffness matrix, how the stress tensor is evaluated, convergence as the mesh size goes to zero, special cases of bending, incompressible materials, and some other quadrature rules for the four-node element.

5.6.1 Eigenvalues and Eigenvectors

We first compute the eigenvalues and eigenvectors of the stiffness matrix for the plane strain case. To do this, let the arbitrary element Ω_e be assumed to be a unit square domain, and let the Jacobi method introduced in Chapter 1 be applied. The results are shown in Figure 5.7. There are three rigid body modes whose eigenvalues are zero. The next level of eigenvalues occurs in the flexual modes, which dominate all the bending types of deformation. Then come the predominantly shear mode and the stretch mode – both associated with the same eigenvalue. The largest eigenvalue arises in the uniform expansive modes that are dominated by hydrostatic pressure. If the modes shown in Figure 5.7 are examined, we can easily realize that these modes correspond to fundamental deformations of a differential element of the body. For example, see the description of the fundamental deformation by Fung (1965, sec. 4.3).

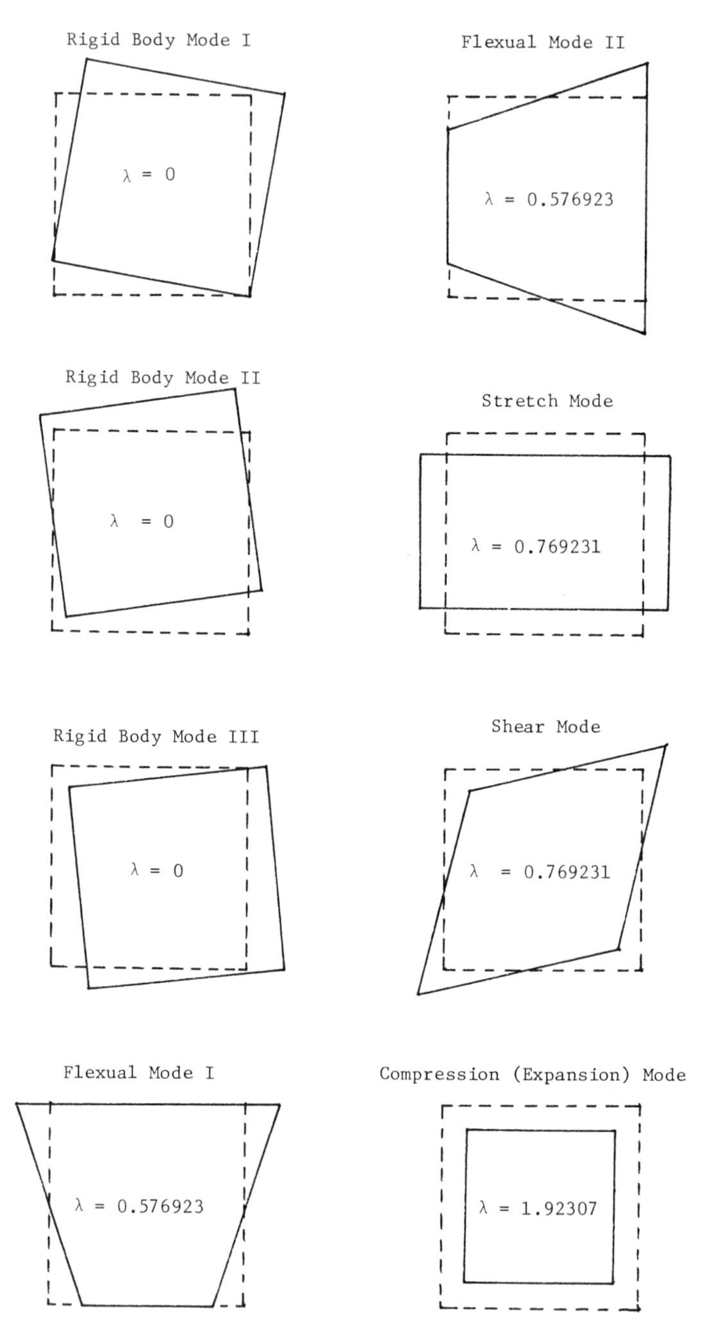

Figure 5.7 Eigenmodes of stiffness matrix obtained by 2 × 2 Gaussian integration rule, plane strain, Young's modulus $E = 1$, Poisson ratio 0.3.

In the above, the element stiffness matrix has been computed using the 2 × 2 Gaussian quadrature rule. The results shown in Figure 5.8 are for the stiffness matrix for plane strain using the one-point Gaussian quadrature rule, which cannot integrate the term $E_{ikjl}N_{\alpha,k}N_{\beta,l}$ *exactly*. Indeed, for constant elasticity E_{ikjl}, since

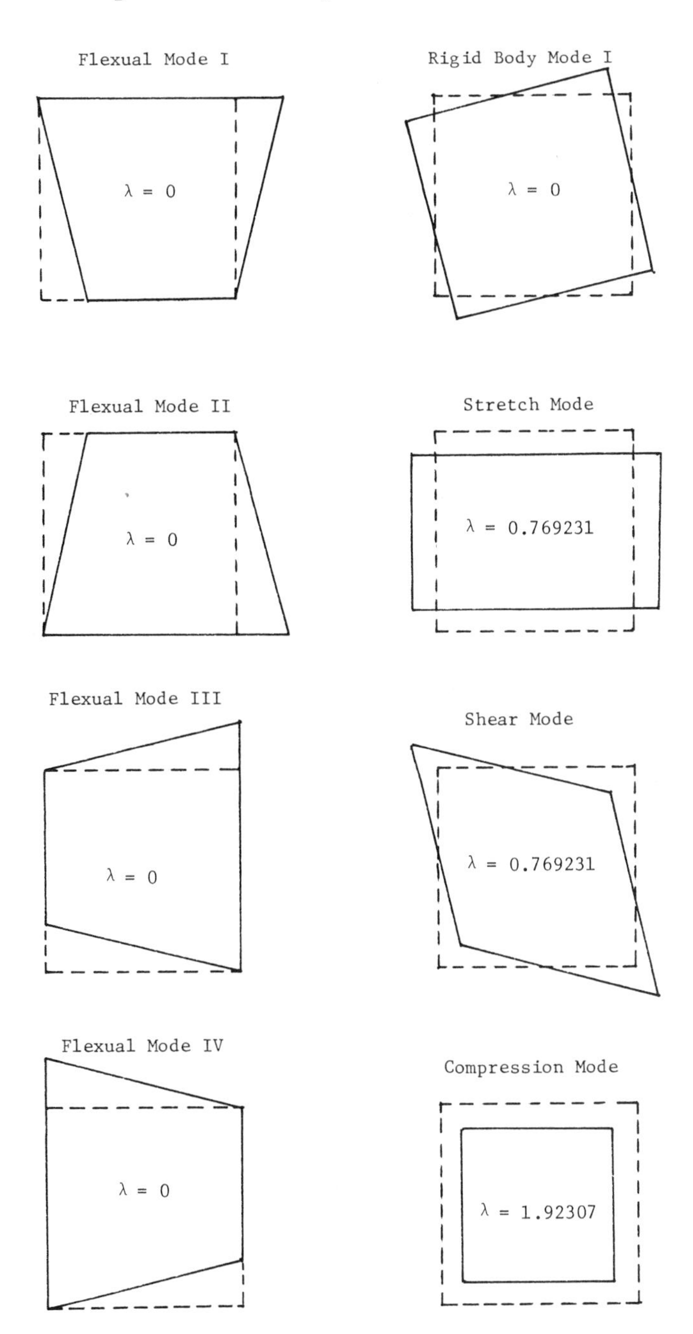

Figure 5.8 Eigenmodes of stiffness matrix obtained by one-point Gaussian integration, plane strain, Young's modulus $E = 1$, Poisson's ratio 0.3.

the first derivatives of shape functions are linear polynomials, at least the 2×2 Gaussian quadrature rule is necessary to integrate exactly. As shown in Figure 5.8, two more zero eigenvalues are obtained if the one-point Gaussian rule is applied. Note that the flexure modes whose deformation is similar to an *hourglass* correspond to zero eigenvalues. This means that the deformation of the body shown

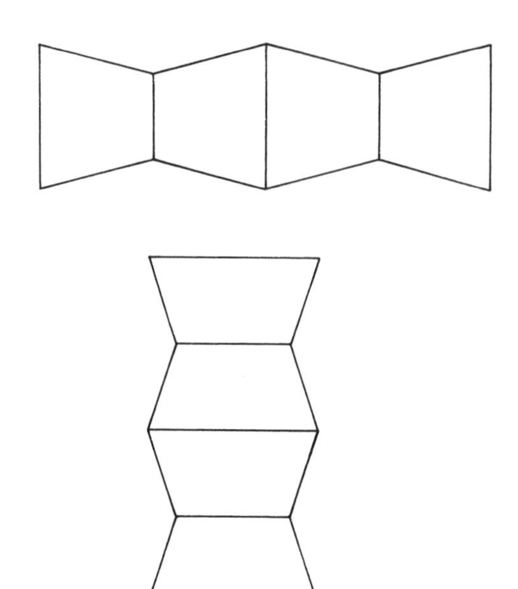

Figure 5.9 Zero-energy modes.

in Figure 5.9 could occur with zero strain energy, although it is not a rigid body motion. This is certainly *not compatible* with the mechanics of deformable bodies, since only rigid body motions give zero strain energy. Thus, the application of the one-point Gaussian quadrature rule to integrate the term $E_{ikjl}N_{\alpha,k}N_{\beta,l}$ must be viewed with caution.

Exercise 5.7: Obtain the eigenvalues and eigenvectors of the stiffness matrix obtained by the four-node element for plane stress problems.

It is also noted that a quadrilateral element can be constructed by several triangular elements, for example, by

Elimination of the degrees of freedom of the fifth node is performed by using the idea of partition of a matrix equation to two parts:

$$\begin{bmatrix} [K^e_{11}] & [K^e_{12}] \\ [K^e_{21}] & [K^e_{22}] \end{bmatrix} \begin{Bmatrix} \{u_1\} \\ \{u_2\} \end{Bmatrix} = \begin{Bmatrix} \{f_1\} \\ \{f_2\} \end{Bmatrix} \tag{5.105}$$

Elimination of $\{u_2\}$ is obtained by solving the second equation by

$$\{u_2\} = -[K^e_{22}]^{-1}[K^e_{21}]\{u_1\} + [K^e_{22}]^{-1}\{f_2\} \tag{5.106}$$

Table 5.1 *Eigenvalues of three different quadrilateral elements for plane problem (plane strain, $E = 1$, $\nu = 0.3$)*

Mode \ Element	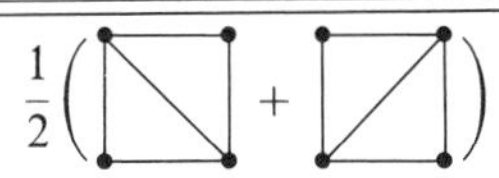 $\frac{1}{2}$(+)	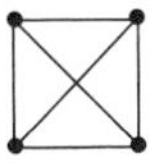	
Rigid body	0	0	0
	0	0	0
	0	0	0
Flexual	1.73076924	0.558690379	0.576923078
	1.73076924	0.558690379	0.576923079
Uniaxial	0.769230777	0.769230769	0.769230769
Shear	0.769230772	0.769230769	0.769230771
Uniform	1.92307693	1.92307693	1.92307693

and by substitution of this into the first equation:

$$([K_{11}^e] - [K_{12}^e][K_{22}^e]^{-1}[K_{21}^e])\{u_1\} = \{f_1\} - [K_{12}^e][K_{22}^e]^{-1}\{f_2\}$$

that is,

$$[\hat{K}^e]\{u_1\} = \{\hat{f}_1\} \tag{5.107}$$

where

$$\begin{aligned} [\hat{K}^e] &= [K_{11}^e] - [K_{12}^e][K_{22}^e]^{-1}[K_{21}^e] \\ \{\hat{f}_1\} &= \{f_1\} - [K_{12}^e][K_{22}^e]^{-1}\{f_2\} \end{aligned} \tag{5.108}$$

We shall obtain eigenvalues and eigenvectors of the above two elements and shall compare them to those of the four-node isotropic quadrilateral element. Summary of results is given in Table 5.1. It is easily understood that both elements provide the same eigenvalues to uniaxial, shear, and uniform deformation modes, while the first one obtained by averaging two three-node elements gives much higher eigenvalues to flexual modes. The quadrilateral element obtained by four three-node elements has almost the same properties as the four-node isoparametric element.

5.6.2 Convergence of the approximation

As seen in the modes of deformations obtained, the four-node element contains the rigid body and constant-strain fields together with continuity of the displacement across the element boundaries. This is sufficient for the convergence of the method as the mesh size goes to zero, *in the sense of Zienkiewicz.** It is intuitively agreeable, since any smooth function can be approximated by piecewise constant functions. Thus, if the modes of deformation by the finite element approximation contain constant strains, the convergence of the strain energy density $U = \frac{1}{2}\sigma_{ij}\varepsilon_{ij}$ can be obtained as the mesh size becomes small. Continuity of displacement is required in the integration by parts used to obtain the weak form: moreover, it is

* If a finite element can reproduce rigid body modes, constant-strain fields, and continuous displacements, then Zienkiewicz (1978) claims convergence is obtained.

a natural requirement, since we are dealing with a *continuum*. There are *hybrid elements* in which displacement fields contain discontinuities. In the author's experience, such techniques frequently lead to difficulties.

We now study some additional convergence issues based on section 2.10.

To investigate convergence of the finite element approximation, recall the total potential energy F, defined by (5.54), and consider its approximation. To simplify the notation, we define

$$a(\mathbf{u}, \mathbf{v}) = \int_{\Omega} E_{ijkl}\varepsilon_{kl}(\mathbf{u})\varepsilon_{ij}(\mathbf{v})\, d\Omega + \int_{\Gamma_{i3}} k_{ij}u_j v_i\, d\Gamma \tag{5.109}$$

and

$$f(\mathbf{v}) = \int_{\Omega} \rho f_i v_i\, d\Omega + \int_{\Gamma_{i2}} h_i v_i\, d\Gamma + \int_{\Gamma_{i3}} k_{ij}u_{\infty j}v_i\, d\Gamma \tag{5.110}$$

Then the weak form (5.53) is expressed as

$$a(\mathbf{u}, \bar{\mathbf{u}}) = f(\bar{\mathbf{u}}), \quad \forall \bar{\mathbf{u}} \ni \bar{\mathbf{u}}_i = 0 \text{ on } \Gamma_{i1} \tag{5.111}$$

and the total potential energy is

$$F(\mathbf{v}) = \tfrac{1}{2}a(\mathbf{v}, \mathbf{v}) - f(\mathbf{v}) \tag{5.112}$$

Similarly, the discretized weak form is

$$a(\mathbf{u}_h, \bar{\mathbf{u}}_h) = f(\bar{\mathbf{u}}_h), \quad \forall \bar{\mathbf{u}}_h \ni \bar{u}_{i\alpha} = 0 \text{ on } \Gamma_{i1} \tag{5.113}$$

for $\mathbf{u}_h = (u_{i\alpha}N_\alpha)\mathbf{i}_i$ in each Ω_e.

The term $a(\mathbf{u}, \mathbf{v})$ physically means the virtual work due to the virtual displacement $\mathbf{v}$ from the equilibrium configuration. Similarly, $f(\mathbf{v})$ is an abstract form of the virtual work done by the applied forces. It is to be noted that $a(\cdot, \cdot)$ is linear in both arguments and clearly $f(\cdot)$ is linear in its argument. Because of this property, $a(\cdot, \cdot)$ and $f(\cdot)$ are called *bilinear* and *linear forms,* respectively. Note that $a(\cdot, \cdot)$ is symmetric, that is, $a(\mathbf{v}, \mathbf{w}) = a(\mathbf{w}, \mathbf{v})$ for any $\mathbf{v}$ and $\mathbf{w}$.

If the elasticity constant $\mathbf{E}$ and the spring constant $\mathbf{k}$ satisfy the conditions for $m \geq 0$ and $\hat{m} \geq 0$,

$$\begin{aligned} E_{ijkl}X_{kl}X_{ij} &\geq mX_{ij}X_{ij}, \quad \forall X_{ij} = X_{ji} \\ k_{ij}x_j x_i &\geq \hat{m}x_i x_i, \quad \forall x_i \end{aligned} \tag{5.114}$$

then it is possible to prove that

$$\tfrac{1}{2}a(\mathbf{v}, \mathbf{v}) - \tfrac{1}{2}a(\mathbf{w}, \mathbf{w}) \geq a(\mathbf{w}, \mathbf{v} - \mathbf{w}) \tag{5.115}$$

for every $\mathbf{w}$ and $\mathbf{v}$ whenever $m \geq 0$ and $\hat{m} \geq 0$. The condition (5.115) is the *convexity* condition on the total strain energy of the body and the potential energy of the spring. More intuitively, the inequalities (5.114) mean that work must be done on the body and the spring in order to deform them. If the elastic body is isotropic, we have

$$E_{ijkl}X_{kl}X_{ij} = \lambda X_{kk}X_{ii} + 2\mu X_{ij}X_{ij}$$

Thus, the condition (5.114) will be satisfied if $\lambda \geq 0$ and $\mu > 0$. Using (5.27), the conditions $\lambda \geq 0$ and $\mu > 0$ will be satisfied if

$$E > 0 \qquad \text{and} \qquad -1 < \nu \leq \tfrac{1}{2}$$

Thus, condition (5.114) is, in general, satisfied.

Exercise 5.8: Prove inequality (5.115) using condition (5.114).

Now let us consider the difference between the total potential energies $F(\mathbf{u}_h)$ and $F(\mathbf{u})$. For any admissible displacement $\mathbf{v}_h$ such that $v_{hi} = g_i$ on Γ_{i1}, we have

$$\begin{aligned} F(\mathbf{u}_h) - F(\mathbf{u}) &= \tfrac{1}{2}a(\mathbf{u}_h, \mathbf{u}_h) - f(\mathbf{u}_h) - \tfrac{1}{2}a(\mathbf{u}, \mathbf{u}) + f(\mathbf{u}) \\ &= \tfrac{1}{2}[a(\mathbf{u}_h, \mathbf{u}_h) - a(\mathbf{v}_h, \mathbf{v}_h) + a(\mathbf{v}_h, \mathbf{v}_h) - a(\mathbf{u}, \mathbf{u})] - f(\mathbf{u}_h) + f(\mathbf{u}) \\ &\leq -a(\mathbf{u}_h, \mathbf{v}_h - \mathbf{u}_h) + \tfrac{1}{2}a(\mathbf{v}_h - \mathbf{u}, \mathbf{v}_h - \mathbf{u}) + \tfrac{1}{2}a(\mathbf{v}_h, \mathbf{u}) \\ &\quad + \tfrac{1}{2}a(\mathbf{u}, \mathbf{v}_h) - a(\mathbf{u}, \mathbf{u}) - f(\mathbf{u}_h) + f(\mathbf{u}) \end{aligned}$$

Since $a(\mathbf{v}, \mathbf{w}) = a(\mathbf{w}, \mathbf{v})$, we have

$$\begin{aligned} F(\mathbf{u}_h) - F(\mathbf{u}) &\leq \tfrac{1}{2}a(\mathbf{v}_h - \mathbf{u}, \mathbf{v}_h - \mathbf{u}) - a(\mathbf{u}_h, \mathbf{v}_h - \mathbf{u}_h) \\ &\quad + a(\mathbf{u}, \mathbf{v}_h - \mathbf{u}) - f(\mathbf{u}_h) + f(\mathbf{u}) \\ &= \tfrac{1}{2}a(\mathbf{v}_h - \mathbf{u}, \mathbf{v}_h - \mathbf{u}) + a(\mathbf{u} - \mathbf{u}_h, \mathbf{v}_h - \mathbf{u}_h) \\ &\quad + a(\mathbf{u}, \mathbf{u}_h - \mathbf{u}) - f(\mathbf{u}_h - \mathbf{u}) \end{aligned}$$

Noting that subtraction of (5.113) from (5.111) with $\bar{\mathbf{u}} = \bar{\mathbf{u}}_h$ yields

$$a(\mathbf{u} - \mathbf{u}_h, \bar{\mathbf{u}}_h) = 0, \quad \forall \bar{\mathbf{u}}_h \ni \bar{u}_{hi} = 0 \text{ on } \Gamma_{i1} \tag{5.116}$$

and using (5.111), we can conclude

$$F(\mathbf{u}_h) - F(\mathbf{u}) \leq \tfrac{1}{2}a(\mathbf{v}_h - \mathbf{u}, \mathbf{v}_h - \mathbf{u}), \quad \forall \mathbf{v}_h$$

where $v_{hi} = g_i$ on Γ_{i1}. Since $\mathbf{v}_h$ is arbitrary, we take $\mathbf{v}_h = \hat{\mathbf{u}}_h$, that is, the interpolation of the true solution $\mathbf{u}$. Then the principle of minimum potential energy yields the inequality

$$0 \leq F(\mathbf{u}_h) - F(\mathbf{u}) \leq \tfrac{1}{2}a(\hat{\mathbf{u}}_h - \mathbf{u}, \hat{\mathbf{u}}_h - \mathbf{u}) \tag{5.117}$$

Assuming the estimate of the interpolation error by four-node isoparametric elements,*

$$\int_\Omega |\nabla \hat{f}_h - \nabla f|^2 \, d\Omega \leq Ch^2 \int_\Omega |\nabla \cdot \nabla f|^2 \, d\Omega \tag{5.118}$$

where $\hat{f}_h$ is the interpolation of a twice continuously differentiable function f, and C is a constant independent of the size h of finite elements, we can estimate an upperbound of (5.117) as follows:

$$0 \leq F(\mathbf{u}_h) - F(\mathbf{u}) \leq \hat{C}h^2 \tag{5.119}$$

where $\hat{C}$ is a positive constant dependent on $\mathbf{u}$ but independent of h. Thus, the total potential energy of the finite element approximation converges to that of the true solution as $h \to 0$.

Using the facts that

$$F(\mathbf{u}_h) = -\tfrac{1}{2}a(\mathbf{u}_h, \mathbf{u}_h), \qquad F(\mathbf{u}) = -\tfrac{1}{2}a(\mathbf{u}, \mathbf{u})$$

inequality (5.117) can also be written as

$$0 \leq a(\mathbf{u}, \mathbf{u}) - a(\mathbf{u}_h, \mathbf{u}_h) \leq a(\hat{\mathbf{u}}_h - \mathbf{u}, \hat{\mathbf{u}}_h - \mathbf{u}) \tag{5.120}$$

This means that the total strain energy of the finite element approximation converges to that of the true solution as $h \to 0$. One may compute $a(\mathbf{u}_h, \mathbf{u}_h)$, and take this as a measure to monitor the quality of the approximations.

* Details of interpolation errors for isoparametric elements can be found in Ciarlet (1978).

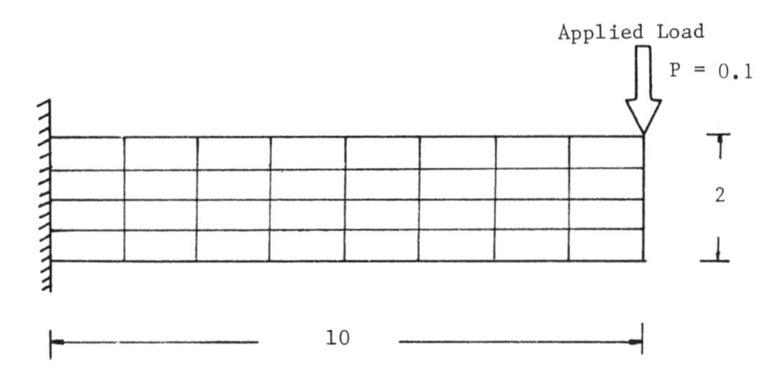

Figure 5.10 Beam-bending problem: 4 × 8 finite elements, Young's modulus $E = 1$, Poisson's ratio 0.3.

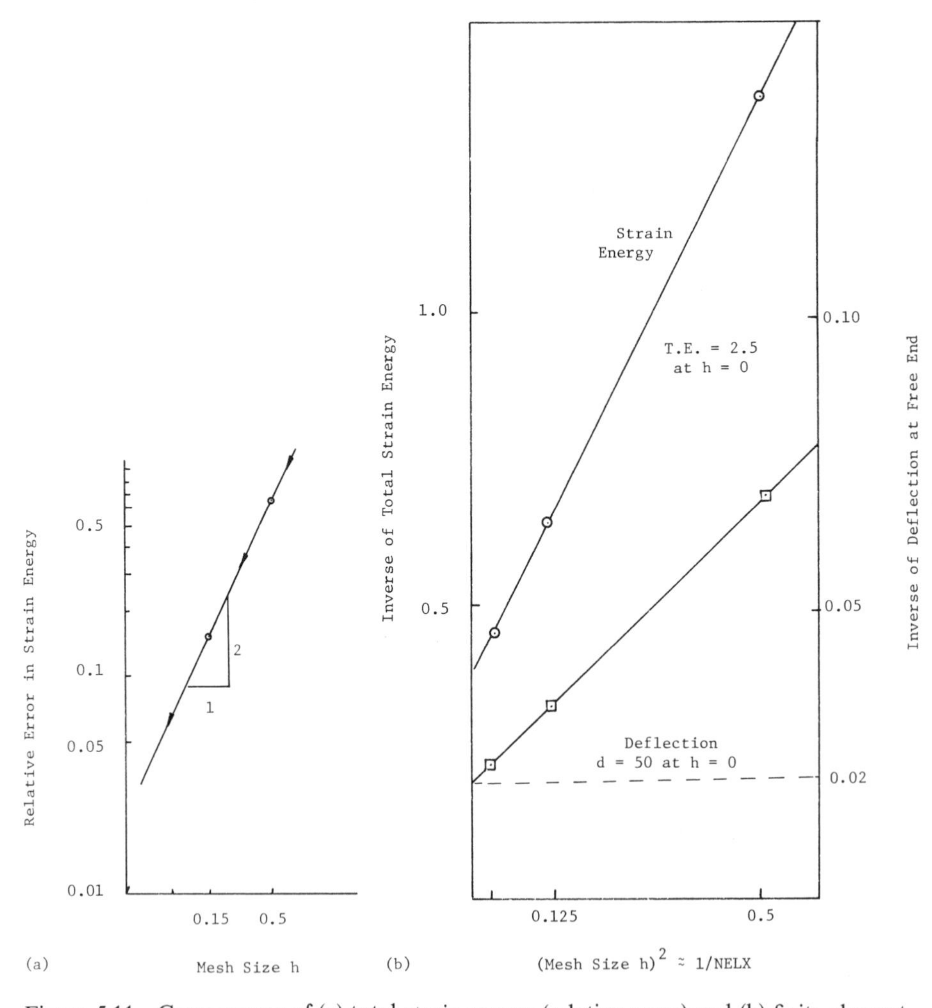

Figure 5.11 Convergence of (a) total strain energy (relative sense) and (b) finite element approximation.

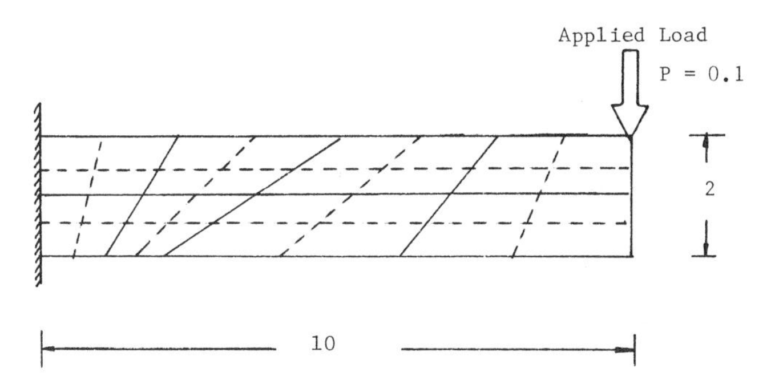

Figure 5.12 Convergence for irregular meshes.

An additional remark on the convergence analysis is that the estimate (5.119) is not necessarily true for all problems, although (5.117) is valid for all. Since the estimate (5.119) is based on the interpolation error estimate (5.118), if (5.118) is not true, then (5.119) is automatically not true. In (5.118) we have assumed that the function f is twice differentiable in Ω. It is certain that the conditions required in (5.118) could be weakened. However, such extensions require the concepts of Lebesgue measure and integration and are best left for a later course. It is, however, noted that the solution $\mathbf{u}$ for a problem involving cracks is not such a function. Thus, (5.118) is not correct. Roughly speaking, for crack problems, $\mathbf{u}$ satisfies

$$\int_\Omega |\nabla \hat{\mathbf{u}}_h - \nabla \mathbf{u}|^2 \, d\Omega \le C(\mathbf{u})h \tag{5.121}$$

where $C(\mathbf{u})$ is a constant dependent on $\mathbf{u}$. This implies convergence of $F(\mathbf{u})$ as $h \to 0$, although the rate of convergence is only h instead of h^2.

Example 5.1: Let us carry out a convergence study for four-node elements using the beam-bending problem shown in Figure 5.10. A cantilever is discretized into 2×1, 4×2, and 8×4 uniform finite elements. Using the generalized plane stress constitutive relation with $E = 1$ and $\nu = 0.3$, the convergence result is shown in Figure 5.11(a) for the applied load $P = 0.1$ at the right end. Relative error in the strain energy is defined by

$$e = \tfrac{1}{2}a(\mathbf{u}_{\hat{h}}, \mathbf{u}_{\hat{h}}) - \tfrac{1}{2}a(\mathbf{u}_h, \mathbf{u}_h) \tag{5.122}$$

where $\hat{h}$ indicates the fixed mesh size and is expected to be the smallest one among the mesh sizes in numerical experiments. Here $\hat{h} = \frac{10}{8} = 1.25$. The rate of convergence in this case is h^2, as shown in Figure 5.11(a), and corresponds to the theoretical result (5.119)). In Figure 5.11(b), inverses of the strain energy and the deflection of the cantilever at the free end are plotted with respect to h^2. This shows that the predicted deflection at the free end as $h \to 0$ is 50, which is also the value of the tip deflection computed using the elementary beam theory.

Exercise 5.9: Perform a convergence study for the meshes shown in Figure 5.12.

5.6.3 Flexural deformations

As shown in example 5.1, the quality of the finite element approximation using the four-node element is rather poor in the beam-bending problem. This is caused by the *high eigenvalues of the flexural modes compared to the case of pure bending deformations.* Suppose that the flexural modes are obtained solely from the bending stress σ_{ss}. Then we have to eliminate all other contributions due to the shear stress σ_{ts}. If the four-node element is applied with the 2×2 Gaussian integration scheme, σ_{ts} contributes to form the flexural modes. Indeed, for the fourth mode in Figure 5.7, the displacement field is

$$u = -a\tfrac{1}{4}(1-s)(1-t) + a\tfrac{1}{4}(1+s)(1-t), \qquad v = 0 \tag{5.123}$$

for some positive constant a. Then

$$\varepsilon_{ts} = -\tfrac{1}{4}as \quad \text{while } \varepsilon_{ss} = \tfrac{1}{2}a(1-t)$$

That is, we have the extra contribution $\frac{1}{2}\sigma_{ts}\varepsilon_{ts}$ from the shear deformation that should not be included in the flexural deformation. In order to exclude these contributions from the shear deformation, we may integrate the stiffness matrix related to the shear deformation ε_{ts} at the centroid by applying the one-point Gaussian quadrature instead of the 2×2 rule, since the shear strain ε_{ts} is linear in s. More precisely, we shall divide the stiffness matrix $E^e_{i\alpha j\beta}$ into two parts, $E^{e,1}_{i\alpha j\beta}$ and $E^{e,2}_{i\alpha j\beta}$. The first part, $E^{e,1}_{i\alpha j\beta}$, is related to the normal stresses σ_{ss} and σ_{tt} and the second part, $E^{e,2}_{i\alpha j\beta}$, is related to only the shear stress. Then, the first part is integrated as before, using the 2×2 Gaussian rule. The second part is integrated by the one-point Gaussian rule, which is a one-order "less" accurate integration rule. This modification by the *selective reduced integration* technique provides much better results for bending-dominated problems than does the usual method. However, it should not be inferred that the reduced integration technique provides better results for all problems.

Remember that the original method, which does not use the reduced integration technique, converges to the correct solution as the mesh size h goes to zero. No matter how bad the quality may be, we are assured that if a fine enough finite element mesh is employed, the quality of the results will be acceptable. The use of reduced integration techniques must be based on thorough analyses in any given problem. The method developed for a specific problem should not be applied blindly to other classes of problems.

Example 5.2 (continuation of example 5.1): Let us apply the selective reduced integration technique to obtain improved approximate solutions for the cantilever problem studied in example 5.1. We shall describe the method stated above as the selective reduced integration I, while the 2×2 Gaussian quadrature rule case is denoted as the full integration scheme. Let us introduce one more selective reduced integration method obtained by the decomposition of the stiffness matrix $E^e_{i\alpha j\beta}$ into $E^{e,1}_{i\alpha j\beta}$, related to the term $\lambda\varepsilon_{kk}\delta_{ij}$, and $E^{e,2}_{i\alpha j\beta}$, related to the term $2\mu\varepsilon_{ij}$ in the constitutive equation $\sigma_{ij} = \lambda\varepsilon_{kk}\delta_{ij} + 2\mu\varepsilon_{ij}$. We shall apply the 2×2 Gaussian rule to the first part and integrate the second part by the one-point rule. This scheme is called the selective reduced integration II. As an extreme case, let us apply the one-point Gaussian

Table 5.2 *Deflection of free end of cantilever*

Mesh	Reduced	Selective reduced II	Selective reduced I	Full integration
2×1	∞	81.3	44.0	14.7
4×2	67.0	55.6	48.7	31.2
8×4	55.1	52.4	50.7	42.3

Note: Deflection at the free end by the beam theory is 50.

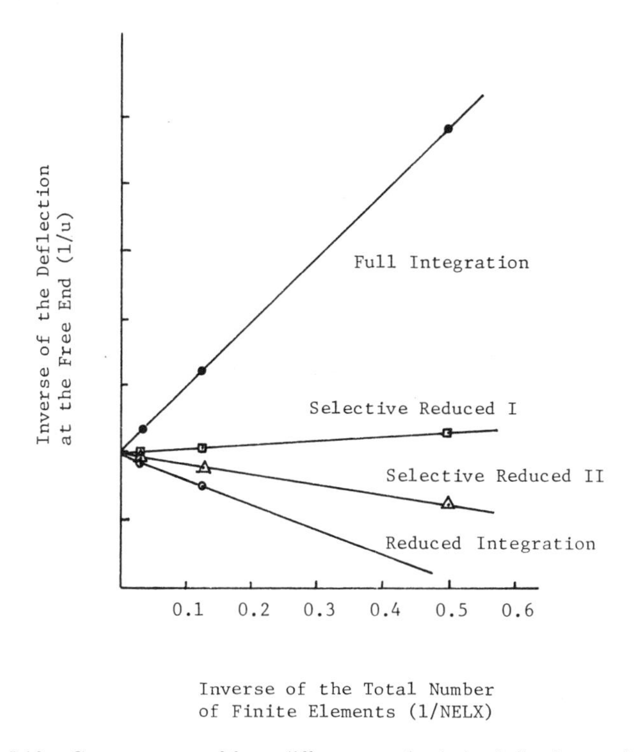

Figure 5.13 Convergence of four different methods in deflection at free end.

quadrature to both terms, and let this be called the *reduced integration* scheme. We shall compare the results obtained by these four different methods. If the deflection of the free end of the cantilever is compared, the results are as shown in Table 5.2 and Figure 5.13. It is understood that each method can predict the limiting value of the deflection as $h \to 0$, although the selective reduced integration scheme I gives results that are very close to those of beam theory for the whole range of mesh sizes.

A final remark on the cantilever problem is that even the best method, the selective reduced integration scheme I, fails to provide reasonable results if the finite element meshes are not uniform. Indeed, for the finite element model shown in Figure 5.14, the deflection at the free end is just 33.8, which is about 77.6% of the deflection determined by

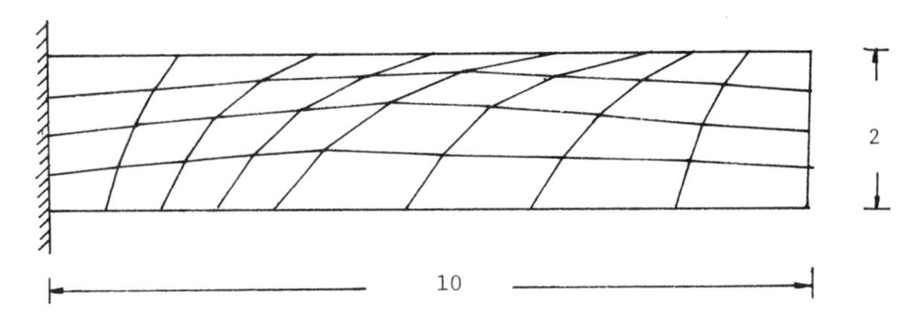

Figure 5.14 Nonuniform finite element model for a convergence test.

beam theory. This indicates the importance of defining a "proper" finite element model in a given problem.*

5.6.4 Nearly incompressible or incompressible materials

When the bulk modulus κ becomes so large that $\varepsilon_{ii} \approx 0$, we have to be very careful in using the four-node isoparametric element. If the fixed boundary condition is imposed on too many nodal points, only the *locked solution* can be obtained by the finite element approximation. That is, no matter how large the applied loads may be, the numerical solution is almost zero because of the large bulk modulus. One way to resolve this difficulty is again the application of the selective reduced integration technique. In this case, we apply the reduced integration scheme to the term $\kappa\varepsilon_{jj}\bar{\varepsilon}_{ii}$ related to the bulk modulus κ in the constitutive equation. More precisely, if the material is isotropic, so that

$$\sigma_{ij} = \kappa\varepsilon_{kk}\delta_{ij} + 2\mu\varepsilon_{ij}^{D} \tag{5.124}$$

then the stiffness matrix $E^{e}_{i\alpha j\beta}$ is separated into two parts, $E^{e,1}_{i\alpha j\beta}$ and $E^{e,2}_{i\alpha j\beta}$, where

$$E^{e,1}_{i\alpha j\beta} = \int_{\Omega_e} \kappa N_{\alpha,i} N_{\beta,j}\, d\Omega$$

and

$$E^{e,2}_{i\alpha j\beta} = \int_{\Omega_e} \mu(N_{\alpha,j}N_{\beta,i} + \delta_{ij}N_{\alpha,k}N_{\beta,k} - \tfrac{2}{3}N_{\alpha,i}N_{\beta,j})\, d\Omega$$

We shall apply the one-point Gaussian integration rule to evaluate the first part, $E^{e,1}_{i\alpha j\beta}$, and apply the 2×2 rule to $E^{e,2}_{i\alpha j\beta}$. If this selective reduced integration technique is used, the locked solution can be avoided for any kind of boundary conditions.

If we consider incompressible materials for which

$$\varepsilon_{kk}(\mathbf{u}) = 0 \quad \text{in } \Omega \tag{5.125}$$

we simply approximate this by saying that the material is nearly incompressible, that is, that κ is large. It is clear that if $\kappa \to \infty$, we can expect $\varepsilon_{ii}(\mathbf{u}) \to 0$ as long as the stress is finite. The mean stress σ_m or the hydrostatic pressure p could be computed by

$$\sigma_m = \kappa\varepsilon_{ii}(u) \qquad \text{or} \qquad p = -\lambda\varepsilon_{ii}(\mathbf{u}) \tag{5.126}$$

* One of possible approaches to obtain a proper finite element model is the adaptive method to relocate the location of nodes (see Diaz et al., 1983).

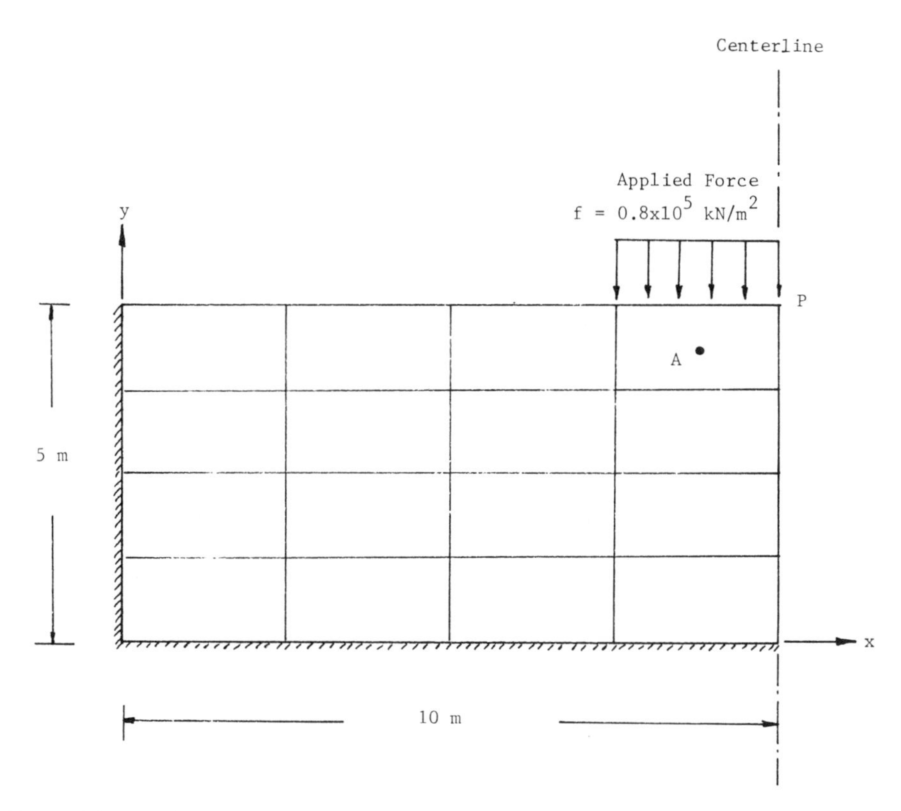

Figure 5.15 Finite element of incompressible elastic foundation: Young's modulus $E = 10^6$ kN/m^2.

after the solution **u** is obtained. It is to be noted that the stress σ_{ij} must be evaluated only on the integration points for the one-point Gaussian rule since the part related to the bulk modulus, which dictates the quality of the solution, is integrated on these points.

Example 5.3: Applying both 2×2 full Gaussian and the selective reduced integration methods, let us solve a problem involving incompressibility to justify the remarks stated above. To do this, let us consider an incompressible material in a fixed container, loaded as shown in Figure 5.15, the cross section of which is rectangular. Under the assumptions of plane strain, finite element equations are obtained using 16 four-node elements for the following three different material constants:

Case	Young's modulus E (kN/m^2)	Poisson's ratio ν	Penalty parameter ε
1	10^6	0.4	0.7
2	10^6	0.49	0.0601
3	10^6	0.499	0.0060

Table 5.3 *Comparison of results using the selective reduced and full integration methods*

	Poisson's ratio	Selective reduced	Full integration
Displacement at P	0.4	−0.26022161	−0.229528124
	0.49	−0.18539858	−0.0806783685
	0.499	−0.17598573	−0.0131050647
Stress σ_{yy} at A	0.4	−0.7081756	−0.644336382
	0.49	−0.717140659	−0.482306832
	0.499	−C.717975897	−0.328824076

where the "penalty" parameter ε is defined by

$$\varepsilon = (1 - 2v)(1 + v)/v \tag{5.127}$$

Let the applied pressure f be given as $f = 0.8 \times 10^5$ kN/m^2 on the part of the top surface as shown in Figure 5.15. Table 5.3 shows the results computed at the two points P and A shown in Figure 5.15. It is clear that if the 2×2 Gaussian quadrature rule is applied to evaluate the stiffness matrix, the size of computed displacement becomes small as Poisson's ratio v approaches 0.5 for the incompressible material in plane strain problems. However, once the selective reduced integration technique is applied, the displacement and stress components converge to reasonable values from the physical point of view as $v \to 0.5$. Such convergence is shown in Figure 5.16 using the relative errors on the displacement and stress defined by

$$e_d = v_\varepsilon - v_{\hat{\varepsilon}} \qquad \text{and} \qquad e_\tau = \sigma_\varepsilon - \sigma_{\hat{\varepsilon}},$$

where the subscripts ε and $\hat{\varepsilon}$ indicate the size of the penalty parameter, and $\hat{\varepsilon} = 0.0060$ is fixed. Values of ε are 0.7 or 0.0601. As shown in Figure 5.16, the rate of convergence is almost 1, that is, the order of errors is $O(\varepsilon)$.

Exercise 5.10: Solve the rigid punch problem for the nearly incompressible material shown in Figure 5.17. Poisson's ratio is $v = 0.499$ and Young's modulus is 10^6 kN/m^2. Using 100 elements, compute the mean stress $\sigma_m = \frac{1}{3}(\sigma_{11} + \sigma_{22} + \sigma_{33})$, plot this along the lines AA' and BB' shown in Figure 5.17, and observe the oscillatory behavior of σ_m.

Another method commonly used for solving incompressible problems is the *mixed finite element method*. This method is based on a modification of the constitutive equation (5.124) for incompressible materials:

$$\sigma_{ij} = -p\delta_{ij} + 2\mu\varepsilon_{ij} \tag{5.128}$$

with the incompressibility constraint (5.125). We treat the displacement **u** and the

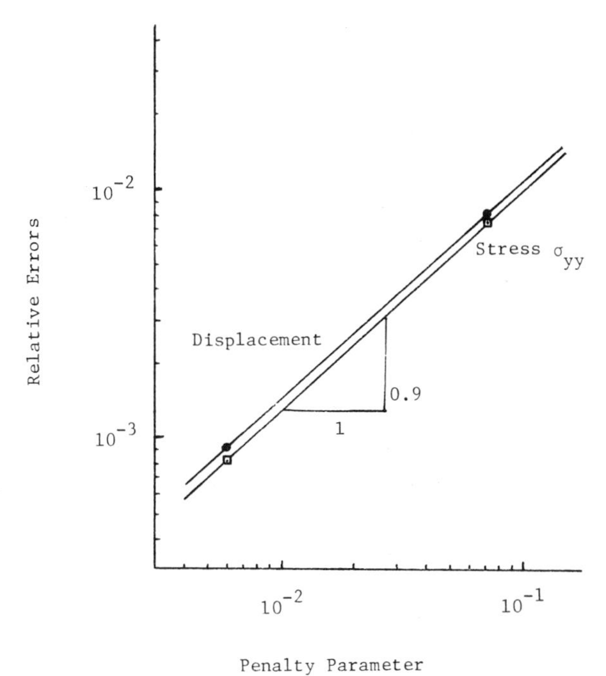

Figure 5.16 Convergence of stress σ_{yy} and displacement in penalty parameter $\varepsilon \to 0$.

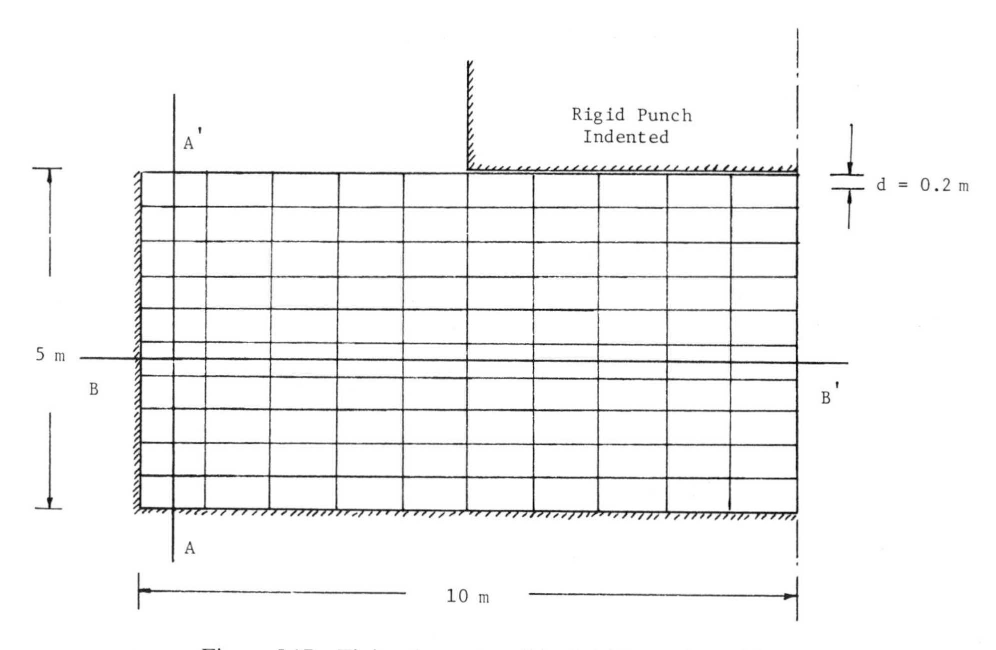

Figure 5.17 Finite element model of rigid punch problem.

"pressure" field p as unrelated variables. The weak form (5.53) is now changed to

$$\int_{\Omega} \{2\mu\varepsilon_{ij}(\mathbf{u})\varepsilon_{ij}(\bar{\mathbf{u}}) - p\varepsilon_{ii}(\bar{\mathbf{u}})\}\, d\Omega + \int_{\Gamma_{i3}} k_{ij}u_j\bar{u}_i\, d\Gamma$$

$$= \int_{\Omega} \rho f_i\bar{u}_i\, d\Omega + \int_{\Gamma_{i2}} h_i\bar{u}_i\, d\Omega + \int_{\Gamma_{i3}} k_{ij}u_{\infty j}\bar{u}_i\, d\Gamma, \quad \forall\bar{\mathbf{u}} \tag{5.129}$$

A weak form for the incompressibility constraint (5.125) is obtained by introducing an arbitrary virtual pressure $\bar{p}$:

$$\int_{\Omega} \bar{p}\varepsilon_{ii}(\mathbf{u})\, d\Omega = 0, \quad \forall\bar{p} \tag{5.130}$$

Let $\mathbf{u}$ and p be discretized by

$$u_j = u_{j\beta}N_\beta \qquad \text{and} \qquad p = p_\delta M_\delta \tag{5.131}$$

respectively. Here the shape functions $\{N_\beta\}$ and $\{M_\delta\}$ might be obtained differently by using different finite element models. For example, a composite element consisting of four four-node elements is taken for the approximation of the pressure field while the displacement is approximated in each four-node element, as shown in Figure 5.18(a). In the most popular method, the pressure field is approximated by a bilinear function of a composite element. In this case, the continuity of the pressure field is preserved.

The weak forms (5.129) and (5.130) are discretized by (5.131):

$$\sum_{e=1}^{E} \bar{u}_{i\alpha} \int_{\Omega_e} \mu(\delta_{ij}N_{\alpha,k}N_{\beta,k} + N_{\alpha,j}N_{\beta,i})\, d\Omega\, u_{j\beta}$$

$$- \sum_{e=1}^{E} \bar{u}_{i\alpha} \int_{\Omega_e} N_{\alpha,i}M_\delta\, d\Omega\, p_\delta + \sum_{e=1}^{E_3} \bar{u}_{i\alpha}E^{3,e}_{i\alpha j\beta}u_{j\beta}$$

$$= \sum_{e=1}^{E} \bar{u}_{i\alpha}f^{1,e}_{i\alpha} + \sum_{e=1}^{E_2} \bar{u}_{i\alpha}f^{2,e}_{i\alpha} + \sum_{e=1}^{E_3} \bar{u}_{i\alpha}f^{3,e}_{i\alpha}, \quad \forall\bar{\mathbf{u}} \tag{5.132}$$

and

$$\sum_{e=1}^{E} \bar{p}_\gamma \int_{\Omega_e} M_\gamma N_{\beta,j}\, d\Omega\, u_{j\beta} = 0, \quad \forall\bar{p} \tag{5.133}$$

where the notation in (5.60) is applied here. Defining

$$E^{1,e}_{i\alpha j\beta} = \int_{\Omega_e} \mu(\delta_{ij}N_{\alpha,k}N_{\beta,k} + N_{\alpha,j}N_{\beta,i})\, d\Omega \tag{5.134}$$

and

$$B^e_{\alpha i\delta} = \int_{\Omega_e} N_{\alpha,i}M_\delta\, d\Omega \tag{5.135}$$

the finite element equation becomes

$$\sum_{e=1}^{E} \begin{bmatrix} E^{1,e}_{i\alpha j\beta} & -B^e_{\alpha i\delta} \\ -B^e_{\beta j\gamma} & 0 \end{bmatrix} \begin{Bmatrix} u_{j\beta} \\ p_\delta \end{Bmatrix} + \sum_{e=1}^{E_3} \begin{bmatrix} E^{3,e}_{i\alpha j\beta} & 0 \\ 0 & 0 \end{bmatrix} \begin{Bmatrix} u_{j\beta} \\ p_\delta \end{Bmatrix}$$

$$= \sum_{e=1}^{E} \begin{Bmatrix} f^{1,e}_{i\alpha} \\ 0 \end{Bmatrix} + \sum_{e=1}^{E_2} \begin{Bmatrix} f^{2,e}_{i\alpha} \\ 0 \end{Bmatrix} + \sum_{e=1}^{E_3} \begin{Bmatrix} f^{3,e}_{i\alpha} \\ 0 \end{Bmatrix} \tag{5.136}$$

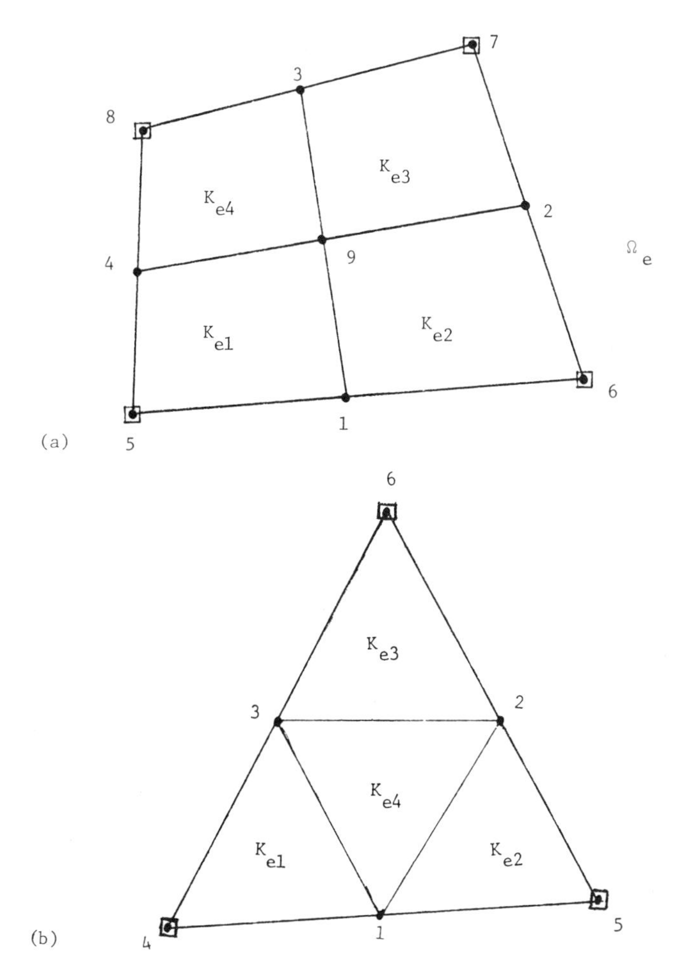

Figure 5.18 Composite elements for mixed methods: (a) four four-node elements, (b) four three-node elements; □, nodes for p; ●, nodes for $\mathbf{u}$.

Note that the sign of (5.133) has been changed to obtain (5.136) so that the symmetry of the stiffness matrix is achieved.

Exercise 5.11: Develop a computer program to form the element stiffness matrix and the element generalized load vector for the composite element Ω_e shown in Figure 5.18(a).

Another composite element that maintains continuity of the pressure field p is the use of four three-node triangular elements, as shown in Figure 5.18(b). Displacement is approximated by a piecewise linear polynomial in each subtriangle and pressure is interpolated linearly on a composite element. This choice is similar to the composite element by four four-node quadrilateral elements. One of the advantages of this four three-node composite element is that the gradient of the displacement vector is constant in each subtriangular element. Thus, the evaluation of element stiffness matrices is much simpler than the four four-node composite element.

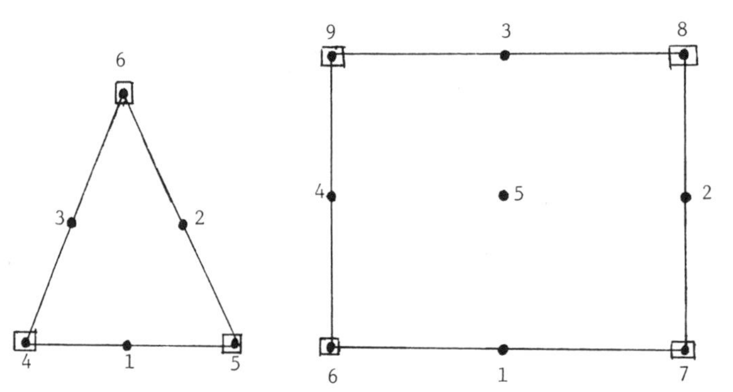

Figure 5.19 Other choices of finite elements for mixed methods: □, pressure; ●, displacement.

Although composite elements have been introduced here, a combination of higher-order interpolation for the displacement and lower-order interpolation for the pressure is more popular in the literature. Popular choices are (a) six-node quadratic elements for the displacement and three-node linear elements for the pressure and (b) nine-node biquadratic elements for the displacement and four-node bilinear elements for the pressure, as shown in Figure 5.19. Another interesting observation is that if the pressure is interpolated by a discontinuous piecewise polynomial, it is possible to eliminate the pressure term p_δ from equation (5.136) in each (composite) finite element Ω_e. Then the element stiffness matrix is represented only by degrees of freedom due to displacement. The use of discontinuous pressure approximations is now popular and is related to the *penalty method* to resolve constraints in the theory of optimization. (See sections 1.4.3 and 5.10.4.)

5.6.5 Evaluation of stresses

Following section 2.10.2, we shall consider the stress evaluation method, which computes the values of the stress tensor at nodal points of the finite element model. In general, we compute the stress tensor at the integration points if Gaussian quadrature rules are applied to evaluate the stiffness matrix. Suppose that one component τ of the stress tensor is evaluated at the 2×2 Gaussian integration points. If a polynomial form of τ is given as

$$\tau = \tau_\alpha \hat{M}_\alpha(s, t) \tag{5.137}$$

where

$$\hat{M}_1 = \frac{1}{4a^2}(a - s)(a - t), \qquad \hat{M}_2 = \frac{1}{4a^2}(a + s)(a - t)$$

$$\hat{M}_3 = \frac{1}{4a^2}(a + s)(a + t), \qquad \hat{M}_4 = \frac{1}{4a^2}(a - s)(a + t) \tag{5.138}$$

and $a = 1/\sqrt{3}$, then it is clear that τ is not continuous along the common boundary of two elements. In order to obtain a continuous stress component $\tilde{\tau}$ from τ, we assume a polynomial form of $\tilde{\tau}$ as

$$\tilde{\tau} = \hat{\tau}_\alpha \hat{N}_\alpha(s, t) \tag{5.139}$$

where $\{\hat{N}_\alpha\}$ are the shape functions for the displacements. The coefficients $\hat{\tau}_\alpha$ are then computed by

$$\int_\Omega \tilde{\tau}\bar{\tau}\, d\Omega = \int_\Omega \tau\bar{\tau}\, d\Omega, \quad \forall\bar{\tau} \tag{5.140}$$

where $\bar{\tau}$ is an arbitrary polynomial whose form is the same as (5.139). Expanding the form (5.140) yields

$$\sum_{e=1}^{E} M^e_{\alpha\beta}\tilde{\tau}_\beta = \sum_{e=1}^{E} f^e_\alpha \tag{5.141}$$

where

$$M^e_{\alpha\beta} = \int_{\Omega_e} \hat{N}_\alpha \hat{N}_\beta\, d\Omega \qquad \text{and} \qquad f^e_\alpha = \left(\int_{\Omega_e} \hat{N}_\alpha \hat{M}_\beta\, d\Omega\right)\tau_\beta \tag{5.142}$$

Solving (5.141), we can obtain the nodal values of the stress component τ and its continuous polynomial form. It may be worthwhile to mention that (5.140) is just the method of least squares! Indeed, $\bar{\tau}$ is the minimizer of the functional to the least-squares best approximation:

$$\min_{\bar{\tau}} F^*(\bar{\tau}) = \tfrac{1}{2}\int_\Omega (\bar{\tau} - \tau)^2\, d\Omega \tag{5.143}$$

for a given function τ.

5.7 Finite element program FEM2 for plane elasticity using FORTRAN

We shall develop a computer program based on the theory studied so far using FORTRAN. The program developed here and shown in Appendix 4 consists of three parts: preprocessing, finite element (FE) processing, and postprocessing. A mesh generator for four-node elements can be obtained by minor modification of the program PRE-FEM1 for three-node elements given in Appendix 2. Similarly, a plotting routine for stress analysis can be easily modified from POST-FEM1 given in Appendix 3.

Preprocessing contains 10 subroutines that generate necessary input data for finite element processing:

PTITLE
HRMESH
PRESKY
MATERL
RBOUN1
RBOUN2
RBOUN3
RBOUN4
RBODYF
RTEMPE

Finite element processing consists of nine subroutines that require the data created in preprocessing:

ASSEMB
  ESTIFO
    TANMOD
    GGRAD4
BOUND1
BOUND2
BOUND3
BOUND4
SKYLIN

Postprocessing (in this program) has three subroutines that provide deformations, stresses, and other quantities needed for stress analysis and design of structures:

DEFORM
STRESS
  PRSTRS

PTITLE: This subroutine reads and outputs the title of the problem to be solved.

HRMESH: The nodal coordinates and element connectivities have to be inputted. Data for this subroutine can be easily obtained by the mesh generator PRE-FEM1 by minor modification in order to generate four-node elements.

PRESKY: In this subroutine the height of the skyline is computed and the diagonal pointers JDIAG are constructed for a skyline solver. This subroutine must be applied at the very end of preprocessing.

MATERL: This subroutine reads and outputs all material properties and defines the characteristics of a problem.

RBOUN1: We shall read input data for the first type of boundary condition,

$$u_i = g_i \quad \text{on } \Gamma_{i1}$$

for $i = 1, 2$. Input data are node numbers, specified directions, and values of constrained displacements g_i.

RBOUN2: This subroutine reads *equivalent nodal forces* computed by the applied traction t_i on Γ_{i2}, using

$$t_{i\alpha} = \sum_{e=1}^{E_2} \int_{\Gamma_{i2}^e} t_i N_\alpha \, d\Gamma \quad (i\text{: no sum}) \tag{5.144}$$

where $t_{i\alpha}$ is the ith component of the equivalent nodal force at the αth node on Γ_{i2} due to the *nonhomogeneous* second type of boundary condition,

$$\sigma_{ij} n_j = t_i \quad \text{on } \Gamma_{i2}$$

If the "free" boundary condition is imposed, we need not input anything since $t_i = 0$ on this boundary. If plane problems are considered, and if the applied trac-

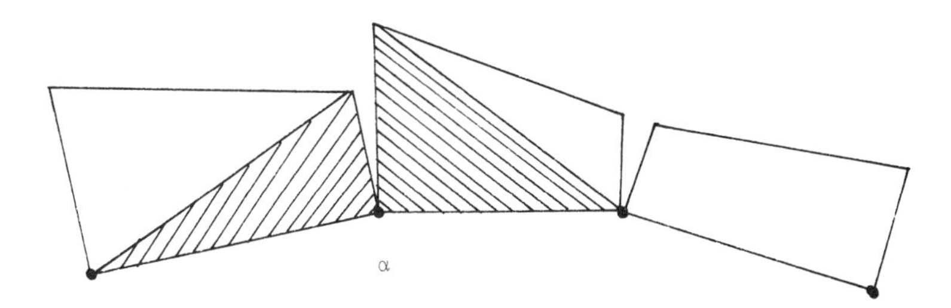

Figure 5.20 Equivalent nodal forces for four-node elements in domain Ω. Distributed traction t_i assumed to be piecewise linear. Summation of two shaded areas yields $t_{i\alpha}$.

tion $\mathbf{t} = t_i \mathbf{i}_i$ is piecewise linear, the equivalent nodal force $t_{i\alpha}$ is computed as the sum of the areas of the two triangles shown in Figure 5.20. This is valid only for three-node triangular or four-node quadrilateral elements whose boundaries are two-node line elements. Input data are thus node numbers and values of equivalent nodal forces.

RBOUN3: Here we consider the generalized third type of boundary conditions in both normal and tangential directions on the boundary Γ:

$$\sigma_n = -k_n(u_n - g_n) + t_n \tag{5.145}$$

and

$$\boldsymbol{\sigma}_T = -k_T(\mathbf{u}_T - \mathbf{g}_T) + \mathbf{t}_T \tag{5.146}$$

where $\sigma_n = \sigma_{ji} n_j n_i$ is the normal traction; $\boldsymbol{\sigma}_T = \boldsymbol{\sigma} - \sigma_n \mathbf{n}$ is the tangential traction; k_n and k_T are spring constants in the normal and tangential directions, respectively; g_n and $\mathbf{g}_T$ are "constrained" normal and tangential displacements, respectively; and t_n and $\mathbf{t}_T$ are applied normal and tangential traction, respectively. If $k_n = 0$ and $t_n \neq 0$, then it is possible to specify the normal traction on the boundary using this generalized third type of boundary condition. Similarly, if $k_n \to +\infty$ with $t_n = 0$, we have

$$u_n - g_n = \sigma_n / k_n \to 0$$

that is, $u_n = g_n$ is approximately satisfied. This means that (5.145) can specify the normal displacement on the boundary.

Input data are element connectivities of two-node line elements and values of k_n, g_n, t_n, k_T, $\mathbf{g}_T$, and $\mathbf{t}_T$, in this subroutine.

RBOUN4: If a point P is on an inclined surface as shown in Figure 5.21, we have to restrict the normal displacement at P. To do this, we input node numbers and the angle of inclined surface from the x axis using this subroutine.

RBODYF: We set up the body force term in this subroutine. To do this, we must input the order of a polynomial LBODY and its coefficients for the components of the body force:

$$\rho f_I(x, y) = \sum_{i,j=1}^{LBODY} c^I_{ij} x^i y^j, \quad I = 1, 2 \tag{5.147}$$

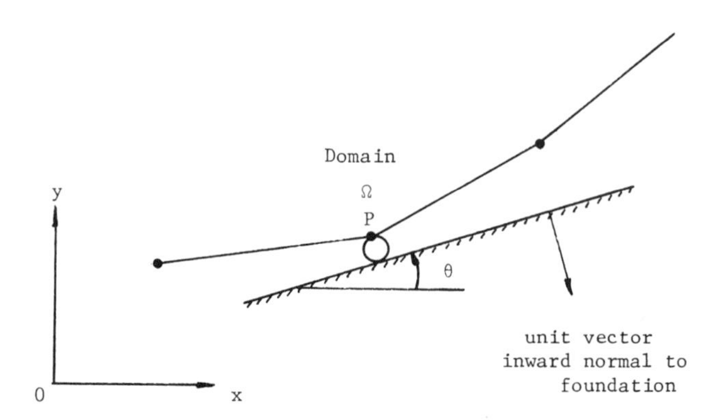

Figure 5.21 Roller support along inclined foundation.

RTEMPE: When thermal stresses have to be considered, we must specify the temperature field. To do this, FEM1 or its modification for heat conduction problems may be used. In this program, we shall assume that the temperature is known at each node of the finite element model and shall input these nodal values.

ASSEMB: This is the main body of the finite element processing. This has three more subroutines, ESTIF0, TANMOD, and GGRAD4. ESTIF0 forms element stiffness matrices and generalized load vectors, TANMOD is used to specify the "tangent" modulus E_{ijkl}, and GGRAD4 computes the gradient of the shape functions with respect to the global coordinate system (x, y).

BOUND1, BOUND2, BOUND3, and BOUND4: According to the data read in subroutines RBOUN1, RBOUN2, RBOUN3, and RBOUN4, we modify the stiffness matrix and the generalized load vector.

SKYLIN: This is a solver of the system of linear equations using the skyline method. Symmetry of the stiffness matrix is assumed here.

DEFORM, STRESS, PRSTRS: These provide output such as displacements, stresses, principal stresses, specific strain energy, and the equivalent (or von Mises) stress

$$\bar{\sigma} = \sqrt{\tfrac{3}{2}\sigma_{ij}^D\sigma_{ij}^D} \tag{5.148}$$

where σ_{ij}^D is the deviator of the stress σ_{ij}.

Example 5.4: As our first example in this section, we shall solve a tension problem involving a thin linearly elastic plate with a hole, as shown in Figure 5.22. Suppose that the plate is a homogeneous isotropic elastic body with the following characteristics:

Young's modulus	$E = 2.1 \times 10^7$ N/cm^2
Poisson's ratio	$\nu = 0.29$
Plate thickness	$t = 1$ cm

Generalized plane stress is assumed in this example problem. To check the quality of the finite element solution, let us compare our solution with that of an infinite plate. For a circular hole in an infinite plate subjected to a pure uniaxial tension or compression σ_0 at ∞, there exists

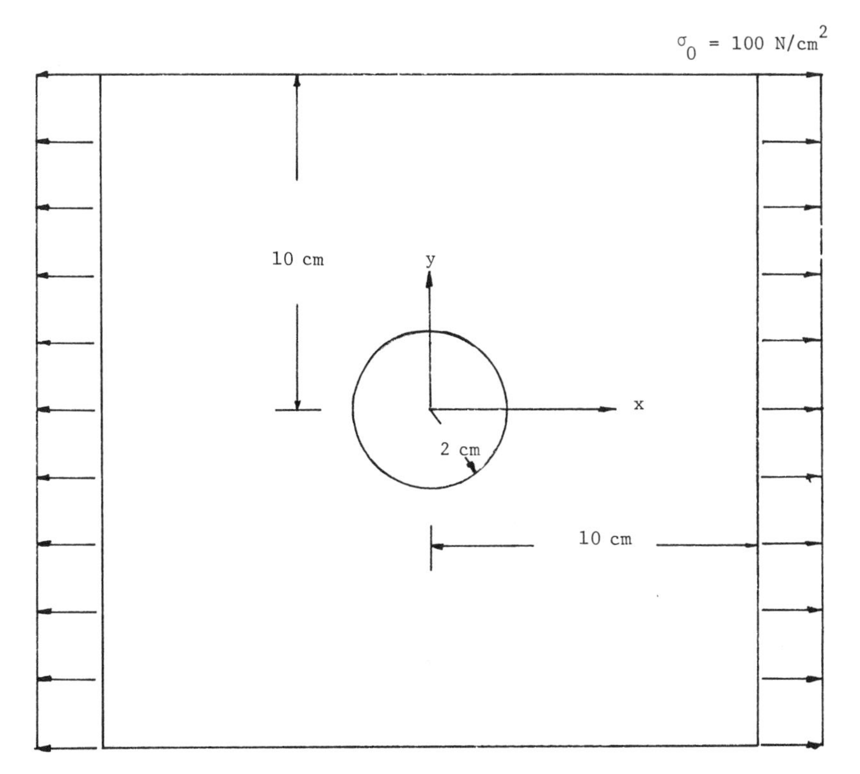

Figure 5.22 Stress analysis of plate with hole: Young's modulus $E = 2.1 \times 10^7$ N/cm^2, Poisson's ratio 0.29.

an exact solution for the stresses (Timoshenko and Goodier, 1970):

$$\sigma_{rr} = \frac{\sigma_0}{2}\left\{\left[1 - \left(\frac{a}{r}\right)^2\right] + \left[1 - 4\left(\frac{a}{r}\right)^2 + 3\left(\frac{a}{r}\right)^4\right]\cos 2\theta\right\}$$

$$\sigma_{\theta\sigma} = \frac{\sigma_0}{2}\left\{\left[1 + \left(\frac{a}{r}\right)^2\right] - \left[1 + 3\left(\frac{a}{r}\right)^4\right]\cos 2\theta\right\} \qquad (5.149)$$

$$\sigma_{r\theta} = -\frac{\sigma_0}{2}\left[1 + 2\left(\frac{a}{r}\right)^2 - 3\left(\frac{a}{r}\right)^4\right]\sin 2\theta$$

where a is the radius of the hole and (r, θ) is a polar coordinate system, the origin of which is located at the center of the hole.

Using symmetry of the problem about both the x and y axes, only a quarter of the whole plate need be considered. It is divided into 48 finite elements, as shown in Figure 5.23. Using the data in Table 5.4, the deformed configuration, the principal stresses (σ_1, σ_2), and the maximum shear stress are obtained as shown in Figures 5.23 and 5.24.

An interesting comparison can be made with the stress pattern obtained by photoelasticity in which fringes correspond to the maximum shear stress $\tau_{max} = \frac{1}{2}(\sigma_1 - \sigma_2)$. It is clear that contour lines of τ_{max} obtained by the finite element approximation are quite similar to the fringe pattern obtained by the photoelasticity shown in Figure 5.25.

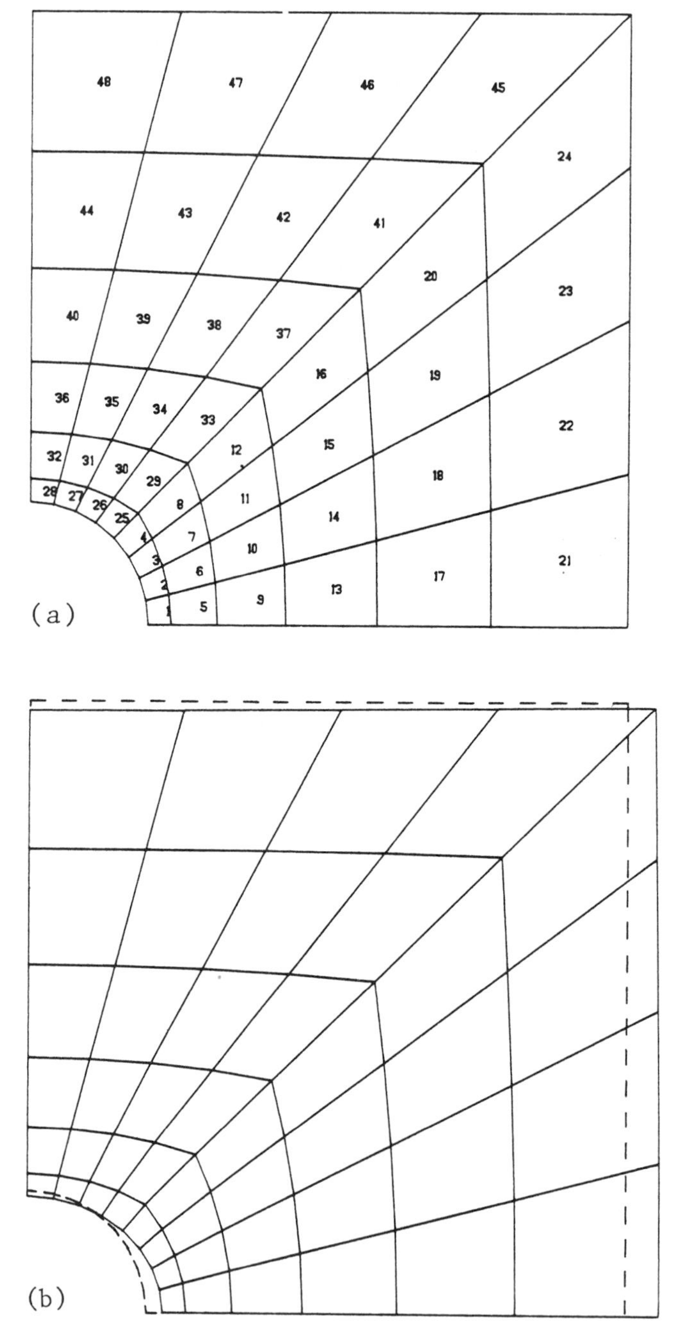

Figure 5.23 (a) Finite element model and (b) deformed configuration.

Table 5.4 *Input data for example 5.4*

Line	Input data
1	TITLE OF THE PROBLEM (MARCH, 1984)
2	THIN PLATE WITH A HOLE
3	CONTROL NUMBERS
4	1 1-ST BOUNDARY CONDITION ON
5	0 2-ND BOUNDARY CONDITION OFF
6	1 3-RD BOUNDARY CONDITION ON
7	0 4-TH BOUNDARY CONDITION OFF
8	0 NO BODY FORCES
9	0 NO TEMPERATURE
10	0 ALL PROCESSINGS
11	FINITE ELEMENT MODEL
12	63 48
13	1 5.8095 0.0000
14	2 4.2857 0.0000
15	3 5.7881 1.3986
16	4 4.2565 0.9981
17	5 7.7143 0.0000
18	6 7.7026 1.8992
19	7 3.1429 −0.0000
20	8 3.1078 0.6977
21	9 5.7298 2.7819
22	10 4.1770 1.9753
23	11 7.6708 3.7901
24	12 3.0124 1.3703
25	13 10.0000 0.0
26	14 10.0000 2.5000
27	15 10.0000 5.0000
28	16 2.3809 −0.0000
29	17 2.3420 0.4974
30	18 2.2360 0.9670
31	19 5.6347 4.1499
32	20 4.0473 2.9317
33	21 7.6189 5.6727
34	22 2.8567 2.0180
35	23 10.0000 7.5000
36	24 2.0630 1.4089
37	25 2.0000 0.0
38	26 1.9591 0.3973
39	27 1.8478 0.7654
40	28 1.6662 1.1044
41	29 5.5027 5.5027
42	30 3.8673 3.8673
43	31 7.5469 7.5469
44	32 2.6407 2.6407
45	33 10.0000 10.0000
46	34 1.8230 1.8230
47	35 1.4142 1.4142
48	36 4.1499 5.6347
49	37 2.9317 4.0473
50	38 5.6727 7.6189
51	39 2.0180 2.8567
52	40 7.5000 10.0000
53	41 1.4089 2.0630
54	42 1.1044 1.6662
55	43 2.7819 5.7298
56	44 1.9753 4.1770
57	45 3.7901 7.6708
58	46 1.3703 3.0124
59	47 5.0000 10.0000
60	48 0.9670 2.2360
61	49 0.7654 1.8478
62	50 1.3986 5.7881
63	51 0.9981 4.2565
64	52 1.8992 7.7026
65	53 0.6977 3.1078
66	54 2.5000 10.0000
67	55 0.4974 2.3420
68	56 0.3973 1.9591
69	57 0.0000 5.8095
70	58 0.0000 4.2857
71	59 0.0000 7.7143
72	60 −0.0000 3.1429
73	61 0.0 10.0000
74	62 −0.0000 2.3809
75	63 0.0 2.0000
76	1 1 25 16 17 26
77	2 1 26 17 18 27
78	3 1 27 18 24 28
79	4 1 28 24 34 35
80	5 1 16 7 8 17
81	6 1 17 8 12 18
82	7 1 18 12 22 24
83	8 1 24 22 32 34
84	9 1 7 2 4 8
85	10 1 8 4 10 12
86	11 1 12 10 20 22
87	12 1 22 20 30 32
88	13 1 2 1 3 4
89	14 1 4 3 9 10
90	15 1 10 9 19 20
91	16 1 20 19 29 30
92	17 1 1 5 6 3
93	18 1 3 6 11 9
94	19 1 9 11 21 19
95	20 1 19 21 31 29
96	21 1 5 13 14 6
97	22 1 6 14 15 11
98	23 1 11 15 23 21
99	24 1 21 23 33 31
100	25 1 35 34 41 42
101	26 1 42 41 48 49
102	27 1 49 48 55 56
103	28 1 56 55 62 63
104	29 1 34 32 39 41
105	30 1 41 39 46 48
106	31 1 48 46 53 55
107	32 1 55 53 60 62
108	33 1 32 30 37 39
109	34 1 39 37 44 46
110	35 1 46 44 51 53
111	36 1 53 51 58 60
112	37 1 30 29 36 37
113	38 1 37 36 43 44
114	39 1 44 43 50 51
115	40 1 51 50 57 58
116	41 1 29 31 38 36
117	42 1 36 38 45 43
118	43 1 43 45 52 50
119	44 1 50 52 59 57
120	45 1 31 33 40 38
121	46 1 38 40 47 45
122	47 1 45 47 54 52
123	48 1 52 54 61 59
124	MATERIAL CONSTANTS
125	1 1 1
126	21000000. 0.29 1.
127	1-ST BOUNDARY CONDITIONS
128	14 100000000000.
129	1 1
130	2 1
131	5 1
132	7 1
133	13 1
134	16 1
135	25 1
136	57 10
137	58 10
138	59 10
139	60 10
140	61 10
141	62 10
142	63 10
143	3-RD BOUNDARY CONDITIONS
144	4
145	1 13 14
146	0.0 0.0 0.0
147	0.0 0.0 100.0
148	2 14 15
149	0.0 0.0 0.0
150	0.0 0.0 100.0
151	3 15 23
152	0.0 0.0 0.0
153	0.0 0.0 100.0
154	4 23 33
155	0.0 0.0 0.0
156	0.0 0.0 100.0

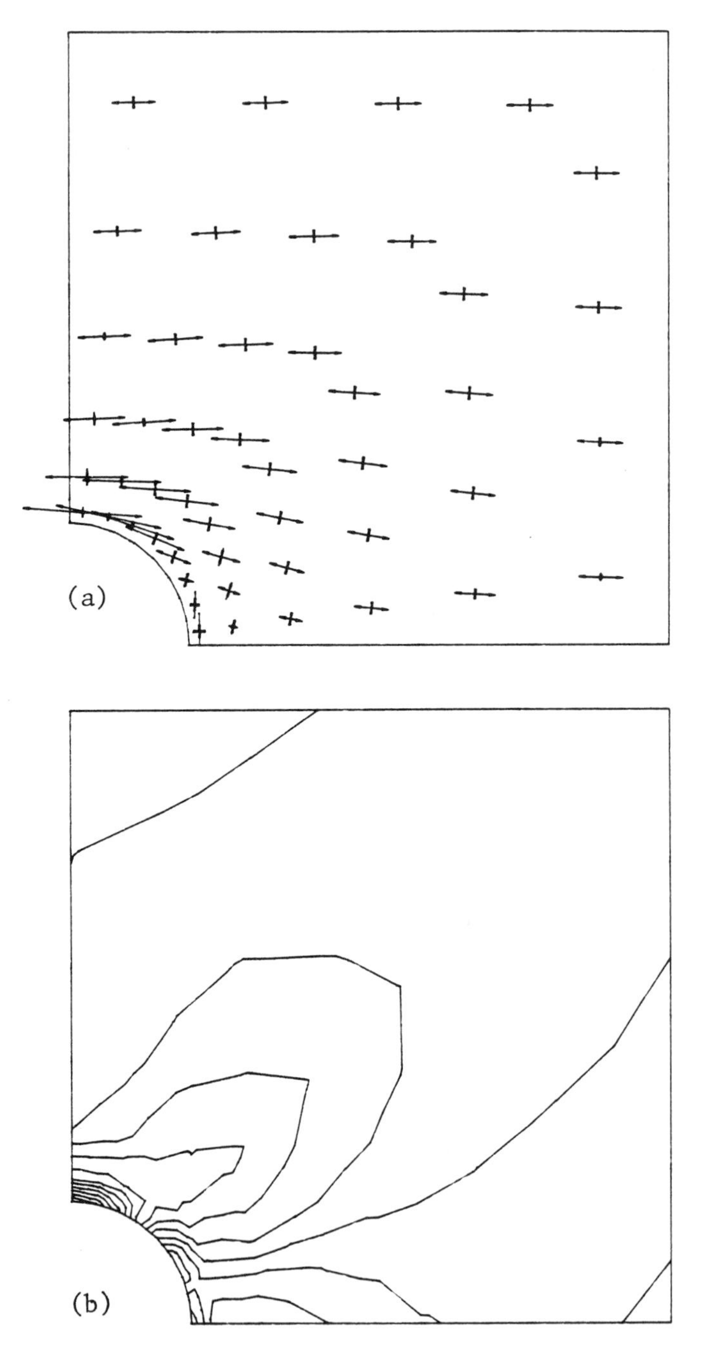

Figure 5.24 Distributions of (a) principal stress and (b) maximum shear stress.

Figure 5.25 Fringe pattern of plate with hole. *From Frocht (1941).*

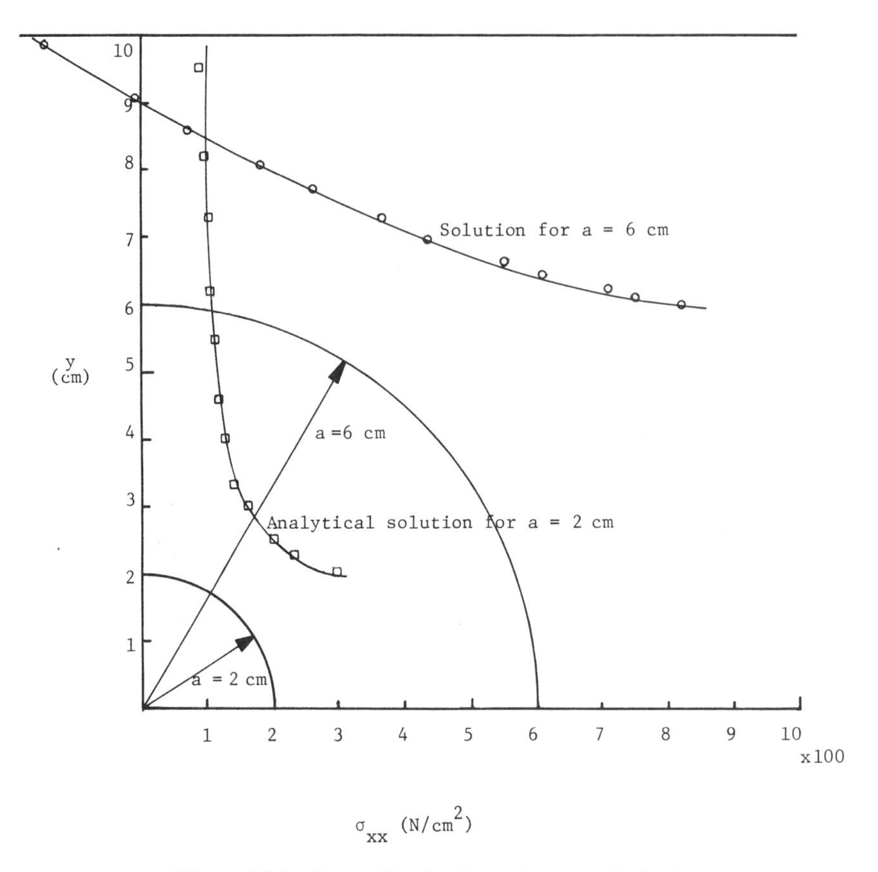

Figure 5.26 Stress distributions along y axis (σ_{xx}).

In Figure 5.26, the stress σ_{xx} along the y axis is plotted. It is clear that the finite element solution predicts σ_{xx} fairly well in comparison to the exact solution for the infinitely wide plate, although the finite element mesh is quite coarse. Another interesting fact is that if the size of the inner hole becomes large enough, compressive stress $\sigma_{xx} < 0$ appears near to the top surface of the plate, as shown in Figure 5.26 although the plate is subjected to the uniform tensile traction at the right edge. This further indicates that the analytical solution (5.149) for the infinite plate is applicable only for small-size holes.

Exercise 5.12: Solve the equilibrium problem of a thin plate with two holes under the uniform tensile field $\sigma_0 = 775$ lb/in.^{2} applied at both end surfaces as shown in Figure 5.27. The plate is a homogeneous isotropic linearly elastic material characterized by

Young's modulus $\quad E = 8.0 \times 10^6$ psi

Poisson's ratio $\quad \nu = 0.21$

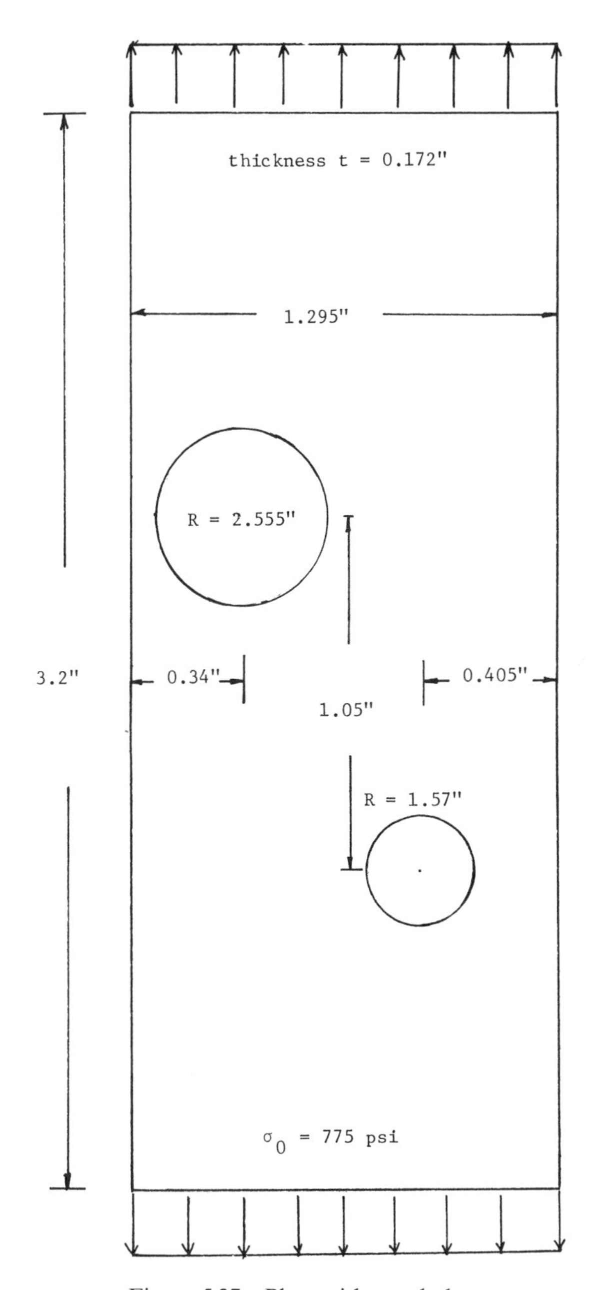

Figure 5.27 Plate with two holes.

Obtain the contour lines of maximum shear stress $\tau_{\max}$ and compare with the photoelastic result. The photoelastic stress pattern is given in Figure 5.28.

Note that the global stiffness matrix must be singular since only the second type of boundary condition is involved in the problem. Thus, to

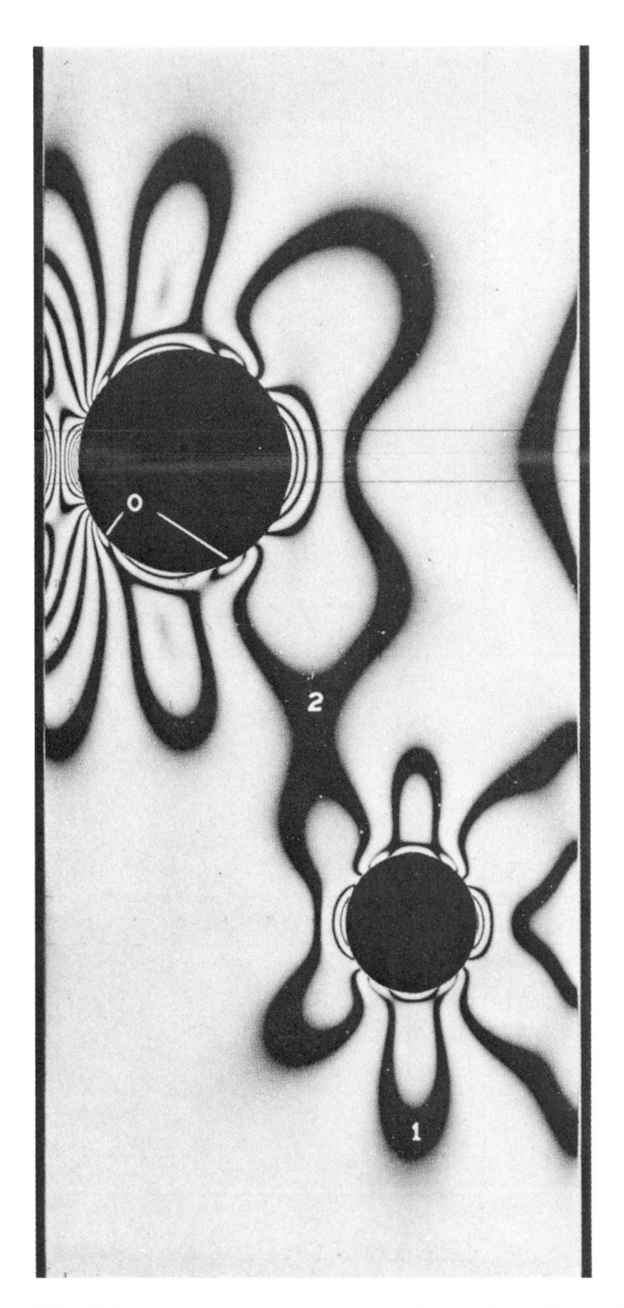

Figure 5.28 Fringe pattern of plate with two holes. *From Frocht (1941).*

avoid singularity, we must add the extra term

$$\varepsilon M_{i\alpha j\beta} = \varepsilon \int_{\Omega_e} \rho \delta_{ij} N_\alpha N_\beta \, d\Omega \tag{5.150}$$

to each element stiffness matrix $E_{i\alpha i\beta}$ for every $i, j = 1, 2$ and $\alpha, \beta = 1, \ldots, 4$, where ε is a very small number such that $\varepsilon = 10^{-3}$–$10^{-5}E$,

ρ is the density of the material, $\{N_\alpha\}$ are the shape functions, and $M_{i\alpha j\beta}$ is the element mass matrix.

It is known that the solution $\mathbf{u}_\varepsilon$ of the problem perturbed by adding the extra term $\varepsilon M_{i\alpha j\beta}$ to the stiffness matrix converges to the original solution $\mathbf{u}$ as $\varepsilon \to 0$. This is called the *regularization method* for singular problems. When this method is applied, we have to check whether or not the applied forces satisfy the condition

$$\int_\Omega \rho f_i u_{Ri} \, d\Omega + \sum_{i=1}^{3} \int_{\Gamma_{i2}} t_i u_{Ri} \, d\Gamma = 0, \quad \forall \mathbf{u}_R \ni \mathbf{u}_R = \mathbf{a} + \mathbf{b} \times \mathbf{x} \tag{5.151}$$

where **a** and **b** are arbitrary constant vectors. If either of the above two equations is not satisfied for the applied body force **f** and traction **t**, the structure will not be in equilibrium since there is no portion of the boundary where the first or third type of boundary conditions are assumed.

In order to see the effect of the regularity parameter ε on the perturbed solution $\mathbf{u}_\varepsilon$ and its corresponding stress field $\boldsymbol{\sigma}_\varepsilon$, solve the problem for $\varepsilon = 10^{-1}$, 10^{-3}, and 10^{-5} and compare the solutions for these different ε values.

Example 5.5: The second example is the *Boussinesq problem* in the theory of linear elasticity. Suppose that a point force P is vertically applied at the center of the top surface of a semiinfinite plate that occupies the domain $\{\mathbf{x} \in \mathbb{R}^2, \mathbf{x} = x\mathbf{i} + y\mathbf{j}, x \geq 0\}$ in $\mathbb{R}^2$. Under the generalized plane stress assumption, the stress field and the corresponding displacement are given (Timoshenko and Goodier, 1970) as

$$\begin{aligned}
\sigma_{rr} &= -\frac{2P}{\pi}\frac{\cos\theta}{r}, \qquad \sigma_{\theta\theta} = 0, \qquad \sigma_{r\theta} = 0 \\
u_r &= -\frac{2P}{\pi E}\cos\theta \log\frac{r}{a} - \frac{(1-\nu)P}{\pi E}\theta \sin\theta \\
u_\theta &= \frac{2\nu P}{\pi E}\sin\theta + \frac{2P}{\pi E}\sin\theta\log\frac{r}{a} - \frac{(1-\nu)P}{\pi E}(\theta\cos\theta - \sin\theta)
\end{aligned} \tag{5.152}$$

where $u_r = 0$ at $r = a$ and $\theta = 0$ is assumed. Then the displacement along the y axis is given by

$$u_x = -\frac{(1+\nu)P}{\pi E} - \frac{2P}{\pi E}\log\frac{y}{a}$$

$$u_y = -\frac{(1-\nu)P}{2E}$$

The stress component σ_{xx} and the displacement u_x along the x axis become

$$\sigma_{xx} = -\frac{2P}{\pi x} \qquad \text{and} \qquad u_x = -\frac{2P}{\pi E}\log\frac{x}{a}$$

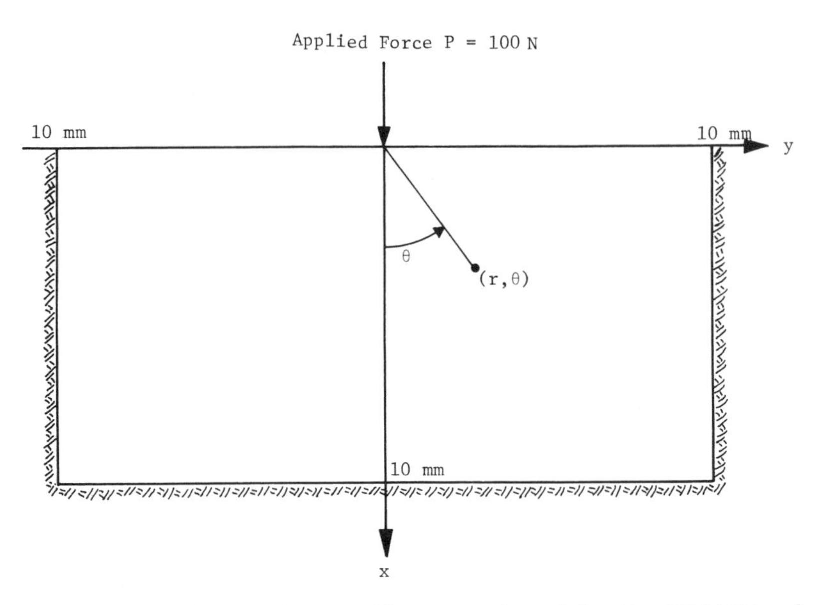

Figure 5.29 Domain for Boussinesq problem: Young's modulus $E = 3200$ kN/mm^2, Poisson's ratio 0.25.

Since infinite domains cannot be treated by the finite element approximations studied so far, we shall make a finite element model by taking only a finite portion of the semiinfinite domain shown in Figure 5.29. Because of symmetry, only the right half portion of the rectangular domain need be discretized. The finite element model is developed using 80 elements.

It is noted that smaller finite elements are allocated near the point at which the point force is applied in order to recover the singular behavior of stresses at this point. At the loading point, triangular elements are assumed. However, these triangular elements could be obtained by merging two nodes of four-node elements. That is, if the element connectivity of the NEL element is, for example,

IJK(1,NEL) = I1
IJK(2,NEL) = I2
IJK(3,NEL) = I3
IJK(4,NEL) = I1

this yields a three-node triangular element, the element connectivity of which is the same as that of the first three connectivities of the four-node element.

Assuming homogeneity and isotropy of the material, let us obtain the displacement u_x along the y axis and the stress σ_{xx} along the x axis using two different integration schemes to compute element stiffness matrices. The first used the 2×2 Gaussian rule, and in the second, the one-point reduced integration scheme is applied. The results are

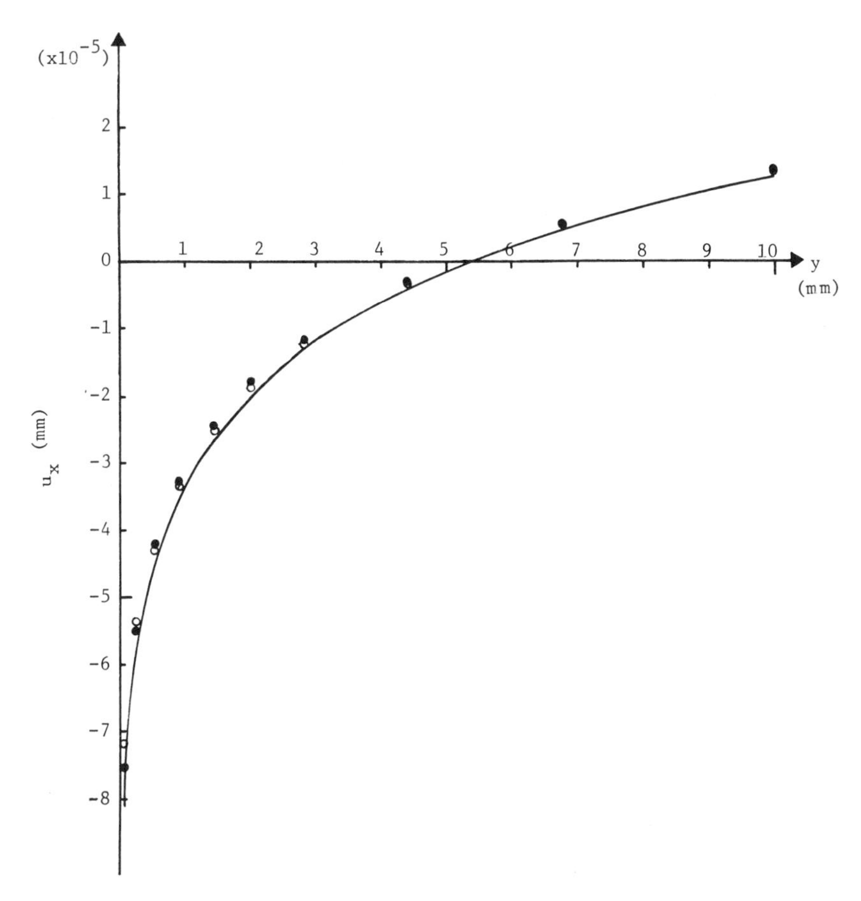

Figure 5.30 Displacement u_x along y axis ($\theta = 90°$): ———, exact solution; ●, 2 × 2-point integration scheme; ○, one-point integration scheme.

shown in Figures 5.30 and 5.31 for the case that

$$E = 3200 \text{ kN/mm}^2, \qquad \text{Poisson's ratio } \nu = 0.25$$

$$\text{Thickness} = 10 \text{ mm} \quad \text{(generalized plane stress)}$$

$$\text{Applied force } P = 100 \text{ N}$$

Here the third type of boundary condition has been assumed along the bottom and right side edges. This is obtained by the given stress field σ_{rr} in (5.152). Figure 5.32 shows the "fringes" obtained by the finite element method and photoelasticity. It is clear that these are very similar.

For this example, the two integration schemes do not produce widely differing numerical results. Another interesting observation concerning the stress distribution obtained from the 2 × 2 integration scheme is that variation of the stress within an element is quite small, even though it changes as a step function at the element boundaries. This suggests that evaluation of the stress tensor element at centroids is sufficient.

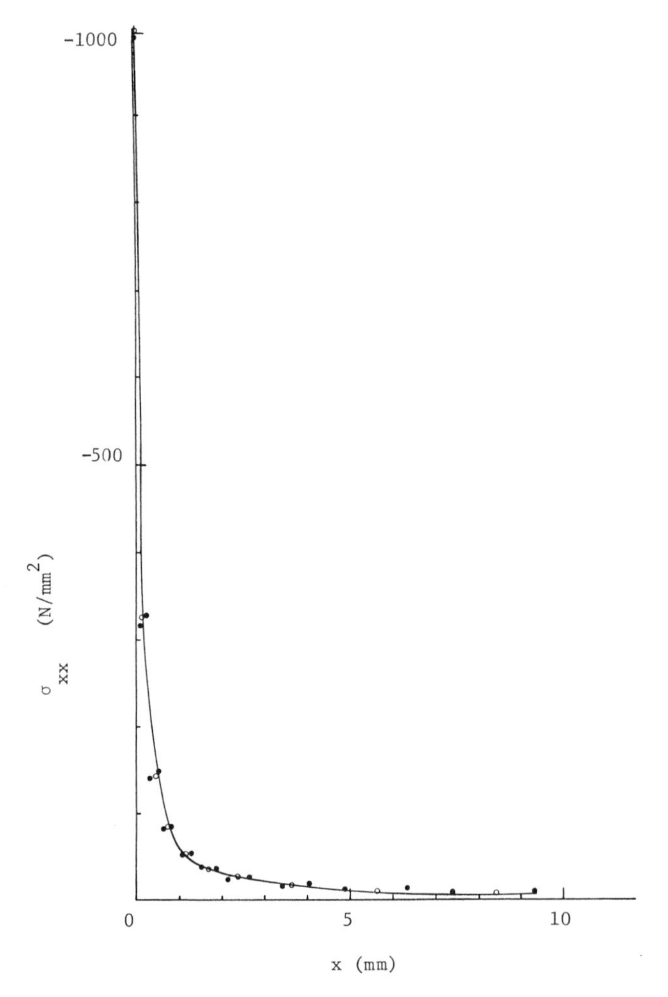

Figure 5.31 Stress σ_{xx} along x axis ($\theta = 0°$): ———, exact solution, $\sigma_{xx} = -2P/\pi x$; ●, 2 × 2-point integration scheme; ○, one-point integration scheme.

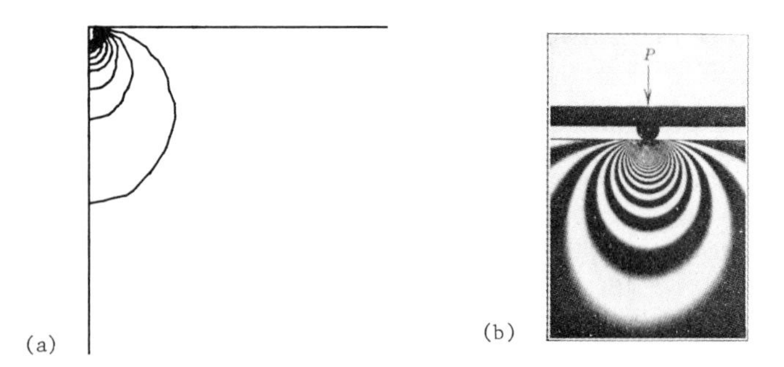

Figure 5.32 Fringes by (a) FEM and (b) photoelasticity. *Part (b) from Hetenyi (1950, chap. 17).*

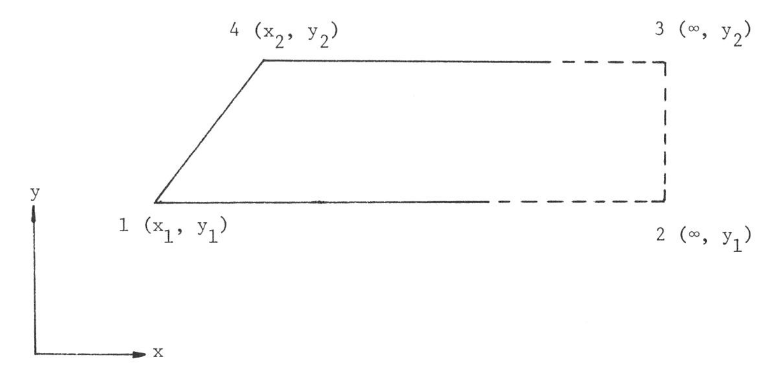

Figure 5.33 Infinite element in $\mathbb{R}^2$.

Exercise 5.13: Explain why the four-node element can be the same as the three-node triangular element if two nodes are emerged into a single node. Can we obtain a similar fact for the case in which three nodes are on the same line?

Exercise 5.14: In example 5.5, an infinite domain has been simulated by considering only a finite portion of the domain. In this case, a difficulty arises as to how large the finite portion should be. The answer depends on the nature of the boundary conditions.

To overcome this difficulty, let us introduce infinitely large finite elements and "consider them" by using a transformation of coordinates. For example, a four-node element that is infinitely long in the x direction is shown in Figure 5.33. If the coordinates of the four nodes are given by (x_1, y_1), (∞, y_1), (∞, y_2), and (x_2, y_2), the transformation to the coordinate system (s, t) in the master element is given by

$$s = 1 - 2\left(\frac{1}{x}\frac{(y_2 - y)x_1 + (y - y_1)x_2}{y_2 - y_1}\right)^m, \qquad t = 1 - 2\frac{y_2 - y}{y_2 - y_1} \tag{5.153}$$

of which the inverse is

$$x = \left(\frac{2}{1 - s}\right)^{1/m}\left(\frac{1 - t}{2}x_1 + \frac{1 + t}{2}x_2\right), \qquad y = \frac{1 - t}{2}y_1 + \frac{1 + t}{2}y_2 \tag{5.154}$$

for some $m \geq 1$.

We shall consider what the above transformation yields to the approximation of the displacement. To do this, let us consider a component u of the displacement vector along the line $y = y_2 = 0$. Because of the approximation of u using the shape functions $\hat{N}_\alpha(s, t)$, we have

$$\begin{aligned} u(x, 0) &= u_\alpha \hat{N}_\alpha(s, 1) \\ &= u_4\tfrac{1}{2}(1 - s) + u_3\tfrac{1}{2}(1 + s) \end{aligned}$$

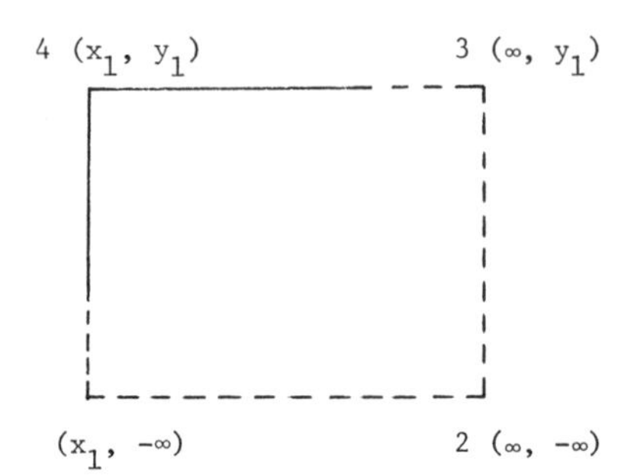

Figure 5.34 Another infinite element in $\mathbb{R}^2$.

where u_α is the value of u at the αth node. For further simplicity, suppose that $u_3 = 0$, that is, u is zero at $x = \infty$. Then

$$u(x, 0) = u_4\tfrac{1}{2}(1 - s)$$

On the other hand, the transformation yields

$$x = \sqrt[m]{\frac{2}{1-s}}\, x_2 \qquad \text{or} \qquad s = 1 - 2\left(\frac{x_2}{x}\right)^m$$

These imply the expression

$$u(x, 0) = u_4\left(\frac{x_2}{x}\right)^m$$

that is, the value of u decays similarly to the function $1/x^m$ in the interval (x_2, ∞). If we expect that u decays similarly to the function $1/(x - x_0)^m$ in the interval (x_2, ∞), the transformation is obtained from the above one by changing x to $x - x_0$:

$$x - x_0 = \left(\frac{2}{1-s}\right)^{1/m}\left(\frac{1-t}{2}x_1 + \frac{1+t}{2}x_2\right)$$

The first derivatives of $\hat{N}_\alpha(s, t)$ are related to those of $N_\alpha(x, y)$ by

$$\begin{Bmatrix} \dfrac{\partial \hat{N}_\alpha}{\partial s} \\[2ex] \dfrac{\partial \hat{N}_\alpha}{\partial t} \end{Bmatrix} = \begin{bmatrix} \dfrac{\partial x}{\partial s} & \dfrac{\partial y}{\partial s} \\[2ex] \dfrac{\partial x}{\partial t} & \dfrac{\partial y}{\partial t} \end{bmatrix} \begin{Bmatrix} \dfrac{\partial N_\alpha}{\partial x} \\[2ex] \dfrac{\partial N_\alpha}{\partial y} \end{Bmatrix}$$

This means that the stiffness matrix for an infinite element could be obtained using the same evaluation method as that used for the four-node element.

Modify the subroutine ESTIF0 in the program FEM2 for the infinite element. Similarly the transformation for the infinite element shown in Figure 5.34 can be obtained as

$$x = \left(\frac{2}{1-s}\right)^{1/m} x_1, \qquad y = \left(\frac{2}{1+t}\right)^{1/m} y_1 \tag{5.155}$$

where $y_1 < 0$. Modify ESTIF0 for this element too.

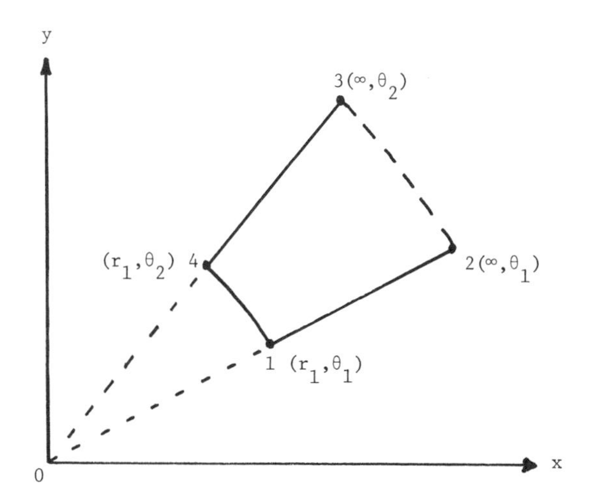

Figure 5.35 Infinite element in polar coordinate system (r, θ).

Exercise 5.15: When an infinite element is considered, it is often convenient to define using the polar coordinate system (r, θ) instead of the Cartesian coordinate system (x, y) as shown in Figure 5.35. Introducing the transformation

$$s = 1 - 2\left(\frac{r_1}{r}\right)^m, \qquad t = 1 - 2\frac{\theta_2 - \theta}{\theta_2 - \theta_1} \tag{5.156}$$

of which the inverse is

$$r = \left(\frac{2}{1 - s}\right)^{1/m} r_1, \qquad \theta = \frac{1 - t}{2}\theta_1 + \frac{1 + t}{2}\theta_2 \tag{5.157}$$

we can approximate a function $u(r, \theta)$ in the infinite element. Assuming that $u(\infty, \theta) = 0$, $u(r, \theta)$ is approximated by finite element methods as

$$\begin{aligned} u(r, \theta) &= u_\alpha \hat{N}_\alpha(s, t) \\ &= u_1 \tfrac{1}{4}(1 - s)(1 - t) + u_4 \tfrac{1}{4}(1 - s)(1 + t) \end{aligned}$$

Applying the transformation, we have

$$u(r, \theta) = \left(\frac{r_1}{r}\right)^m \left(u_1 \frac{\theta_2 - \theta}{\theta_2 - \theta_1} + u_4 \frac{\theta - \theta_1}{\theta_2 - \theta_1}\right)$$

This means that the infinite transformation simulates the decay of $u(r, \theta)$ at $r = \infty$ similar to the function $1/r^m$.

Find the element stiffness matrix using the above transformation in the polar coordinate system (r, θ). Furthermore, noting that the displacement components (u_r, u_θ) are related to those in the Cartesian system (u_x, u_y) by

$$\begin{Bmatrix} u_r \\ u_\theta \end{Bmatrix} = \begin{bmatrix} \cos\theta & \sin\theta \\ -\sin\theta & \cos\theta \end{bmatrix} \begin{Bmatrix} u_x \\ u_y \end{Bmatrix} \tag{5.158}$$

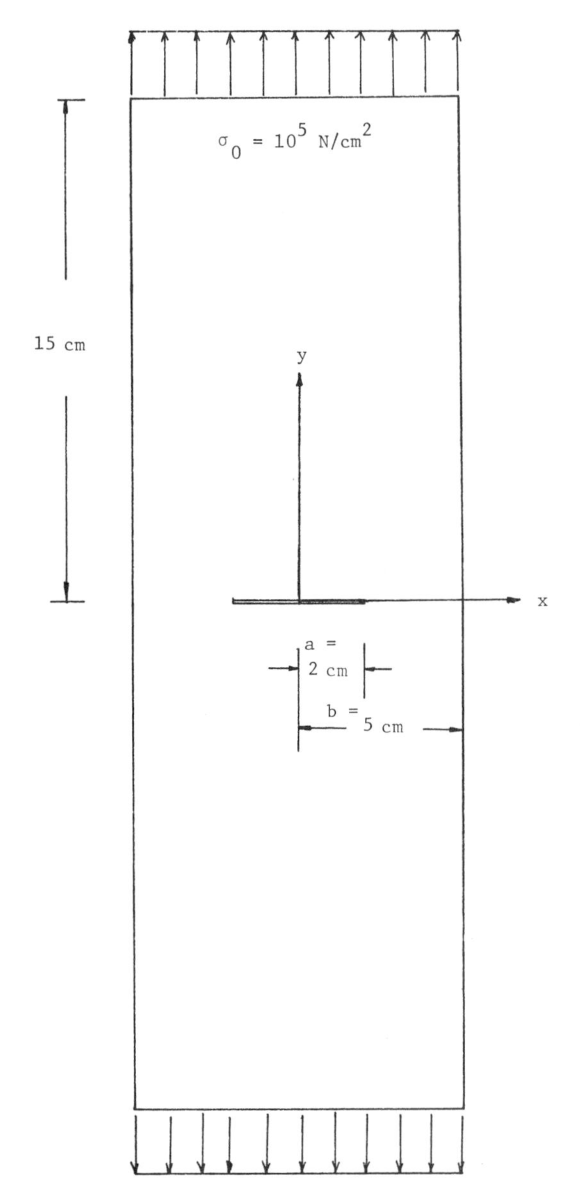

Figure 5.36 Stress analysis near crack tips: Young's modulus $E = 2.1 \times 10^7$ N/cm^2, Poisson's ratio 0.29.

transform the element stiffness matrix in the polar coordinate system (r, θ) to the one in the Cartesian system (x, y).*

Example 5.6: Let us consider another singular problem involving cracks for in-plane deformation of a thin plate. As a model problem, let us consider the plate shown in Figure 5.36 subjected to the tensile

* More detailed and systematic treatment of infinite elements can be found in Okabe (1981).

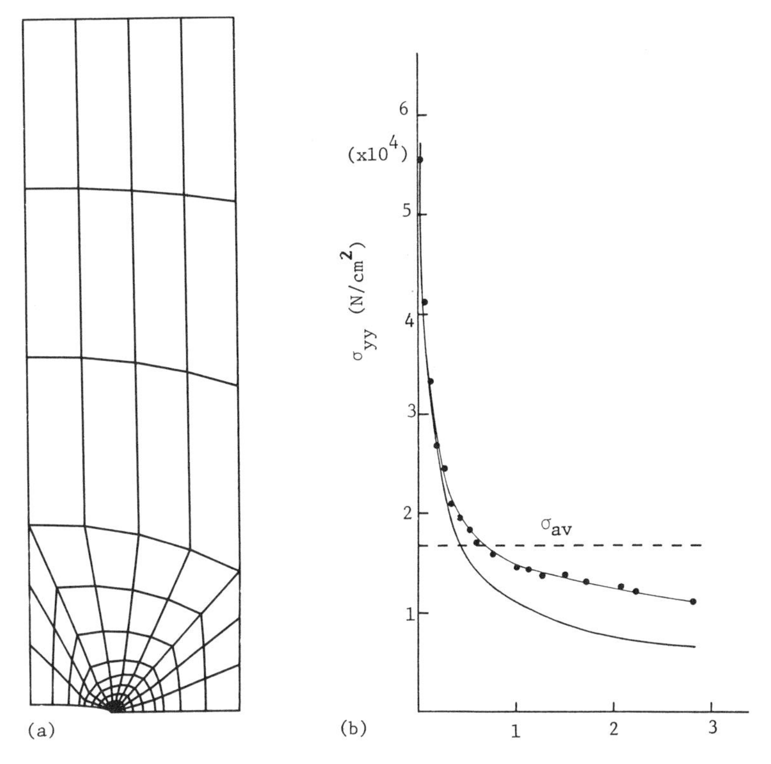

Figure 5.37 (a) Deformation and (b) stress distribution along x axis at $\theta = 0°$ (σ_{yy}): ———, solution by $(K_I)a$ (defined in Table 5.5); ●, solution by FEM; $\sigma_{av} = 16{,}667\ \mathrm{N/cm^2}$.

traction $\sigma_0 = 100\ \mathrm{kN/cm^2}$. Suppose that the thickness of the plate is 1 cm and that the material is homogeneous and isotropic:

$$E = 2.1 \times 10^4\ \mathrm{kN/cm^2}, \qquad \nu = 0.29$$

Using the symmetry of the problem, we shall solve only a quarter part of the plate. The deformed configuration, the principal stresses, and the maximum shear stress τ_{max} are shown in Figures 5.37(a), 5.38, and 5.39.

To assess the quality of the approximate solution, let us plot the stress σ_{yy} along the x [see Figure 5.37(b)] together with the analytical expressions for the stresses around the crack tip (Timoshenko and Goodier, 1970):

$$\sigma_{xx} = \frac{K_I}{(2\pi r)^{1/2}} \cos \tfrac{1}{2}\theta(1 - \sin \tfrac{1}{2}\theta \sin \tfrac{3}{2}\theta)$$

$$\sigma_{yy} = \frac{K_I}{(2\pi r)^{1/2}} \cos \tfrac{1}{2}\theta(1 + \sin \tfrac{1}{2}\theta \sin \tfrac{3}{2}\theta) \tag{5.159}$$

$$\sigma_{xy} = \frac{K_I}{(2\pi r)^{1/2}} \sin \tfrac{1}{2}\theta \cos \tfrac{1}{2}\theta \cos \tfrac{3}{2}\theta$$

where $K_I = \sigma_0(\pi a)^{1/2}$, σ_0 is the stress σ_{yy} at ∞, and a is the half size of

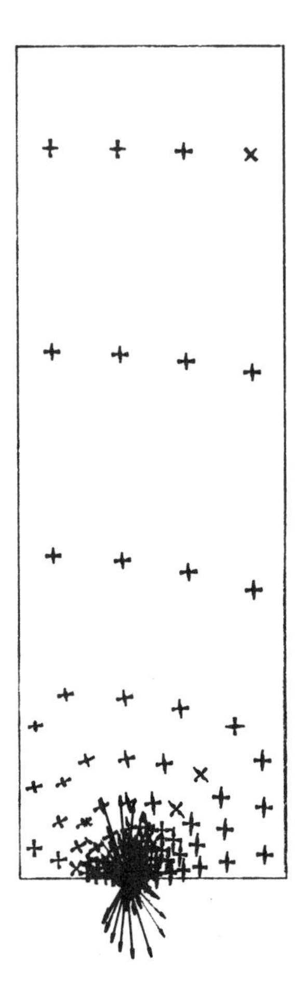

Figure 5.38 Principal stresses.

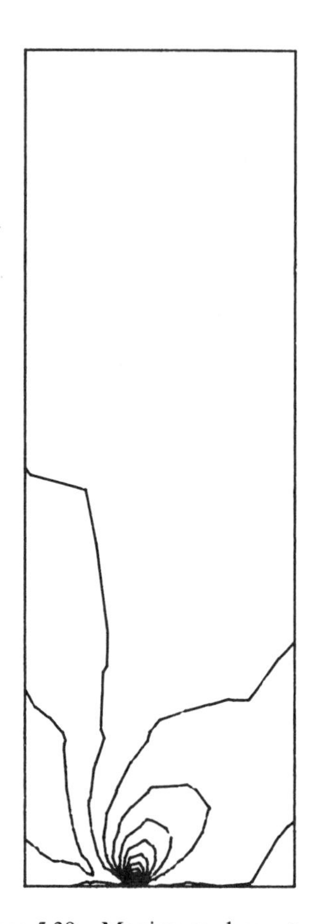

Figure 5.39 Maximum shear stress.

the crack. Here the analytical expressions correspond to the case of an infinite plate with a crack, the length of which is $2a$.

The value of K_I in the analytical solution is said to be the *stress intensity factor* of the first mode (or the opening type) in fracture mechanics and depends on the shape of the domain and the applied load. If the applied load is perpendicular to the crack and if the domain is infinite, then K_I is obtained as $K_I = \sigma_0\sqrt{\pi a}$. If the domain of finite width is similar to the one in Figure 5.36, we have

$$(K_I)_a = \sigma_0\sqrt{\pi a}\,\sqrt{\frac{2b}{\pi a}\tan\frac{\pi a}{2b}} \tag{5.160}$$

according to Lawn and Wilshaw (1975).

There are several methods to find K_I using finite element methods. For example, using the solution near the crack tip yields

$$K_I = \sqrt{2\pi r}\sigma_{yy}\Big/\cos\frac{\theta}{2}\left(1 + \sin\frac{\theta}{2}\sin\frac{3\theta}{2}\right) \tag{5.161}$$

Table 5.5 *Stress intensity factors computed by stress, displacement, and energy method*

	Stress method	Displacement method	Energy method
K_I ($N/cm^{3/2}$)	269327	248857	274371
(r, θ)(cm, rad)	(0.0372, 0)	(0.0318, π)	—
Error (%)	0.08	7.67	−1.79

Note: Error $= [(K_I)_a - K_I]/(K_I)a$, where $(K_I)_a = 269{,}544$ $N/cm^{3/2}$, which is defined in (5.160).

By finding σ_{yy} along the x axis corresponding to $\theta = 0$ using finite element solutions, it is possible to evaluate the value of K_I. Similarly, the analytical method for the infinite domain implies the displacement

$$u_x = \frac{K_I}{2G}\sqrt{\frac{r}{2\pi}}\cos\tfrac{1}{2}\theta(\kappa - 1 + 2\sin^2\tfrac{1}{2}\theta)$$

$$u_y = \frac{K_I}{2G}\sqrt{\frac{r}{2\pi}}\sin\tfrac{1}{2}\theta(\kappa + 1 - 2\cos^2\tfrac{1}{2}\theta)$$

where $G = \mu$ is the shear modulus and κ is defined as

$$\kappa = \begin{cases} 3 - 4v, & \text{for plane strain} \\ (3 - v)/(1 + v), & \text{for plane stress} \end{cases}$$

Thus, if $\theta = \pi$ is taken, the value of u_y along the x axis implies

$$K_I = \frac{2G\sqrt{2\pi}}{(\kappa + 1)\sqrt{r}}\,u_y \tag{5.162}$$

This also provides a method to evaluate the stress intensity factor K_I. Another popular method is obtained by computing the strain energy $U + \Delta U$ for the case of the crack with the length $a + \Delta a$, where U is the strain energy for the crack with the length a. Applying the relation between the rate of energy release $G = \partial U/\partial a \approx \Delta U/\Delta a$ and the stress intensity factor K_I, we have

$$K_I = \sqrt{\frac{\Delta U}{\Delta a}\frac{E}{b}} \tag{5.163}$$

where $b = 1 - v^2$ for plane strain, and $b = 1$ for plane stress. Thus, solving the similar problem for the crack length $a + \Delta a$ for a sufficiently small Δa, the stress intensity factor K_I can be obtained.

We compute K_I in Table 5.5 using the above three methods for the finite element model shown in Figure 5.36. It is clear that the stress method based on (5.161) is the best in this case.

In order to trace the rapid growth of the stress near the crack tip, very small size finite elements are allocated there. An interesting question is design of finite element mesh near crack tips to simulate the

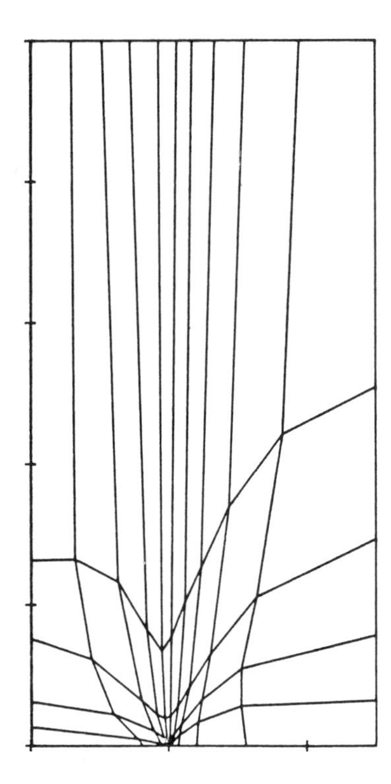

Figure 5.40 Improved finite element model using mesh optimization scheme.

singular behavior adequately. One such method is the technique of mesh *optimization* discussed in section 2.10. Introducing a proper measure of error in a finite element approximation, an improved mesh is obtained so that the error is distributed equally to each finite element. In Figure 5.40, an improved finite element model from the uniform mesh using 10×5 four-node elements is shown. It is clear that very small elements are allocated near the crack tip. The stress distribution $\sigma_{yy}(x, 0)$ along the x axis is given in Figure 5.41 for both the (initial) uniform and the improved finite element models. Rapid growth of the stress near the crack tip is simulated well by the improved finite element model, although the initial uniform mesh does not give impressive results. As a measure of error to improve the mesh, the value of stress discontinuity across the element boundaries has been used here.

Exercise 5.16: Solve the problem shown in Figure 5.42 for a homogeneous isotropic material characterized by

Young's modulus $E = 2.1 \times 10^4$ kN/cm^2

Poisson's ratio $\nu = 0.29$

Thickness $t = 1$ cm

Generalized plane stress

Compute the quantity

$$J = \int_{\Gamma_{0i}} \left(U\, dy - \boldsymbol{\sigma} \frac{\partial \mathbf{u}}{\partial x}\, ds \right) \tag{5.164}$$

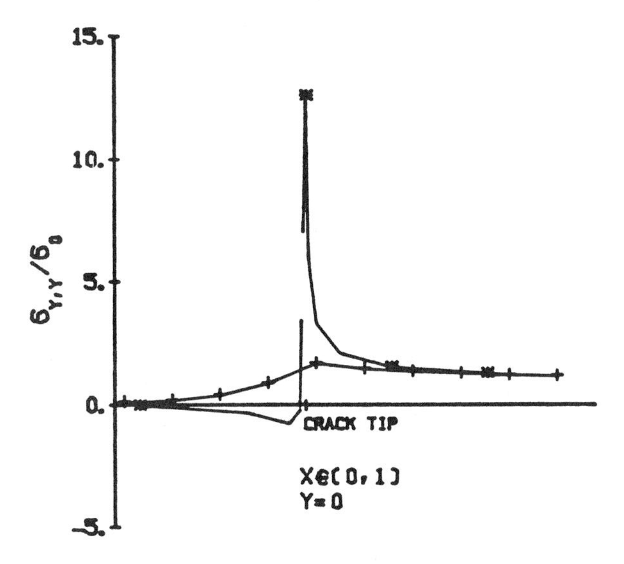

Figure 5.41 Stress distribution along x axis: $+$, initial grid; $*$, final grid.

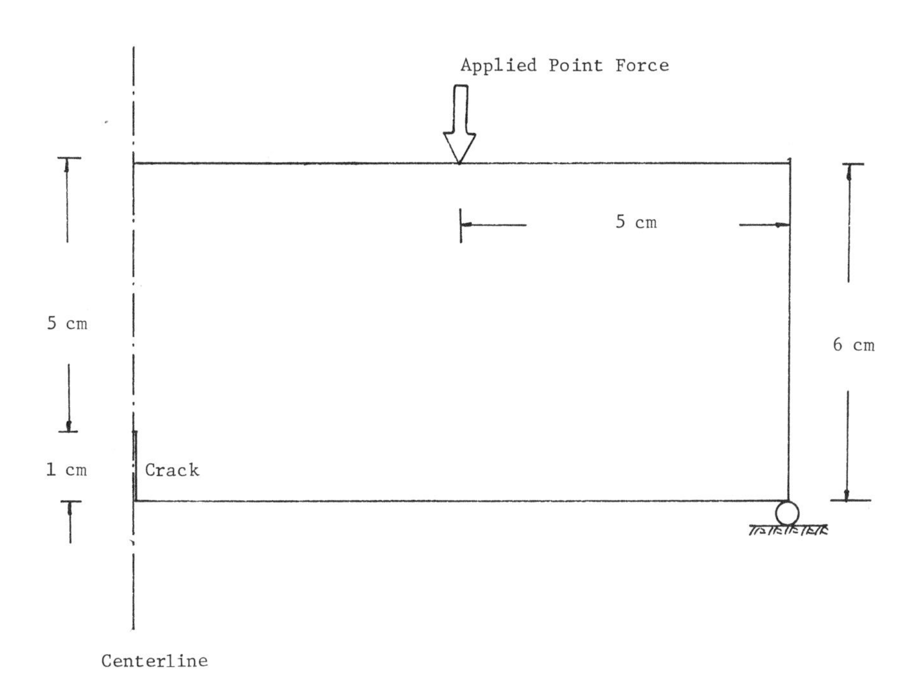

Figure 5.42 Bending test of plate with surface crack.

along the curve Γ_{0i}, $i = 1, 2, 3$, where U is the specific strain energy defined by

$$U = \tfrac{1}{2}\sigma_{ij}\varepsilon_{ij}$$

and $\boldsymbol{\sigma}$ is the traction vector defined by

$$\sigma_i = \sigma_{ji} n_i$$

The quantity J is called *Rice's J integral* and is the same as the strain energy release rate per unit thickness (defined by $\partial U/\partial a$, where a is the crack length) for linear elasticity. The stress intensity factor K_I is then related to the value of J by

$$K_I = \begin{cases} (JE)^{1/2}, & \text{for plane stress} \\ (JE/(1-\nu^2))^{1/2}, & \text{for plane strain} \end{cases} \tag{5.165}$$

Thus, it is important to compute the value of J along a curve surrounding the crack tip in fracture mechanics.

Example 5.7 (comparison to photoelasticity): We shall provide qualitative comparison of stress patterns from finite element solutions to fringes by photoelasticity for several cases shown in the handbook of experimental stress analysis (Hetény, 1950). Since qualitative comparison is aimed here, we shall describe only results without details of information about finite element analyses. Results are obtained using FEM2 listed above. The material is assumed to be isotropic and homogeneous and is characterized by

Young's modulus	$E = 2.1 \times 10^4$ kN/cm^2
Poisson's ratio	$\nu = 0.29$
Thickness	$t = 1$ cm

without regarding the material property for photoelasticity. Plane stress is assumed in analysis. Examples follow:

(a) Stress concentrations at semicircular notches in a tension bar,
(b) bakelite model taken through lucite holders,
(c) contact stresses with a hollow roller,
(d) localized stresses in the fillets of a gear tooth, and
(e) stresses in model of radial section of railway-car wheel.

Comparisons are shown in Figures 5.43–5.47.

It is clear that finite element solutions provide very similar stress patterns to those of photoelasticity. The only difference observed is that patterns near the boundaries are not the same as the models for photoelasticity since boundary conditions are not the same.

Exercise 5.17: Let us consider the stress–strain relations for plane stress in an orthotropic material. For a lamina in the $\hat{x}_1\hat{x}_2$ plane as shown in Figure 5.48, plane stress is represented by putting

$$\hat{\sigma}_{33} = \hat{\sigma}_{32} = \hat{\sigma}_{31} = 0$$

in the three-dimensional stress–strain relations, where $\hat{\sigma}_{IJ}$ is the IJ component of the stress tensor in the $(\hat{x}_1, \hat{x}_2, \hat{x}_3)$ coordinate system. Thus, from the three-dimensional stress–strain relation, we have either

$$\begin{Bmatrix} \hat{\sigma}_{11} \\ \hat{\sigma}_{22} \\ \hat{\sigma}_{12} \end{Bmatrix} = \begin{bmatrix} \hat{c}_{11} & \hat{c}_{12} & 0 \\ \hat{c}_{21} & \hat{c}_{22} & 0 \\ 0 & 0 & 2\hat{c}_{66} \end{bmatrix} \begin{Bmatrix} \hat{\varepsilon}_{11} \\ \hat{\varepsilon}_{22} \\ \hat{\varepsilon}_{12} \end{Bmatrix} \tag{5.166}$$

(a)

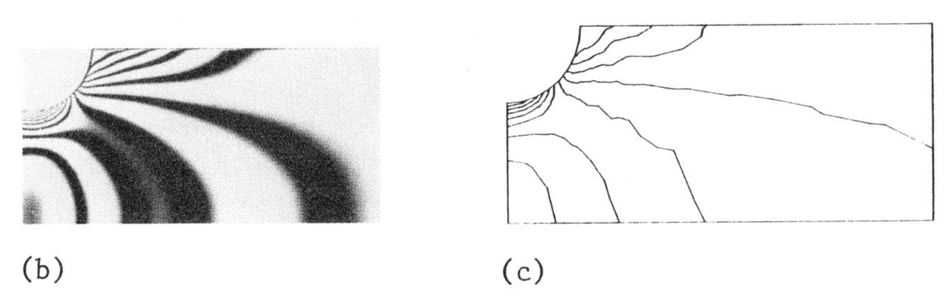

(b) (c)

Figure 5.43 Stress concentrations at semicircular notches in a tension bar: (a) fringe photograph; (b) a quarter of (a); (c) maximum shear stress lines by FEM. *Photograph in (a) and (b) by M. M. Frocht; from Hetenyi (1950).*

or

$$\begin{Bmatrix} \hat{\varepsilon}_{11} \\ \hat{\varepsilon}_{22} \\ \hat{\varepsilon}_{12} \end{Bmatrix} = \begin{bmatrix} \hat{s}_{11} & \hat{s}_{12} & 0 \\ \hat{s}_{21} & \hat{s}_{22} & 0 \\ 0 & 0 & \frac{1}{2}\hat{s}_{66} \end{bmatrix} \begin{Bmatrix} \hat{\sigma}_{11} \\ \hat{\sigma}_{22} \\ \hat{\sigma}_{12} \end{Bmatrix} \tag{5.167}$$

where $[\hat{s}_{IJ}]$ is the inverse of $[\hat{c}_{IJ}]$:

$$\begin{aligned} \hat{c}_{11} &= \hat{s}_{22}/(\hat{s}_{11}\hat{s}_{22} - \hat{s}_{12}\hat{s}_{21}) \\ \hat{c}_{22} &= \hat{s}_{11}/(\hat{s}_{11}\hat{s}_{22} - \hat{s}_{12}\hat{s}_{21}) \\ \hat{c}_{12} &= -\hat{s}_{12}/(\hat{s}_{11}\hat{s}_{22} - \hat{s}_{12}\hat{s}_{21}) \\ \hat{c}_{21} &= -\hat{s}_{21}/(\hat{s}_{11}\hat{s}_{22} - \hat{s}_{12}\hat{s}_{21}) \\ \hat{c}_{33} &= 1/\hat{s}_{66} \end{aligned} \tag{5.168}$$

For a unidirectional fiber-reinforced lamina, we have

$$\begin{aligned} &\hat{s}_{11} = 1/\hat{E}_1, \qquad \hat{s}_{22} = 1/\hat{E}_2, \qquad \hat{s}_{66} = 1/\hat{G}_{12} \\ &\hat{s}_{12} = -\hat{\nu}_{21}/\hat{E}_2, \qquad \hat{s}_{21} = -\hat{\nu}_{12}/\hat{E}_1 \end{aligned} \tag{5.169}$$

where $\hat{E}_I$ is Young's modulus in the $\hat{x}_I$ direction, $\hat{\nu}_{IJ}$ is Poisson's ratio for transverse strain in the $\hat{x}_J$ direction when stressed in the $\hat{x}_I$ direction,

(a)

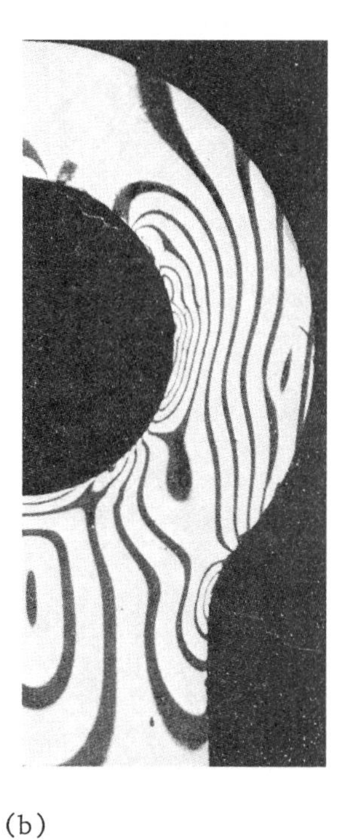

(b)

(c)

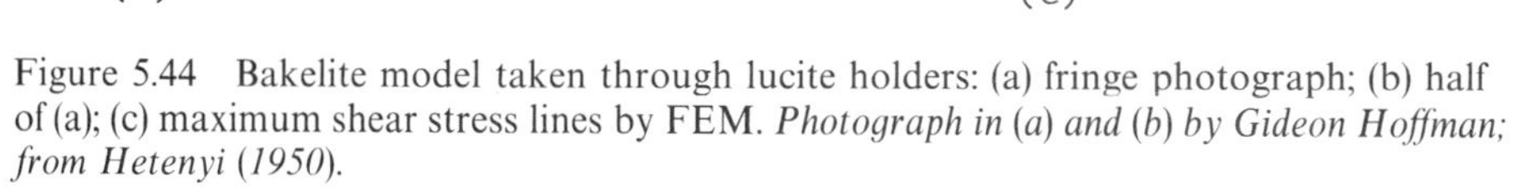

Figure 5.44 Bakelite model taken through lucite holders: (a) fringe photograph; (b) half of (a); (c) maximum shear stress lines by FEM. *Photograph in (a) and (b) by Gideon Hoffman; from Hetenyi (1950).*

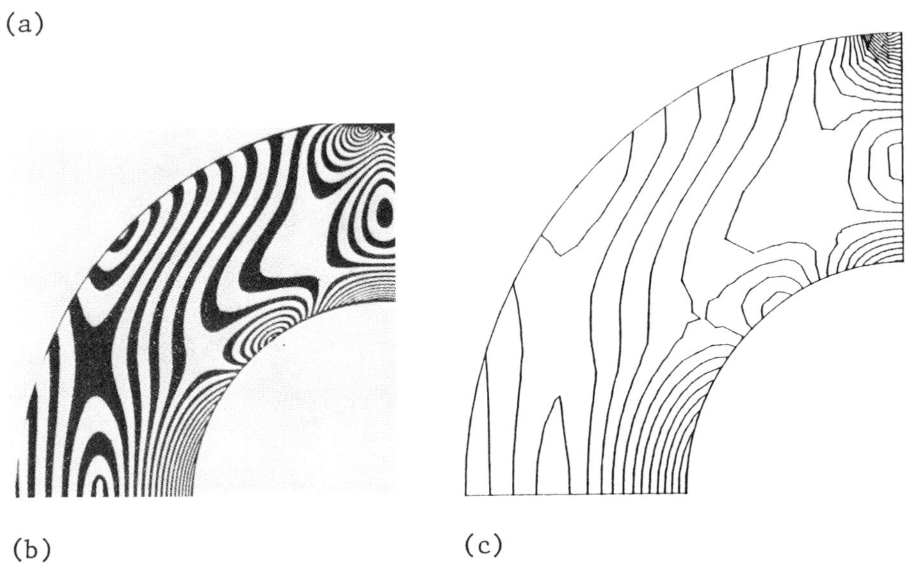

Figure 5.45 Contact stress with hollow roller: (a) fringe pattern; (b) a quarter of (a); (c) maximum shear stress lines by FEM. *From Hetenyi (1950).*

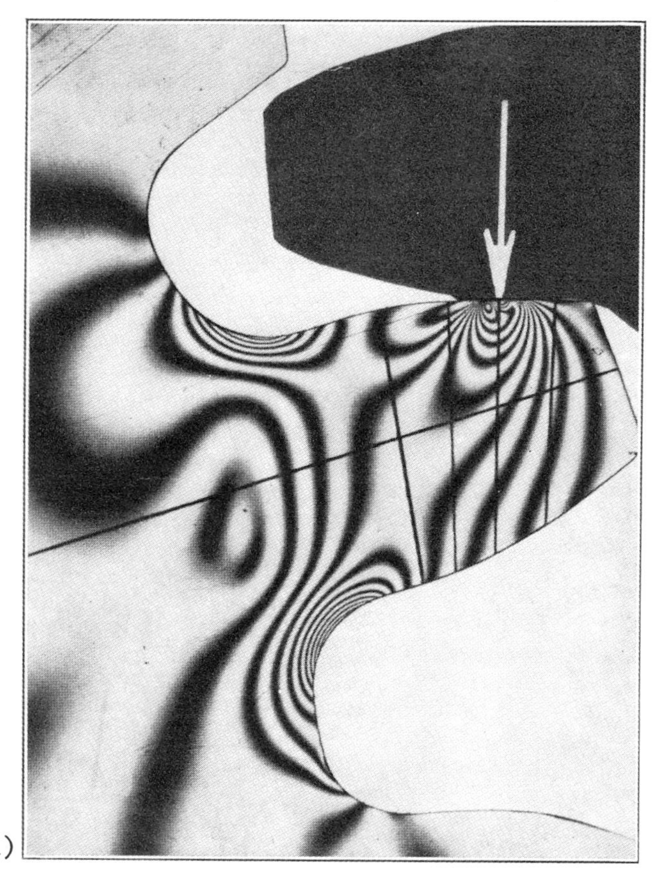
(a)

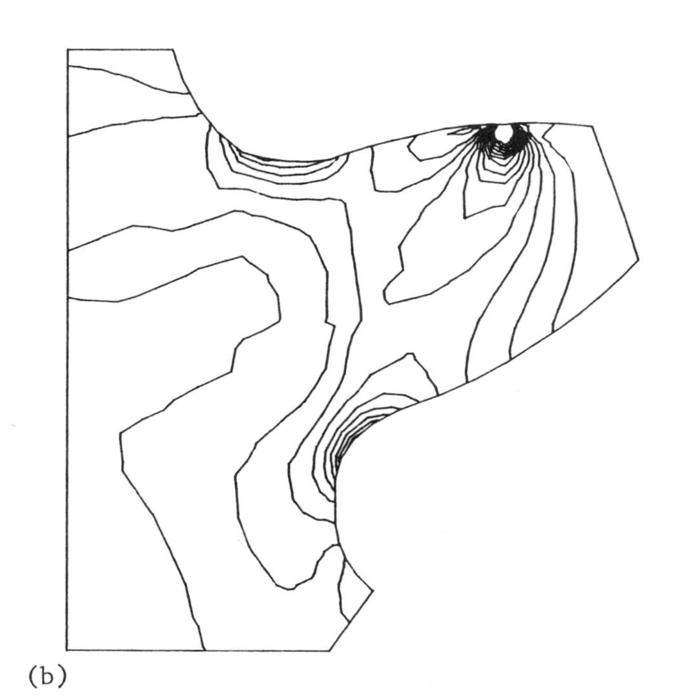
(b)

Figure 5.46 Localized stresses in fillets of gear tooth: (a) fringe photograph; (b) maximum shear stress lines by FEM. *From Hetenyi (1950).*

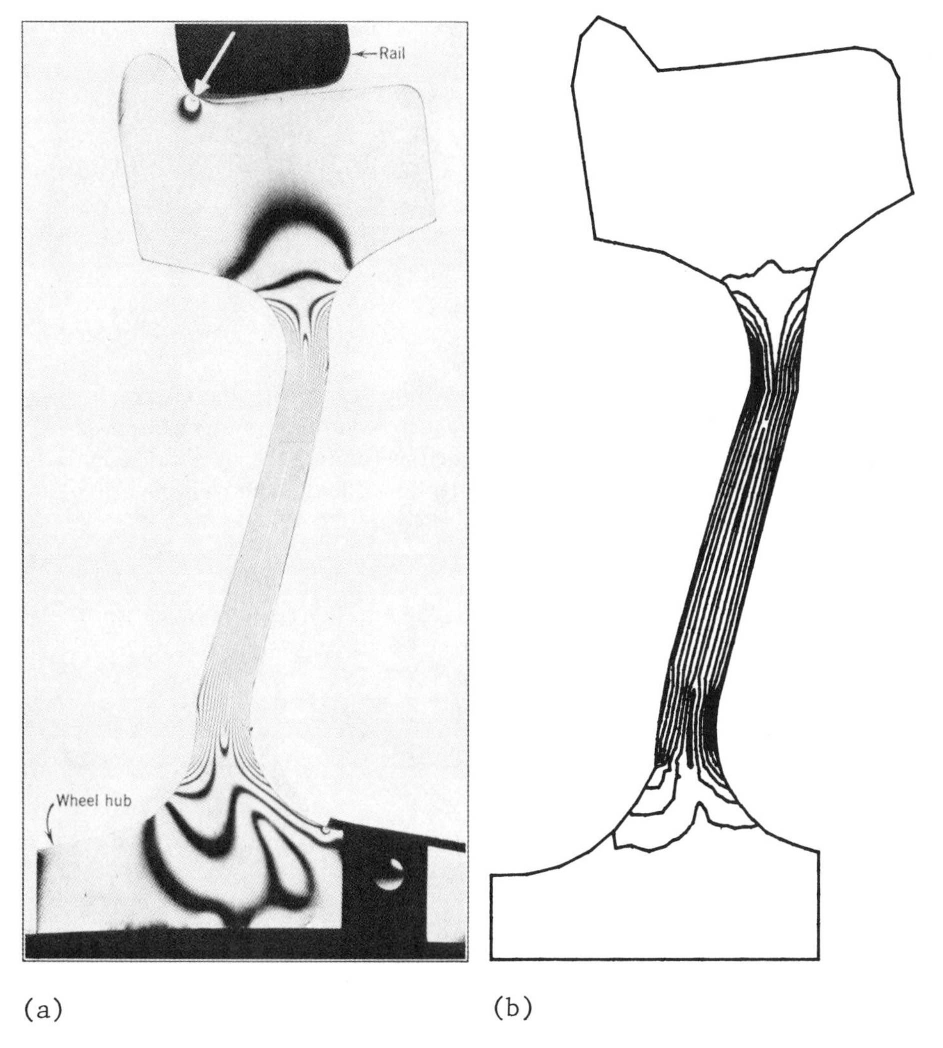

Figure 5.47 Stress in model of radial section of railway-car wheel: (a) fringe photograph; (b) maximum shear stress lines by FEM. *From Hetenyi (1950).*

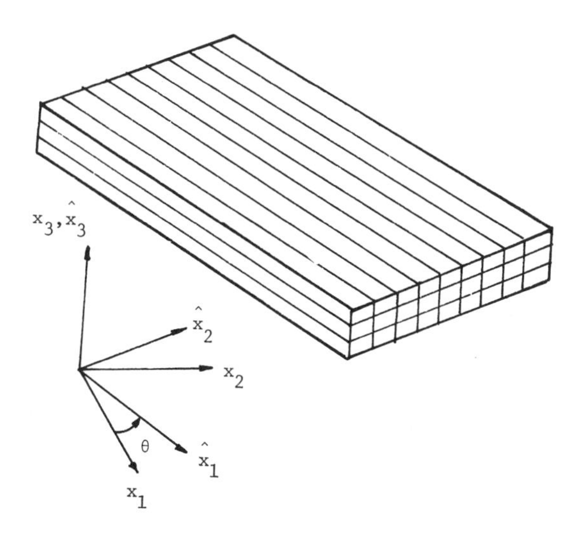

Figure 5.48 Orthotropic fiber-reinforced lamina.

and $\hat{G}_{IJ}$ is the shear modulus in the $\hat{x}_I\hat{x}_J$ plane. Thus,

$$\begin{aligned}
&\hat{c}_{11} = \hat{E}_1/(1 - \hat{v}_{12}\hat{v}_{21}), \qquad \hat{c}_{22} = \hat{E}_2/(1 - \hat{v}_{12}\hat{v}_{21}) \\
&\hat{c}_{12} = \hat{c}_{21} = \hat{v}_{12}\hat{E}_2/(1 - \hat{v}_{12}\hat{v}_{21}) = \hat{v}_{21}\hat{E}_1/(1 - \hat{v}_{12}\hat{v}_{21}) \\
&\hat{c}_{66} = \hat{G}_{12}
\end{aligned} \tag{5.170}$$

Note that the relation

$$\hat{v}_{12}/\hat{E}_1 = \hat{v}_{21}/\hat{E}_2 \tag{5.171}$$

has to be satisfied.

Now let us transform the stress–strain relation in the $(\hat{x}_1, \hat{x}_2, \hat{x}_3)$ coordinate system to that in the (x_1, x_2, x_3) coordinate system. Suppose that $x_3 = \hat{x}_3$ and the $\hat{x}_1$ axis is obtained by rotating the x_1 axis through an angle θ in the counterclockwise direction. Then, as shown in equation (1.16b),

$$\hat{\sigma}_{IJ} = \beta_{Ii}\sigma_{ij}\beta_{Jj} \qquad \text{and} \qquad \hat{\varepsilon}_{IJ} = \beta_{Ii}\varepsilon_{ij}\beta_{Jj}$$

where $\beta_{Ii} = \cos(\theta_{Ii})$ is the direction cosine of the angle between the I and i axes. Using the angle θ, we have for $c = \cos\theta$ and $s = \sin\theta$

$$\begin{Bmatrix} \sigma_{11} \\ \sigma_{22} \\ \sigma_{12} \end{Bmatrix} = \begin{bmatrix} c^2 & s^2 & -2cs \\ s^2 & c^2 & 2cs \\ cs & -cs & c^2 - s^2 \end{bmatrix} \begin{Bmatrix} \hat{\sigma}_{11} \\ \hat{\sigma}_{22} \\ \hat{\sigma}_{12} \end{Bmatrix} \tag{5.172}$$

and

$$\begin{Bmatrix} \hat{\varepsilon}_{11} \\ \hat{\varepsilon}_{22} \\ \hat{\varepsilon}_{12} \end{Bmatrix} = \begin{bmatrix} c^2 & s^2 & 2cs \\ s^2 & c^2 & -2cs \\ -cs & cs & c^2 - s^2 \end{bmatrix} \begin{Bmatrix} \varepsilon_{11} \\ \varepsilon_{22} \\ \varepsilon_{12} \end{Bmatrix} \tag{5.173}$$

Thus, the matrix $[C_{ij}]$ in the (x_1, x_2, x_3) coordinate system is obtained as

$$\begin{aligned}
&\begin{bmatrix} c_{11} & c_{12} & 2c_{16} \\ c_{21} & c_{22} & 2c_{26} \\ 2c_{16} & 2c_{26} & 2c_{66} \end{bmatrix} \\
&\quad = \begin{bmatrix} c^2 & s^2 & -2cs \\ s^2 & c^2 & 2cs \\ cs & -cs & c^2 - s^2 \end{bmatrix} \begin{bmatrix} \hat{c}_{11} & \hat{c}_{12} & 0 \\ \hat{c}_{21} & \hat{c}_{22} & 0 \\ 0 & 0 & 2\hat{c}_{66} \end{bmatrix} \begin{bmatrix} c^2 & s^2 & 2cs \\ s^2 & c^2 & -2cs \\ -cs & cs & c^2 - s^2 \end{bmatrix}
\end{aligned} \tag{5.174}$$

that is,

$$\begin{aligned}
c_{11} &= \hat{c}_{11}c^4 + 2(\hat{c}_{12} + 2\hat{c}_{66})s^2c^2 + \hat{c}_{22}s^4 \\
c_{12} &= (\hat{c}_{11} + \hat{c}_{22} - 4\hat{c}_{66})s^2c^2 + \hat{c}_{12}(s^4 + c^4) \\
c_{22} &= \hat{c}_{11}s^4 + 2(\hat{c}_{12} + 2\hat{c}_{66})s^2c^2 + \hat{c}_{22}c^4 \\
c_{16} &= (\hat{c}_{11} - \hat{c}_{12} - 2\hat{c}_{66})sc^3 + (\hat{c}_{12} - \hat{c}_{22} + 2\hat{c}_{66})s^3c \\
c_{26} &= (\hat{c}_{11} - \hat{c}_{12} - 2\hat{c}_{66})s^3c + (\hat{c}_{12} - \hat{c}_{22} + 2\hat{c}_{66})sc^3 \\
c_{66} &= (\hat{c}_{11} + \hat{c}_{22} - 2\hat{c}_{12} - 2c_{66})s^2c^2 + \hat{c}_{66}(s^4 + c^4)
\end{aligned} \tag{5.175}$$

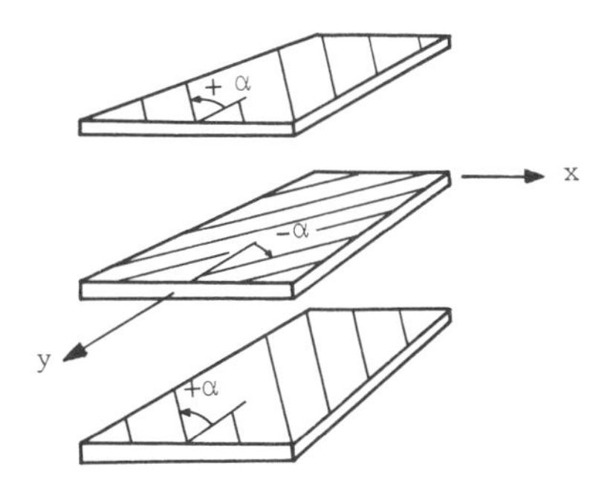

Figure 5.49 Layered symmetric angle-ply laminate.

The stress–strain relation in the (x_1, x_2, x_3) coordinate system becomes

$$\begin{Bmatrix} \sigma_{11} \\ \sigma_{22} \\ \sigma_{12} \end{Bmatrix} = \begin{bmatrix} c_{11} & c_{12} & 2c_{16} \\ c_{21} & c_{22} & 2c_{26} \\ 2c_{16} & 2c_{26} & 2c_{66} \end{bmatrix} \begin{Bmatrix} \varepsilon_{11} \\ \varepsilon_{22} \\ \varepsilon_{12} \end{Bmatrix} \tag{5.176}$$

Modify the subroutines MATPRO and TANMOD in the program FEM2 so that the above constitutive relations for an orthotropic material can be considered as well as for an isotropic material.

Exercise 5.18: Let us consider a layered symmetric angle-ply laminate shown in Figure 5.49 consisting of orthotropic layers that are symmetrically disposed about the middle surface so that no coupling between bending and extension is exhibited. More specifically, using orthotropic laminae of equal thicknesses, a laminate is composed so that the adjacent laminae have opposite signs of the angle of orientation of the principal material properties with respect to the laminate axes, for example, $+\alpha/-\alpha/+\alpha$. Assuming the generalized plane stress, obtain the relation of stress–strain for the laminate shown in Figure 5.49, assuming the material property described in the previous exercise.

5.8 BASIC program for microcomputers using hourglass control for reduced integration methods

Let us transform FEM2 in FORTRAN to BASIC-FEM2 using BASIC, a standard language in microcomputers such as IBM PC and Apple II series. Since computation is slow in today's microcomputers, it is necessary to introduce an algorithm that forms element stiffness matrices in a relatively short time without losing accuracy of finite element approximations. As discussed in section 5.6, the selective reduced integration technique sometimes provides better results than the one by full integration, especially for bending-dominated problems or for incompressible materials. A disadvantage of the selective reduced integration technique is that there are no ways to substantially reduce computer time. One of the methods to overcome this disadvantage is to use the reduced integration technique, which integrates all the terms in the integrand of element stiffness matrices using less accurate lower-order quadrature rules. However, this yields singularity of the global stiffness matrix for certain boundary conditions because of the existence of hourglass (zero-energy) modes, see Figure 5.9. Recently, Belytschko et al. (1983) show a consistent manner that eliminates hourglass modes without losing the advantage of the reduced integration scheme. This development enables us to modify the program FEM2 written in FORTRAN for microcomputers using BASIC, where speed of computation is considerably slower than in large-size computers. The program BASIC-FEM2 listed below is developed using IBM PC-XT, although it is designed to be machine independent so that it can be executed by microcomputers with BASIC.

Let N_α, $\alpha = 1, \ldots, n$, be the shape functions of the four-node quadrilateral isoparametric element defined by

$$N_\alpha(s, t) = \tfrac{1}{4}(1 + s_\alpha s)(1 + t_\alpha t)$$

where (s_α, t_α) are the coordinates of the four corner nodes in the normalized coordinate system (s, t). Because of the isoparametric relation, the global coordinate (x, y) is related to the normalized (s, t) by

$$\begin{aligned} x &= \sum_{\alpha=1}^{4} x_\alpha N_\alpha(s, t) \\ &= \mathbf{a} \cdot \mathbf{x} + \tfrac{1}{4}(x_{31} + x_{24})s + \tfrac{1}{4}(x_{31} + x_{42})t + (\mathbf{h} \cdot \mathbf{x})st \\ y &= \sum_{\alpha=1}^{4} y_\alpha N_\alpha(s, t) \\ &= \mathbf{a} \cdot \mathbf{y} + \tfrac{1}{4}(y_{31} + y_{24})s + \tfrac{1}{4}(y_{31} + y_{42})t + (\mathbf{h} \cdot \mathbf{y})st \end{aligned} \tag{5.177}$$

where (x_α, y_α) are the coordinates of the corresponding four corner nodes in the global coordinate system (x, y),

$$\mathbf{a} = \frac{1}{4}\begin{Bmatrix} 1 \\ 1 \\ 1 \\ 1 \end{Bmatrix}, \qquad \mathbf{h} = \frac{1}{4}\begin{Bmatrix} 1 \\ -1 \\ 1 \\ -1 \end{Bmatrix}, \qquad \mathbf{x} = \begin{Bmatrix} x_1 \\ x_2 \\ x_3 \\ x_4 \end{Bmatrix}, \qquad \mathbf{y} = \begin{Bmatrix} y_1 \\ y_2 \\ y_3 \\ y_4 \end{Bmatrix} \tag{5.178}$$

where $x_{\alpha\beta} = x_\alpha - x_\beta$ and $y_{\alpha\beta} = y_\alpha - y_\beta$. From (5.177), we have

$$\begin{bmatrix} x_{31} + x_{24} & x_{31} + x_{42} \\ x_{31} + y_{24} & y_{31} + y_{42} \end{bmatrix} \begin{Bmatrix} s \\ t \end{Bmatrix} = \begin{Bmatrix} 4x - 4\mathbf{a} \cdot \mathbf{x} - 4(\mathbf{h} \cdot \mathbf{x})st \\ 4y - 4\mathbf{a} \cdot \mathbf{y} - 4(\mathbf{h} \cdot \mathbf{y})st \end{Bmatrix}$$

Solving this matrix equation yields

$$\begin{Bmatrix} s \\ t \end{Bmatrix} = \frac{1}{A}\begin{bmatrix} y_{31} + y_{42} & x_{13} + x_{24} \\ y_{13} + y_{42} & x_{31} + x_{24} \end{bmatrix} \begin{Bmatrix} x - \mathbf{a} \cdot \mathbf{x} - (\mathbf{h} \cdot \mathbf{x})st \\ y - \mathbf{a} \cdot \mathbf{y} - (\mathbf{h} \cdot \mathbf{y})st \end{Bmatrix} \tag{5.179}$$

where A is the area of the four-node element given by

$$A = \tfrac{1}{2}(x_{13}y_{24} + x_{24}y_{31}) \tag{5.180}$$

Substituting (5.179) into the relation

$$u = \sum_{\alpha=1}^{4} u_\alpha N_\alpha(s, t)$$

where u is the x component of the displacement vector and u_α, $\alpha = 1, \ldots, 4$, are its values at the four corner nodes, we can obtain

$$\begin{aligned} u &= \mathbf{a} \cdot \mathbf{u} + \tfrac{1}{4}(u_{31} + u_{24})s + \tfrac{1}{4}(u_{31} + u_{42})t + (\mathbf{h} \cdot \mathbf{u})st \\ &= [\mathbf{a} - (\mathbf{a} \cdot \mathbf{x})\mathbf{b}_x - (\mathbf{a} \cdot \mathbf{y})\mathbf{b}_y] \cdot \mathbf{u} + (\mathbf{b}_x \cdot \mathbf{u})x \\ &\quad + (\mathbf{b}_y \cdot \mathbf{u})y + st[\mathbf{h} - (\mathbf{h} \cdot \mathbf{x})\mathbf{b}_x - (\mathbf{h} \cdot \mathbf{y})\mathbf{b}_y] \cdot \mathbf{u} \end{aligned} \tag{5.181}$$

where

$$\mathbf{b}_x = \frac{1}{2A}\begin{Bmatrix} y_{24} \\ y_{31} \\ y_{42} \\ y_{13} \end{Bmatrix}, \qquad \mathbf{b}_y = \frac{1}{2A}\begin{Bmatrix} x_{42} \\ x_{13} \\ x_{24} \\ x_{31} \end{Bmatrix}, \qquad \mathbf{u} = \begin{Bmatrix} u_1 \\ u_2 \\ u_3 \\ u_4 \end{Bmatrix} \tag{5.182}$$

Similarly, the y component of the displacement vector is written by

$$\begin{aligned} v &= [\mathbf{a} - (\mathbf{a} \cdot \mathbf{x})\mathbf{b}_x - (\mathbf{a} \cdot \mathbf{y})\mathbf{b}_y] \cdot \mathbf{v} + (\mathbf{b}_x \cdot \mathbf{v})x \\ &\quad + (\mathbf{b}_y \cdot \mathbf{v})y + st[\mathbf{h} - (\mathbf{h} \cdot \mathbf{x})\mathbf{b}_x - (\mathbf{h} \cdot \mathbf{y})\mathbf{b}_y] \cdot \mathbf{v} \end{aligned} \tag{5.183}$$

It follows from (5.181) and (5.183) that the first derivatives of u and v in the global coordinate system *at the centroid of each four-node element* become

$$\begin{aligned} \frac{\partial u}{\partial x} &= \mathbf{b}_x \cdot \mathbf{u}, & \frac{\partial u}{\partial y} &= \mathbf{b}_y \cdot \mathbf{u} \\ \frac{\partial v}{\partial x} &= \mathbf{b}_x \cdot \mathbf{v}, & \frac{\partial v}{\partial y} &= \mathbf{b}_y \cdot \mathbf{v} \end{aligned} \tag{5.184}$$

It is noted that the displacement vectors that provide zero strains at the centroid of each element are given by

$$\mathbf{u} = a_u\mathbf{a} + h_u\mathbf{h} + c\mathbf{y} \qquad \text{and} \qquad \mathbf{v} = a_v\mathbf{a} + h_v\mathbf{h} - c\mathbf{x} \tag{5.185}$$

where a_u, a_v, h_u, h_v, and c are arbitrary real numbers. This means that there are five zero strain modes if the reduced integration technique, that is, the one-point Gaussian quadrature rule, is applied, although only three physically consistent zero strain modes must be related to a_u, a_v, and c. The modes characterized by $\mathbf{h}$ are called *hourglass modes* in the literature.

Now, let

$$\mathbf{g} = \mathbf{h} - (\mathbf{h} \cdot \mathbf{x})\mathbf{b}_x - (\mathbf{h} \cdot \mathbf{y})\mathbf{b}_y \qquad \text{and} \qquad \eta = st \tag{5.186}$$

Then the strain tensor $\boldsymbol{\varepsilon}$ becomes

$$\begin{aligned} \varepsilon_{xx} &= \mathbf{b}_x \cdot \mathbf{u} + \mathbf{g} \cdot \mathbf{u}\frac{\partial \eta}{\partial x} \\ \varepsilon_{xy} &= \varepsilon_{yx} = \tfrac{1}{2}(\mathbf{b}_y \cdot \mathbf{u} + \mathbf{b}_x \cdot \mathbf{v}) + \frac{1}{2}\left(\mathbf{g} \cdot \mathbf{u}\frac{\partial \eta}{\partial y} + \mathbf{g} \cdot \mathbf{v}\frac{\partial \eta}{\partial x}\right) \\ \varepsilon_{yy} &= \mathbf{b}_y \cdot \mathbf{v} + \mathbf{g} \cdot \mathbf{v}\frac{\partial \eta}{\partial y} \end{aligned} \tag{5.187}$$

Recalling that:

stiffness due to the shear deformation must be obtained by the reduced integration scheme in order to obtain "soft" enough stiffness for "bending" (see section 5.6) and

stiffness due to the volumetric deformation must be obtained by the reduced integration scheme in order to avoid the "locking" phenomenon for nearly incompressible materials (see section 5.6),

the constitutive relation for isotropic materials may be decomposed to two parts:

$$\begin{Bmatrix} \sigma_{xx} \\ \sigma_{yy} \\ \sigma_{xy} \end{Bmatrix} = \begin{bmatrix} \lambda & \lambda & 0 \\ \lambda & \lambda & 0 \\ 0 & 0 & 2\mu \end{bmatrix} \begin{Bmatrix} \varepsilon_{xx} \\ \varepsilon_{yy} \\ \varepsilon_{xy} \end{Bmatrix} + \begin{bmatrix} 2\mu & 0 & 0 \\ 0 & 2\mu & 0 \\ 0 & 0 & 0 \end{bmatrix} \begin{Bmatrix} \varepsilon_{xx} \\ \varepsilon_{yy} \\ \varepsilon_{xy} \end{Bmatrix} \tag{5.188}$$

where

$$\lambda = \begin{cases} Ev/(1-2v)(1+v), & \text{plane strain} \\ Ev/(1-v^2), & \text{plane stress} \end{cases}, \qquad \mu = \frac{E}{2(1+v)}$$

where E is Young's modulus and v is Poisson's ratio. The first part of the constitutive relation is evaluated by the reduced integration scheme, while the second part must be evaluated by either the full integration scheme or its approximations. Since the constant and linear terms in the integrand of each element stiffness matrix can be exactly integrated by the reduced integration scheme (i.e., the one-point Gaussian quadrature), we need to consider the integration

$$\int_{\Omega_e} 2\mu g_\alpha g_\beta \delta_{ij} \frac{\partial \eta}{\partial x_i} \frac{\partial \eta}{\partial x_j} \, d\Omega$$

$$= 2\mu g_\alpha g_\beta \delta_{ij} \int_{\Omega_e} \frac{\partial \eta}{\partial x_i} \frac{\delta \eta}{\partial x_j} \, d\Omega \qquad \alpha, \beta = 1, \ldots, 4; i, j = 1, 2 \tag{5.189}$$

by either the full integration scheme or its approximation to form the element stiffness matrix.

Noting that the inverse of the transpose of the Jacobian matrix evaluated at the centroid is given by

$$\mathbf{J}^{-\mathrm{T}} = \frac{1}{A} \begin{bmatrix} t_\beta y_\beta & -s_\beta y_\beta \\ -t_\beta x_\beta & s_\beta x_\beta \end{bmatrix} \tag{5.190}$$

we can approximate

$$\int_{\Omega_e} \frac{\partial \eta}{\partial x} \frac{\partial \eta}{\partial x} \, \partial\Omega \approx \tfrac{1}{3} A \frac{1}{A^2} (y_{24}^2 + y_{31}^2 + y_{42}^2 + y_{13}^2) = \tfrac{16}{12} A \mathbf{b}_x \cdot \mathbf{b}_x \tag{5.191}$$

and

$$\int_{\Omega_e} \frac{\partial \eta}{\partial y} \frac{\partial \eta}{\partial y} \, d\Omega \approx \tfrac{16}{12} A \mathbf{b}_y \cdot \mathbf{b}_y \tag{5.192}$$

Thus, the additional element stiffness matrix to the one by the one-point Gaussian integration rule can approximately be computed by (5.189) using (5.191) and (5.192), while the stiffness matrix by the one-point Gaussian integration rule is obtained

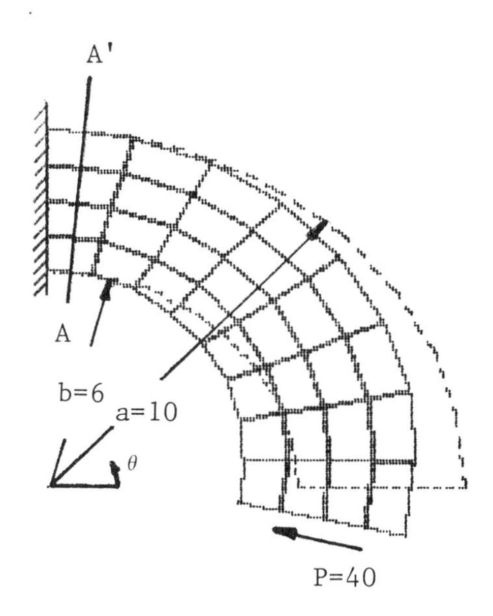

Figure 5.50 Initial and deformed configuration of example 5.8 for BASIC-FEM2.

using the constitutive relation

$$\begin{Bmatrix} \sigma_{xx} \\ \sigma_{yy} \\ \sigma_{xy} \end{Bmatrix} = \begin{bmatrix} \lambda + 2\mu & \lambda & 0 \\ \lambda & \lambda + 2\mu & 0 \\ 0 & 0 & 2\mu \end{bmatrix} \begin{Bmatrix} \varepsilon_{xx} \\ \varepsilon_{yy} \\ \varepsilon_{xy} \end{Bmatrix} \tag{5.193}$$

where

$$\varepsilon_{xx} = \mathbf{b}_x \cdot \mathbf{u}, \qquad \varepsilon_{yy} = \mathbf{b}_y \cdot \mathbf{v}, \qquad \text{and} \qquad \varepsilon_{xy} = \tfrac{1}{2}(\mathbf{b}_y \cdot \mathbf{u} + \mathbf{b}_x \cdot \mathbf{v}) \tag{5.194}$$

It is clear that the above method coincides with the selective reduced integration method, which integrates the first part of (5.188) by the one-point Gaussian rule and the second part of (5.188) by the 2×2 rule, if all elements are rectangular.

Example 5.8: Let us solve a bending problem of the curved cantilever beam shown in Figure 5.50. Suppose that the outer and inner radii are $a = 10$ cm and $b = 6$ cm, respectively, and that the thickness of the curved beam is 1 cm. Assuming isotropic and homogeneous material such that Young's modulus $E = 21{,}000$ kN/cm^2 and Poisson's ratio $\nu = 0.29$. Let us obtain the displacement and stresses of the curved beam subjected to the horizontal force $P = 40$ kN at the bottom surface using 32 uniform four-node quadrilateral elements, and let us compare the result obtained by the reduced integration scheme with the hourglass control described above to the analytical solution given, for example, in Fung (1965):

$$\sigma_{rr} = \frac{P}{N}\left(r + \frac{a^2b^2}{r^3} - \frac{a^2 + b^2}{r}\right)\sin\theta$$

$$\sigma_{\theta\theta} = \frac{P}{N}\left(3r - \frac{a^2b^2}{r^3} - \frac{a^2 + b^2}{r}\right)\sin\theta$$

$$\sigma_{r\theta} = -\frac{P}{N}\left(r + \frac{a^2b^2}{r^3} - \frac{a^2 + b^2}{r}\right)\cos\theta$$

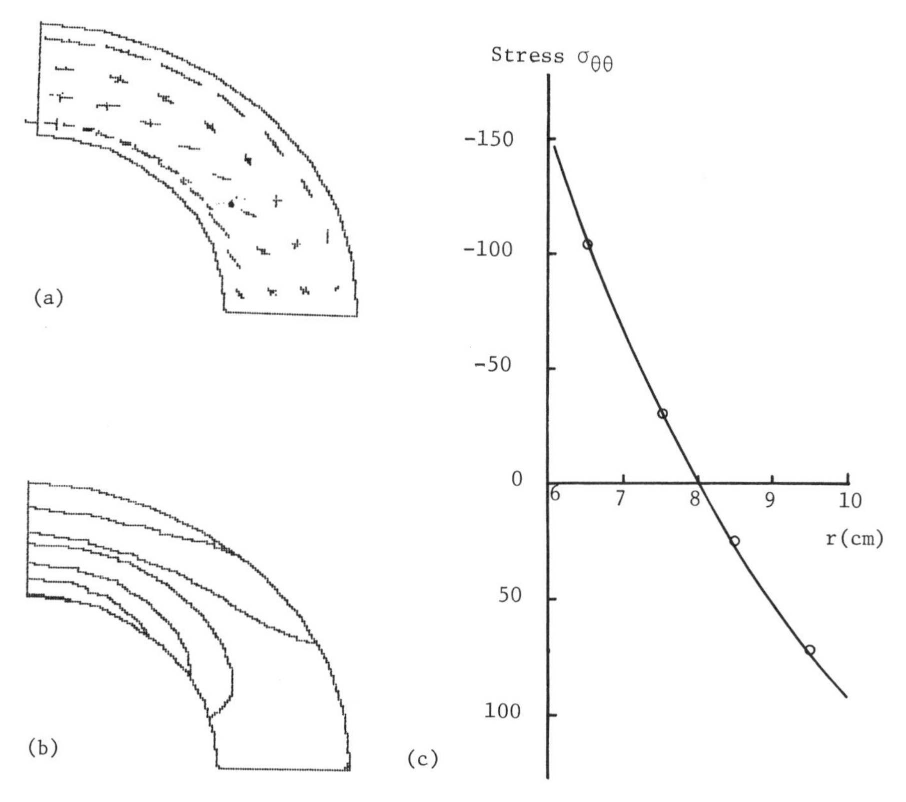

Figure 5.51 Results of example 5.8 for BASIC-FEM2: (a) principal stresses; (b) contour lines of $\bar{\sigma}$; (c) stress distribution along AA', ———, exact solution; ○, BASIC-FEM2.

where (r, θ) is the polar coordinate system, the origin of which coincides with that of the circular curved beam, and

$$N = b^2 - a^2 + (a^2 + b^2)\ln\frac{a}{b}$$

Figure 5.50 shows the initial configuration and the deformed finite element grids after magnifying the size of displacement 10 times. Figure 5.51 describes the principal stresses and contour lines of the von Mises equivalent stress together with the stress distribution of the cross section AA'. It is clear that the values of the stress $\sigma_{\theta\theta}$ along the line AA' computed by the finite element method agree very well to the analytical ones. Figures shown in this example are plotted using Apple IIe with a program that is separately developed for plotting the result computed by BASIC-FEM2.

Computing time to form an element stiffness matrix and its assembling to the global one is only 13.9 s, whereas if the 2×2 integration rule is applied, about 55 s are required. This indicates that the reduced integration method with the hourglass control is 4 times faster than the ordinary full integration scheme. Furthermore, because of possession of the effect by the selective reduced integration technique, this four-node element provides much accurate numerical results not only for bending dominant problems but also for nearly incompressible materials. The system of linear equations with 90 unknowns is solved in 472 s using the skyline solver for symmetric coefficient matrices.

While the basic program for stress analysis is listed for IBM PC, XT, and AT series, a similar but more general program is prepared in the lab manual for this textbook. The program for IBM can solve not only isotropic but also anisotropic problems such as for monotropic, orthotropic, and transverse isotropic materials. Furthermore, if the BASIC compiler (IBM-Microsoft) is applied, the speed of computation is improved by more than ten times. For example, the compiled version required only 0.8 s to form and assemble an element stiffness matrix, while 9 s is necessary for the IBM version BASIC program without compilation.

```
1000 PRINT "   *******************************"
1010 PRINT "          PROGRAM  :  FEM2"
1020 PRINT "   *******************************"
1030 PRINT "                           Version 1.0 / Fall 1984"
1040 PRINT : PRINT
1050 PRINT "        This program is for stress analysis of linearly elastic"
1060 PRINT "        planer structures using 4-node quadrilateral isoparametric"
1070 PRINT "        elements and the reduced integration technique with"
1080 PRINT "        hourglass control introduced by Professor Ted Belytschko."
1090 PRINT : PRINT : PRINT
1100 DIM SKE(8,8),FE(8),DX(4),DY(4),GH(4)
1110 EPI=10000000000#
1120 PRINT "Preprocessing : Input data of a finite element model"
1130 PRINT "----------------------------------------------------------------" : PRINT
1140 INPUT "Is the GEOMETRIC data of the finite element model in a file ?
     (YES/NO) = ";L$
1150 IF L$="NO" THEN GOTO 1240
1160 PRINT : INPUT "   File name (e.g. A:DATA ) = ";F$
1170 OPEN F$ FOR INPUT AS #1
1180 INPUT#1,NX,NELX
1190 DIM X(NX),Y(NX),MPE(NELX),IJK(4,NELX),F(2*NX),TEM(NX),JDIAG(2*NX)
1200 FOR I=1 TO NX : INPUT#1,X(I),Y(I) : NEXT I
1210 FOR NEL=1 TO NELX :
     INPUT#1,MPE(NEL),IJK(1,NEL),IJK(2,NEL),IJK(3,NEL),IJK(4,NEL) : NEXT NEL
1220 CLOSE #1
1230 GOTO 1320
1240 PRINT
1250 INPUT "The total number of nodes in the model   NX = ";NX
1260 INPUT "The total number of finite element    NELX = ";NELX
1270 DIM X(NX),Y(NX),MPE(NELX),IJK(4,NELX),F(2*NX),TEM(NX),JDIAG(2*NX)
1280 PRINT : PRINT "GEOMETRY .......... NODAL COORDINATES .........." : PRINT
1290 FOR I=1 TO NX : PRINT " node number ";I;TAB(20);" < x , y > = "; :
     INPUT X(I),Y(I) : NEXT I
1300 PRINT : PRINT "GEOMETRY .......... ELEMENT CONNECTIVITIES .........." : PRINT
1310 FOR NEL=1 TO NELX : PRINT " element ";NEL;TAB(16);"<MPE>, <IJK> = ";:
     INPUT MPE(NEL),IJK(1,NEL),IJK(2,NEL),IJK(3,NEL),IJK(4,NEL) : NEXT NEL
1320 PRINT :
     PRINT "MATERIAL PROPERTIES ..........................................." : PRINT
1330 INPUT "Is it plane strain/ stress ?   (STRAIN/STRESS) = ";AE$
1340 PRINT "Is the structure" :
     PRINT "           isotropic    (I)," :
     PRINT "           transverse isotropic   (TI)," :
     PRINT "           orthotropic    (O)," :
     PRINT "           anisotropic    (A) ?" ;
1350 INPUT "      Which one ";IOA$
1360 INPUT "Is the structure homogeneous or nonhomogeneous (H/N) = ";HN$
1370 MPX=1 : IF HN$="N"THEN INPUT "How many different homogeneities ?   mpx = ";MPX
1380 DIM ELAST(MPX,7),THERS(MPX,3),PZX(MPX),PZY(MPX),PXY(MPX)
1390 FOR I=1 TO MPX : PRINT : PRINT "MATERIAL ..... ";I : PRINT
```

```
1400 IF IOA$<>"I" THEN GOTO 1540
1410 INPUT "   Thickness of plate    TH = ";TH : ELAST(I,7)=TH
1420 INPUT "   Young's modulus    E = ";E : E=E*TH
1430 INPUT "   Poisson's ratio    v = ";V
1440 INPUT "   Thermal expansion coefficient    a = ";AL
1450 IF AE$="STRESS" THEN GOTO 1490
1460 D11=E*(1-V)/((1-2*V)*(1+V)) : D12=E*V/((1-2*V)*(1+V)) : D33=.5*E/(1+V)
1470 ELAST(I,1)=D11 : ELAST(I,2)=D12 : ELAST(I,3)=0 : ELAST(I,4)=D11 :
     ELAST(I,5)=0 : ELAST(I,6)=D33
1480 GOTO 1520
1490 D11=E/(1-V*V) : D12=E*V/(1-V*V) : D33=.5*E/(1+V)
1500 PZX(I)=V : PZY(I)=V : PXY(I)=0
1510 ELAST(I,1)=D11 : ELAST(I,2)=D12 : ELAST(I,3)=0 : ELAST(I,4)=D11 :
     ELAST(I,5)=0 : ELAST(I,6)=D33
1520 THERS(I,1)=(D11+D12)*AL : THERS(I,2)=(D12+D11)*AL : THERS(I,3)=0
1530 GOTO 2340
1540 IF IOA$<>"TI" THEN GOTO 1810
1550 INPUT "   Thickness of plate   TH = ";TH : ELAST(I,7)=TH
1560 INPUT "   Young's modulus on the isotropic plane   E = ";E : E=E*TH
1570 IF AE$="STRAIN" THEN INPUT "   Poisson's ratio in the isotropic plane   v = ";V
1580 INPUT "   Young's modulus in the transverse direction   E3 = ";E3 : E3=E3*TH
1590 INPUT "   Shear modulus    G3 = ";G3
1600 INPUT "   Poisson's ratio   v13 = ";V13 : V31=V13*E3/E
1610 INPUT "   Thermal expansion coefficient    a = ";AL1
1620 INPUT "                                   a3 = ";AL3
1630 INPUT "   Angle of rotation of the isotropic plane   ANG = ";ANG :
     ANG=3.141529*ANG/180
1640 IF AE$<>"STRAIN" THEN GOTO 1670
1650 Q11=E*(1-V13*V31)/((1+V)*(1-V-2*V13*V31)) : Q12=E*V31/(1-V-2*V13*V31) :
     Q22=E3*(1-V)/(1-V-2*V13*V31) : Q33=G3
1660 GOTO 1680
1670 Q11=E/(1-V13*V31) : Q12=E*V31/(1-V13*V31) : Q22=E3/(1-V13*V31) : Q33=G3
1680 S=SIN(ANG) : C=COS(ANG) : S2=S*S : C2=C*C : SC=S*C : S4=S2*S2 : C4=C2*C2
1690 D11=Q11*C4+2*(Q12+2*Q33)*SC*SC+Q22*S4
1700 D12=(Q11+Q22-4*Q33)*SC*SC+Q12*(S4+C4)
1710 D13=(Q11-Q12-2*Q33)*SC*C2+(Q12-Q22+2*Q33)*SC*S2
1720 D22=Q11*S4+2*(Q12+2*Q33)*SC*SC+Q22*C4
1730 D23=(Q11-Q12-2*Q33)*SC*S2+(Q12-Q22+2*Q33)*SC*C2
1740 D33=(Q11+Q22-2*Q12-2*Q33)*SC*SC+Q33*(S4+C4)
1750 A11=AL1*C2+AL3*S2 : A22=AL1*S2+AL3*C2 : A12=(AL1-AL3)*SC
1760 B11=D11*A11+D12*A22+D13*A12 : B22=D12*A11+D22*A22+D23*A12 :
     B12=D13*A11+D23*A22+D33*A12
1770 THERS(I,1)=B11 : THERS(I,2)=B22 : THERS(I,3)=B12
1780 ELAST(I,1)=D11 : ELAST(I,2)=D12 : ELAST(I,3)=D13 : ELAST(I,4)=D22 :
     ELAST(I,5)=D23 : ELAST(I,6)=D33
1790 PZX(I)=V*C*C+V13*S*S : PZY(I)=V13*C*C+V*S*S : PXY(I)=(V-V13)*C*S
1800 GOTO 2340
1810 IF IOA$<>"O" THEN GOTO 2170
1820 INPUT "   Number of laminates   NL = ";NL : PRINT
1830 IF AE$="STRAIN" THEN NL=1
1840 FOR L=1 TO NL : PRINT "  Laminate ";L
1850 INPUT "   Thickness of the laminate   TH = ";THI : TH=TH+THI
1860 INPUT "   Young's modulus in the 1-st direction  E1 = ";E1
1870 INPUT "   Poisson's ratio in the 1-st direction v12 = ";V12
1880 INPUT "   Young's modulus in the 2-nd direction  E2 = ";E2 : V21=V12*E2/E1
1890 INPUT "   Shear modulus in the 1-2 direction    G12 = ";G12
1900 IF AE$<>"STRAIN" THEN GOTO 1940
1910 INPUT "   Young's modulus in the 3-rd direction  E3 = ";E3
1920 INPUT "   Poisson's ratio                       V13 = ";V13 : V31=V13*E3/E1
1930 INPUT "                                         V23 = ";V23 : V32=V23*E3/E2
1940 INPUT "   Thermal expansion coefficient in the 1-st direction   a1 = ";AL1
1950 INPUT "                                 in the 2-nd direction   a2 = ";AL2
1960 INPUT "   The angle of the 1-st direction  ANG = ";ANG : ANG=ANG*3.141529/180
1970 IF AE$<>"STRAIN" GOTO 2010
1980 DET=1-V23*V32-V21*V13*V32-V12*V21-V31*V13-V31*V12*V23
1990 Q11=E1*(1-V23*V32)/DET : Q12=E1*(V31*V23+V21)/DET : Q22=E2*(1-V31*V13)/DET :
     Q66=G12
2000 GOTO 2020
2010 Q11=E1/(1-V12*V21) : Q12=V21*Q11 : Q22=E2/(1-V12*V21) : Q66=G12
2020 S=SIN(ANG) : C=COS(ANG) : S2=S*S : C2=C*C : SC=S*C : S4=S2*S2 : C4=C2*C2
2030 D11= THI*(Q11*C4+2*(Q12+2*Q66)*SC*SC+Q22*S4)
```

```
2040 D12= THI*((Q11+Q22-4*Q66)*SC*SC+Q12*(S4+C4))
2050 D13= THI*((Q11-Q12-2*Q66)*SC*C2+(Q12-Q22+2*Q66)*SC*S2)
2060 D22= THI*(Q11*S4+2*(Q12+2*Q66)*SC*SC+Q22*C4)
2070 D23= THI*((Q11-Q12-2*Q66)*SC*S2+(Q12-Q22+2*Q66)*SC*C2)
2080 D33= THI*((Q11+Q22-2*Q12-2*Q66)*SC*SC+Q66*(S4+C4))
2090 A11=AL1*C2+AL2*S2 : A22=AL1*S2+AL2*C2 : A12=(AL1-AL2)*SC
2100 B11=D11*A11+D12*A22+D13*A12 : B22=D12*A11+D22*A22+D23*A12 :
     B12=D13*A11+D23*A22+D33*A12
2110 PZX(I)=V31*C*C+V32*S*S : PZY(I)=V32*C*C+V31*S*S : PXY(I)=(V31-V32)*C*S
2120 ELAST(I,1)=ELAST(I,1)+D11 : ELAST(I,2)=ELAST(I,2)+D12 :
     ELAST(I,3)=ELAST(I,3)+D13
2130 ELAST(I,4)=ELAST(I,4)+D22 : ELAST(I,5)=ELAST(I,5)+D23 :
     ELAST(I,6)=ELAST(I,6)+D33
2140 THERS(I,1)=THERS(I,1)+B11 : THERS(I,2)=THERS(I,2)+B22 :
     THERS(I,3)=THERS(I,3)+B12
2150 NEXT L
2160 GOTO 2340
2170 IF IOA$<>"A" THEN GOTO 2340
2180 PRINT "   Only plane stress problems are considered for both anisotropic" :
     PRINT "   and monotropic materials." : PRINT
2190 INPUT "   Thickness of the plate    TH = ";TH : ELAST(I,7)=TH
2200 INPUT "   Young's modulus in the 1-st direction    E1 = ";E1 : E1=E1*TH
2210 INPUT "                          2-nd              E2 = ";E2 : E2=E2*TH
2220 INPUT "   Shear modulus in the 1-2 plane          G12 = ";G12 : G12=G12*TH
2230 INPUT "   Poisson's ratio                          v12 = ";V12 : V21=V12*E2/E1
2240 INPUT "                                           y112 = ";Y112 : Y121=Y112*G12/E1
2250 INPUT "                                           y212 = ";Y212 : Y122=Y212*G12/E2
2260 INPUT "   Thermal expansion coefficient             a1 = ";AL1
2270 INPUT "                                             a2 = ";AL2
2280 INPUT "                                            a12 = ";AL3
2290 DET=1-V12*V21-Y112*Y121-Y112*Y122*V21-Y212*V12*Y121-Y212*Y122
2300 D11=E1*(1-Y212*Y122)/DET : D12=E1*(Y212*Y121+V21)/DET :
     D13=-E1*(Y122*V21+Y121)/DET
2310 D22=E2*(1-Y112*Y121)/DET : D23=-E2*(Y122+V12*Y121)/DET :
     D33=G12*(1-V12*V21)/DET
2320 ELAST(I,1)=D11 : ELAST(I,2)=D12 : ELAST(I,3)=D13 : ELAST(I,4)=D22 :
     ELAST(I,5)=D23 : ELAST(I,6)=D33
2330 THERS(I,1)=D11*AL1+D12*AL2+D13*AL3 : THERS(I,2)=D12*AL1+D22*AL2+D23*AL3 :
     THERS(I,3)=D13*AL1+D23*AL2+D33*AL3
2340 NEXT I
2350 PRINT :
     PRINT "COMPUTE THE SKYLINE HEIGHT  ....................................." : PRINT
2360 FOR NEL=1 TO NELX
2370 FOR IA=1 TO 4 : IJKIA=IJK(IA,NEL) : FOR I=1 TO 2 : IJKIAI=2*(IJKIA-1)+I
2380 FOR JB=1 TO 4 : IJKJB=IJK(JB,NEL) : FOR J=1 TO 2 : IJKJBJ=2*(IJKJB-1)+J
2390 IAIJBJ=IJKJBJ-IJKIAI+1 : IF IAIJBJ>JDIAG(IJKJBJ) THEN JDIAG(IJKJBJ)=IAIJBJ
2400 NEXT J : NEXT JB : NEXT I : NEXT IA
2410 NEXT NEL
2420 PRINT :
     PRINT "CONSTRAINED CONDITIONS BETWEEN TWO-POINTS ......................" : PRINT
2430 INPUT "Are there constrained equations between two nodal points? (YES/NO) = ";L$
2440 IF L$="YES" THEN GOTO 2490
2450 REM SET THE DIAGONAL POINTOR  JDIAG
2460 FOR I=2 TO 2*NX : JDIAG(I)=JDIAG(I-1)+JDIAG(I) : NEXT I
2470 DIM SK(JDIAG(2*NX))
2480 GOTO 2960
2490 INPUT "How many sets of constrained equations ?   NB4 = ";NB4
2500 IF NB4<=0 THEN GOTO 2960
2510 DIM NOD4(2,NB4)
2520 PRINT
2530 FOR I=1 TO NB4
2540 PRINT " set ";I;TAB(10);" node1, node2 = "; : INPUT N1,N2
2550 IF N1<N2 THEN GOTO 2600
2560 I1=2*(N1-N2)+1 : I2=I1+1
2570 IF JDIAG(2*N1-1)<I1 THEN JDIAG(2*N1-1)=I1
2580 IF JDIAG(2*N1)<I2 THEN JDIAG(2*N1)=I2
2590 GOTO 2630
2600 I1=2*(N2-N1)+1 : I2=I1+1
2610 IF JDIAG(2*N2-1)<I1 THEN JDIAG(2*N2-1)=I1
2620 IF JDIAG(2*N2)<I2 THEN JDIAG(2*N2)=I2
2630 NOD4(1,I)=N1 : NOD4(2,I)=N2
```

```
2640 NEXT I
2650 REM  SET THE JDIAG AS THE DIAGONAL POINTER
2660 FOR I=2 TO 2*NX : JDIAG(I)=JDIAG(I-1)+JDIAG(I) : NEXT I
2670 DIM SK(JDIAG(2*NX))
2680 PRINT : PRINT "          CONSTRAINED EQUATION" : PRINT
2690 PRINT "          AC1*Ux(1)+AC2*Uy(1)+AC3*Ux(2)+AC4*Uy(2)+AC5 = 0" : PRINT
2700 PRINT "             AC1,...,AC5 = given coefficients"
2710 PRINT "             Ux(1),Uy(1) = displacement of the 1-st node"
2720 PRINT "             Ux(2),Uy(2) = displacement of the 2-nd node" : PRINT
2730 FOR I=1 TO NB4
2740 PRINT " set ";I;TAB(10);" AC1,AC2,AC3,AC4,AC5 = "; : INPUT AC1,AC2,AC3,AC4,AC5
2750 N1=NOD4(1,I) : N2=NOD4(2,I)
2760 J1=JDIAG(2*N1-1) : J2=JDIAG(2*N1) : J3=JDIAG(2*N2-1) : J4=JDIAG(2*N2)
2770 SK(J1)=SK(J1)+EPI*AC1*AC1 : SK(J2)=SK(J2)+EPI*AC2*AC2
2780 SK(J3)=SK(J3)+EPI*AC3*AC3 : SK(J4)=SK(J4)+EPI*AC4*AC4
2790 IF N1<N2 THEN GOTO 2870
2800 SK(J1-2*(N1-N2))=SK(J1-2*(N1-N2))+EPI*AC1*AC3
2810 SK(J1-2*(N1-N2)+1)=SK(J1-2*(N1-N2)+1)+EPI*AC1*AC4
2820 SK(J2-2*(N1-N2)-1)=SK(J2-2*(N1-N2)-1)+EPI*AC2*AC3
2830 SK(J2-2*(N1-N2)=SK(J2-2*(N1-N2))+EPI*AC2*AC4
2840 SK(J2-1)=SK(J2-1)+EPI*AC1*AC2
2850 SK(J4-1)=SK(J4-1)+EPI*AC3*AC4
2860 GOTO 2930
2870 SK(J3-2*(N2-N1))=SK(J3-2*(N2-N1))+EPI*AC1*AC3
2880 SK(J3-2*(N2-N1)+1)=SK(J3-2*(N2-N1)+1)+EPI*AC2*AC3
2890 SK(J4-2*(N2-N1))=SK(J4-2*(N2-N1))+EPI*AC2*AC4
2900 SK(J4-2*(N2-N1)-1)=SK(J4-2*(N2-N1)-1)+EPI*AC1*AC4
2910 SK(J2-1)=SK(J2-1)+EPI*AC1*AC2
2920 SK(J4-1)=SK(J4-1)+EPI*AC3*AC4
2930 F(2*N1-1)=F(2*N1-1)-EPI*AC5*AC1 : F(2*N1)=F(2*N1)-EPI*AC5*AC2
2940 F(2*N2-1)=F(2*N2-1)-EPI*AC5*AC3 : F(2*N2)=F(2*N2)-EPI*AC5*AC4
2950 NEXT I
2960 PRINT :
     PRINT "DISPLACEMENT BOUNDARY CONDITION ................................" : PRINT
2970 INPUT "How many nodes are specified their displacement components ?  NB1 =";NB1
2980 IF NB1<=0 THEN GOTO 3150
2990 PRINT "   constrained in both directions x  and  y     .....  DIR = 11"
3000 PRINT "                   in the x direction           .....  DIR = 10"
3010 PRINT "                   in the y direction           .....  DIR =  1"
3020 PRINT
3030 FOR I=1 TO NB1
3040 INPUT " node number,  DIR, values of displacements UX,UY = ";NODE,NDIR,UX,UY
3050 IF NDIR<=10 THEN GOTO 3090
3060 N1=2*NODE-1 : N2=N1+1 : F(N1)=F(N1)+EPI*UX : F(N2)=F(N2)+EPI*UY
3070 SK(JDIAG(N1))=SK(JDIAG(N1))+EPI : SK(JDIAG(N2))=SK(JDIAG(N2))+EPI
3080 GOTO 3140
3090 IF NDIR<=1 THEN GOTO 3120
3100 N1=2*NODE-1 : SK(JDIAG(N1))=SK(JDIAG(N1))+EPI : F(N1)=F(N1)+EPI*UX
3110 GOTO 3140
3120 IF NDIR<=0 THEN GOTO 3140
3130 N2=2*NODE : SK(JDIAG(N2))=SK(JDIAG(N2))+EPI : F(N2)=F(N2)+EPI*UY
3140 NEXT I
3150 PRINT : INPUT "Can the structure be translated/rotated rigidly ? (YES/NO) =";L$
3160 IF L$<>"YES" THEN GOTO 3260
3170 PRINT "   translation in the x direction        .... INDEX = 1"
3180 PRINT "   translation in the y direction        .... INDEX = 2"
3190 PRINT "   translation/rotation                  .... INDEX = 3"
3200 PRINT
3210 INPUT "What is the INDEX ?    INDEX = ";INDEX
3220 DEL11=0 : DEL22=0
3230 IF INDEX=1 THEN DEL11=1
3240 IF INDEX=2 THEN DEL22=1
3250 IF INDEX=3 THEN DEL11=1 : DEL22=1
3260 PRINT :
     PRINT "TRACTION BOUNDARY CONDITION ...................................." : PRINT
3270 INPUT "How many nodal points are subject to applied forces ?  NB2 = ";NB2
3280 IF NB2<=0 THEN GOTO 3340
3290 PRINT
3300 FOR I=1 TO NB2
3310 INPUT " node number, applied forces FX,FY = ";NODE,FX,FY
3320 I1=2*NODE-1 : I2=I1+1 : F(I1)=F(I1)+FX : F(I2)=F(I2)+FY
```

```
3330 NEXT I
3340 PRINT :
     PRINT "THIRD TYPE BOUNDARY CONDITION (element) ........................" : PRINT
3350 INPUT "How many line elements are on the 3-rd type boundary ?   NB3 = ";NB3
3360 IF NB3<=0 THEN GOTO 3620
3370 DIM IJB(2,NB3),VB3(6,NB3)
3380 PRINT : PRINT " (normal)        Sn = - Kn(Un-Gn) + Tn"
3390 PRINT " (tangential)   St = - Kt(Ut-Gt) + Tt" : PRINT
3400 FOR I=1 TO NB3
3410 INPUT " node1, node2, Kn, Gn, Tn, Kt, Gt, and Tt = ";
     NODE1,NODE2,KN,GN,TN,KT,GT,TT
3420 X1=X(NODE1) : X2=X(NODE2) : Y1=Y(NODE1) : Y2=Y(NODE2)
3430 X21=X2-X1 : Y21=Y2-Y1 : HE=SQR(X21*X21+Y21*Y21) :
     N1=(Y21*Y21)/(HE*HE) : N2=-Y21*X21/(HE*HE) : N3=(X21*X21)/(HE*HE) :
     T1=N3 : T2=-N2 : T3=N1
3440 N1=KN*N1+KT*T1 : N2=KN*N2+KT*T2 : N3=KN*N3+KT*T3 : K1=HE/3 : K2=HE/6
3450 J1=JDIAG(2*NODE1-1) : J2=JDIAG(2*NODE1) : J3=JDIAG(2*NODE2-1) :
     J4=JDIAG(2*NODE2)
3460 SK(J1)=SK(J1)+N1*K1 : SK(J2)=SK(J2)+N3*K1 : SK(J3)=SK(J3)+N1*K1 :
     SK(J4)=SK(J4)+N3*K1
3470 IF NODE1<NODE2 THEN GOTO 3530
3480 SK(J4-1)=SK(J4-1)+N2*K1 : SK(J2-1)=SK(J2-1)+N2*K1
3490 J3=J1-2*(NODE1-NODE2) : J4=J2-2*(NODE1-NODE2)
3500 SK(J3)=SK(J3)+N1*K2 : SK(J4)=SK(J4)+N3*K2
3510 SK(J3+1)=SK(J3+1)+N2*K2 : SK(J4-1)=SK(J4-1)+N2*K2
3520 GOTO 3570
3530 SK(J2-1)=SK(J2-1)+N2*K1 : SK(J4-1)=SK(J4-1)+N2*K1
3540 J1=J3-2*(NODE2-NODE1) : J2=J4-2*(NODE2-NODE1)
3550 SK(J1)=SK(J1)+N1*K2 : SK(J2)=SK(J2)+N3*K2
3560 SK(J1+1)=SK(J1+1)+N2*K2 : SK(J2-1)=SK(J2-1)+N2*K2
3570 F1=.5*(KN*GN+TN) : F2=.5*(KT*GT+TT)
3580 J1=2*NODE1-1:J2=J1+1 : J3=2*NODE2-1 : J4=J3+1
3590 F(J1)=F(J1)+F1*(Y2-Y1)+F2*(X2-X1) : F(J3)=F(J3)+F1*(Y2-Y1)+F2*(X2-X1)
3600 F(J2)=F(J2)+F1*(X1-X2)+F2*(Y2-Y1) : F(J4)=F(J4)+F1*(X1-X2)+F2*(Y2-Y1)
3610 NEXT I
3620 PRINT :
     PRINT "DISTRIBUTED BODY FORCES (node) ..................................." : PRINT
3630 INPUT "Are there any distributed body forces applied ?   (YES/NO) = ";L$
3640 IF L$<>"YES" THEN GOTO 3760
3650 INPUT "Are equivalent nodal forces in a file ?   (YES/NO) = ";L$
3660 IF L$<>"NO" THEN GOTO 3720
3670 FOR I=1 TO NX
3680 PRINT "node ";I;TAB(12);"Fx, Fy = "; : INPUT FX,FY
3690 I1=2*I-1 : I2=I1+1 : F(I1)=F(I1)+FX : F(I2)=F(I2)+FY
3700 NEXT I
3710 GOTO 3760
3720 INPUT "What is the file name ?   (e.g. A:FORCE) F2$ = ";F2$
3730 OPEN F2$ FOR INPUT AS #2
3740 FOR I=1 TO 2*NX : INPUT#2,FX : F(I)=F(I)+FX : NEXT I
3750 CLOSE #2
3760 PRINT :
     PRINT "TEMPERATURE DISTRIBUTION (node) ................................" : PRINT
3770 INPUT "Is the thermal stress considered ?   (YES/NO) = ";L$
3780 IF L$<>"YES" THEN GOTO 3890
3790 INPUT "Is the temperature field in a  file ?   (YES/NO) = ";L$
3800 IF L$<>"NO" THEN GOTO 3850
3810 FOR I=1 TO NX
3820 PRINT " node ";I;TAB(12);"temperature = "; : INPUT TEM(I)
3830 NEXT I
3840 GOTO 3890
3850 INPUT "What is the file name ?   (e.g. A:TEMPER)   F3$ = ";F3$
3860 OPEN F3$ FOR INPUT AS #2
3870 FOR I=1 TO NX : INPUT#2,TEM(I) : NEXT I
3880 CLOSE#2
3890 REM ........................................................................
3900 PRINT : PRINT : PRINT
3910 PRINT "Finite element processing / Forming stiffness matrix and solve"
3920 PRINT "---------------------------------------------------------------------"
3930 PRINT
3940 PRINT "FORMING STIFFNESS MATRIX ....................................." : PRINT
3950 FAC=1
```

```
3960 NEL=0
3970 NEL=NEL+1 : PRINT NEL;
3980 IJK1=IJK(1,NEL) : IJK2=IJK(2,NEL) : IJK3=IJK(3,NEL) : IJK4=IJK(4,NEL)
3990 X1=X(IJK1) : X2=X(IJK2) : X3=X(IJK3) : X4=X(IJK4)
4000 Y1=Y(IJK1) : Y2=Y(IJK2) : Y3=Y(IJK3) : Y4=Y(IJK4)
4010 X13=X1-X3 : X24=X2-X4 : Y13=Y1-Y3 : Y24=Y2-Y4
4020 X5=X1-X2+X3-X4 : Y5=Y1-Y2+Y3-Y4 : AREA=.5*(X13*Y24-X24*Y13) : ARI=1/AREA
4030 T1=TEM(IJK1) : T2=TEM(IJK2) : T3=TEM(IJK3) : T4=TEM(IJK4)
4040 T5=.25*(T1+T2+T3+T4) : T6=-T1+T2+T3-T4 : T7=-T1-T2+T3+T4
4050 DX(1)=.5*Y24*ARI : DX(2)=-.5*Y13*ARI : DX(3)=-.5*Y24*ARI : DX(4)=.5*Y13*ARI
4060 DY(1)=-.5*X24*ARI : DY(2)=.5*X13*ARI : DY(3)=.5*X24*ARI : DY(4)=-.5*X13*ARI
4070 C11=(-Y13-Y24)*ARI : C12=(Y13-Y24)*ARI : C21=(X13+X24)*ARI : C22=(-X13+X24)*ARI
4080 GH(1)=.25*(1-X5*DX(1)-Y5*DY(1)) : GH(2)=.25*(-1-X5*DX(2)-Y5*DY(2))
4090 GH(3)=.25*(1-X5*DX(3)-Y5*DY(3)) : GH(4)=.25*(-1-X5*DX(4)-Y5*DY(4))
4100 F11=(C11*C11+C12*C12)/3 : F22=(C21*C21+C22*C22)/3 : F12=(C11*C21+C12*C22)/3
4110 I=MPE(NEL) : D11=ELAST(I,1) : D12=ELAST(I,2) : D13=ELAST(I,3) :
     D22=ELAST(I,4) : D23=ELAST(I,5) : D33=ELAST(I,6)
4120 EP11=D11/16000 : EP22=D22/16000
4130 B11=THERS(I,1) : B22=THERS(I,2) : B12=THERS(I,3)
4140 IF AE$<>"STRAIN" OR D11/ABS(D11-D12)<50 THEN GOTO 4160
4150 F11=F11*(D11-D12)/D11 : F22=F22*(D22-D12)/D22 : F12=0
4160 FOR IA=1 TO 4 : DXIA=DX(IA) : DYIA=DY(IA) : IA1=2*IA-1 : IA2=2*IA : GHIA=GH(IA)
4170 FE(IA1)=(B11*DXIA+B12*DYIA)*T5+GHIA*B11*(C12*T6+C11*T7)/3
4180 FE(IA2)=(B22*DYIA+B12*DXIA)*T5+GHIA*B22*(C22*T6+C21*T7)/3
4190 FOR JB=IA TO 4 : DXJB=DX(JB) : DYJB=DY(JB) : JB1=2*JB-1 : JB2=2*JB : GHJB=GH(JB)
4200 SKE(IA1,JB1)=DXIA*(D11*DXJB+D13*DYJB)+DYIA*(D13*DXJB+D33*DYJB)
              +GHIA*F11*D11*GHJB+DEL11*EP11
4210 SKE(IA1,JB2)=DXIA*(D12*DYJB+D13*DXJB)+DYIA*(D23*DYJB+D33*DXJB)
              +GHIA*F12*D12*GHJB
4220 SKE(IA2,JB1)=DYIA*(D12*DXJB+D23*DYJB)+DXIA*(D13*DXJB+D33*DYJB)
              +GHIA*F12*D12*GHJB
4230 SKE(IA2,JB2)=DYIA*(D22*DYJB+D23*DXJB)+DXIA*(D23*DYJB+D33*DXJB)
              +GHIA*F22*D22*GHJB+DEL22*EP22
4240 SKE(JB1,IA1)=SKE(IA1,JB1)  : SKE(JB2,IA1)=SKE(IA1,JB2) :
     SKE(JB1,IA2)=SKE(IA2,JB1)  : SKE(JB2,IA2)=SKE(IA2,JB2)
4250 NEXT JB : NEXT IA
4260 FOR II=1 TO 8 : I=2-II+INT(II/2)*2 : IA=INT((II-I)/2)+1
4270 IJKIA=IJK(IA,NEL) : IJKIAI=2*(IJKIA-1)+I
4280 F(IJKIAI)=F(IJKIAI)+FE(II)*AREA
4290 FOR JJ=1 TO 8 : J=2-JJ+INT(JJ/2)*2 : JB=INT((JJ-J)/2)+1
4300 IJKJB=IJK(JB,NEL) : IJKJBJ=2*(IJKJB-1)+J
4310 IAIJBJ=IJKIAI-IJKJBJ
4320 IF IAIJBJ<0 THEN GOTO 4340
4330 JPOS=JDIAG(IJKIAI)-IAIJBJ : SK(JPOS)=SK(JPOS)+SKE(JJ,II)*AREA
4340 NEXT JJ : NEXT II
4350 IF NEL<NELX THEN GOTO 3970
4360 PRINT : PRINT :
     PRINT "SKYLINE SOLVER ......................................................"
4370 PRINT
4380 NEQ=2*NX
4390 GOSUB 4940
4400 PRINT : PRINT "Postprocessing / Output of displacements and stresses"
4410 PRINT "----------------------------------------------------------------------"
4420 PRINT :
     INPUT "Are the computed results stored in the file ?
          If yes, then input the file name.   (e.g. A:RESULT)   F4$ = ";F4$
4430 OPEN F4$ FOR OUTPUT AS #2
4440 PRINT#2,NX : PRINT#2,NELX
4450 FOR I=1 TO NX : PRINT#2,X(I) : PRINT#2,Y(I) : NEXT I
4460 FOR NEL=1 TO NELX : PRINT#2,IJK(1,NEL) : PRINT#2,IJK(2,NEL) :
     PRINT#2,IJK(3,NEL) : PRINT#2,IJK(4,NEL) : NEXT NEL
4470 PRINT
4480 PRINT "NODE       Ux             Uy               X                 Y"
4490 FOR I=1 TO NX
4500 UX=F(2*I-1) : UY=F(2*I) : XI=X(I)+UX : YI=Y(I)+UY
4510 PRINT I;TAB(7);INT(10000000#*UX)/10000000#;TAB(24);INT(10000000#*UY)/10000000#;
     TAB(42);INT(1000*XI)/1000;TAB(60);INT(1000*YI)/1000
4520 PRINT#2,UX : PRINT#2,UY
4530 NEXT I
4540 PRINT : PRINT : PRINT
```

```
4550 NEL=0
4560 NEL=NEL+1
4570 IJK1=IJK(1,NEL) : IJK2=IJK(2,NEL) : IJK3=IJK(3,NEL) : IJK4=IJK(4,NEL)
4580 X1=X(IJK1) : X2=X(IJK2) : X3=X(IJK3) : X4=X(IJK4)
4590 Y1=Y(IJK1) : Y2=Y(IJK2) : Y3=Y(IJK3) : Y4=Y(IJK4)
4600 X13=X1-X3 : X24=X2-X4 : Y13=Y1-Y3 : Y24=Y2-Y4 : AREA=.5*(X13*Y24-X24*Y13) :
     ARI2=.5/AREA
4610 DX(1)=Y24*ARI2 : DX(2)=-Y13*ARI2 : DX(3)=-Y24*ARI2 : DX(4)=Y13*ARI2
4620 DY(1)=-X24*ARI2 : DY(2)=X13*ARI2 : DY(3)=X24*ARI2 : DY(4)=-X13*ARI2
4630 U1=F(2*IJK1-1) : U2=F(2*IJK2-1) : U3=F(2*IJK3-1) : U4=F(2*IJK4-1)
4640 V1=F(2*IJK1) : V2=F(2*IJK2) : V3=F(2*IJK3) : V4=F(2*IJK4)
4650 TNEL=.25*(TEM(IJK1)+TEM(IJK2)+TEM(IJK3)+TEM(IJK4))
4660 EXX=DX(1)*U1+DX(2)*U2+DX(3)*U3+DX(4)*U4
4670 EYY=DY(1)*V1+DY(2)*V2+DY(3)*V3+DY(4)*V4
4680 EXY=DX(1)*V1+DX(2)*V2+DX(3)*V3+DX(4)*V4+DY(1)*U1+DY(2)*U2+DY(3)*U3+DY(4)*U4
4690 I=MPE(NEL) : D11=ELAST(I,1) : D12=ELAST(I,2) : D13=ELAST(I,3) :
     D22=ELAST(I,4) : D23=ELAST(I,5) : D33=ELAST(I,6)
4700 B11=THERS(I,1) : B22=THERS(I,2) : B12=THERS(I,3)
4710 SXX=D11*EXX+D12*EYY+D13*EXY-B11*TNEL
4720 SYY=D12*EXX+D22*EYY+D23*EXY-B22*TNEL
4730 SXY=D13*EXX+D23*EYY+D33*EXY-B12*TNEL
4740 P1=.5*(SXX+SYY)+SQR(.25*(SXX-SYY)*(SXX-SYY)+SXY*SXY)
4750 P2=.5*(SXX+SYY)-SQR(.25*(SXX-SYY)*(SXX-SYY)+SXY*SXY)
4760 THETA=90*ATN(2*SXY/(SXX-SYY))/3.141529 : IF SXX<SYY THEN THETA=90+THETA
4770 SZZ=0
4780 IF AE$="STRAIN" THEN SZZ=PZX(I)*SXX+PZY(I)*SYY+PXY(I)*SXY
4790 PRES=(SXX+SYY+SZZ)/3 : SDXX=SXX-PRES : SDYY=SYY-PRES : SDZZ=SZZ-PRES
4800 MISE=SQR(1.5*(SDXX*SDXX+SDYY*SDYY+SDZZ*SDZZ+2*SXY*SXY))
4810 STRE=.5*(SXX*EXX+SYY*EYY+SXY*EXY)
4820 TSE=TSE+STRE*AREA
4830 PRINT "ELEMENT ";NEL
4840 PRINT "   STRAIN";TAB(15);"EXX ";EXX;TAB(35);"EYY ";EYY;TAB(55);"EXY ";EXY
4850 PRINT "   STRESS";TAB(15);"SXX ";SXX;TAB(35);"SYY ";SYY;TAB(55);"SXY ";SXY
4860 PRINT "   P. STRESS";TAB(15);"P1 ";P1;TAB(35);"P2 ";P2;TAB(55);"THETA ";THETA
4870 PRINT "   MISES STRESS";TAB(20);"SS ";MISE;TAB(40);"STRAIN E.D. ";STRE
4880 PRINT#2,EXX : PRINT#2,EYY : PRINT#2,EXY : PRINT#2,SXX : PRINT#2,SYY :
     PRINT#2,SXY : PRINT#2,P1 : PRINT#2,P2 : PRINT#2,THETA : PRINT#2,MISE :
     PRINT#2,STRE
4890 SK(NEL)=P1 : SK(NELX+NEL)=P2 : SK(2*NELX+NEL)=THETA
4900 IF NEL<NELX THEN GOTO 4560
4910 CLOSE #2
4920 PRINT : PRINT " TOTAL STRAIN ENERGY IN THIS SYSTEM   TSE = ";TSE
4930 GOTO 5320
4940 REM  subroutine skyline(s)
4950 REM        developed by  Professor Taylor in University of California
4960 REM ------------------------------------------------------------------------
4970 JR=0
4980 FOR J=1 TO NEQ : JE=JDIAG(J) : JH=JE-JR : IS=J-JH+2 : PRINT J;
4990 IF JH<2 THEN GOTO 5180
5000 IF JH=2 THEN GOTO 5110
5010 IE=J-1 : K=JR+2
5020 I1=IS-1 : ID=JDIAG(I1)
5030 FOR I=IS TO IE : IR=ID : ID=JDIAG(I) :IH=ID-IR-1 : I1=I-IS+1
5040 IF IH>I1 THEN IH=I1
5050 IF IH<=0 THEN GOTO 5090
5060 AK=0
5070 FOR L=1 TO IH : AK=AK+SK(K-IH+L-1)*SK(ID-IH+L-1) : NEXT L
5080 SK(K)=SK(K)-AK
5090 K=K+1
5100 NEXT I
5110 IR=JR+1 : IE=JE-1 : K=J-JE
5120 FOR I=IR TO IE : ID=JDIAG(K+I) : IF SK(ID)=0 THEN GOTO 5140
5130 D=SK(I) : SK(I)=D/SK(ID) : SK(JE)=SK(JE)-D*SK(I)
5140 NEXT I
5150 FJ=0 : LE=JH-1
5160 FOR L=1 TO LE : FJ=FJ+SK(JR+L)*F(IS+L-2) : NEXT L
5170 F(J)=F(J)-FJ
5180 JR=JE
5190 NEXT J
5200 FOR I=1 TO NEQ : ID=JDIAG(I) : IF SK(ID)=0 THEN GOTO 5220
```

```
5210 F(I)=F(I)/SK(ID)
5220 NEXT I
5230 REM ..... BACK-SUBSTITUTION .....
5240 J=NEQ : JE=JDIAG(J)
5250 D=F(J) : J=J-1 : IF J<=0 THEN GOTO 5310
5260 JR=JDIAG(J) : JS=JE-JR : IF JS<=1 THEN GOTO 5300
5270 IS=J-JE+JR+2 : K=JR-IS+1
5280 FOR I=IS TO J : F(I)=F(I)-SK(K+I)*D
5290 NEXT I
5300 JE=JR : GOTO 5250
5310 PRINT : RETURN
5320 REM  plotting routine
5330 REM ------------------------------------------------------------------------
5340 PRINT :
     PRINT : INPUT "Shall we look at the deformed configuration ?   (YES/NO) = ";L$
5350 IF L$="NO" THEN GOTO 5850
5360 INPUT "Will the initial configuration be plotted ?   (YES/NO) = ";IN$
5370 INPUT "What is the magnification factor of displacements ?   MFC = ";MFC
5380 INPUT "Will the principal stresses be plotted ?  (YES/NO) = ";PS$
5390 IF PS$<>"NO" THEN INPUT "What is the size of P1/P2 ?   ARR = ";ARR
5400 INPUT "The 1-st and last element numbers   (NEL1,NEL2) = ";NEL1,NEL2
5410 XMIN=1E+10 : XMAX=-1E+10 : YMIN=1E+10 : YMAX=-1E+10 : PMAX=-1E+10
5420 IF NEL2>NELX THEN NEL2=NELX
5430 FOR NEL=NEL1 TO NEL2
5440 P1=SK(NEL) : P2=SK(NELX+NEL)
5450 IF PMAX<ABS(P1) THEN PMAX=ABS(P1)
5460 IF PMAX<ABS(P2) THEN PMAX=ABS(P2)
5470 FOR IA=1 TO 4 : IJKIA=IJK(IA,NEL)
5480 XIA=X(IJKIA)+MFC*F(2*IJKIA-1) : YIA=Y(IJKIA)+MFC*F(2*IJK2)
5490 IF XIA<XMIN THEN XMIN=XIA
5500 IF XIA>XMAX THEN XMAX=XIA
5510 IF YIA<YMIN THEN YMIN=YIA
5520 IF YIA>YMAX THEN YMAX=YIA
5530 NEXT IA
5540 NEXT NEL
5550 R=2.16
5560 SX=600/(R*(XMAX-XMIN)) : SY=170/(YMAX-YMIN) : IF SY<SX THEN SX=SY
5570 SCREEN 2 : CLS
5580 FOR NEL=NEL1 TO NEL2
5590 IJK1=IJK(1,NEL) : IJK2=IJK(2,NEL) : IJK3=IJK(3,NEL) : IJK4=IJK(4,NEL)
5600 X1=15+SX*R*(X(IJK1)-XMIN) : Y1=185-SX*(Y(IJK1)-YMIN)
5610 X2=15+SX*R*(X(IJK2)-XMIN) : Y2=185-SX*(Y(IJK2)-YMIN)
5620 X3=15+SX*R*(X(IJK3)-XMIN) : Y3=185-SX*(Y(IJK3)-YMIN)
5630 X4=15+SX*R*(X(IJK4)-XMIN) : Y4=185-SX*(Y(IJK4)-YMIN)
5640 DX1=X1+SX*R*MFC*F(2*IJK1-1) : DY1=Y1-SX*MFC*F(2*IJK1)
5650 DX2=X2+SX*R*MFC*F(2*IJK2-1) : DY2=Y2-SX*MFC*F(2*IJK2)
5660 DX3=X3+SX*R*MFC*F(2*IJK3-1) : DY3=Y3-SX*MFC*F(2*IJK3)
5670 DX4=X4+SX*R*MFC*F(2*IJK4-1) : DY4=Y4-SX*MFC*F(2*IJK4)
5680 IF IN$="YES" THEN LINE (X1,Y1)-(X2,Y2) : LINE (X2,Y2)-(X3,Y3) :
     LINE (X3,Y3)-(X4,Y4) : LINE (X4,Y4)-(X1,Y1)
5690 LINE (DX1,DY1)-(DX2,DY2) : LINE (DX2,DY2)-(DX3,DY3) :
     LINE (DX3,DY3)-(DX4,DY4) : LINE (DX4,DY4)-(DX1,DY1)
5700 IF PS$="NO" THEN GOTO 5800
5710 P1=SK(NEL) : P2=SK(NELX+NEL) : TH=3.141529*SK(2*NELX+NEL)/180
5720 AX=ARR*ABS(P1)*.5*COS(TH)*R/PMAX : AY=ARR*ABS(P1)*.5*SIN(TH)/PMAX
5730 TH=3.141529*(SK(2*NELX+NEL)+90)/180
5740 BX=ARR*ABS(P2)*.5*COS(TH)*R/PMAX : BY=ARR*ABS(P2)*.5*SIN(TH)/PMAX
5750 XC=.25*(DX1+DX2+DX3+DX4) : YC=.25*(DY1+DY2+DY3+DY4)
5760 X1=XC-AX : X2=XC+AX : Y1=YC+AY : Y2=YC-AY : LINE (X1,Y1)-(X2,Y2)
5770 X3=XC-BX : X4=XC+BX : Y3=YC+BY : Y4=YC-BY : LINE (X3,Y3)-(X4,Y4)
5780 IF P1>0 THEN LINE (X1,Y1)-(X1+2,Y1) : LINE (X1+2,Y1)-(X1+2,Y1+1) :
     LINE (X1+2,Y1+1)-(X1,Y1+1) : LINE (X1,Y1+1)-(X1,Y1)
5790 IF P2>0 THEN LINE (X3,Y3)-(X3+2,Y3) : LINE (X3+2,Y3)-(X3+2,Y3+1) :
     LINE (X3+2,Y3+1)-(X3,Y3+1) : LINE (X3,Y3+1)-(X3,Y3)
5800 NEXT NEL
5810 INPUT "Shall we redraw ?   (YES/NO) = ";L$
5820 IF L$="YES" THEN GOTO 5360
5830 INPUT "Push the return key to continue ... ";L$
5840 SCREEN 0
5850 END
```

5.9 Families of isoparametric elements

We shall consider some generalizations of the four-node element studied so far. As in the family of triangular elements, many higher-order quadrilateral elements are popular in practice. Roughly speaking, there are two families of *isoparametric elements.* The first is the *Lagrangian family of quadrilateral* (*or brick*) *elements and the other is the serendipity family* of quadrilateral (or brick) elements.

5.9.1 Lagrange interpolation on a line

As a basis of interpolation on a finite element, we need to extend the idea of piecewise linear interpolation of a function f defined on the real line $\mathbb{R}$. To do this, we shall briefly review *Lagrange interpolation* on the real line $\mathbb{R}$.

Lagrange interpolation is based on the existence of a unique polynomial $f^I(x)$ of degree not exceeding $(n - 1)$ such that

$$f^I(x_i) = f(x_i), \quad i = 1, \dots, n \tag{5.195}$$

for a given function f on an interval $I = [a, b]$, where $a = x_1 < x_2 < x_3 < \cdots < x_{n-1} < x_n = b$. It is noted that the existence of such a polynomial f^I is equivalent to the existence of the set of polynomials $\{L_i(x)\}$, $i = 1, \dots, n$, which possess the following properties:

(i) $\qquad$ Degree of $L_i(x) \leq n - 1$

(ii) $\qquad L_i(x_j) = \delta_{ij}, \quad \text{for } i = 1, \dots, n \qquad (5.196)$

Indeed, if the set of polynomials $\{L_i(x)\}$ exists, the polynomial f^I is given by

$$f^I(x) = f_i L_i(x) \tag{5.197}$$

where $f_i = f(x_i)$ is the value of f at point i.

The polynomials $L_i(x)$ that satisfy properties (i) and (ii) are

$$\begin{aligned} L_i(x) &= \frac{(x - x_1)\cdots(x - x_{i-1})(x - x_{i+1})\cdots(x - x_n)}{(x_i - x_1)\cdots(x_i - x_{i-1})(x_i - x_{i+1})\cdots(x_i - x_n)} \\ &= \prod_{j=1, j\neq i}^{n} \frac{x - x_j}{x_i - x_j} \end{aligned} \tag{5.198}$$

The first derivative of $L_i(x)$ is easily found by logarithmic differentiation, and the result is

$$L_i'(x) = L_i(x) \sum_{j=1, j\neq i}^{n} \frac{1}{x - x_j} \tag{5.199}$$

Using the property that any interval $I = [a, b]$ in $\mathbb{R}$ can be transformed from a normalized interval $[-1, 1]$ using the mapping

$$x = \tfrac{1}{2}(1 - s)a + \tfrac{1}{2}(1 + s)b \tag{5.200}$$

for $x \in I$ and $s \in [-1, 1]$, the set of polynomials $\{L_i(x)\}$ may be normalized to the set $\{L_i(s)\}$ such that

$$L_i(s) = \prod_{j=1, j\neq i}^{n} \frac{s - s_j}{s_i - s_j}, \qquad s_j = -1 + \frac{2j}{n} \tag{5.201}$$

The first derivative of $L_i(s)$ with respect to x becomes

$$\frac{dL_i}{dx} = \frac{ds}{dx}\frac{dL_i}{ds} = \frac{2}{b-a} L_i(s) \sum_{j=1, j\neq i}^{n} \frac{1}{s - s_j} \tag{5.202}$$

where s and x are related by (5.200). Thus, once we have a routine for $L_i(s)$ and $dL_i(s)/ds$ for a given value $s \in [-1, 1]$ and n, it is possible to define the interpolation of f on any interval I as

$$f^I(x) = f_i L_i(s), \qquad f_i = f(x_i) \tag{5.203}$$

Example 5.9: Let us obtain Lagrange interpolations of a function

$$f(x) = 1/(1 + 10x^2)$$

in an interval $I = [-1, 1]$. Using a general algorithm to create the normalized basis polynomials $\{L_i(s)\}$ and its derivatives–

```
820    FOR I = 1 TO ND
830 XI = -1 + 2 * (I - 1)/(ND - 1)
840 SN = 1:SD = 1:DS = 0
850    FOR J = 1 TO ND
860    IF J = I THEN GOTO 970
870 XJ = -1 + 2 * (J - 1)/(ND - 1)
880 SN = SN * (XL - XJ)
890 SD = SD * (XI - XJ)
900 DJ = 0
910    FOR K = 1 TO ND
920    IF K = J OR K = I THEN GOTO 950
930 XK = -1 + 2 * (K - 1)/(ND - 1)
940 DJ = DJ * (XL - XK)
950    NEXT K
960 DS = DS + DJ
970    NEXT J
980 SH(I) = SN / SD
990 DH(I) = DS / SD
1000   NEXT I
```
(5.204)

where XL is the given coordinate s, SH(I) = $L_i(s)$, DH(I) = $dL_i(s)/ds$, and ND is the number of points n – we obtain the results shown in Figures 5.52–5.54. It is clear that the quality of higher-order polynomials is not very good in this case. Especially in Figures 5.53 and 5.54, the Lagrange interpolations by 4- and 10-degree polynomials are very oscillatory. The graphs in the figures were obtained using the subroutine GRAPH written in BASIC. This program is developed based on Dwyer and Critchfield (1978, chap. 7).

Now how can we overcome the large oscillatory behavior of higher-order Lagrange interpolation? One very effective way to eliminate oscillations is to use the idea of finite element methods. That is, we first divide the interval $I = [-1, 1]$ to several finite elements (i.e., subinter-

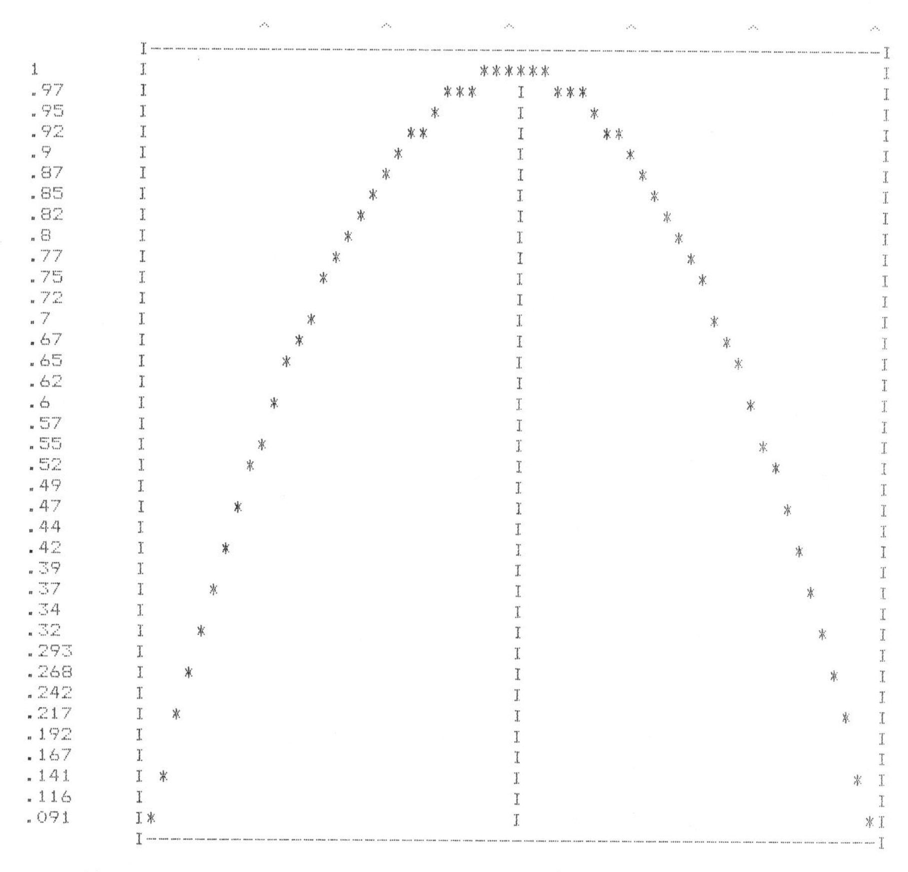

Figure 5.52 Interpolation of $f(x) = 1/(1 + 10x^2)$ by quadratic polynomial.

vals) and interpolate the function on each finite element using Lagrange interpolation. An example is shown in Figure 5.55. Here two finite elements $\Omega_1 = [-1, 0]$ and $\Omega_2 = [0, 1]$ were used together with a fifth-degree polynomial for interpolation.

Exercise 5.19: Interpolate a function

$$f(x) = 1/(1 + 10x^2)$$

on an interval $I = [-1, 1]$ using two, four, and eight equal-size finite elements for the following two cases: On each finite element, the function f is interpolated by (a) a quadratic polynomial and (b) a cubic polynomial. Compare the first derivative of the interpolation with that of the original function. Find those points at which the quality of the first derivative of the interpolation is good (1%, say).

Next we shall introduce the concept of isoparametric representation and its generalization. As has been noted, if the geometry of the domain in which a function

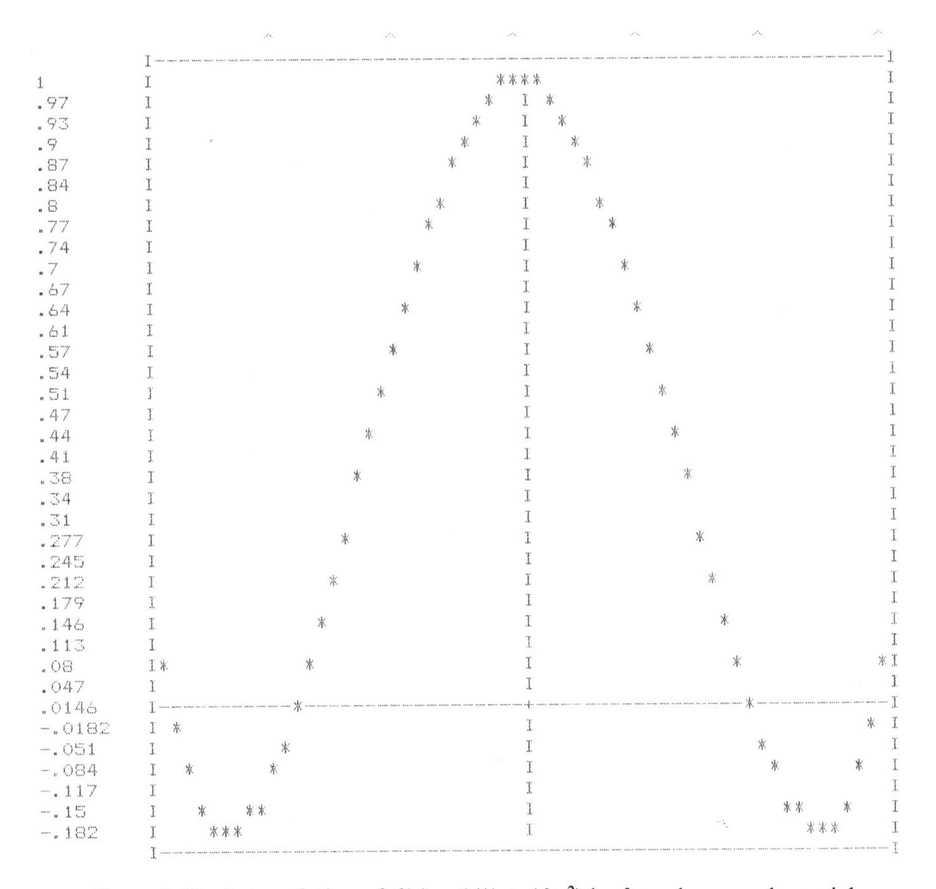

Figure 5.53 Interpolation of $f(x) = 1/(1 + 10x^2)$ by four-degree polynomial.

is defined is approximated by the same "shape" functions [for Lagrange interpolation on a real line, the set of polynomials $L_i(s)$] that are used for the interpolation of the function itself, it is said that *isoparametric* representation is being employed for the approximation (or interpolation). For example, consider a function f defined on a curved line in $\mathbb{R}^3$ as shown in Figure 5.56.

Suppose that this line is approximated by piecewise linear segments and that the function is also interpolated by a piecewise linear polynomial. That is, a typical segment $\Omega_e = [\xi_e, \xi_{e+1}]$ is a line element identified with the coordinates ξ_e and ξ_{e+1} along the curved line, and the function f is interpolated by a linear polynomial on Ω_e. Let Ω_e be considered as the image of a line segment $\hat{\Omega}_e$ normalized so that $\hat{\Omega}_e = [-1, 1]$ along some straight line with coordinate s. Then any point P on Ω_e, which may be represented by the coordinates (x, y, z) in the Cartesian coordinate system, as well as by ξ, the coordinate along the curve, is related to the *master segment* Ω_e by

$$x = x_i L_i(s), \qquad y = y_i L_i(s), \qquad z = z_i L_i(s), \quad i = 1, 2 \tag{5.205}$$

where (x_i, y_i, z_i), $i = 1, 2$, are the coordinates of two end points of a segment Ω_e.

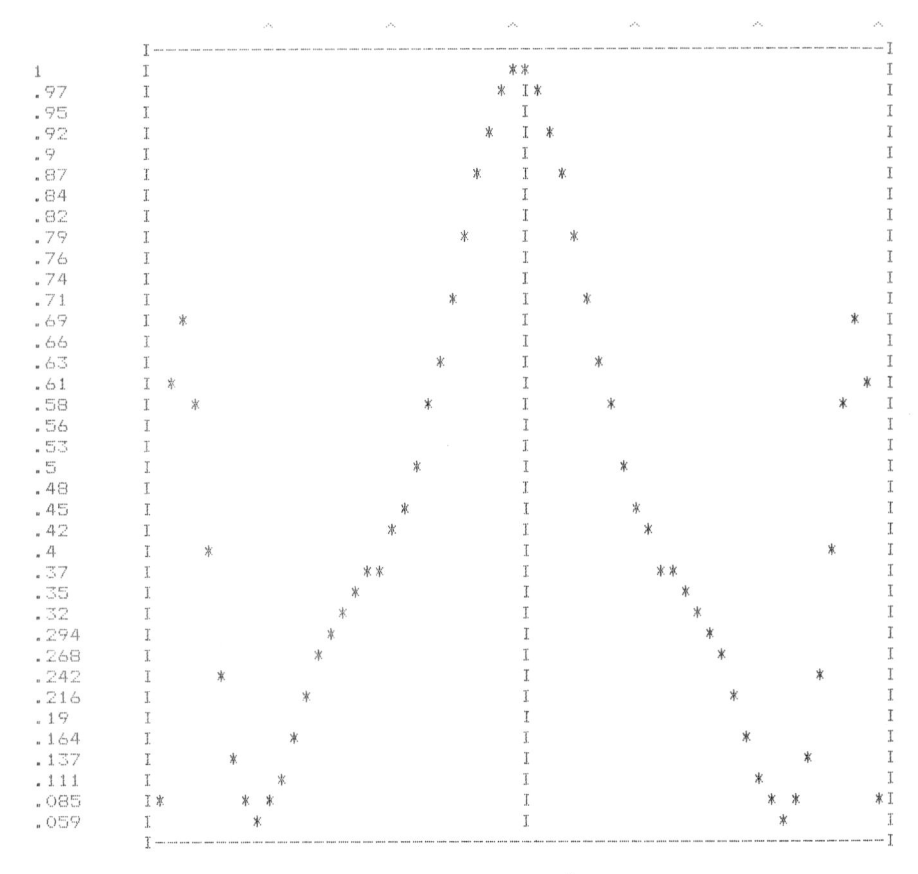

Figure 5.54 Interpolation of $f(x) = 1/(1 + 10x^2)$ by 10-degree polynomial.

Similarly, f is interpolated by

$$f^I = f_i L_i(s) \quad \text{on } \Omega_e, \qquad f_i = f(x_i, y_i, z_i) = f(\xi_i) \tag{5.206}$$

Then the differential line element $d\xi$ along the curve is given by

$$\delta\xi = \sqrt{dx^2 + dy^2 + dz^2} = \sqrt{\left(x_i \frac{dL_i}{ds}\right)^2 + \left(y_i \frac{dL_i}{ds}\right)^2 + \left(z_i \frac{dL_i}{ds}\right)^2}\, ds \tag{5.207}$$

Putting

$$J = \sqrt{\left(x_i \frac{dL_i}{ds}\right)^2 + \left(y_i \frac{dL_i}{ds}\right)^2 + \left(z_i \frac{dL_i}{ds}\right)^2} \tag{5.208}$$

we have

$$\frac{d\xi}{ds} = J \qquad \text{and} \qquad \frac{ds}{d\xi} = J^{-1} \tag{5.209}$$

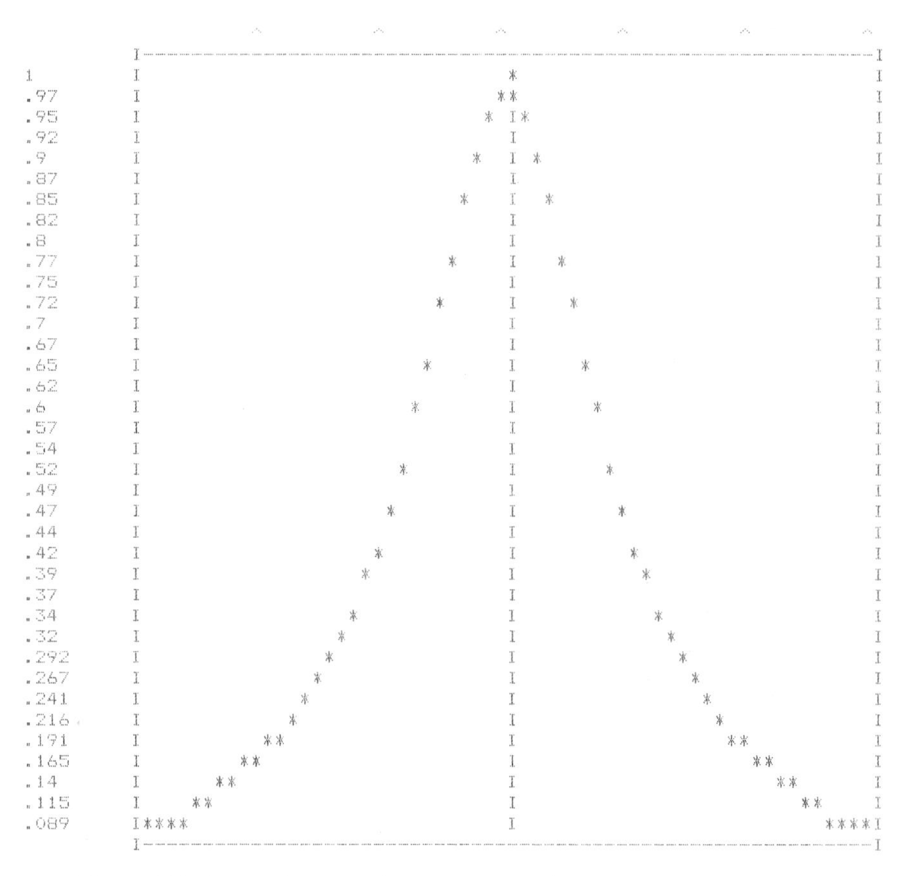

Figure 5.55 Interpolation of $f(x) = 1/(1 + 10x^2)$ using two finite elements.

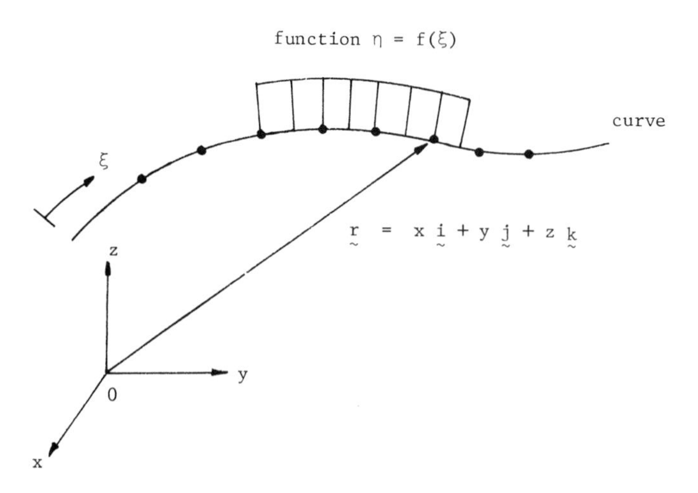

Figure 5.56 Function defined on a curve in $\mathbb{R}^3$.

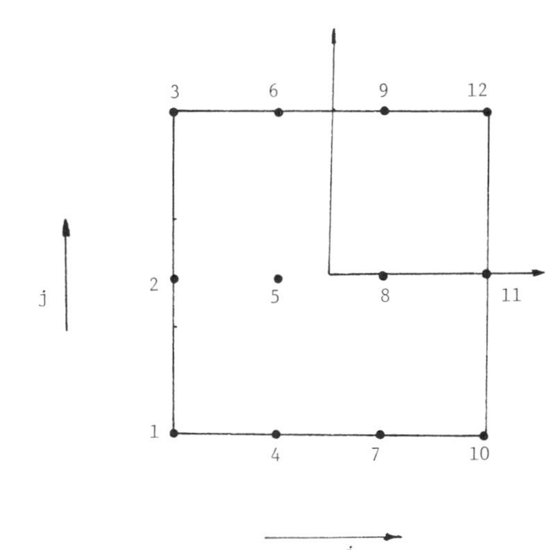

Figure 5.57 Numbering system of element constructed by tensor product: $n_s = 4$, $n_t = 3$; $\alpha = 8$ when $(i, j) = (3, 2)$.

Thus, the first derivative of f^I is given as

$$\frac{df^I}{d\xi} = f_i \frac{dL_i}{d\xi} = f_i \frac{ds}{d\xi} \frac{dL_i}{ds} = J^{-1} f_i \frac{dL_i}{ds} \tag{5.210}$$

on each element Ω_e. The integration over Ω_e of some function $g(\xi)$ is given by

$$\int_{\Omega_e} g(\xi)\, d\xi = \int_{-1}^{1} \hat{g}(s) J\, ds \tag{5.211}$$

where $\hat{g}(s) \equiv g(\xi(s))$.

It is noted that the expressions (5.205)–(5.211) are valid for other higher-order approximations of a curved line and a function defined on an interval as well as for piecewise linear approximation.

5.9.2 Lagrange interpolation by tensor product

Let us use the normalized polynomials (5.201) of Lagrange interpolation defined on the normalized interval $\hat{I} = [-1, 1]$ to construct interpolations of functions defined on $\mathbb{R}^2$ or $\mathbb{R}^3$. To do this, we shall define the *shape functions* $\hat{N}_\alpha(s, t)$ and $\hat{N}_\alpha(r, s, t)$ from L_i using the idea of tensor product. We shall first consider $\hat{N}_\alpha(s, t)$, which are appropriate for $\mathbb{R}^2$.

It is noted that the degree of the polynomial is decided by the number of nodes at which the value of the interpolation is to agree with that of the function being interpolated. In fact, if n nodes are taken in a subinterval (i.e., an element on $\mathbb{R}$), the degree of the interpolating polynomial is $n - 1$ on a real line $\mathbb{R}$. Now, for $\mathbb{R}^2$, let us take n_s and n_t points in the directions s and t on the normalized domain $\hat{\Omega}_e = [-1, 1] \times [-1, 1]$, which is called the master element. That is, the value of the interpolation is to agree with that of the function at $n_s n_t$ nodes in $\hat{\Omega}_e$ as shown in Figure 5.57 (for $n_s = 4$, $n_t = 3$). If the numbering of the nodes in $\hat{\Omega}_e$ is as shown in Figure 5.57, we have the rule

$$\alpha = n_t(i - 1) + j, \quad i = 1, \ldots, n_s; j = 1, \ldots, n_t \tag{5.212}$$

where (i, j) indicates the location of the α node using the grid numbers i and j in the s and t directions, respectively. This further yields a compact definition of the shape functions by tensor product of $\{L_i(s)\}$ and $\{L_j(t)\}$ defined by (5.201) in each direction:

$$\hat{N}_\alpha(s, t) = L_i(s)L_j(t), \quad \alpha = n_t(i - 1) + j \tag{5.213}$$

For the case $n_s = 4$ and $n_t = 3$ as in Figure 5.57, $\{L_i(s)\}$ and $\{L_j(t)\}$ are defined as

$$\begin{aligned} L_1(s) &= -\tfrac{9}{16}(s + \tfrac{1}{3})(s - \tfrac{1}{3})(s - 1) \\ L_2(s) &= \tfrac{27}{16}(s + 1)(s - \tfrac{1}{3})(s - 1) \\ L_3(s) &= -\tfrac{27}{16}(s + 1)(s + \tfrac{1}{3})(s - 1) \\ L_4(s) &= \tfrac{9}{16}(s + 1)(s + \tfrac{1}{3})(s - \tfrac{1}{3}) \end{aligned} \tag{5.214}$$

and

$$L_1(t) = \tfrac{1}{2}t(t - 1), \qquad L_2(t) = -(t - 1)(t + 1), \qquad L_3(t) = \tfrac{1}{2}t(t + 1) \tag{5.215}$$

Thus, we have

$$\begin{aligned} \hat{N}_1(s, t) &= -\tfrac{9}{32}(s + \tfrac{1}{3})(s - \tfrac{1}{3})(s - 1)t(t - 1) \\ \hat{N}_2(s, t) &= \tfrac{9}{16}(s + \tfrac{1}{3})(s - \tfrac{1}{3})(s - 1)(t - 1)(t + 1) \\ &\vdots \\ \hat{N}_4(s, t) &= \tfrac{27}{32}(s + 1)(s - \tfrac{1}{3})(s - 1)t(t - 1) \\ &\vdots \\ \hat{N}_{12}(s, t) &= \tfrac{9}{32}(s + 1)(s + \tfrac{1}{3})(s - \tfrac{1}{3})t(t + 1) \end{aligned} \tag{5.216}$$

The first derivatives of $\hat{N}_\alpha$ with respect to the master coordinates s and t are given by

$$\frac{\partial \hat{N}_\alpha}{\partial s} = L_i'(s)L_j(t) \qquad \text{and} \qquad \frac{\partial \hat{N}_\alpha}{\partial t} = L_i(s)L_j'(t) \tag{5.217}$$

As for the four-node element, finite elements $\{\Omega_e\}$ in some domain $\Omega \subset \mathbb{R}^2$ are "defined" by the mapping that uses the shape function $\{\hat{N}_\alpha(s, t)\}$ on the master element $\hat{\Omega}_e$. Indeed, if the coordinates (x_α, y_α) are specified at the node α of a finite element Ω_e in the "global" coordinate system (x, y), and α is the corresponding node number in the master element $\hat{\Omega}_e$, the relation between (s, t) in $\hat{\Omega}_e$ and (x, y) in Ω_e is given by

$$x = x_\alpha \hat{N}_\alpha(s, t), \qquad y = y_\alpha \hat{N}_\alpha(s, t) \tag{5.218}$$

If the shape functions $\{N_\alpha(x, y)\}$ are introduced for each Ω_e using the coordinate system (x, y), the relation (5.218) yields

$$\hat{N}_\alpha(s, t) \triangleq N_\alpha(x(s, t), y(s, t)) \tag{5.219}$$

as shown in (5.72). Then, as shown in section 5.4, we have

$$\begin{Bmatrix} \dfrac{\partial N_\alpha}{\partial x} \\ \dfrac{\partial N_\alpha}{\partial y} \end{Bmatrix} = \frac{1}{J} \begin{bmatrix} \dfrac{\partial y}{\partial t} & -\dfrac{\partial y}{\partial s} \\ -\dfrac{\partial x}{\partial t} & \dfrac{\partial x}{\partial s} \end{bmatrix} \begin{Bmatrix} \dfrac{\partial \hat{N}_\alpha}{\partial s} \\ \dfrac{\partial \hat{N}_\alpha}{\partial t} \end{Bmatrix} \tag{5.220}$$

where the Jacobian J is given by

$$J = \frac{\partial x}{\partial s}\frac{\partial y}{\partial t} - \frac{\partial x}{\partial t}\frac{\partial y}{\partial s} \tag{5.221}$$

The area elements $d\Omega$ and $d\hat{\Omega}$ are related by

$$d\Omega = J\, d\hat{\Omega} \tag{5.222}$$

Extension to elements in $\mathbb{R}^3$ is straightforward. Indeed, the numbering (5.212) is changed by

$$\begin{aligned} &\alpha = n_t n_s(i-1) + n_s(j-1) + k \\ &i = 1, \ldots, n_r;\ j = 1, \ldots, n_s;\ k = 1, \ldots, n_t \end{aligned} \tag{5.223}$$

where (n_r, n_s, n_t) are the number of nodes in the r, s, and t direction, respectively. Then the shape functions are defined as

$$\hat{N}_\alpha(r, s, t) = L_i(r)L_j(s)L_k(t) \tag{5.224}$$

The relation (5.220) between the first derivative becomes

$$\begin{Bmatrix} \dfrac{\partial N_\alpha}{\partial x} \\ \dfrac{\partial N_\alpha}{\partial y} \\ \dfrac{\partial N_\alpha}{\partial z} \end{Bmatrix} = \begin{bmatrix} \dfrac{\partial x}{\partial r} & \dfrac{\partial y}{\partial r} & \dfrac{\partial z}{\partial r} \\ \dfrac{\partial x}{\partial s} & \dfrac{\partial y}{\partial s} & \dfrac{\partial z}{\partial s} \\ \dfrac{\partial x}{\partial t} & \dfrac{\partial y}{\partial t} & \dfrac{\partial z}{\partial t} \end{bmatrix}^{-1} \begin{Bmatrix} \dfrac{\partial \hat{N}_\alpha}{\partial r} \\ \dfrac{\partial \hat{N}_\alpha}{\partial s} \\ \dfrac{\partial \hat{N}_\alpha}{\partial t} \end{Bmatrix} \tag{5.225}$$

The simplest brick element is the eight-node element whose shape functions are given by

$$\begin{aligned} \hat{N}_1(r, s, t) &= \tfrac{1}{8}(1-r)(1-s)(1-t) & \hat{N}_5(r, s, t) &= \tfrac{1}{8}(1+r)(1-s)(1-t) \\ \hat{N}_2(r, s, t) &= \tfrac{1}{8}(1-r)(1-s)(1+t) & \hat{N}_6(r, s, t) &= \tfrac{1}{8}(1+r)(1-s)(1+t) \\ \hat{N}_3(r, s, t) &= \tfrac{1}{8}(1-r)(1+s)(1-t) & \hat{N}_7(r, s, t) &= \tfrac{1}{8}(1+r)(1+s)(1-t) \\ \hat{N}_4(r, s, t) &= \tfrac{1}{8}(1-r)(1+s)(1+t) & \hat{N}_8(r, s, t) &= \tfrac{1}{8}(1+r)(1+s)(1+t) \end{aligned} \tag{5.226}$$

where the numbering is as shown in Figure 5.58. Although the numbering is arbitrary, we use the one in Figure 5.58 because of the numbering rule (5.223).

Exercise 5.20: Obtain the shape functions for the nine-node quadrilateral element in $\mathbb{R}^2$.

Exercise 5.21: Obtain the shape functions for the 27-node brick element in $\mathbb{R}^3$.

Exercise 5.22: Modify subroutines ESTIF0 and GGRAD4 in FEM2 for the $n_s n_t$-node quadrilateral element in $\mathbb{R}^2$. Specify the proper number of integration points in each direction for the Gaussian quadrature rules.

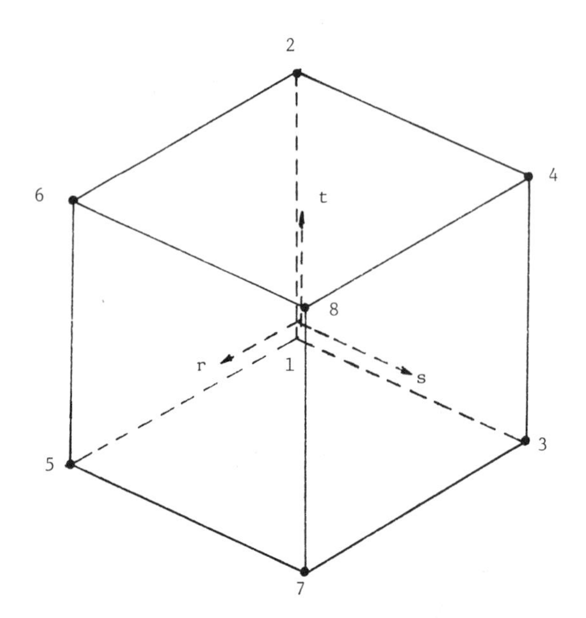

Figure 5.58 Numbering for eight-node brick element in $\mathbb{R}^3$.

5.9.3 Serendipity elements

As a variation of Lagrange interpolation using the tensor product, there is a family consisting of the *serendipity elements.* Although the quality of the approximate solutions obtained using serendipity elements is sometimes quite poor in comparison to the above method, these elements are very popular.

The eight-node quadrilateral element shown in Figure 5.59 is characterized by the shape functions as follows:

$$\begin{aligned}
\hat{N}_1(s,t) &= \tfrac{1}{4}(1-s)(1-t)(-1-s-t) & \hat{N}_5(s,t) &= \tfrac{1}{2}(1-s^2)(1-t)\\
\hat{N}_2(s,t) &= \tfrac{1}{4}(1+s)(1-t)(-1+s-t) & \hat{N}_6(s,t) &= \tfrac{1}{2}(1+s)(1-t^2)\\
\hat{N}_3(s,t) &= \tfrac{1}{4}(1+s)(1+t)(-1+s+t) & \hat{N}_7(s,t) &= \tfrac{1}{2}(1-s^2)(1+t)\\
\hat{N}_4(s,t) &= \tfrac{1}{4}(1-s)(1+t)(-1+s+t) & \hat{N}_8(s,t) &= \tfrac{1}{2}(1-s)(1-t^2)
\end{aligned}$$

There are two ways to explain how to obtain these shape functions. The first is based on the conditions to be satisfied by the shape functions:

$$\hat{N}_\alpha(s_\beta, t_\beta) = \delta_{\alpha\beta}, \qquad \sum_{\alpha=1}^{N_e} \hat{N}_\alpha(s,t) = 1 \tag{5.227}$$

Drawing the lines passing through nodal points as shown in Figure 5.59, we can define

$$\begin{aligned}
\hat{N}_1(s,t) &= \tfrac{1}{4}(1-s)(1-t)(-1-s-t)\\
\hat{N}_5(s,t) &= \tfrac{1}{2}(1-s^2)(1-t), \quad \text{etc.}
\end{aligned}$$

This intuitive approach cannot be extended to higher-order serendipity elements. The second approach is more systematic. We first define the shape functions $\hat{N}_5$,

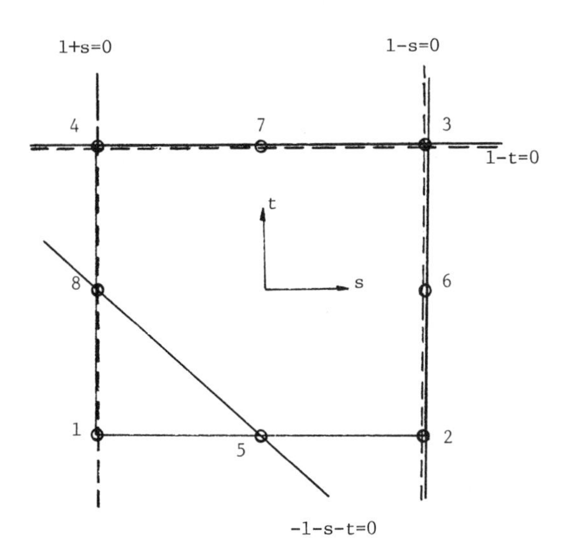

Figure 5.59 Eight-node serendipity element: solid lines for node 1, dashed for node 5.

$\hat{N}_6$, $\hat{N}_7$, and $\hat{N}_8$ for midnodes using the condition (5.227). Now we must find the shape functions for the corner nodes. There are several ways to do this, each of which utilizes some of the previously determined $\hat{N}$'s. For example, for the first node we may construct, using the fifth node,

$$\hat{N}_1^5(s, t) = \tfrac{1}{4}(1 - s)(1 - t) - \tfrac{1}{2}\hat{N}_5(s, t) = \tfrac{1}{4}s(s - 1)(1 - t)$$

The part $\frac{1}{2}s(s - 1)$ is nothing but the shape function along the line 1–5–2, and the part $\frac{1}{2}(1 - t)$ is similar to $\hat{N}_5$. It is clear that $\hat{N}_1^5$ is zero at 5, 2, 6, 3, 7, and 4. However, $\hat{N}_1^5 \neq 0$ at node 8. To remedy this, we modify $\hat{N}_1^5$ using $\hat{N}_8$ already defined:

$$\begin{aligned}\hat{N}_1(s, t) &= \hat{N}_1^5(s, t) - \tfrac{1}{2}\hat{N}_8(s, t)\\ &= \tfrac{1}{4}s(s - 1)(1 - t) - \tfrac{1}{4}(1 - s)(1 - t)^2\\ &= \tfrac{1}{4}(1 - s)(1 - t)(-1 - s - t)\end{aligned}$$

We can apply similar steps to obtain the remaining shape functions.

For the 12-node quadrilateral element shown in Figure 5.60, we have

$$\begin{aligned}\hat{N}_5(s, t) &= \tfrac{9}{32}(1 - 3s)(1 - s^2)(1 - t)\\ &\vdots\\ \hat{N}_{11}(s, t) &= \tfrac{9}{32}(1 - s)(1 + 3t)(1 - t^2)\\ \hat{N}_{12}(s, t) &= \tfrac{9}{32}(1 - s)(1 - 3t)(1 - t^2)\end{aligned}$$

and then

$$\begin{aligned}\hat{N}_1 &= \tfrac{1}{4}(1 - s)(1 - t) - \tfrac{2}{3}\hat{N}_5 - \tfrac{1}{3}\hat{N}_6 - \tfrac{2}{3}\hat{N}_{12} - \tfrac{1}{3}\hat{N}_{11}\\ &= \tfrac{1}{32}(1 - s)(1 - t)(-10 + 9s^2 + 9t^2)\end{aligned}$$

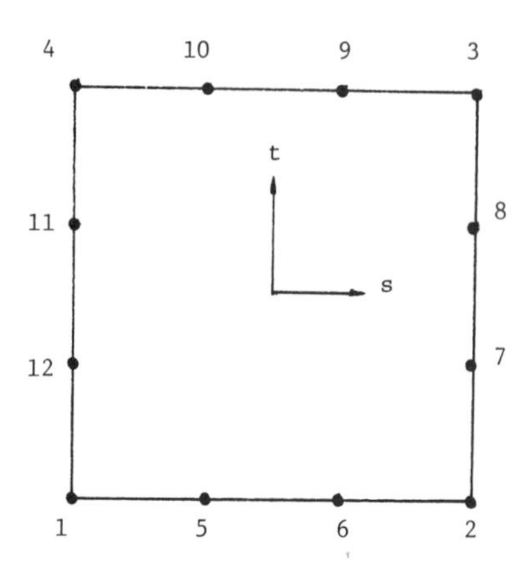

Figure 5.60 Twelve-node serendipity element.

The complete list of these shape functions is given as follows.

$$\hat{N}_1(s,t) = \tfrac{1}{32}(1-s)(1-t)(-10+9s^2+9t^2) \qquad \hat{N}_7(s,t) = \tfrac{9}{32}(1+s)(1-3t)(1-t^2)$$
$$\hat{N}_2(s,t) = \tfrac{1}{32}(1+s)(1-t)(-10+9s^2+9t^2) \qquad \hat{N}_8(s,t) = \tfrac{9}{32}(1+s)(1+3t)(1-t^2)$$
$$\hat{N}_3(s,t) = \tfrac{1}{32}(1+s)(1+t)(-10+9s^2+9t^2) \qquad \hat{N}_9(s,t) = \tfrac{9}{32}(1-3s)(1-s^2)(1+t)$$
$$\hat{N}_4(s,t) = \tfrac{1}{32}(1-s)(1+t)(-10+9s^2+9t^2) \qquad \hat{N}_{10}(s,t) = \tfrac{9}{32}(1-3s)(1-s^2)(1+t)$$
$$\hat{N}_5(s,t) = \tfrac{9}{32}(1-3s)(1-s^2)(1-t) \qquad \hat{N}_{11}(s,t) = \tfrac{9}{32}(1-s)(1+3t)(1-t^2)$$
$$\hat{N}_6(s,t) = \tfrac{9}{32}(1+3s)(1-s^2)(1-t) \qquad \hat{N}_{12}(s,t) = \tfrac{9}{32}(1-s)(1-3t)(1-t^2)$$

The 20-node brick element in $\mathbb{R}^3$ has the following shape functions:

$$\hat{N}_\alpha(r,s,t) = \tfrac{1}{8}(1+r_\alpha r)(1+s_\alpha s)(1+t_\alpha t)(-2+r_\alpha r+s_\alpha s+t_\alpha t) \quad (\text{no sum on } \alpha) \tag{5.228}$$

for $\alpha = 1, 2, \ldots, 8$, and

$$\begin{aligned}
\hat{N}_9(r,s,t) &= \tfrac{1}{4}(1-r^2)(1-s)(1-t) & \hat{N}_{15}(r,s,t) &= \tfrac{1}{4}(1-r^2)(1+s)(1+t)\\
\hat{N}_{10}(r,s,t) &= \tfrac{1}{4}(1+r)(1-s^2)(1-t) & \hat{N}_{16}(r,s,t) &= \tfrac{1}{4}(1-r)(1-s^2)(1+t)\\
\hat{N}_{11}(r,s,t) &= \tfrac{1}{4}(1-r^2)(1+s)(1-t) & \hat{N}_{17}(r,s,t) &= \tfrac{1}{4}(1-r)(1-s)(1-t^2)\\
\hat{N}_{12}(r,s,t) &= \tfrac{1}{4}(1-r)(1-s^2)(1-t) & \hat{N}_{18}(r,s,t) &= \tfrac{1}{4}(1+r)(1-s)(1-t^2)\\
\hat{N}_{13}(r,s,t) &= \tfrac{1}{4}(1-r^2)(1-s)(1+t) & \hat{N}_{19}(r,s,t) &= \tfrac{1}{4}(1+r)(1+s)(1-t^2)\\
\hat{N}_{14}(r,s,t) &= \tfrac{1}{4}(1+r)(1-s^2)(1+t) & \hat{N}_{20}(r,s,t) &= \tfrac{1}{4}(1-r)(1+s)(1-t^2)
\end{aligned} \tag{5.229}$$

where $(r_\alpha, s_\alpha, t_\alpha)$ are the coordinates of the α node, and the numbering system is shown in Figure 5.61.

Exercise 5.23: Change subroutines ESTIF0 and GGRAD4 in FEM2 using eight-node serendipity quadrilateral elements.

In the above, both the geometry of an element and a function are approximated by using the same shape functions $\hat{N}_\alpha(s,t)$, and we called such an element an *isoparametric element*. A variant of such an element is the *subparametric element*, such as

$$x = x_\alpha \hat{N}_\alpha^4(s,t), \qquad y = y_\alpha \hat{N}_\alpha^4(s,t), \qquad f = f_\alpha \hat{N}_\alpha^8(s,t) \tag{5.230}$$

where $\hat{N}_\alpha^4$ and $\hat{N}_\alpha^8$ are, respectively, the shape functions for the four- and eight-node serendipity quadrilateral elements. In this case, the geometry of an element is ap-

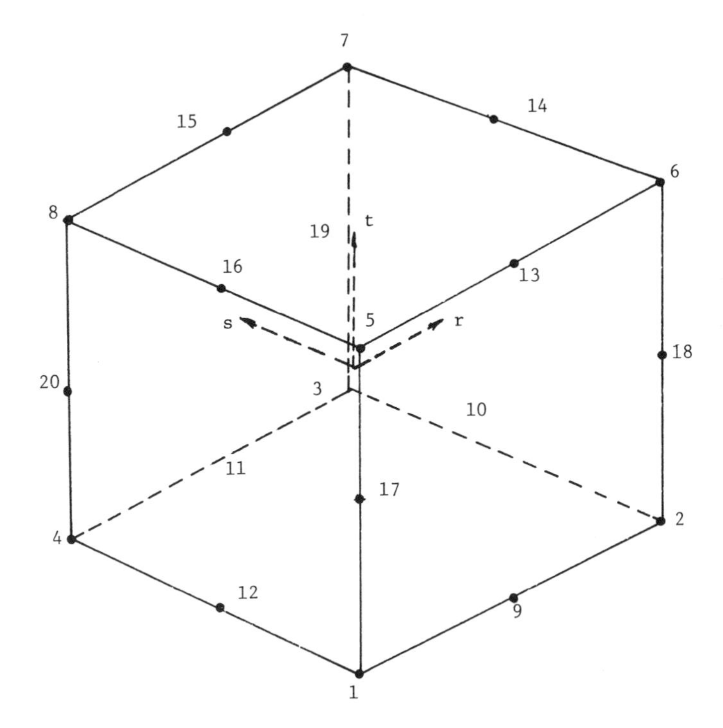

Figure 5.61 Twenty-node serendipity brick element.

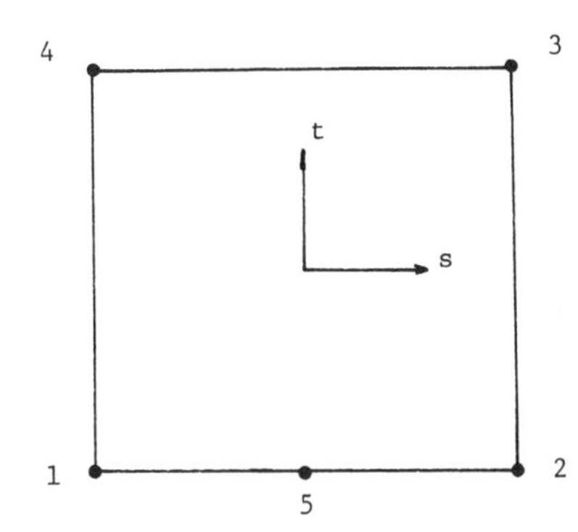

Figure 5.62 Five-node transition quadrilateral element in $\mathbb{R}^2$.

proximated by polynomials of degree lower than those used to approximate the function. If the "geometry polynomials" are of higher degrees than the "function polynomials," then we say that *superparametric* elements are being used.

In certain finite element models, we need to combine two different types of element. To do this, the five-node element shown in Figure 5.62 is introduced. Applying the first method to define the shape functions for the corner nodes of a serendipity element, the shape functions of this five-node element are obtained as

$$\begin{aligned}
&\hat{N}_1(s,t) = \tfrac{1}{4}(1-s)(1-t) - \tfrac{1}{2}\hat{N}_5(s,t) = \tfrac{1}{4}s(s-1)(1-t)\\
&\hat{N}_2(s,t) = \tfrac{1}{4}(1+s)(1-t) - \tfrac{1}{2}\hat{N}_5(s,t) = \tfrac{1}{4}s(s+1)(1-t)\\
&\hat{N}_3(s,t) = \tfrac{1}{4}(1+s)(1+t) \qquad \hat{N}_4(s,t) = \tfrac{1}{4}(1-s)(1+t)\\
&\hat{N}_5(s,t) = \tfrac{1}{2}(1-s^2)(1-t)
\end{aligned} \tag{5.231}$$

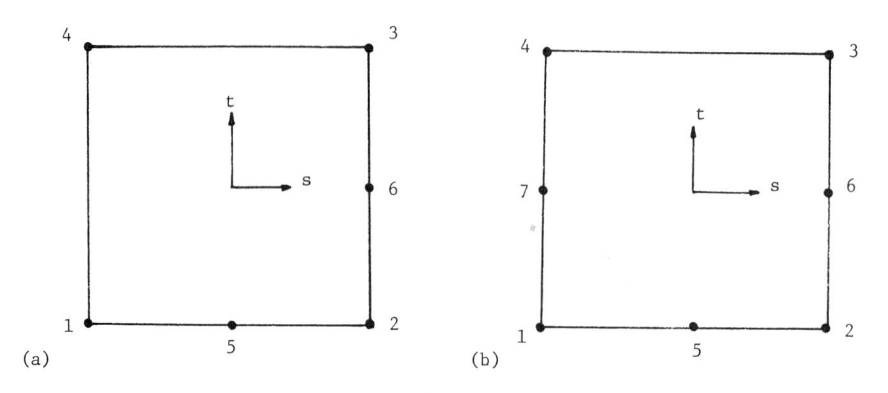

Figure 5.63 (a) Six- and (b) seven-node transition elements in $\mathbb{R}^2$.

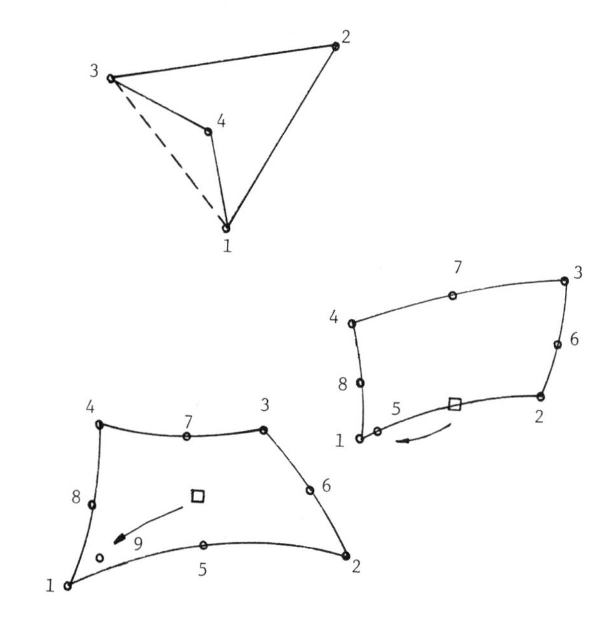

Figure 5.64 Location of midnodes and center node prohibited.

Exercise 5.24: Determine the shape functions for the six- and seven-node transition elements shown in Figure 5.63.

For higher-order elements, the location of the interior nodes is important. For example, the finite elements in the global coordinate system shown in Figure 5.64 may cause serious trouble when the element stiffness matrix is formed. Let us first look at the one-dimensional case. Suppose that three nodes are taken to form a line element and that the shape functions in the master element (segment) are

$$\begin{aligned} \hat{N}_1(s) &= \tfrac{1}{2}s(s-1) \\ \hat{N}_2(s) &= 1 - s^2 \qquad -1 \leq s \leq 1 \\ \hat{N}_3(s) &= \tfrac{1}{2}s(s+1) \end{aligned}$$

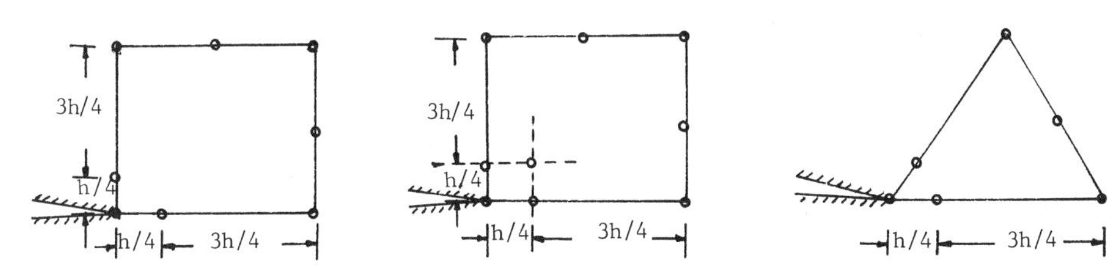

Figure 5.65 Location of nodes around crack tip.

Let the coordinates of these three nodes along the line in the global system be given as 0, ah, and h, where a is a parameter such that $0 < a < 1$, and h is the length of the element. If the line element is straight, then

$$d\xi = \xi_\alpha \frac{d\hat{N}_\alpha}{ds}(s)\,ds = (-2as + s + \tfrac{1}{2})h\,ds$$

Thus, the Jacobian of the transformation between the element in the global system and the master element is

$$J = (-2as + s + \tfrac{1}{2})h \tag{5.232}$$

Note that if a $> \frac{3}{4}$, there will be a point at which the Jacobian becomes negative. Thus, it is clear that the midnode should be located near the center of the element. With a corner node as a vertex, construct a square of side $\frac{1}{2}$. No other node should appear in or on this square. The first derivative of the interpolation of a function f with respect to the coordinate ξ along the (curved) line becomes

$$\frac{df^I}{d\xi} = J^{-1}\frac{df^I}{ds} = \frac{1}{[(1-2a)s + \frac{1}{2}]h} f_\alpha N'_\alpha(s) \tag{5.233}$$

This shows that as $a \to \frac{3}{4}$, $(df^I/d\xi)_{s=1} \to +\infty$. Thus, by taking $a = \frac{3}{4}$, we can simulate singular behavior of the first derivative of f at a point such that $df^I/d\xi \approx 0(1/\sqrt{1 - \xi/h})$. In this example, the first derivative of f is infinite at $s = 1$, that is, at the third node. What is the first derivative in the current stress analysis? It is related to the strain and stress tensor. From this, we suggest that if there is a reason to suspect singular behavior of the stress tensor at a corner node of an element, we may shift the location of the midnodes so that their distances from the corner exceeds one-fourth the element length by an arbitrary small amount. Thus, for example, at a crack tip, we may use six-, eight-, and nine-node elements with the special location of the midnodes as shown in Figure 5.65.

5.9.4 Interpolation with singularity

Suppose that a four-node quadrilateral element is considered for stress analysis in which singular behavior can be observed in the stress field. For simplicity, let the first node of the four-node quadrilateral element be the location of singularity, and let a component u of the displacement vector be approximately represented by $u = a_0 r^m$, for a constant a_0, where r is the distance of a point from the first node and m is a given power less than unity. If $m = \frac{1}{2}$, then application of the idea of shifting midnodes toward the singular point for quadratic elements (i.e., six-node triangular, eight- and nine-node quadrilateral elements) can reproduce the adequate singular behavior as shown in the above. Now we shall consider the case of an arbitrary power $m < 1$.

Applying the method described in Okabe (1981), we shall obtain interpolation functions for the geometry of the element while shape functions $\hat{N}_\alpha(s, t), \alpha = 1, \ldots, 4$, are the same as the one already introduced for displacement components. That is, we shall obtain shape functions $\hat{M}_\alpha(s, t)$ to interpolate coordinates x and y so that singular behavior can be reproduced, while displacements are interpolated by

$$\hat{N}_\alpha(s, t) = \tfrac{1}{4}(1 + s_\alpha t)(1 + t_\alpha t)$$

where (s_α, t_α) are coordinates of the four corner nodes in the normalized coordinate system.

Let us first introduce a nondimensionalized distance function ρ corresponding to the distance r from the first node:

$$\rho = 1 - \hat{M}_1(s, t) \tag{5.234}$$

where $\hat{M}_1(s, t)$ is the shape function of the first node such that

$$\hat{M}_1(s_\alpha, t_\alpha) = 0, \quad \alpha = 2, 3, 4, \qquad \text{and} \qquad \hat{M}_1(s_1, t_1) = 1$$

Then ρ is zero at the first node and unity along the opposite edges of the element to the first node.

Now, how can we provide the singularity $u = a_0 r^m$ in the vicinity of the first node? Under the assumption that ρ is proportional to r, this may be obtained by imposing the condition

$$\rho^m = \sum_{\alpha=2}^{4} \hat{N}_\alpha(s, t) \tag{5.235}$$

Indeed, if $u_1 = 0$ and $u_2 = u_3 = u_4 = a_1$ are assumed, we have

$$u = u_\alpha \hat{N}_\alpha(s, t) = a_1 \sum_{\alpha=2}^{4} \hat{N}_\alpha(s, t)$$

Then we have $u = a_1 \rho^m$. If ρ is proportional to r, then $u = a_0 r^m$, that is, this has reproduced the desired singularity. Equations (5.234) and (5.235) yield

$$\begin{aligned} \hat{M}_1(s, t) &= 1 - \left[\sum_{\alpha=2}^{4} \hat{N}_\alpha(s, t) \right]^{1/m} = 1 - [1 - \hat{N}_1(s, t)]^{1/m} \\ &= 1 - [\tfrac{1}{4}(3 + s + t - st)]^{1/m} \end{aligned} \tag{5.236}$$

Shape functions $\hat{M}_\alpha(s, t), \alpha = 2, 3, 4$, for other nodes can be obtained by the relation

$$\frac{\hat{M}_\alpha}{\rho} = \frac{\hat{N}_\alpha}{\rho^m}, \quad \alpha = 2, 3, 4 \tag{5.237}$$

These reflect the fact that the displacement interpolation should reproduce the singular behavior ρ^m at points whose distance is ρ. Since the distance from the singularity is interpolated by $M_\alpha, \alpha = 2, 3, 4$, the relation (5.237) can be introduced.

Thus, we have

$$\begin{aligned}\hat{M}_\alpha(s, t) &= \rho^{1-m}\hat{N}_\alpha(s, t)\\ &= [\tfrac{1}{4}(3 + s + t - st)]^{1/m-1}\hat{N}_\alpha(s, t)\end{aligned} \tag{5.238}$$

for $\alpha = 2, 3, 4$. It is clear that

$$\hat{M}_\alpha(s_\beta, t_\beta) = \delta_{\alpha\beta} \qquad \text{and} \qquad \sum_{\alpha=1}^{4} \hat{M}_\alpha(s, t) = 1$$

Using $\hat{M}_\alpha(s, t)$, $\alpha = 1, \ldots, 4$, we interpolate geometry by

$$x = x_\alpha \hat{M}_\alpha(s, t) \qquad \text{and} \qquad y = y_\alpha \hat{M}_\alpha(s, t)$$

Since

$$\frac{\partial \hat{M}_\alpha}{\partial s} = O(\rho^{1-m}) \qquad \text{and} \qquad \frac{\partial \hat{M}_\alpha}{\partial t} = O(\rho^{1-m})$$

for $\alpha = 1, \ldots, 4$, and since the following relation holds:

$$\begin{Bmatrix} \dfrac{\partial \hat{N}_\alpha}{\partial x} \\ \dfrac{\partial \hat{N}_\alpha}{\partial y} \end{Bmatrix} = \begin{bmatrix} \dfrac{\partial x}{\partial s} & \dfrac{\partial y}{\partial s} \\ \dfrac{\partial x}{\partial t} & \dfrac{\partial y}{\partial t} \end{bmatrix}^{-1} \begin{Bmatrix} \dfrac{\partial \hat{N}_\alpha}{\partial s} \\ \dfrac{\partial \hat{N}_\alpha}{\partial t} \end{Bmatrix}$$

we have

$$\frac{\partial u}{\partial x} = O(\rho^{m-1}) \qquad \text{and} \qquad \frac{\partial u}{\partial y} = O(\rho^{m-1})$$

That is, the first derivatives of displacement components have singularity ρ^{m-1}. Since $m < 1$ is already assumed, this implies that each component of the stress will be infinity at the first node, that is, stress singularity of $O(r^{m-1})$ can be reproduced by the above interpolation.

5.9.5 Hermite interpolation of a function

Another class of function interpolations is the *Hermite interpolation,* which uses nodal values of both the function *and* its first derivatives.

Let a function f be defined on an interval $I = [x_1, \ldots, x_n]$, and let

$$f_\alpha = f(x_\alpha), \qquad \theta_\alpha = f'(x_\alpha), \quad \alpha = 1, \ldots, n \tag{5.239}$$

Then we define Hermite interpolation by

$$f^I = f_\alpha H_\alpha^0(x) + \theta_\alpha H_\alpha^1(x) \tag{5.240}$$

where

$$\begin{aligned} H_\alpha^0(x_\beta) &= \delta_{\alpha\beta}, \qquad (H_\alpha^0)'(x_\beta) = 0 \\ H_\alpha^1(x_\beta) &= 0, \qquad (H_\alpha^1)'(x_\beta) = \delta_{\alpha\beta} \end{aligned} \tag{5.241}$$

If the basis polynomials L_α of Lagrange interpolation are known, then we have

$$\begin{aligned} H_\alpha^0(x) &= \{1 - 2(x - x_\alpha)L_\alpha'(x)\}\{L_\alpha(x)\}^2 \\ H_\alpha^1(x) &= (x - x_\alpha)\{L_\alpha(x)\}^2 \end{aligned} \qquad \text{(no sum on } \alpha) \tag{5.242}$$

Deflection at A and flexural stress at B in various two-element cantilever beams of aspect ratio 10/1 (not drawn to scale). Results are reported as the ratio of computed value to the value obtained by elementary beam theory (shear deformation neglected). Point B is a Gauss point of a 2×2 rule. Poisson's ratio = 0·30

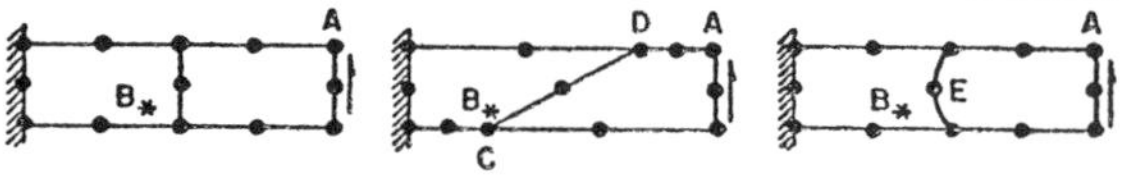

Element type	Gauss rule	e	β	v_A	σ_B	v_A	σ_B	v_A	σ_B
8 node	3×3	0	0	0·9301	1·1292	0·1610	0·0475	0·2212	0·0499
8 node	2×2	0	0	0·9678	0·9999	0·3623	0·0507	0·4301	−0·0476
9 node	3×3	0	0	0·9541	1·1405	0·7913	0·6867	0·7370	0·7051
9 node	2×2	0	0	1·0058	0·9999	1·1086	1·1252	0·9549	0·9583

Notes: 1. Points C and D are, respectively, one-quarter and three-quarters along the beam from the support. Side nodes are evenly spaced.

2. Point E is positioned left of centre an amount equal to one-twentieth of the length of the beam.

Figure 5.66 Comparison of eight- and nine-node elements in $\mathbb{R}^2$. *From Cook (1982).*

Note that we applied Hermite interpolation (5.240)–(5.242) to approximate the transverse deflection of the beam in Chapter 4. The idea of a tensor product may be to form shape functions for finite elements in $\mathbb{R}^2$ and $\mathbb{R}^3$. For example, the shape functions of the four-node element in $\mathbb{R}^2$ are

$$\begin{aligned} \hat{N}^1_\alpha(s,t) &= H^0_\alpha(s)H^0_\alpha(t) \\ \hat{N}^2_\alpha(s,t) &= H^1_\alpha(s)H^0_\alpha(t) \\ \hat{N}^3_\alpha(s,t) &= H^0_\alpha(s)H^1_\alpha(t) \\ \hat{N}^4_\alpha(s,t) &= H^1_\alpha(s)H^1_\alpha(t) \end{aligned} \qquad \text{(no sum on } \alpha) \tag{5.243}$$

for $\alpha = 1, \ldots, 4$. In this case, a function $f(x, y)$ is approximated by the form

$$f = f_\alpha \hat{N}^1_\alpha + \theta_{x\alpha}\hat{N}^2_\alpha + \theta_{y\alpha}\hat{N}^3_\alpha + \theta_{xy\alpha}\hat{N}^4_\alpha \tag{5.244}$$

where θ_x, θ_y, and θ_{xy} are abbreviations for $\partial f/\partial x$, $\partial f/\partial y$, and $\partial^2 f/\partial x\, \partial y$, respectively, and, for example, $\theta_{x\alpha} = (\partial f/\partial x)(x_\alpha, y_\alpha)$, and so on.

5.9.6 Speciality of higher-order elements

We have developed higher-order elements based on tensor product Lagrange interpolation, serendipity interpolation, and Hermite interpolation. One remark concerning these higher-order elements, especially the serendipity elements, is that if the geometry of the elements is significantly distorted from a rectangle in $\mathbb{R}^2$ (or a brick in $\mathbb{R}^3$), the quality of approximation deteriorates significantly while tensor product Lagrange elements can maintain their accuracy reasonably well, even for such distorted cases. An example that illustrates this fact can be found in the bending of a cantilever. If uniform, rectangular, eight-node serendipity elements are used in the problem, the numerical results are very accurate, as shown in Figure 5.66. However, if the elements are distorted as in Figure 5.66, the accuracy of the numerical solution drastically reduces to about 30%, whereas nine-node tensor product Lagrange elements still maintain an accuracy of more than 70%. This suggests that eight-node serendipity elements must be used with caution.

Table 5.6 *Weight and integration points of the Gaussian quadrature rule from Zienkiewicz (1978)*

$$\int_{-1}^{1} f(x)\,dx = \sum_{j=1}^{n} H_i f(a_j)$$

$\pm a$		H
	$n = 1$	
0		2.00000 00000 00000
	$n = 2$	
0.57735 02691 89626		1.00000 00000 00000
	$n = 3$	
0.77459 66692 41483		0.55555 55555 55556
0.00000 00000 00000		0.88888 88888 88889
	$n = 4$	
0.86113 63115 94053		0.34785 48451 37454
0.33998 10435 84856		0.65214 51548 62546
	$n = 5$	
0.90617 98459 38664		0.23692 68850 56189
0.53846 93101 05683		0.47862 86704 99366
0.00000 00000 00000		0.56888 88888 88889
	$n = 6$	
0.93246 95142 03152		0.17132 44923 79170
0.66120 93864 66265		0.36076 15730 48139
0.23861 91860 83197		0.46791 39345 72691
	$n = 7$	
0.94910 79123 42759		0.12948 49661 68870
0.74153 11855 99394		0.27970 53914 89277
0.40584 51513 77397		0.38183 00505 05119
0.00000 00000 00000		0.41795 91836 73469
	$n = 8$	
0.96028 98564 97536		0.10122 85362 90376
0.79666 64774 13627		0.22238 10344 53374
0.52553 24099 16329		0.31370 66458 77887
0.18343 46424 95650		0.36268 37833 78362
	$n = 9$	
0.96816 02395 07626		0.08127 43883 61574
0.83603 11073 26636		0.18064 81606 94857
0.61337 14327 00590		0.26061 06964 02935
0.32425 34234 03809		0.31234 70770 40003
0.00000 00000 00000		0.33023 93550 01260
	$n = 10$	
0.97390 65285 17172		0.06667 13443 08688
0.86506 33666 88985		0.14945 13491 50581
0.67940 95682 99024		0.21908 63625 15982
0.43339 53941 29247		0.26926 67193 09996
0.14887 43389 81631		0.29552 42247 14753

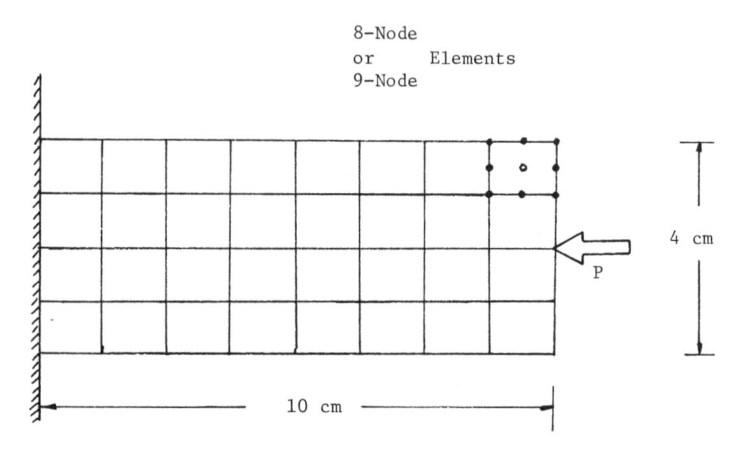

Figure 5.67 Test problem for reduced integration schemes: $E = 2.1 \times 10^4$ kN/cm^2, $\nu = 0.29$, $t = 1$ cm.

5.9.7 Gaussian quadrature rules

The final remark on higher-order elements and generalization of elements to $\mathbb{R}^3$ problems concerns the integration rule used to compute element stiffness matrices:

$$K^e_{i\alpha j\beta} = \int_{\Omega_e} E_{ikjl} N_{\alpha,k} N_{\beta,l}\, d\Omega$$

$$= \begin{cases} \displaystyle\sum_{p=1}^{M} \sum_{q=1}^{N} w_p w_q (E_{ikjl} N_{\alpha,k} N_{\beta,l} J)(s_p, t_q) \\ \quad \text{for } \mathbb{R}^2 \text{ problems} \\ \displaystyle\sum_{p=1}^{M} \sum_{q=1}^{N} \sum_{a=1}^{L} w_p w_q w_a (E_{ikjl} N_{\alpha,k} N_{\beta,l} J)(r_p, s_q, t_a) \\ \quad \text{for } \mathbb{R}^3 \text{ problems} \end{cases}$$

where w_p, w_q, and w_a are weights; (s_p, t_q) and (r_p, s_q, t_a) are integration points; and M, N, and L are the numbers of integration points in each direction (see Table 5.6). As shown in section 5.4, for four-node elements in plane problems, $M = N = 2$, that is, the 2×2 Gaussian integration rule was applied. For eight-node elements in $\mathbb{R}^3$ problems, $M = N = L = 2$, that is, the $2 \times 2 \times 2$ Gaussian rule must be applied. In practice, the number of integration points in each direction is the same as the number of nodes in that direction. Thus, if the element shown in Figure 5.57 is used for $\mathbb{R}^2$ problems, $M = 4$ and $N = 3$, that is, the 4×3 integration should be applied.

Exercise 5.25: Apply the reduced integration scheme to the eight-node serendipity and the nine-node tensor product Lagrange elements in order to evaluate stiffness matrices. That is, apply the 2×2 Gaussian integration rule instead of the 3×3 Gaussian rule. Then obtain the deformation of the body shown in Figure 5.67.

5.10 Navier–Stokes flow problems

Although fluid mechanics is quite different from solid mechanics, they are quite close to each other as far as finite element approximations are concerned. Indeed, if we introduce the correspondences:

Solids	Fluids
Displacement $\mathbf{u}$	Velocity $\mathbf{u}$
Strain tensor $\boldsymbol{\varepsilon}$	Rate of deformation $\mathbf{d}$
Mean stress σ_m	Hydrostatic pressure p

it is easy to understand how a computer program for a finite element approximation of a Navier–Stokes problem can be obtained by modifying the program FEM2. To see this, we first write the *Navier–Stokes problem* for the flow of an incompressible fluid:

$$\left.\begin{aligned} &\rho \dot{u}_i = \frac{\partial \sigma_{ji}}{\partial x_j} + \rho f_i \\ &\sigma_{ij} = -p\delta_{ij} + 2\mu d_{ij}, \qquad d_{ij} = \frac{1}{2}\left(\frac{\partial u_i}{\partial x_j} + \frac{\partial u_j}{\partial x_i}\right) \\ &\frac{\partial u_i}{\partial x_i} = 0 \end{aligned}\right\} \text{ in } \Omega \tag{5.245}$$

$$\begin{aligned} u_i &= g_i \quad \text{on } \Gamma_{i1} \\ \sigma_{ji} n_j &= h_i \quad \text{on } \Gamma_{i2} \\ \sigma_{ji} n_j &= -k_{ij}(u_j - u_{\infty j}) \quad \text{on } \Gamma_{i3} \end{aligned}$$

with the initial condition

$$u_i = u_{0i} \quad \text{in } \Omega \text{ at } t = 0 \tag{5.246}$$

Here the acceleration $\dot{u}_i$ is given by

$$\dot{u}_i = \frac{\partial u_i}{\partial t} + u_j \frac{\partial u_i}{\partial x_j} \tag{5.247}$$

If we assume a slight compressibility,* then the continuity equation $\partial u_i/\partial x_i = 0$ (or div $\mathbf{u} = 0$) should be modified to read

$$\frac{1}{\lambda} p + \frac{\partial u_i}{\partial x_i} = 0 \tag{5.248}$$

or

$$\frac{1}{\lambda}\frac{\partial p}{\partial t} + \frac{\partial u_i}{\partial x_i} = 0 \tag{5.249}$$

If the form (5.248) is assumed for a sufficiently large $\lambda > 0$, the constitutive and the continuity equation in (5.245) can be combined to give

$$\sigma_{ij} = \lambda d_{kk}\delta_{ij} + 2\mu d_{ij} \tag{5.250}$$

* Only small Mach number flows are considered in this book so that there is no possibility of shock forming.

In this case, the problem has the exact same form as the isotropic linear elastic solid studied above.

5.10.1 Newton–Raphson method

Because of the *convective term* $u_j \, \partial u_i/\partial x_j$ in the material derivative, our problem is nonlinear even for Newtonian fluid flow. Thus, we need to introduce a method for solving nonlinear problems. In this section, let us describe the *Newton–Raphson method* for solving the problem:

$$F(x) = 0 \tag{5.251}$$

The method involves three basic steps:

(i) Assume an initial approximation x^0.
(ii) Knowing the nth approximation of the solution x^n, compute the increment Δx^{n+1} by

$$F(x^n + \Delta x^{n+1}) \doteq F(x^n) + F'(x^n)\,\Delta x^{n+1} = 0$$

that is

$$F'(x^n)\,\Delta x^{n+1} = -F(x^n) \tag{5.252}$$

(iii) If $\|\Delta x^{n+1}\|$ is small enough, $x^{n+1} = x^n + \Delta x^{n+1}$ will be the solution. Otherwise, repeat step (ii) until a sufficiently small increment is obtained.

For the steady-state problem, the Navier–Stokes problem (5.245) becomes

$$\rho u_j \frac{\partial u_i}{\partial x_j} = \frac{\partial}{\partial x_j} \sigma_{ji}(\mathbf{u}) + \rho f_i$$

where $\sigma_{ji}(\mathbf{u})$ is the stress tensor evaluated at the level $\mathbf{u}$ of the velocity. Replacing $\mathbf{u}$ by $\mathbf{u}^n + \Delta\mathbf{u}^{n+1}$ yields

$$\rho(u_j^n + \Delta u_j^{n+1}) \frac{\partial}{\partial x_j}(u_i^n + \Delta u_i^{n+1}) = \frac{\partial}{\partial x_j} \sigma_{ji}(\mathbf{u}^n + \Delta\mathbf{u}^{n+1}) + \rho f_i$$

and then

$$\rho\left(u_j^n \frac{\partial}{\partial x_j} \Delta u_i^{n+1} + \Delta u_j^{n+1} \frac{\partial}{\partial x_j} u_i^n + \Delta u_j^{n+1} \frac{\partial}{\partial x_j} \Delta u_i^{n+1}\right)$$
$$= \frac{\partial}{\partial x_j} \sigma_{ji}(\Delta\mathbf{u}^{n+1}) - \rho u_j^n \frac{\partial}{\partial x_j} u_i^n + \frac{\partial}{\partial x_j} \sigma_{ji}(\mathbf{u}^n) + \rho f_i$$

Neglecting the higher-order term, $\Delta u_j^{n+1}\, \partial\, \Delta u_i^{n+1}/\partial x_j$, we have

$$\rho\left(u_j^n \frac{\partial}{\partial x_j} \Delta u_i^{n+1} + \Delta u_j^{n+1} \frac{\partial}{\partial x_j} u_i^n\right) = \frac{\partial}{\partial x_j} \sigma_{ji}(\Delta\mathbf{u}^{n+1}) + R_i \tag{5.253}$$

where

$$R_i = -\rho u_j^n \frac{\partial}{\partial x_j} u_i^n + \frac{\partial}{\partial x_j} \sigma_{ji}(\mathbf{u}^n) + \rho f_i$$

After solving the *linearized* boundary value problem for $\Delta\mathbf{u}^{n+1}$, we can obtain the solution of the original nonlinear boundary value problem as the limit of the sequence $\mathbf{u}^n = \mathbf{u}^{n-1} + \Delta\mathbf{u}^n$. If the initial approximation $\mathbf{u}^0$ of the velocity is taken to be the solution of the problem

$$0 = \frac{\partial}{\partial x_j}[\sigma_{ji}(\mathbf{u}^0)] + \rho f_i$$
$$u_i^0 = g_i \quad \text{on } \Gamma_{i1}, \qquad \sigma_{ji}(\mathbf{u}^0)n_j = h_i \quad \text{on } \Gamma_{i2} \tag{5.254}$$
$$\sigma_{ji}(\mathbf{u}^0)n_j = -k_{ij}(u_j^0 - u_{\infty j}) \quad \text{on } \Gamma_{i3}$$

then the linearized problem for the increment Δu_i^{n+1} becomes

$$\rho\left(u_j^n \frac{\partial\, \Delta u_i^{n+1}}{\partial x_j} + \Delta u_j^{n+1}\frac{\partial u_i^n}{\partial x_j}\right) = \frac{\partial}{\partial x_i}(\sigma_{ji}(\Delta\mathbf{u}^{n+1})) + R_i$$
$$\Delta u_i^{n+1} = 0 \quad \text{on } \Gamma_{i1} \tag{5.255}$$
$$\sigma_{ji}(\Delta u_k^{n+1})n_j = 0 \quad \text{on } \Gamma_{i2}$$
$$\sigma_{ji}(\Delta u_k^{n+1})n_j = -k_{ij}\,\Delta u_j^{n+1} \quad \text{on } \Gamma_{i3}$$

for $n = 0, 1, 2, 3, \ldots$. Here $\mathbf{R} = R_i\mathbf{i}_i$ is the residual of the approximation defined by (5.253).

5.10.2 Finite element approximation

Let us only consider a finite element approximation of the linearized problem (5.255), since the problem (5.254) for the initial approximation $\mathbf{u}^0$ is basically the linear elasticity problem considered already. To simplify the notation, let us change (5.255) to read

$$\rho\left(w_j\frac{\partial u_i}{\partial x_j} + u_j\frac{\partial w_i}{\partial x_j}\right) = \frac{\partial}{\partial x_j}[\sigma_{ji}(\mathbf{u})] + R_i$$
$$u_i = 0 \quad \text{on } \Gamma_{i1}, \qquad \sigma_{ji}(\mathbf{u})n_j = 0 \quad \text{on } \Gamma_{i2} \tag{5.256}$$
$$\sigma_{ji}(\mathbf{u})n_j = -k_{ij}u_j \quad \text{on } \Gamma_{i3}$$

by making the identifications

$$\mathbf{u}^n = \mathbf{w}, \qquad \Delta\mathbf{u}^{n+1} = \mathbf{u}$$

Then the weak form is obtained as follows:

$$u_i = 0 \quad \text{on } \Gamma_{i1}$$
$$\int_\Omega \rho(w_ju_{i,j}\bar{u}_i + u_jw_{i,j}\bar{u}_i)\,d\Omega + \int_\Omega \sigma_{ji}(u)\bar{u}_{i,j}\,d\Omega + \int_{\Gamma_{i3}} k_{ij}u_j\bar{u}_i\,d\Gamma \tag{5.257}$$
$$= \int_\Omega R_i\bar{u}_i\,d\Omega, \quad \forall\bar{\mathbf{u}} \ni \bar{u}_i = 0 \text{ on } \Gamma_{i1}$$

Applying the constitutive equation (5.250), the term $\sigma_{ji}\bar{u}_{i,j}$ becomes

$$\sigma_{ji}(\mathbf{u})\bar{u}_{i,j} = (\lambda u_{k,k}\delta_{ij} + \mu u_{i,j} + \mu u_{j,i})\bar{u}_{i,j}$$
$$= \lambda u_{j,j}\bar{u}_{i,i} + \mu(u_{i,j}\bar{u}_{i,j} + u_{j,i}\bar{u}_{i,j})$$

The next step is the discretization of the domain and the velocity. Suppose that

$$u_j = u_{j\beta}N_\beta(\mathbf{x}), \qquad \bar{u}_i = \bar{u}_{i\alpha}N_\alpha(\mathbf{x}) \tag{5.258}$$

within an element Ω_e. Then

$$\int_\Omega \rho(w_ju_{i,j}\bar{u}_i + u_jw_{i,j}\bar{u}_i)\,d\Omega = \sum_{e=1}^{E}\bar{u}_{i\alpha}(C_{i\alpha j\beta}^{1,e} + C_{i\alpha j\beta}^{2,e})u_{j\beta}$$

where

$$C^{1,e}_{i\alpha j\beta} = \int_{\Omega_e} \rho \delta_{ij} w_k N_\alpha N_{\beta,k}\, d\Omega$$
$$C^{2,e}_{i\alpha j\beta} = \int_{\Omega_e} \rho w_{i,j} N_\alpha N_\beta\, d\Omega \tag{5.259}$$

The known terms w_k and $w_{i,j}$ may be computed by

$$w_k = w_{k\gamma} N_\gamma \quad \text{and} \quad w_{i,j} = w_{i\gamma} N_{\gamma,j}$$

If the matrices $[E^e_{i\alpha j\beta}]$ and $[E^{e,3}_{i\alpha j\beta}]$ and the vector $\{f^e_{i\alpha}\}$ are defined by

$$E^e_{i\alpha j\beta} = \int_{\Omega_e} \{\lambda N_{\alpha,i} N_{\beta,j} + \mu(N_{\alpha,j} N_{\beta,i} + \delta_{ij} N_{\alpha,k} N_{\beta,k})\}\, d\Omega$$
$$E^{3,e}_{i\alpha j\beta} = \int_{\Gamma^e_{i3}} k_{ij} N_\alpha N_\beta\, d\Gamma, \qquad f^e_{i\alpha} = \int_{\Omega_e} R_i N_\alpha\, d\Omega \tag{5.260}$$

the weak form is discretized as

$$u_{i\alpha} = 0 \quad \text{on } \Gamma_{i1}$$
$$\sum_{e=1}^{E} \bar{u}_{i\alpha}(C^{1,e}_{i\alpha j\beta} + C^{2,e}_{i\alpha j\beta} + E^e_{i\alpha j\beta}) u_{j\beta} + \sum_{e=1}^{E_3} \bar{u}_{i\alpha} E^{3,e}_{i\alpha j\beta} u_{j\beta}$$
$$= \sum_{e=1}^{E} \bar{u}_{i\alpha} f^e_{i\alpha}, \quad \forall \bar{u}_i \ni \bar{u}_{i\alpha} = 0 \text{ on } \Gamma_{i1} \tag{5.261}$$

Then the finite element equation becomes

$$u_{i\alpha} = 0 \quad \text{on } \Gamma_{i1}$$
$$\sum_{e=1}^{E} (C^{1,e}_{i\alpha j\beta} + C^{2,e}_{i\alpha j\beta} + E^e_{i\alpha j\beta}) u_{j\beta} + \sum_{e=1}^{E_3} E^{3,e}_{i\alpha j\beta} u_{j\beta} = \sum_{e=1}^{E} f^e_{i\alpha} \tag{5.262}$$

There is a difference between (5.262) and the finite element equation (5.62) for linear elasticity. Here, we have nonsymmetric element stiffness matrices due to the convective term. Thus, we cannot use the symmetric skyline solver since it requires symmetric matrices.

A special remark regarding the above formulation is that it may be necessary to introduce the upwind technique when the convective term dominates the flow field. In this case, in order to reduce the significant effect of the term

$$\int_\Omega \rho w_j u_{i,j} \bar{u}_i\, d\Omega$$

to the quality of approximate solutions, we add an artificial viscosity term to the left side of the second equation in (5.257) (see section 3.1). For the four-node element, the introduction of artificial viscosity yields the additional term

$$E^{av,e}_{i\alpha j\beta} = \int_{-1}^{1} \int_{-1}^{1} \mu^{av,e}_{kl} \frac{\partial N_\alpha}{\partial s_k} \frac{\partial N_\beta}{\partial s_l} J\, d\hat{\Omega}\, \delta_{ij} \tag{5.263}$$

where $s_1 = s$, $s_2 = t$, and $\mu^{av,e}_{kl}$ is the artificial viscosity defined by

$$\mu^{av,e}_{kl} = \alpha_0 \rho w_k w_l / |\mathbf{w}| \tag{5.264}$$

Here $|\mathbf{w}|$ is the magnitude of the velocity vector $\mathbf{w}$, and α_0 is a parameter to be defined by numerical experiments.

5.10.3 Time-dependent cases

If transient problems must be analyzed, we may "deal with" the nonlinear convection term by applying a sort of θ-method (see section 3.2). Indeed, the nonlinear equation

$$\rho\left(\frac{\partial u_i}{\partial t} + u_j \frac{\partial u_i}{\partial x_j}\right) = \frac{\partial}{\partial x_j}\,[\sigma_{ji}(\mathbf{u})] + f_i \tag{5.265}$$

may be discretized in the time direction as

$$\rho\,\frac{u_i^{n+1} - u_i^n}{\Delta t} + \theta u_j^{n+1}\frac{\partial u_i^n}{\partial x_j} + (1-\theta)u_j^n \frac{\partial u_i^{n+1}}{\partial x_j}$$

$$= \hat{\theta}\frac{\partial}{\partial x_j}\,[\sigma_{ji}(\mathbf{u}^{n+1})] + (1-\hat{\theta})\frac{\partial}{\partial x_j}\,[\sigma_{ji}(\mathbf{u}^n)] + f_i \tag{5.266}$$

where the two parameters θ and $\hat{\theta}$ are such that

$$0 \le \theta \le 1 \qquad \text{and} \qquad 0 \le \hat{\theta} \le 1 \tag{5.267}$$

Thus, each time step involves the solution of linear equations.

Exercise 5.26: In the formulation, described traction is specified on the boundary Γ_2. Now we shall consider the boundary condition

$$p = \hat{p} \qquad \text{and} \qquad \frac{\partial \mathbf{u}}{\partial n} = 0 \quad \text{on } \Gamma_2 \tag{5.268}$$

where $\hat{p}$ is a specified hydrostatic pressure and $\partial/\partial n$ is the normal derivative along the boundary Γ_2. It is clear that the above condition reflects the situation that the flow field becomes constant in the normal direction at the boundary Γ_2 and is a representation of the boundary condition for flow problems defined in an infinite domain but modeled in a finite domain such as Couette's flow. Assuming that μ *is constant* and

$$\mathbf{u} = \mathbf{g} \quad \text{on } \Gamma_1 \tag{5.269}$$

derive the weak form

$$\int_\Omega \left(\rho u_j \frac{\partial u_i}{\partial x_j}\bar{u}_i - p\bar{u}_{i,i} + \mu u_{i,j}\bar{u}_{i,j}\right) d\Omega$$

$$= \int_\Omega \rho f_i \bar{u}_i \, d\Omega + \int_{\Gamma_2} (-\hat{p}) n_i \bar{u}_i \, d\Gamma \tag{5.270}$$

for every $\bar{\mathbf{u}}$ such that $\bar{\mathbf{u}} = \mathbf{0}$ on Γ_1. Here stationary problems are considered together with the continuity equation

$$u_{i,i} = 0 \quad \text{in } \Omega$$

due to incompressibility of fluid.

5.10.4 Remarks on the penalty method

In the above, incompressibility $[u_{i,i}\,(=\text{div}\,\mathbf{u}) = 0]$ is treated by the penalty method, which multiplies the modified continuity equation (5.248) or (5.249) with a sufficiently large parameter λ. The advantage of this method is that the hydrostatic pressure p is eliminated from the governing equations. Thus, the only unknown in the formulation is the velocity vector $\mathbf{u}$. As shown in section 5.6.4, the technique of selective reduced integration must be used to form element stiffness matrices. Indeed, (5.260) is divided into two parts:

$$\begin{aligned} E^e_{i\alpha j\beta} &= E^{e,1}_{i\alpha j\beta} + E^{e,2}_{i\alpha j\beta} \\ E^{e,1}_{i\alpha j\beta} &= \int_{\Omega_e} \lambda N_{\alpha,i} N_{\beta,j}\, d\Omega \\ E^{e,2}_{i\alpha j\beta} &= \int_{\Omega_e} \mu (N_{\alpha,j} N_{\beta,i} + \delta_{ij} N_{\alpha,k} N_{\beta,k})\, d\Omega \end{aligned} \tag{5.271}$$

Then the first part, $E^{e,1}_{i\alpha j\beta}$, is integrated by a Gaussian quadrature rule whose order is one less than that used for the second part, $E^{e,2}_{i\alpha j\beta}$. More precisely, if four-node quadrilateral elements are used for plane problems, $E^{e,1}_{i\alpha j\beta}$ is integrated by one-point Gaussian quadrature, while $E^{e,2}_{i\alpha j\beta}$ is computed by the 2×2 Gaussian rule. If an eight- or nine-node quadrilateral element is used, $E^{e,1}_{i\alpha j\beta}$ is integrated via the 2×2 Gaussian rule, whereas $E^{e,2}_{i\alpha j\beta}$ is obtained by using 3×3 Gaussian quadrature. If six-node triangular elements are used for plane problems, we need not use the reduced integration for the first parts, $E^{e,1}_{i\alpha j\beta}$, since the number of constraints due to the incompressibility $u_{i,i} = 0$ is, in general, considerably less than the number of degrees of freedom in a finite element model. In fact, a six-node element has three constraints due to the incompressibility compared to a total of 12 degrees of freedom, while seven and eight constraints are introduced for an eight- and a nine-node element, respectively, if both parts are integrated by 3×3 Gaussian quadrature without using reduced integration. Applying the selective reduced integration, we reduce the number of constraints to four in an eight- and a nine-node element.

It is noted that the distribution of the hydrostatic pressure p computed by (5.248) or (5.249) at the reduced integration points (after the velocity field $\mathbf{u}$ is obtained) may have small local oscillations that, in general, decay as the mesh size becomes small. To "deal with" these oscillations, we may apply the scheme (5.140), which computes the "average" value of the stress at nodes as its value at the element's centroid.

If the least-squares method (5.140) is not applied, the following method may be applicable to obtain smooth distribution of the pressure. According to mathematical theories of mixed finite element approximations of Stokes flow problems (Johnson and Pickaranta, 1982), the composite element of four four-node elements shown in Figure 5.18 has convergence property under the constraint on the pressure distribution p on the element:

$$p_1 A_1 - p_2 A_2 + p_3 A_3 - p_4 A_4 = 0 \tag{5.272}$$

where p_α and A_α are the value of the pressure and the area of the αth subelement. Since the pressure p_λ computed by the penalty method

$$p_\lambda = -\lambda u_{i,i}$$

may not satisfy the condition (5.272), the following modified pressure $\hat{p}_\lambda$ will be identified as the finite element approximation:

$$\hat{p}_{\lambda 1} = p_{\lambda 1} - B/A_1, \quad \hat{p}_{\lambda 2} = p_{\lambda 2} + B/A_2, \quad \hat{p}_{\lambda 3} = p_{\lambda 3} - B/A_3, \quad \hat{p}_{\lambda 4} = p_4 + B/A_4$$

when

$$B = p_{\lambda 1} A_1 - p_{\lambda 2} A_2 + p_{\lambda 3} A_3 - p_{\lambda 4} A_4$$

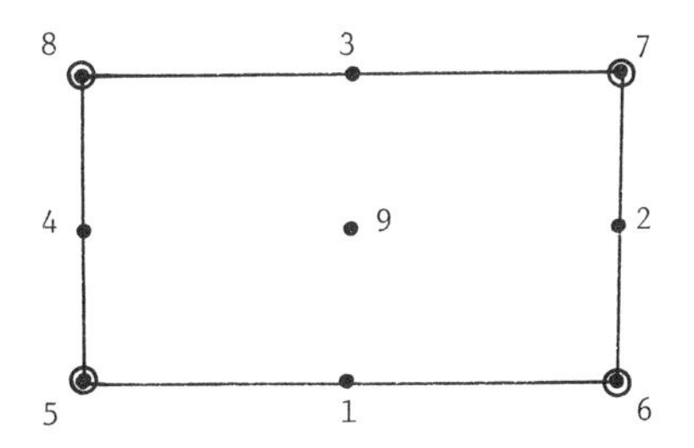

Figure 5.68 Element for mixed method: ○, pressure; ●, velocity.

It is clear that the modified pressure $\hat{p}_\lambda$ satisfies the condition (5.272) for the convergence of the mixed formulation, which is "equivalent" to the penalty formulation described above.*

Exercise 5.27: Modify the program FEM2 so that steady-state Navier–Stokes flow problems in the plane can be solved using four-node elements.

Exercise 5.28: Applying the mixed method introduced in section 5.6.4, develop a computer code that solves steady-state Navier–Stokes flow problems in the plane in terms of both the velocity $\mathbf{u}$ and the pressure p. Use the element shown in Figure 5.68 in which the velocity $\mathbf{u}$ and the pressure p are approximated by biquadratic and bilinear polynomials, respectively.

A final remark to Navier–Stokes flow problems: The use of the above finite element approximations should be restricted to rather low-velocity viscous flow problems, although they provide approximate solutions even for a large Reynolds number if artificial viscosity is applied. Also, some modification of the Newton–Raphson procedure, using the idea of incremental methods, may be necessary to obtain convergence of the iteration method for a large Reynolds number. In any case, physics of fluid mechanics has to be understood before using finite element approximations developed on the analogy of linear elasticity.

Example 5.10: Let stationary Navier–Stokes flow be considered on the two-dimensional domain shown in Figure 5.69. We shall apply penalty finite element methods using four-node elements to discretize the Navier–Stokes equations, the form of which is

$$\mathbf{u} \cdot \nabla \mathbf{u} = -\frac{1}{\rho} \nabla p + \nu \nabla \cdot \nabla \mathbf{u}, \qquad \nabla \cdot \mathbf{u} = 0$$

where $\nu = \mu/\rho$ is assumed to be a constant.

Let the entrance velocity be unity, that is, $\mathbf{u} = u_x\mathbf{i} + u_y\mathbf{j}$, $u_x = 1$ m/s, and $u_y = 0$, and let the stress-free condition be assumed at the exit. The velocity fields shown in Figures 5.70–5.73 are obtained for the finite element model of the flow domain given in Figure 5.69. As the viscosity ν becomes smaller, the number of iterations of the Newton–Raphson method increases to obtain its convergence. Here the artificial viscosity is not applied.

* Details of equivalence relations of penalty and mixed formulations are discussed in Reddy (1982), which also contains many mathematical developments of penalty methods.

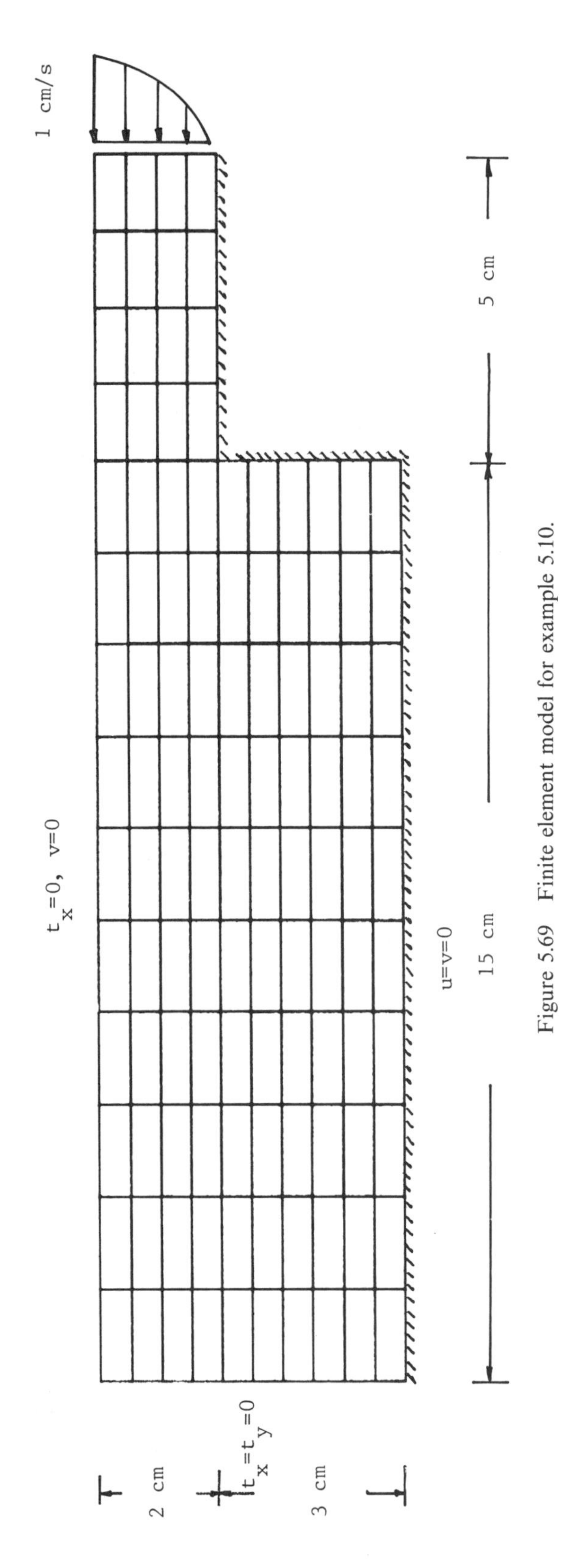

Figure 5.69 Finite element model for example 5.10.

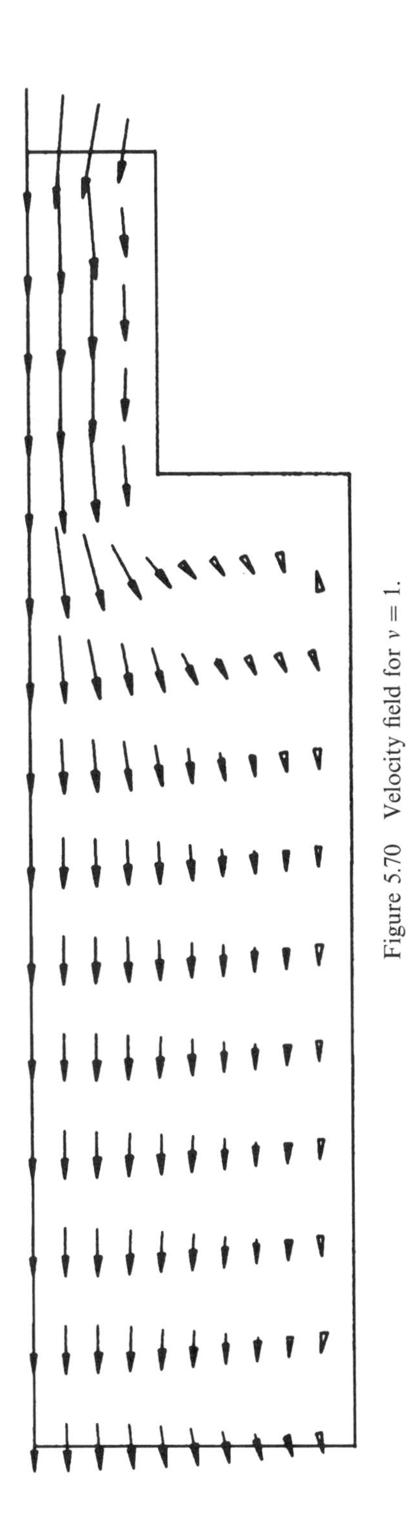

Figure 5.70 Velocity field for $\nu = 1$.

Figure 5.71 Velocity field for $\nu = 0.1$.

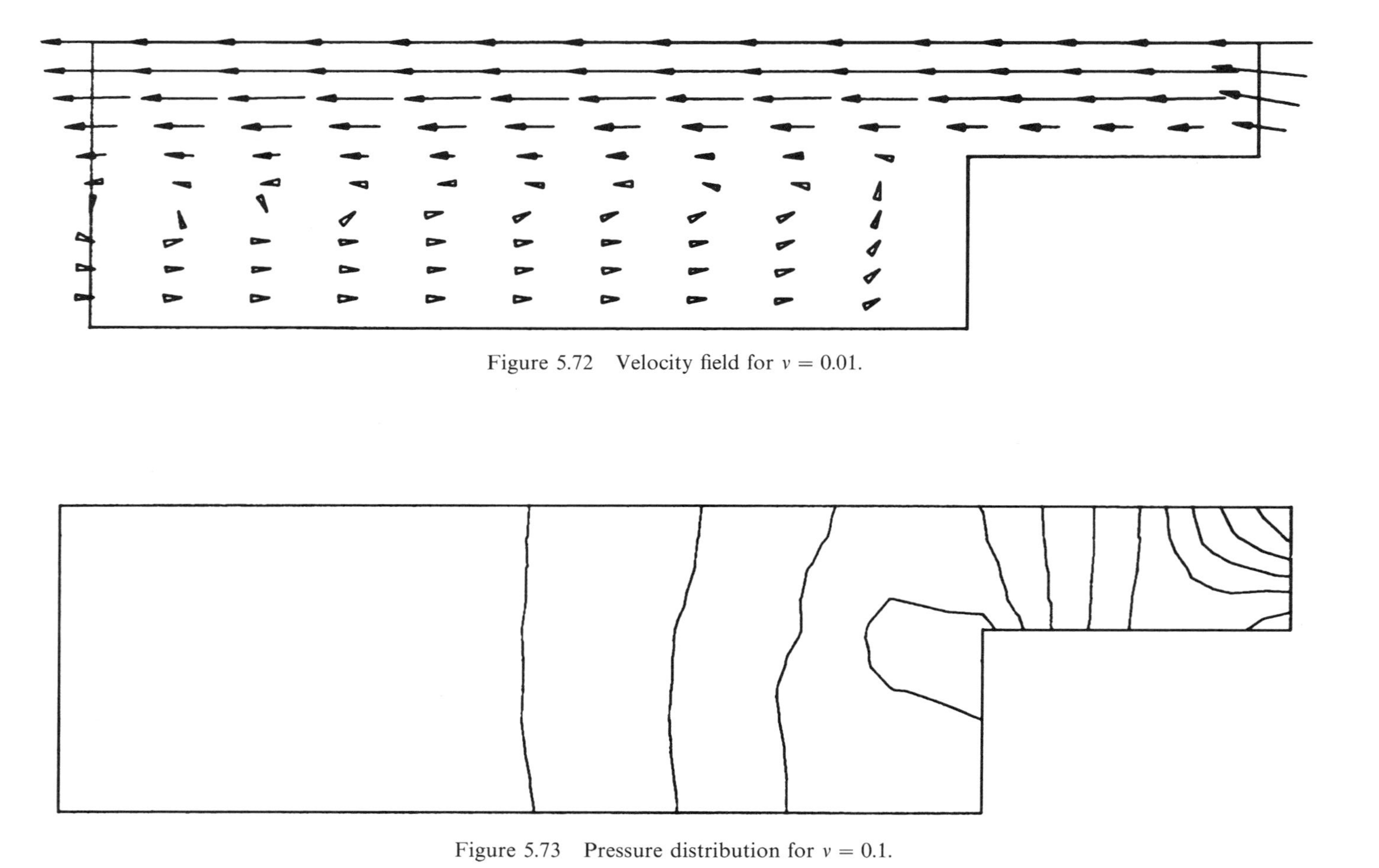

Figure 5.72 Velocity field for $\nu = 0.01$.

Figure 5.73 Pressure distribution for $\nu = 0.1$.

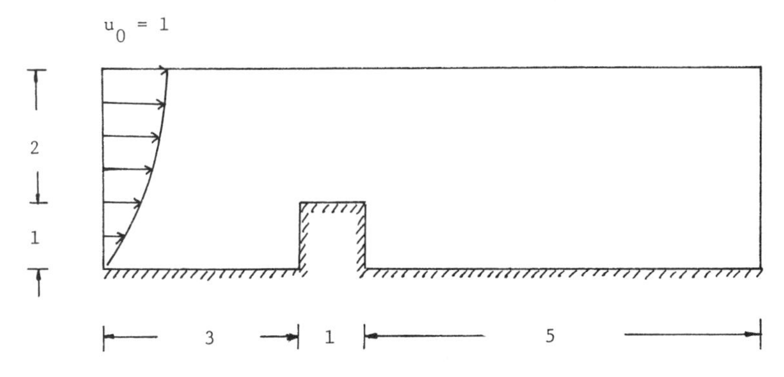

Figure 5.74 Flow domain for exercise 5.29.

Exercise 5.29: Solve the same problem of example 5.10 on the domain shown in Figure 5.74. Discuss if the artificial viscosity must be applied.

References

Belytschko, T., Liu, W. K., and Kennedy, J. M., "Hourglass control in linear and nonlinear problems," in Atluri, S., and Perrone, N. (eds.), *Computer methods for nonlinear solids and structures,* ASME/AMD vol. 54, American Society of Mechanical Engineers, New York, 1983, pp. 37–63.

Ciarlet, P. G., *The finite element method of elliptic problems,* North-Holland, Amsterdam, 1978, chap. 2.

Cook, R. D., and Feng, Z. H., "Control of spurious modes in the nine-node quadrilateral element," *International Journal of Numerical Methods and Engineering* 18: 1576–80, 1982.

Diaz, A. R., Kikuchi, N., and Taylor, J. E., "A method of grid optimization for finite element methods," *Computer Methods in Applied Mechanics and Engineering* 41: 29–45, 1983.

Dwyer, T., and Critchfield, M., *BASIC and the personal computer,* Addison-Wesley, Reading, Massachusetts, 1978.

Fung, Y. C., *Foundations of solids mechanics,* Prentice-Hall, Englewood Cliffs, New Jersey, 1965.

Hetény, M., *Handbook of Experimental stress analysis,* Wiley, New York, 1950. Johnson, C., and Pickaranta, J., "Analysis of some mixed methods related to reduced integration," *Mathematics of Computation* 38: 375–400, 1982.

Johnson, C., and Pickaranta, J., "Analysis of some mixed finite element methods related to reduced integration," *Mathematics of Computation* 38: 375–400, 1982.

Lawn, B. R., and Wilshaw, T. R., *Fracture of brittle solids,* Cambridge University Press, Cambridge, England, 1975, p. 59.

Okabe, M., "Fundamental theory of the semi-radial singularity mapping with applications to fracture mechanics," *Computer Methods in Applied Mechanics and Engineering* 26: 53–73, 1981.

Reddy, J. N. (ed.), *Penalty-finite element methods,* ASME/AMD vol. 51, American Society of Mechanical Engineers, New York, 1982.

Timoshenko, S. P., and Goodier, J. N., *Theory of elasticity,* 3rd ed., McGraw-Hill, New York, 1970.

Zienkiewicz, O. C., *The finite element method,* 3rd ed., McGraw-Hill, London, 1978.

6 PLATE-BENDING PROBLEMS

Another popular application of finite element methods in classical elasticity is the bending of thin plate structures. A thin plate is a solid body bounded by a cylindrical surface (or surfaces), called the edge (or edges) of the plate and by two parallel planes (normal to the generators of the cylindrical surface), called the faces of the plate. The distance h between the faces is called the thickness of the plate and is relatively small compared to lateral dimensions. In the following discussion, we shall study plates of an isotropic, linearly elastic material, under the assumption of small deformation such that $\|u\| \ll h$.

6.1 Basic plate equations

Let the xy plane be the middle surface of the plate (the plane midway between the faces), and let the z axis be normal to this surface. We will consider two formulations of the *plate equations*. The first involves the Kirchhoff approximations in which it is assumed that normals to the plate's midsurface, before deformation, remain normal to *this* surface after deformation and experience no length changes. Exaggerated views (in the xz and yz planes) of the deformation are shown in Figures 6.1(a), (b). Consider a normal through $(x, y, 0)$ and a point on this normal at (x, y, z) (shown as a solid circle in the following figures). In the deformed state, the projections of this normal into the xz and yz coordinate planes are induced to the vertical by the (small) angles $\partial w/\partial x$ and $\partial w/\partial y$ as shown. The displacement components of the typical point at x, y, z are

$$
\begin{aligned}
u_x(x, y, z) &\doteq u - z\frac{\partial w}{\partial x} \\
u_y(x, y, z) &\doteq v - z\frac{\partial w}{\partial y} \qquad (6.1)\\
u_z(x, y, z) &\doteq w
\end{aligned}
$$

where $u(x, y)$, $v(x, y)$ and $w(x, y)$ are the displacement components of the points at $(x, y, 0)$, in the x, y, and z directions, respectively. From (6.1), we find the strain components to be

$$
\begin{aligned}
\varepsilon_{xx} &= \frac{\partial u_x}{\partial x} = \frac{\partial u}{\partial x} - z\frac{\partial^2 w}{\partial x^2} \\
\varepsilon_{yy} &= \frac{\partial u_y}{\partial y} = \frac{\partial v}{\partial y} - z\frac{\partial^2 w}{\partial y^2}
\end{aligned}
$$

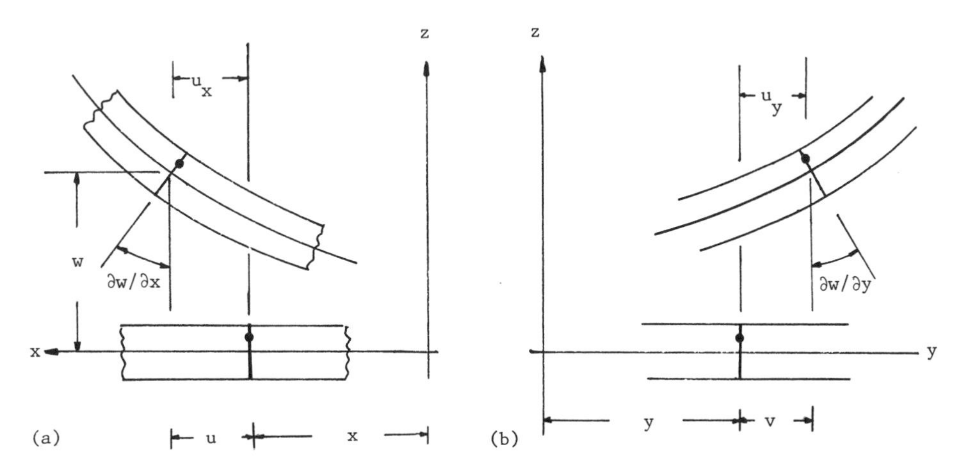

Figure 6.1 Cross sections of Kirchhoff's plate.

$$
\begin{aligned}
\varepsilon_{zz} &= \frac{\partial u_z}{\partial z} \equiv 0 \\
\varepsilon_{xy} &= \frac{1}{2}\left(\frac{\partial u_x}{\partial y} + \frac{\partial u_y}{\partial x}\right) = \frac{1}{2}\left(\frac{\partial u}{\partial y} + \frac{\partial v}{\partial x}\right) - z\frac{\partial^2 w}{\partial x\,\partial y} \\
\varepsilon_{xz} &= \frac{1}{2}\left(\frac{\partial u_x}{\partial z} + \frac{\partial u_z}{\partial x}\right) \equiv 0 \\
\varepsilon_{yz} &= \frac{1}{2}\left(\frac{\partial u_y}{\partial z} + \frac{\partial u_z}{\partial y}\right) \equiv 0
\end{aligned} \tag{6.2}
$$

Hooke's law and the last two equations of (6.2) give

$$\sigma_{xz} = \sigma_{yz} \equiv 0$$

It is customary to neglect the normal stress σ_{zz} on the grounds that it is small compared to the other normal stresses. We shall do this. Then from Hooke's law again, we find that the nonzero stress components are

$$
\begin{aligned}
\sigma_{xx} &= \frac{E}{1-\nu^2}(\varepsilon_{xx} + \nu\varepsilon_{yy}) \\
&= \frac{E}{1-\nu^2}\left[\frac{\partial u}{\partial x} + \nu\frac{\partial v}{\partial y} - z\left(\frac{\partial^2 w}{\partial x^2} + \nu\frac{\partial^2 w}{\partial y^2}\right)\right] \\
\sigma_{yy} &= \frac{E}{1-\nu^2}(\nu\varepsilon_{xx} + \varepsilon_{yy}) \\
&= \frac{E}{1-\nu^2}\left[\nu\frac{\partial u}{\partial x} + \frac{\partial v}{\partial y} - z\left(\nu\frac{\partial^2 w}{\partial x^2} + \frac{\partial^2 w}{\partial y^2}\right)\right] \\
\sigma_{xy} &= 2\mu\varepsilon_{xy} \\
&= \mu\left(\frac{\partial u}{\partial y} + \frac{\partial v}{\partial x} - 2z\frac{\partial^2 w}{\partial x\,\partial y}\right)
\end{aligned} \tag{6.3}
$$

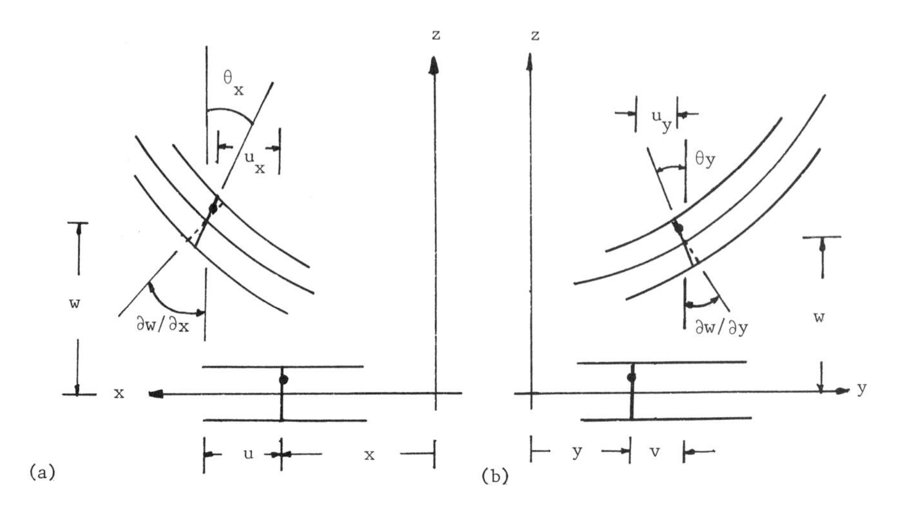

Figure 6.2 Cross sections of Mindlin's plate.

Before making any further use of (6.3), we describe another formulation of the plate problem. This leads to the equations that govern the behavior of the *Mindlin plate.* As in the Kirchhoff plate, it is assumed that normals (to the plate's midsurface) remain straight and experience no length changes. The Mindlin formulation differs from that of Kirchhoff in that these normals are not assumed to remain normal. Instead they are allowed an "additional rotation" that, at the expense of a more complicated theory, results in the shear stresses σ_{xz} and σ_{yz} being nonzero. In other words, in a Mindlin plate the effects of shear are taken into account. Figures 6.2(a), (b) similar to those in Figure 6.1, show an exaggerated view of the deformation. In this case, the displacement components are

$$\begin{aligned} u_x(x, y, z) &\doteq u - z\theta_x \\ u_y(x, y, z) &\doteq v - z\theta_y \\ u_z(x, y, z) &\doteq w \end{aligned} \tag{6.4}$$

where $u(x, y)$, $v(x, y)$, and $w(x, y)$ have the same interpretation as they did in the Kirchhoff formulation. The angles $\theta_x(x, y)$ and $\theta_y(x, y)$ are as shown in the figures.

From (6.4), we find the strain components to be

$$\begin{aligned} &\varepsilon_{xx} = \frac{\partial u}{\partial x} - z\frac{\partial \theta_x}{\partial x}, \qquad \varepsilon_{yy} = \frac{\partial v}{\partial y} - z\frac{\partial \theta_y}{\partial y}, \qquad \varepsilon_{zz} \equiv 0 \\ &\varepsilon_{xy} = \frac{1}{2}\left[\frac{\partial u}{\partial y} + \frac{\partial v}{\partial x} - z\left(\frac{\partial \theta_x}{\partial y} + \frac{\partial \theta_y}{\partial x}\right)\right], \qquad \varepsilon_{xz} = \frac{1}{2}\left(\frac{\partial w}{\partial x} - \theta_x\right) \\ &\varepsilon_{yz} = \frac{1}{2}\left(\frac{\partial w}{\partial y} - \theta_y\right) \end{aligned} \tag{6.5}$$

We assume, as before, that σ_{zz} is negligible, and then (6.5) and Hooke's law give

$$\begin{aligned}
\sigma_{xx} &= \frac{E}{1-\nu^2}\left[\frac{\partial u}{\partial x} + \nu\frac{\partial v}{\partial y} - z\left(\frac{\partial \theta_x}{\partial x} + \nu\frac{\partial \theta_y}{\partial y}\right)\right] \\
\sigma_{yy} &= \frac{E}{1-\nu^2}\left[\nu\frac{\partial u}{\partial x} + \frac{\partial v}{\partial y} - z\left(\nu\frac{\partial \theta_x}{\partial x} + \frac{\partial \theta_y}{\partial y}\right)\right] \\
\sigma_{xy} &= \mu\left[\frac{\partial u}{\partial y} + \frac{\partial v}{\partial x} - z\left(\frac{\partial \theta_x}{\partial y} + \frac{\partial \theta_y}{\partial x}\right)\right] \\
\sigma_{xz} &= \mu\left(\frac{\partial w}{\partial x} - \theta_x\right), \qquad \sigma_{yz} = \mu\left(\frac{\partial w}{\partial y} - \theta_y\right)
\end{aligned} \tag{6.6}$$

We introduce the *in-plane* resultants of the stresses σ_{xx}, σ_{xy}, and σ_{yy} by defining

$$\begin{aligned}
N_{xx} &= \int \sigma_{xx}\, dz = \frac{Eh}{1-\nu^2}\left(\frac{\partial u}{\partial x} + \nu\frac{\partial v}{\partial y}\right) \\
N_{xy} &= \int \sigma_{xy}\, dz = \mu h\left(\frac{\partial u}{\partial y} + \frac{\partial v}{\partial x}\right) \\
N_{yy} &= \int \sigma_{yy}\, dz = \frac{Eh}{1-v^2}\left(\nu\frac{\partial u}{\partial x} + \frac{\partial v}{\partial y}\right)
\end{aligned} \tag{6.7}$$

Similarly, we introduce moment resultants defined by

$$\begin{aligned}
M_{xx} &= \int z\sigma_{xx}\, dx = -D\left(\frac{\partial \theta_x}{\partial x} + \nu\frac{\partial \theta_y}{\partial y}\right) \\
M_{yy} &= \int z\sigma_{yy}\, dz = -D\left(\nu\frac{\partial \theta_x}{\partial x} + \frac{\partial \theta_y}{\partial y}\right) \\
M_{xy} &= \int z\sigma_{xy}\, dz = -D\frac{1-\nu}{2}\left(\frac{\partial \theta_x}{\partial y} + \frac{\partial \theta_y}{\partial x}\right)
\end{aligned} \tag{6.8}$$

where M_{xx} and M_{yy} are called the *bending moments*, M_{xy} the *twisting moment*, and D the *bending rigidity* defined by

$$D = \frac{Eh^3}{12(1-\nu^2)} \tag{6.9}$$

Finally, we introduce the transverse shear force resultants Q_x and Q_y by

$$\begin{aligned}
Q_x &= \int \sigma_{xz}\, dz = k\mu h\left(\frac{\partial w}{\partial x} - \theta_x\right) \\
Q_y &= \int \sigma_{yz}\, dz = k\mu h\left(\frac{\partial w}{\partial y} - \theta_y\right)
\end{aligned} \tag{6.10}$$

The factor k in (6.10) is a "fudge factor" that attempts to take into account the actual nonuniform distribution of shear stress in the thickness direction. [Note from the last two of (6.6) that these stresses are independent of z and hence are uniform in the thickness direction.] Reissner suggests $\frac{5}{6}$ as a reasonable value for k. Mindlin's choice is $k = \pi^2/12$.

If we neglect the effect of the transverse shear forces by taking $Q_x = Q_y = 0$, then (6.10) implies that $\theta_x = \partial w/\partial x$ and $\theta_y = \partial w/\partial y$. Then (6.4) reduces to (6.1) and we recover the Kirchhoff formulation.

We now recall the equilibrium equations of a deformable body:

$$\sigma_{ji,j} + \rho \hat{f}_i = 0 \tag{6.11}$$

where $\hat{\mathbf{f}} = \hat{f}_i \mathbf{i}_i$ is the body force per unit mass.

If we integrate (6.11) in the thickness direction, we find

$$\begin{aligned} \frac{\partial N_{xx}}{\partial x} + \frac{\partial N_{yx}}{\partial y} + f_x = 0 \\ \frac{\partial N_{xy}}{\partial x} + \frac{\partial N_{yy}}{\partial y} + f_y = 0 \end{aligned} \tag{6.12}$$

and

$$\frac{\partial Q_x}{\partial x} + \frac{\partial Q_y}{\partial y} + q = 0 \tag{6.13}$$

where

$$\begin{aligned} f_x = \int \rho \hat{f}_x \, dz + [\sigma_{zx}]_{-h/2}^{h/2} \\ f_y = \int \rho \hat{f}_y \, dz + [\sigma_{zy}]_{-h/2}^{h/2} \end{aligned} \tag{6.14}$$

and

$$q = \int \rho \hat{f}_z \, dz + [\sigma_{zz}]_{-h/2}^{h/2} \tag{6.15}$$

Similarly, if we multiply the first two of (6.11) by z and then integrate over the thickness, we obtain

$$\begin{aligned} \frac{\partial M_{xx}}{\partial x} + \frac{\partial M_{yx}}{\partial y} - Q_x = -m_x \\ \frac{\partial M_{xy}}{\partial x} + \frac{\partial M_{yy}}{\partial y} - Q_y = -m_y \end{aligned} \tag{6.16}$$

where

$$\begin{aligned} m_x = \int z\rho \hat{f}_x \, dz + \tfrac{1}{2}h[\sigma_{zx}(\tfrac{1}{2}h) + \sigma_{zx}(-\tfrac{1}{2}h)] \\ m_y = \int z\rho \hat{f}_y \, dz + \tfrac{1}{2}h[\sigma_{zy}(\tfrac{1}{2}h) + \sigma_{zy}(-\tfrac{1}{2}h)] \end{aligned} \tag{6.17}$$

It is noted that the equations in (6.12) are decoupled from (6.13) and (6.16) and constitute a plane stress problem. Thus, these must be solved independently of the others. The governing equations for plate bending are therefore (6.13) and (6.16).

We now state the boundary conditions, which are analogous to those that arise in beam bending. One such boundary condition is the clamped-edge condition

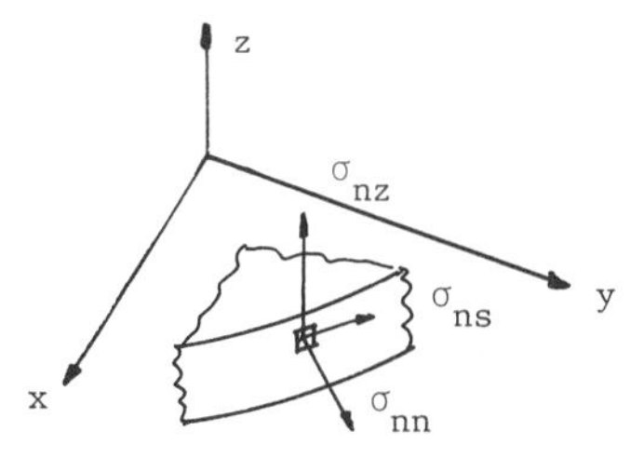

Figure 6.3 Stress distribution on edge.

described by

$$u_x = 0, \qquad u_y = 0, \qquad \text{and} \qquad u_z = 0 \tag{6.18a}$$

on the edge. Applying the approximations (6.4) yields

$$w = \theta_x = \theta_y = 0 \qquad \text{and} \qquad u = v = 0 \tag{6.18b}$$

As in beam-bending theory, the deflection w and the rotations (or slopes) θ_x and θ_y are the primitive variables whose prescription forms the first type (i.e., essential) boundary condition because of (6.18a). If the boundary Γ of the middle surface is an arbitrary closed curve in $\mathbb{R}^2$, the clamped-edge condition on Γ must be given with respect to the normal direction. Indeed, let $\mathbf{n} = n_i \mathbf{i}_i$ be the unit vector outward normal to the boundary Γ. Then, in addition to $u_z = 0$, the condition requires $\mathbf{n} \cdot \mathbf{u} = 0$, which in view of (6.4) reads

$$n_x u + n_y v - z(n_x \theta_x + n_y \theta_y) = 0$$

Since this must hold for all z in $(-\frac{1}{2}h, \frac{1}{2}h)$, we see that both $n_x u + n_y v$ and $n_x \theta_x + n_y \theta_y$ must vanish. Thus, (6.18b) is replaced by

$$n_x u + n_y v = w = n_x \theta_x + n_y \theta_y = 0 \tag{6.18c}$$

For the *simply supported* case, the deflection w must be zero and the edge must be *moment free*. To explain the concept of a moment-free edge, we first introduce the normal and shear stresses acting on an edge. Let these stresses be denoted by σ_{nn}, σ_{ns}, and σ_{nz}, as shown in Figure 6.3. We define the resultants M_{nn}, M_{ns}, and, for future reference, Q_n as follows:

$$M_{nn} = \int_{-h/2}^{h/2} z\sigma_{nn}\, dz, \qquad M_{ns} = \int_{-h/2}^{h/2} z\sigma_{ns}\, dz, \qquad Q_n = \int_{-h/2}^{h/2} \sigma_{nz}\, dz \tag{6.19}$$

The view looking down onto the piece of edge material is as shown in Figure 6.3, and an enlarged view of the state of stress "at" the shaded edge point is shown in Figure 6.4. The usual Mohr's-circle-type relations then give

$$\begin{aligned} \sigma_{nn} &= \sigma_{xx} n_x^2 + \sigma_{yy} n_y^2 + 2\sigma_{xy} n_x n_y \\ &= n_i \sigma_{ij} n_j \\ \sigma_{ns} &= -(\sigma_{xx} - \sigma_{yy}) n_x n_y + \sigma_{xy}(n_x^2 - n_y^2) \end{aligned} \tag{6.20}$$

where $n_x = \cos\theta$ and $n_y = \sin\theta$. In view of (6.20), the first two of (6.19) give

$$\begin{aligned} M_{nn} &= M_{xx} n_x^2 + M_{yy} n_y^2 + 2M_{xy} n_x n_y \\ &= n_i M_{ij} n_j \\ M_{ns} &= -(M_{xx} - M_{yy}) n_x n_y + M_{xy}(n_x^2 - n_y^2) \end{aligned} \tag{6.21}$$

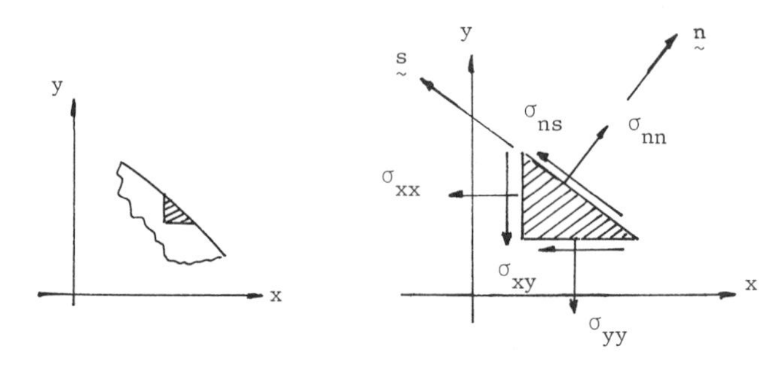

Figure 6.4 Stress on edge.

By moment free, we mean that the left sides of (6.21) are to be zero. Hence, for a simply supported edge, we require

$$w = M_{nn} = M_{ns} = 0 \tag{6.22}$$

Clearly, the relation between σ_{nz} and the shear stresses σ_{xz} and σ_{yz} is

$$\sigma_{nz} = \sigma_{xz}n_x + \sigma_{yz}n_y$$

so the last of (6.19) gives

$$Q_n = Q_x n_x + Q_y n_y \tag{6.23}$$

Now we can state that for a *free edge*, we must have

$$M_{nn} = M_{ns} = Q_n = 0 \tag{6.24}$$

It should be noted that if there is a corner point on Γ, the decomposition into normal and tangential moment components is not unique, since the normal vector n is not defined at such a point.*

6.2 Weak form for the Mindlin plate

Let us use a combination of index notation and regular notation. To do this, we make the identifications

$$\begin{array}{llll} M_{xx} = M_{11}, & M_{xy} = M_{12}, & M_{yx} = M_{21}, & M_{yy} = M_{22} \\ Q_x = Q_1, & Q_y = Q_2, & m_x = m_1, & m_y = m_2 \\ n_x = n_1, & n_y = n_2, & \dfrac{\partial}{\partial x}(\ ) = (\ )_{,1}, & \dfrac{\partial}{\partial y}(\ ) = (\ )_{,2} \end{array}$$

Then the equilibrium equations (6.16) and (6.13) become

$$\begin{aligned} M_{ji,j} - Q_i &= -m_i \\ -Q_{i,i} &= q \end{aligned} \tag{6.25}$$

* At corner points, *corner forces* often arise. If P is the magnitude of such a force at an xy corner, then $P = 2[M_{xy}]_{\text{corner}}$.

Let us also introduce the rotation "vector" $\boldsymbol{\theta} = \mathbf{i}\theta_x + \mathbf{j}\theta_y$. Then the boundary conditions suggested from the standard choices (6.18), (6.19), (6.22), and (6.24) might be given by

$$\begin{aligned} w &= w^1, & \theta_n &= \theta_n^1, & \theta_s &= \theta_s^1 \quad \text{on } \Gamma_1 \\ w &= w^2, & M_{nn} &= M_{nn}^2, & M_{ns} &= M_{ns}^2 \quad \text{on } \Gamma_2 \\ \theta_n &= \theta_n^3, & \theta_s &= \theta_s^3, & \theta_n &= Q_n^3 \quad \text{on } \Gamma_3 \\ M_{nn} &= M_{nn}^4, & M_{ns} &= M_{ns}^4, & Q_n &= Q_n^4 \quad \text{on } \Gamma_4 \end{aligned} \tag{6.26}$$

where the functions with superscripts are given on each part of the boundary Γ, $\theta_n = \boldsymbol{\theta} \cdot \mathbf{n} = \theta_i n_i$, and $\theta_s = \boldsymbol{\theta} \cdot \mathbf{s} = -\theta_x n_y + \theta_y n_x$.

The weak form of (6.25) and (6.26) is obtained by multiplying the virtual rotation $\bar{\theta}_i$ and the virtual deflection $\bar{w}$ to (6.25) and by integrating over the domain Ω. Indeed, applying the integration by parts, we have

$$\begin{aligned} &-\int_\Omega M_{ji}\bar{\theta}_{i,j}\, d\Omega + \int_\Gamma (M_{ji}n_j)\bar{\theta}_i\, d\Gamma - \int_\Omega Q_i\bar{\theta}_i\, d\Omega \\ &+ \int_\Omega Q_i\bar{w}_{,i}\, d\Omega - \int_\Gamma Q_i n_i \bar{w}\, d\Gamma = -\int_\Omega m_i\bar{\theta}_i\, d\Omega + \int_\Omega q\bar{w}\, d\Omega \end{aligned}$$

Using the decomposition rule,

$$(M_{ji}n_j)\bar{\theta}_i = M_{nn}\bar{\theta}_n + M_{ns}\bar{\theta}_s$$

where $\bar{\theta}_n = \bar{\theta}_i n_i$, $\bar{\boldsymbol{\theta}}_s = \bar{\boldsymbol{\theta}} - \bar{\theta}_n\mathbf{n}$, and $\bar{\boldsymbol{\theta}} = \bar{\theta}_i\mathbf{i}_i$, and applying the boundary conditions (6.26), we have

$$\begin{aligned} &-\int_\Omega M_{ji}\bar{\theta}_{i,j}\, d\Omega - \int_\Omega Q_i\bar{\theta}_i\, d\Omega + \int_\Omega Q_i\bar{w}_{,i}\, d\Omega \\ &\quad = -\int_\Omega m_i\bar{\theta}_i\, d\Omega + \int_\Omega q\bar{w}\, d\Omega - \int_{\Gamma_2} (M_{nn}^2\bar{\theta}_n + M_{ns}^2\bar{\theta}_s)\, d\Gamma \\ &\quad + \int_{\Gamma_3} Q_n^3\bar{w}\, d\Gamma - \int_{\Gamma_4} (M_{nn}^4\bar{\theta}_n + M_{ns}^4\bar{\theta}_s - Q_n^4\bar{w})\, d\Gamma \end{aligned} \tag{6.27}$$

for every $\bar{w}$ and $\bar{\boldsymbol{\theta}}$ such that

$$\begin{aligned} &\bar{w} = 0 \quad \text{on } \Gamma_1 \text{ and } \Gamma_2, \qquad \bar{\theta}_n = 0 \quad \text{on } \Gamma_1 \text{ and } \Gamma_3 \\ &\bar{\theta}_s = 0 \quad \text{on } \Gamma_1 \text{ and } \Gamma_3 \end{aligned} \tag{6.28}$$

Let moduli D_{ijkl} be defined by

$$\begin{aligned} &D_{1111} = D, \qquad D_{1122} = D\nu, \qquad D_{2211} = D\nu, \qquad D_{2222} = D \\ &D_{1212} = D_{1221} = D_{2112} = D_{2121} = \tfrac{1}{2}D(1-\nu) \end{aligned} \tag{6.29}$$

and all other terms are zero. Then (6.8) may be written as

$$M_{ij} = -D_{ijkl}\theta_{k,l} \tag{6.30}$$

The relation (6.10), in index notation, reads

$$Q_i = k\mu h(w_{,i} - \theta_i) \tag{6.31}$$

Substitution of (6.30) and (6.31) into (6.27) yields

$$\begin{aligned}\int_\Omega D_{jikl}\theta_{k,l}\bar{\theta}_{i,j}\,d\Omega - \int_\Omega k\mu h(w_{,i}-\theta_i)\bar{\theta}_i\,d\Omega \\ + \int_\Omega k\mu h(w_{,i}-\theta_i)\bar{w}_{,i}\,d\Omega \\ = -\int_\Omega m_i\bar{\theta}_i\,d\Omega - \int_{\Gamma_2}(M_{nn}^2\bar{\theta}_n + M_{ns}^2\bar{\theta}_s)\,d\Gamma \\ - \int_{\Gamma_4}(M_{nn}^4\bar{\theta}_n + M_{ns}^4\bar{\theta}_s)\,d\Gamma + \int_\Omega q\bar{w}\,d\Omega \\ + \int_{\Gamma_3} Q_n^3\bar{w}\,d\Gamma + \int_{\Gamma_4} Q_n^4\bar{w}\,d\Gamma \quad (n, s\text{: no sum})\end{aligned} \tag{6.32}$$

Since the right side of the weak form (6.32) depends on the boundary and load conditions, let us introduce two linear functionals, $L_1(\bar{\boldsymbol{\theta}})$ and $L_2(\bar{w})$, in terms of which arbitrary boundary and load conditions can be described. Then the weak form (6.32) can be written as

$$\begin{aligned}\int_\Omega D_{jikl}\theta_{k,l}\bar{\theta}_{i,j}\,d\Omega - \int_\Omega k\mu h(w_{,i}-\theta_i)\bar{\theta}_i\,d\Omega \\ + \int_\Omega k\mu h(w_{,i}-\theta_i)\bar{w}_{,i}\,d\Omega = L_1(\bar{\boldsymbol{\theta}}) + L_2(\bar{w})\end{aligned} \tag{6.33}$$

If (6.26) is considered, we have

$$\begin{aligned}L_1(\bar{\boldsymbol{\theta}}) = -\int_\Omega m_i\bar{\theta}_i\,d\Omega - \int_{\Gamma_2}(M_{nn}^2\bar{\theta}_n + M_{ns}^2\bar{\theta}_s)\,d\Gamma \\ - \int_{\Gamma_4}(M_{nn}^4\bar{\theta}_n + M_{ns}^4\bar{\theta}_s)\,d\Gamma \quad (n, s\text{: no sum})\end{aligned} \tag{6.34a}$$

$$L_2(\bar{w}) = \int_\Omega q\bar{w}\,d\Omega + \int_{\Gamma_3} Q_n^3\bar{w}\,d\Gamma + \int_{\Gamma_4} Q_n^4\bar{w}\,d\Gamma \quad (n, s\text{: no sum}) \tag{6.34b}$$

for the two functionals. For other choices of the boundary conditions, L_1 and L_2 will have definitions differing from (6.34).

6.3 Finite element approximations

It is noted that the first-order derivatives of θ_i, w, $\bar{\theta}_i$, and $\bar{w}$ are used in the weak form (6.33). This means that the continuity on θ_i, w, $\bar{\theta}_i$, and $\bar{w}$ is assumed in the form (6.33) and that the continuity of the normal derivatives of w and $\bar{w}$ across the line in Ω need not be required. This setting yields the choice of Lagrangian interpolation for both the deflection and the rotations, while the Kirchhoff plate theory requires using a sort of Hermitian interpolation for the deflection w in order to satisfy the continuity of the normal derivative $\partial w/\partial n$. It is difficult to find a finite element that satisfies the continuity of the normal derivative of the deflection. Thus, the Mindlin plate theory has an advantage for the choice of finite elements, since only the continuity of the functions is assumed.

Let the domain (i.e., the middle surface) be divided into E finite elements Ω_e in each of which the deflection w and the rotation $\boldsymbol{\theta}$ are interpolated by

$$\begin{aligned} w &= w_\alpha N_\alpha(x, y) \\ \theta_i &= \theta_{i\alpha} M_\alpha(x, y), \quad i = 1, 2\end{aligned} \tag{6.35}$$

where N_α and M_α are, respectively, the shape functions of the α node in the element for the deflection w and the rotations θ_i. If the same interpolation is used for the virtual deflection $\bar{w}$ and the virtual rotation $\bar{\boldsymbol{\theta}}$, the weak form (6.33) is discretized as

$$\sum_{e=1}^{E} \bar{\theta}_{i\alpha}(K_{i\alpha j\beta}^{1,e}\theta_{j\beta} - K_{i\alpha\beta}^{2,e}w_\beta) + \sum_{e=1}^{E} \bar{w}_\alpha(-K_{\alpha j\beta}^{3,e}\theta_{j\beta} + K_{\alpha\beta}^{4,e}w_\beta)$$
$$= \sum_{e=1}^{E} \bar{\theta}_{i\alpha}M_{i\alpha} + \bar{w}_\alpha P_\alpha \tag{6.36}$$

or

$$\sum_{e=1}^{E} \{\bar{\theta}_{i\alpha}, \bar{w}_\alpha\}\begin{bmatrix} K_{i\alpha j\beta}^{1,e} & -K_{i\alpha\beta}^{2,e} \\ -K_{\alpha j\beta}^{3,e} & K_{\alpha\beta}^{4,e} \end{bmatrix}\begin{Bmatrix} \theta_{j\beta} \\ w_\beta \end{Bmatrix} = \sum_{e=1}^{E} \{\bar{\theta}_{i\alpha}, \bar{w}_\alpha\}\begin{Bmatrix} M_{i\alpha} \\ P_\alpha \end{Bmatrix} \tag{6.37}$$

where

$$K_{i\alpha j\beta}^{1,e} = \int_{\Omega_e} (D_{kijl}M_{\alpha,k}M_{\beta,l} + \delta_{ij}k\mu hM_\alpha M_\beta)\, d\Omega \tag{6.38}$$

$$K_{i\alpha\beta}^{2,e} = \int_{\Omega_e} k\mu hM_\alpha N_{\beta,i}\, d\Omega, \qquad K_{\alpha j\beta}^{3,e} = \int_{\Omega_e} k\mu hM_\beta N_{\alpha,j}\, d\Omega \tag{6.39}$$

$$K_{\alpha\beta}^{4,e} = \int_{\Omega_e} k\mu hN_{\alpha,i}N_{\beta,i}\, d\Omega \tag{6.40}$$

and $M_{i\alpha}$ and P_α are defined from

$$\bar{\theta}_{i\alpha}M_{i\alpha} = L_1(\bar{\boldsymbol{\theta}}) \quad \text{and} \quad \bar{w}_\alpha P_\alpha = L_2(\bar{w}) \tag{6.41}$$

6.3.1 Choices of the shape functions

A standard choice of the shape functions $\{M_\alpha\}$ and $\{N_\alpha\}$ for the Mindlin plate is

$$\hat{N}_\alpha = \hat{M}_\alpha = \tfrac{1}{4}(1 + s_\alpha s)(1 + t_\alpha t), \quad \alpha = 1, 2, 3, 4 \tag{6.42}$$

for the master element of the four-node quadrilateral isoparametric element in the normalized coordinate system, where (s_α, t_α) is the set of coordinates of the αth node in the master element. The choice (6.42) yields the approximation of the rotations and the deflection by bilinear polynomials in each element Ω_e. The relation of N_α to $\hat{N}_\alpha$ is obtained by the rule of isoparametric representation such that

$$x = x_\alpha\hat{N}_\alpha(s, t), \qquad y = y_\alpha\hat{N}_\alpha(s, t) \tag{6.43}$$

by using the set of coordinates (x_α, y_α) of the αth node in the global coordinate system. Indeed,

$$\hat{N}_\alpha(s, t) = N_\alpha(x_\beta\hat{N}_\beta(s, t), y_\beta\hat{N}_\beta(s, t))$$

Details for such a relation have been discussed in the previous chapter.

It must be noted that a blind application of (6.42) leads to absurd results for the case of thin plates. That is, if the thickness of the plate is quite small, then the stiffness matrix obtained by (6.42) becomes too stiff and provides almost zero solutions. This phenomenon is called *locking*, as in the case of incompressible materials. To avoid this, we may apply the technique of selective reduced integration once more.

The coupling term $(w_{,i} - \theta_i)$ involving the first derivatives of the deflection w and the rotations is present in the weak form (6.33). If the thickness of the plate $h \to 0$, the Kirchhoff hypothesis is preferable to that of Mindlin, so $\theta_i = w_{,i}$ is called

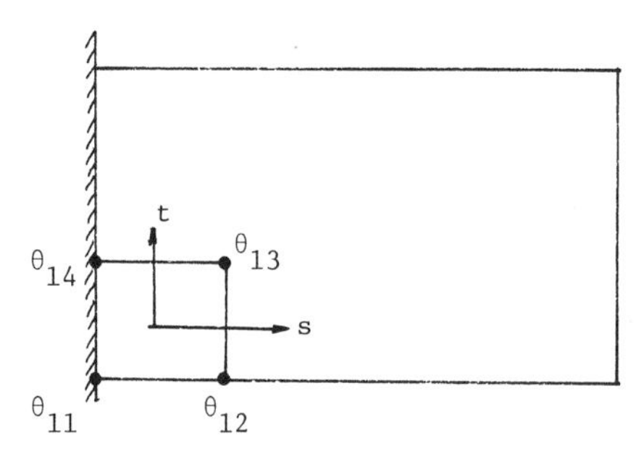

Figure 6.5 Clamped plate.

for. If the finite element approximation by (6.42) is applied to the discretized weak form, we have

$$\frac{\partial w}{\partial s} = a_1 + a_3 t, \qquad \frac{\partial w}{\partial t} = a_2 + a_3 s$$

$$\theta_1 = b_0 + b_1 s + b_2 t + b_3 st, \qquad \theta_2 = c_0 + c_1 s + c_2 t + c_3 st$$

Then $\theta_i = w_{,i}$ for the approximate θ_i and w can be satisfied only if

$$\begin{aligned} &a_1 = b_0, \quad a_3 = b_2, \quad a_2 = c_0, \quad a_3 = c_1 \\ &b_1 = b_3 = c_2 = c_3 = 0 \end{aligned} \tag{6.44}$$

If the conditions (6.44) are imposed, along with the given boundary conditions, we merely obtain a zero solution. For example, if a plate clamped along its left edge, as shown in Figure 6.5, is considered, the boundary condition requires

$$\theta_{11} = \theta_{14} = 0$$

Moreover, the relations (6.44) yield

$$\begin{aligned} -\theta_{11} + \theta_{12} + \theta_{13} - \theta_{14} = 0 \\ \theta_{11} - \theta_{12} + \theta_{13} - \theta_{14} = 0 \end{aligned}$$

Combining these, we see that

$$\theta_{12} = \theta_{13} = 0$$

This means that the plate does not bend in the s direction. Thus, if we expect $\theta_i = w_{,i}$ for the approximation by (6.42), the plate becomes so stiff that no deflection can be achieved as the thickness of the plate goes to zero. What is the next step? Recall that the relation $\theta_i = w_{,i}$ involves the first derivative of the deflection. Since the first derivatives are one order less polynomials than the function itself, the first derivatives $w_{,i}$ are, roughly speaking, constants in the i direction. This suggests that the requirement of the Kirchhoff hypothesis for very thin plates is assigned just on the centroid of the element. This means that the evaluation of the terms $K^{2,e}_{i\alpha\beta}$, $K^{3,e}_{\alpha j\beta}$, and $K^{4,e}_{\alpha\beta}$ in the stiffness matrix is performed by the one-point Gaussian quadrature rule, whereas the term $K^{1,e}_{i\alpha j\beta}$ related to the bending is computed by the 2×2 Gaussian rule. This is the selective reduced integration technique introduced by Hughes et al. (1977).

Even for thick plates, which need not satisfy the Kirchhoff hypothesis, the selective reduced integration technique provides better results than the case of full inte-

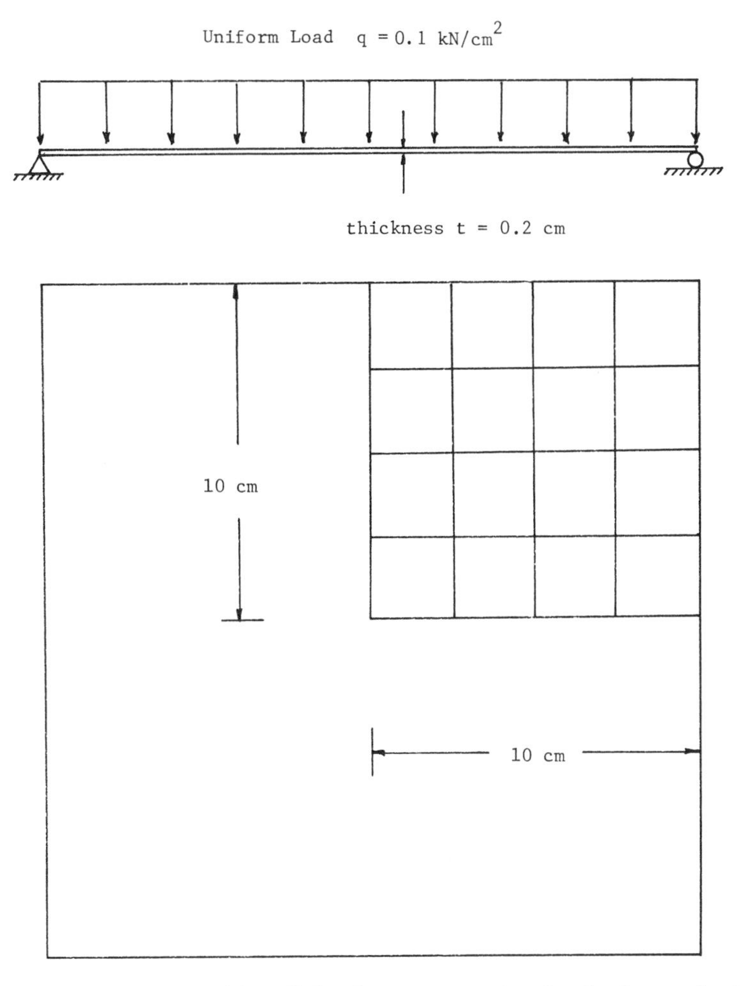

Figure 6.6 Plate-bending problem (finite element approximation by four-node elements): $E = 2.1 \times 10^4$ kN/cm^2, $\nu = 0.29$, analytical solution $w_{\max} = 4.249776$ cm.

gration, since the "degree" of the coupling term $\theta_i - w_{,i}$ is equivalent to θ_i and w if the reduced integration is applied.

Example 6.1: Suppose that a square plate is simply supported and subjected to a uniform transverse force as shown in Figure 6.6. Let the plate be a homogeneous isotropic linearly elastic material characterized by

Young's modulus	$E = 2.1 \times 10^4$ kN/cm^2
Poisson's ratio	$\nu = 0.29$
Width	$2a = 20$ cm

Let us solve the problem for $q = 10$ kN/cm^2 and three different thicknesses: $t = 0.2, 0.5$, and 1 cm, using 16 four-node elements for a quadrant of the plate. Numerical results are shown in Figure 6.7, which displays the deflection of the plate at its center. Figure 6.8 shows convergence

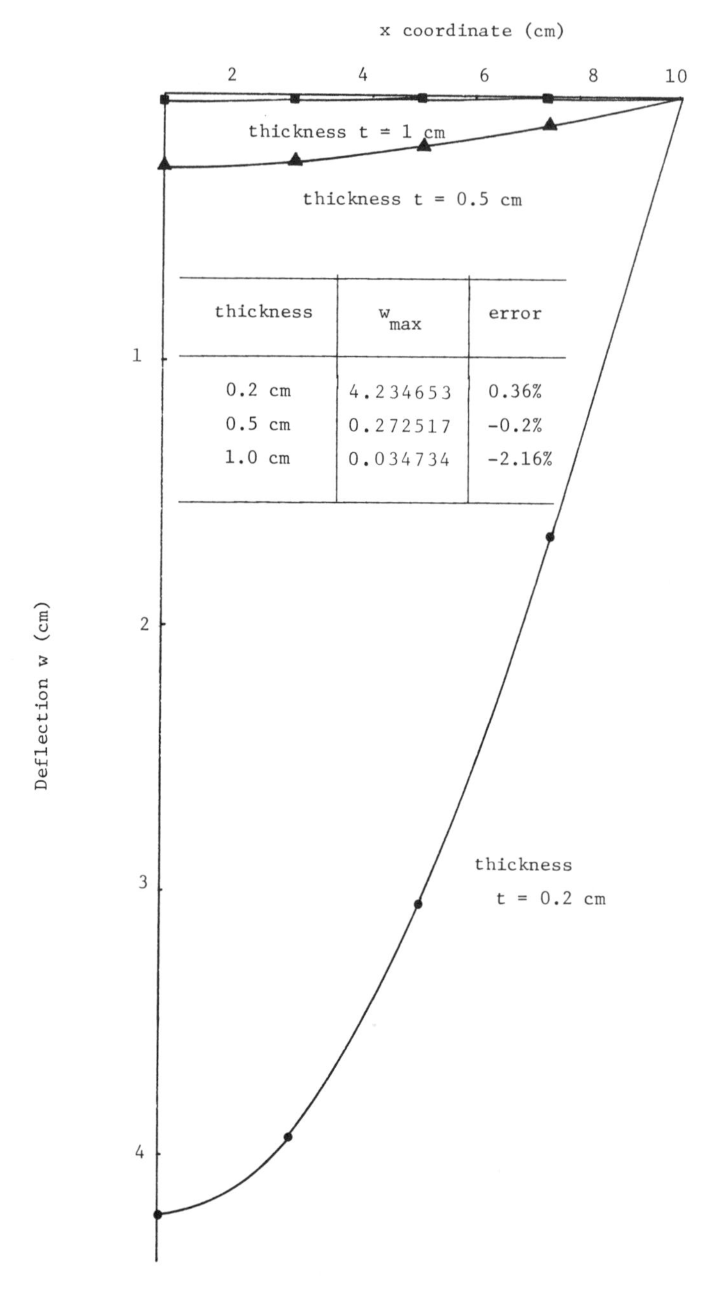

thickness	w_{max}	error
0.2 cm	4.234653	0.36%
0.5 cm	0.272517	-0.2%
1.0 cm	0.034734	-2.16%

Figure 6.7 Deflection of plate along x axis. Error $= (w_{exact} - w_{max})/w_{exact}$.

of the finite element approximation as the size of finite elements goes to zero for the case $t = 0.2$ cm. Figure 6.9 shows an irregular finite element model and its results. It is clear that the quality of the approximation does not depend on the modeling too much.

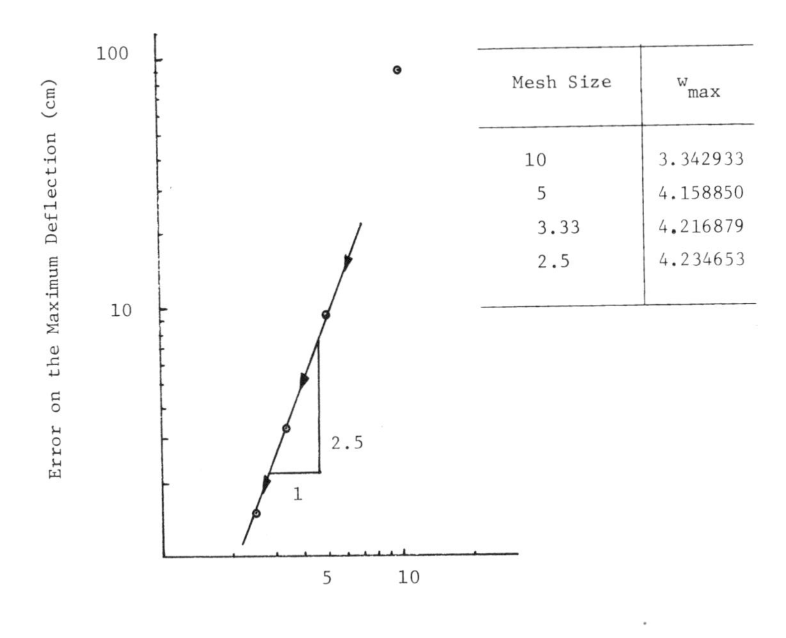

Mesh Size	w_{max}
10	3.342933
5	4.158850
3.33	4.216879
2.5	4.234653

Figure 6.8 Convergence of finite element approximation as $h \to 0$: $w_{exact} = 4.249776$ cm.

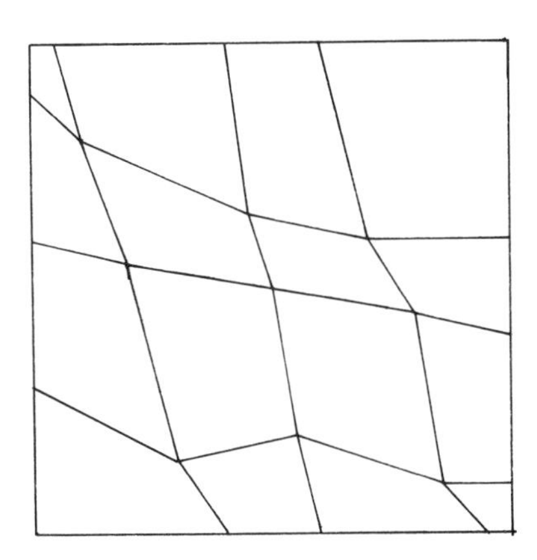

Figure 6.9 Irregular finite element model for plate-bending problem: $w_{max} = 3.860944$ cm, error $= 9.15\%$.

Exercise 6.1: Derive a routine to compute element stiffness matrices using:

1. for the deflection w, the shape functions $\{N_\alpha\}$, $\alpha = 1, \ldots, 9$, of the nine-node tensor product Lagrange element and
2. for the slope $\boldsymbol{\theta}$, the shape functions $\{M_\alpha\}$, $\alpha = 1, \ldots, 4$, of the four-node element, as shown in Figure 6.10.

Solve the problem of example 6.1, and discuss the quality of this element.

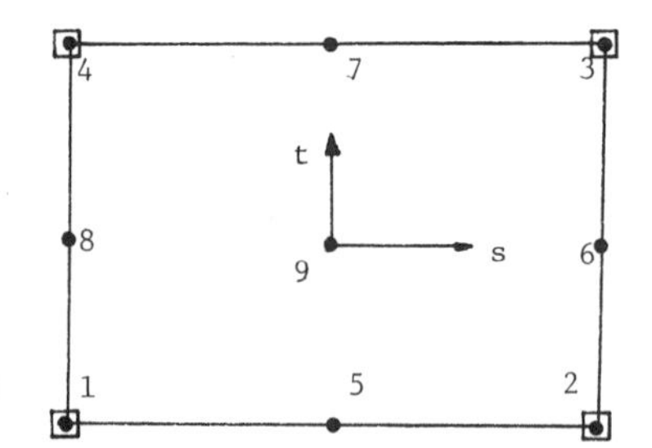

Figure 6.10 Quadratic Mindlin plate element.

Exercise 6.2: If both the deflection w and the slope $\boldsymbol{\theta}$ are approximated by the nine-node tensor product Lagrange element, the selective reduced integration technique must be applied to compute element stiffness matrices. Indeed, the pure bending term is integrated with a 3×3 Gaussian quadrature rule and the other terms related to the shear force are "underintegrated" using a 2×2 Gaussian rule. What happens if the reduced integration 2×2 Gaussian rule is applied to all the terms? Are spurious modes present in the deflection?

6.4 Finite elements for the Kirchhoff plate

The Mindlin plates studied so far did not assume the Kirchhoff condition

$$\theta_x = \frac{\partial w}{\partial x} \quad \text{and} \quad \theta_y = \frac{\partial w}{\partial y} \tag{6.45}$$

a priori and are approximated by Lagrange-type finite elements, which assume only continuity of the variables themselves. In this section, let us study the plate assuming the Kirchhoff condition (6.45). To do this, we first look at the equilibrium (6.16). Differentiating the first of (6.16) with respect to x and the second of (6.16) with respect to y, we have

$$\frac{\partial^2 M_{xx}}{\partial x^2} + \frac{\partial^2 M_{yx}}{\partial x\, \partial y} - \frac{\partial Q_x}{\partial x} = -\frac{\partial m_x}{\partial x}$$

$$\frac{\partial^2 M_{xy}}{\partial x\, \partial y} + \frac{\partial^2 M_{yy}}{\partial y^2} - \frac{\partial Q_y}{\partial y} = -\frac{\partial m_y}{\partial y}$$

Adding these two equations and substituting (6.13) yields

$$-\frac{\partial^2 M_{xx}}{\partial x^2} - 2\frac{\partial^2 M_{xy}}{\partial x\, \partial y} - \frac{\partial^2 M_{yy}}{\partial y^2} = q + \frac{\partial m_x}{\partial x} + \frac{\partial m_y}{\partial y} \tag{6.46}$$

Here $M_{xy} = M_{yx}$ has been used. For simplicity, let us assume that $m_x = m_y = 0$ in the subsequent study in this section. Then the equilibrium equation of a plate becomes

$$-\frac{\partial^2 M_{xx}}{\partial x^2} - 2\frac{\partial^2 M_{xy}}{\partial x\, \partial y} - \frac{\partial^2 M_{yy}}{\partial y^2} = q \quad \text{in } \Omega \tag{6.47}$$

Now we shall obtain the weak form of (6.47). Multiplying both sides of (6.47) by a virtual deflection $\bar{w}$, integrating over the domain Ω, and applying integration by parts, we have

$$\int_\Omega \left\{ M_{xx}\left(-\frac{\partial^2 \bar{w}}{\partial x^2}\right) + 2M_{xy}\left(-\frac{\partial^2 \bar{w}}{\partial x\,\partial y}\right) + M_{yy}\left(-\frac{\partial^2 \bar{w}}{\partial y^2}\right)\right\} d\Omega$$

$$= \int_\Omega q\bar{w}\, d\Omega - \int_\Gamma \left\{\left(-\frac{\partial M_{xx}}{\partial x} - \frac{\partial M_{yx}}{\partial y}\right) n_x + \left(-\frac{\partial M_{xy}}{\partial x} - \frac{\partial M_{yy}}{\partial y}\right) n_y\right\} \bar{w}\, d\Gamma$$

$$- \int_\Gamma \left\{(M_{xx}n_x + M_{yx}n_y)\frac{\partial \bar{w}}{\partial x} + (M_{xy}n_x + M_{yy}n_y)\frac{\partial \bar{w}}{\partial y}\right\} d\Gamma$$

Defining

$$Q_x = \frac{\partial M_{xx}}{\partial x} + \frac{\partial M_{yx}}{\partial y}, \qquad Q_y = \frac{\partial M_{xy}}{\partial x} + \frac{\partial M_{yy}}{\partial y} \qquad \text{on } \Gamma \tag{6.48}$$
$$M_x = M_{xx}n_x + M_{yx}n_y, \qquad M_y = M_{xy}n_x + M_{yy}n_y$$

and

$$\bar{K}_{xx} = -\frac{\partial^2 \bar{w}}{\partial x^2}, \qquad \bar{K}_{xy} = \bar{K}_{yx} = -\frac{\partial^2 \bar{w}}{\partial x\,\partial y}, \qquad \bar{K}_{yy} = -\frac{\partial^2 \bar{w}}{\partial y^2} \tag{6.49}$$

we have

$$\int_\Omega (M_{xx}\bar{K}_{xx} + M_{xy}\bar{K}_{xy} + M_{yx}\bar{K}_{yx} + M_{yy}\bar{K}_{yy})\, d\Omega$$
$$= \int_\Omega q\bar{w}\, d\Omega + \int_\Gamma (Q_x n_x + Q_y n_y)\bar{w}\, d\Gamma - \int_\Gamma \left(M_x \frac{\partial \bar{w}}{\partial x} + M_y \frac{\partial \bar{w}}{\partial y}\right) d\Gamma, \quad \forall \bar{w} \tag{6.50a}$$

If index notation is applied, (6.50a) can be written as

$$\int_\Omega M_{ij}\bar{K}_{ij}\, d\Omega = \int_\Omega q\bar{w}\, d\Omega + \int_\Gamma Q_i n_i \bar{w}\, d\Gamma - \int_\Gamma M_i \bar{w}_{,i}\, d\Gamma, \quad \forall \bar{w} \tag{6.50b}$$

Here we have assumed that an *arbitrary virtual deflection* $\bar{w}$ *and its normal derivative along any smooth curve in* Ω *are continuous* in order to justify the two integrations by parts used to obtain the weak form (6.50). Note that only continuity has been needed in the integration by parts used for other problems studied in previous chapters.

Now the boundary conditions must be specified to obtain the final form of the weak form. As an example, let us assume the boundary conditions

$$w = g, \qquad \frac{\partial w}{\partial n} = g_n, \qquad M_{ns} = 0 \quad \text{on } \Gamma_1$$
$$Q_n = Q_i n_i = h_n, \qquad M_i = h_i \quad \text{on } \Gamma_2 \tag{6.51}$$

Then (6.50) becomes

$$\int_\Omega M_{ij}\bar{K}_{ij}\, d\Omega = \int_\Omega q\bar{w}\, d\Omega + \int_{\Gamma_2} h_n \bar{w}\, d\Gamma - \int_{\Gamma_2} h_i \bar{w}_{,i}\, d\Gamma \tag{6.52}$$

for every $\bar{w}$ such that $\bar{w} = \partial\bar{w}/\partial n = 0$ on Γ_1. Since $w = g$ and $\partial w/\partial n = g_n$ on Γ_1 are essential boundary conditions (i.e., first type), a virtual deflection $\bar{w}$ must be restricted by $\bar{w} = \partial\bar{w}/\partial n = 0$ on Γ_1.

Exercise 6.3: Find the weak form when the boundary conditions are

(a) $w = 0, \quad M_{nn} = 0 \quad \text{on } \Gamma_1$ (simply supported) (n: no sum)

(b) $\dfrac{\partial w}{\partial n} = 0, \quad M_{ns} = 0, \quad Q_n = 0 \quad \text{on } \Gamma_2$ (free)

6.4.1 Nonconforming triangular element (Zienkiewicz)

Let three-node triangular elements be considered, and let three degrees of freedom w, $\partial w/\partial x$, and $\partial w/\partial y$ be assumed at each node. Since nine degrees of freedom are present in such an element, the deflection w can be approximated by a cubic polynomial, but not a complete cubic polynomial. Suppose that the deflection w is approximated by

$$w = \sum_{i=1}^{9} \alpha_i S_i(\xi_1, \xi_2, \xi_3; \beta) \tag{6.53}$$

where S_i are the basis functions; ξ_1, ξ_2, ξ_3 are the area coordinates defined in section 2.11; and β is a constant to be chosen later. Recalling the shape functions for 10-node triangular elements introduced in section 2.11, we shall define the basis functions S_i by

$$\begin{aligned}
&S_1 = \xi_1, \qquad S_2 = \xi_2, \qquad S_3 = \xi_3,\\
&S_4 = \xi_1^2\xi_2 + \beta\xi_1\xi_2\xi_3, \qquad S_5 = \xi_1\xi_2^2 + \beta\xi_1\xi_2\xi_3\\
&S_6 = \xi_2^2\xi_3 + \beta\xi_1\xi_2\xi_3, \qquad S_7 = \xi_2\xi_3^2 + \beta\xi_1\xi_2\xi_3\\
&S_8 = \xi_3^2\xi_1 + \beta\xi_1\xi_2\xi_3, \qquad S_9 = \xi_3\xi_1^2 + \beta\xi_1\xi_2\xi_3
\end{aligned} \tag{6.54}$$

According to Zienkiewicz, the parameter β must be $\frac{1}{2}$ in order to reproduce complete quadratic polynomials within an element.

Noting that

$$\frac{\partial \xi_i}{\partial x} = b_i \quad \text{and} \quad \frac{\partial \xi_i}{\partial y} = c_i \tag{6.55}$$

where

$$\begin{aligned}
&b_1 = \frac{y_2 - y_3}{2\Delta}, \qquad b_2 = \frac{y_3 - y_1}{2\Delta}, \qquad b_3 = \frac{y_1 - y_2}{2\Delta}\\
&c_1 = \frac{x_3 - x_2}{2\Delta}, \qquad c_2 = \frac{x_1 - x_3}{2\Delta}, \qquad c_3 = \frac{x_2 - x_1}{2\Delta}
\end{aligned} \tag{6.56}$$

$$2\Delta = x_2y_3 - x_3y_2 + x_3y_1 - x_1y_3 + x_1y_2 - x_2y_1$$

and using the nodal coordinates (x_α, y_α), $\alpha = 1, 2, 3$, of the three nodes, the shape functions N_α, $N_{x\alpha}$, and $N_{y\alpha}$ satisfying

$$\begin{aligned}
&N_\alpha(x_\beta, y_\beta) = \delta_{\alpha\beta}, \qquad \frac{\partial N_\alpha}{\partial x}(x_\beta, y_\beta) = 0, \qquad \frac{\partial N_\alpha}{\partial y}(x_\beta, y_\beta) = 0\\
&N_{x\alpha}(x_\beta, y_\beta) = 0, \qquad \frac{\partial N_{x\alpha}}{\partial x}(x_\beta, y_\beta) = \delta_{\alpha\beta}, \qquad \frac{\partial N_{x\alpha}}{\partial y}(x_\beta, y_\beta) = 0\\
&N_{y\alpha}(x_\beta, y_\beta) = 0, \qquad \frac{\partial N_{y\alpha}}{\partial x}(x_\beta, y_\beta) = 0, \qquad \frac{\partial N_{y\alpha}}{\partial y}(x_\beta, y_\beta) = \delta_{\alpha\beta}
\end{aligned} \tag{6.57}$$

are obtained as

$$
\begin{aligned}
N_\alpha &= \xi_\alpha + \xi_\alpha\xi_\beta(\xi_\alpha - \xi_\beta) + \xi_\alpha\xi_\gamma(\xi_\alpha - \xi_\gamma) \\
N_{x\alpha} &= 2\Delta[c_\gamma(\xi_\alpha^2\xi_\beta + \tfrac{1}{2}\xi_1\xi_2\xi_3) - c_\beta(\xi_\alpha^2\xi_\gamma + \tfrac{1}{2}\xi_1\xi_2\xi_3)] \\
N_{y\alpha} &= 2\Delta[b_\beta(\xi_\alpha^2\xi_\gamma + \tfrac{1}{2}\xi_1\xi_2\xi_3) - b_\gamma(\xi_\alpha^2\xi_\beta + \tfrac{1}{2}\xi_1\xi_2\xi_3)]
\end{aligned} \tag{6.58}
$$

where the summation convention is not applied in (6.58), and (α, β, γ) is *the permutation (1, 2, 3)*. Then the deflection w is approximated by

$$
w = w_\alpha N_\alpha + \left.\frac{\partial w}{\partial x}\right|_\alpha N_{x\alpha} + \left.\frac{\partial w}{\partial y}\right|_\alpha N_{y\alpha} \tag{6.59}
$$

It is clear that w is continuous along the element boundaries, but $\partial w/\partial n$ is not necessarily continuous. Thus, this element is *nonconforming* because it cannot produce continuous w and $\partial w/\partial n$, which are required to obtain the weak form.

Reassigning the numbering for the degrees of freedom and the shape functions,

$$
(a_1, a_2, a_3, a_4, \ldots, a_9) = \left(w_1, \left.\frac{\partial w}{\partial x}\right|_1, \left.\frac{\partial w}{\partial y}\right|_1, w_2, \ldots, \left.\frac{\partial w}{\partial y}\right|_3\right)
$$

$$
(\hat{N}_1, \hat{N}_2, \hat{N}_3, \hat{N}_4, \ldots, \hat{N}_9) = (N_1, N_{x1}, N_{y1}, N_2, \ldots, N_{yz})
$$

(6.59) is written as

$$
w = a_q \hat{N}_q \tag{6.60}
$$

Similarly, let $\bar{w} = \bar{a}_p \hat{N}_p$. Substituting these into the left side of the weak form (6.52) yields

$$
\begin{aligned}
\int_\Omega M_{ij}\bar{K}_{ij}\, d\Omega &= \sum_{e=1}^{E} \int_{\Omega_e} M_{ij}\bar{K}_{ij}\, d\Omega \\
&= \sum_{e=1}^{E} \int_{\Omega_e} D_{ijkl}K_{kl}\bar{K}_{ij}\, d\Omega \\
&= \sum_{e=1}^{E} \bar{a}_p K_{pq}^e a_q
\end{aligned} \tag{6.61}
$$

where

$$
K_{pq}^e = \int_{\Omega_e} D_{ijkl} \frac{\partial^2 \hat{N}_q}{\partial x_k \partial x_l} \frac{\partial^2 \hat{N}_p}{\partial x_i \partial x_j}\, d\Omega \tag{6.62}
$$

The matrix $[K_{pq}^e]$ is called the element stiffness matrix. The right side of the weak form (6.52) can be discretized similarly. The remaining procedures are exactly the same as in other problems studied so far.

Exercise 6.4: Let a simply supported square plate be subjected to a uniform load q. Solve this problem using the finite element model shown in Figure 6.11 for

$$
E = 2.1 \times 10^4\ \text{kN/cm}^2, \qquad \nu = 0.29
$$

$$
\text{Thickness } t = 0.2\ \text{cm}, \qquad \text{Load } q = 10\ \text{kN/cm}^2
$$

Use the symmetry of the problem and consider only one quadrant.

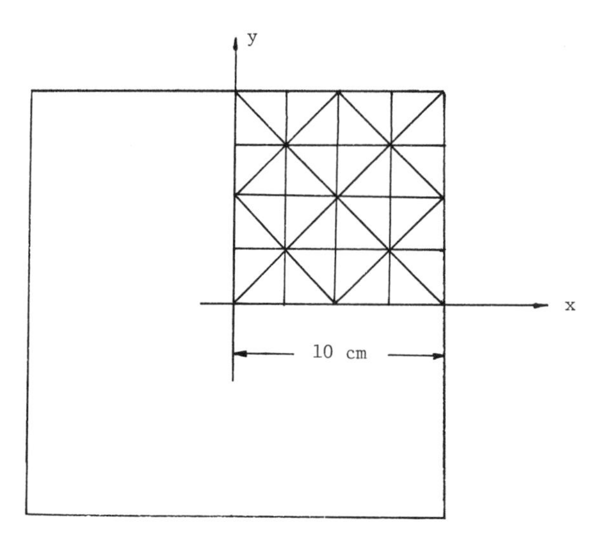

Figure 6.11 Plate-bending problem using Zienkiewicz nonconforming element.

6.4.2 Nonconforming triangular element (Fujino–Morley)

Another possibility is to use a complete quadratic polynomial for approximating the deflection w. This can be obtained by introducing three values of the deflection w at corner nodes of a triangular element and three values of the normal derivative $\partial w/\partial n$ at midnodes as six degrees of freedom. Introducing the notation

$$S_\alpha = 2\Delta\sqrt{b_\alpha^2 + c_\alpha^2}$$
$$l_\alpha = -b_\alpha/\sqrt{b_\alpha^2 + c_\alpha^2}, \qquad m_\alpha = -c_\alpha/\sqrt{b_\alpha^2 + c_\alpha^2} \quad \text{(no sum)} \tag{6.63}$$

we can define the shape functions

$$N_\alpha = -\frac{2\Delta}{S_\alpha}\,\xi_\alpha(1 - \xi_\alpha), \quad \text{when } \alpha = 1, 2, 3$$
$$N_\alpha = \xi_{\alpha-3} - \sum_{\beta=4}^{6}(b_{\alpha-3}l_{\beta-3} + c_{\alpha-3}m_{\beta-3})N_{\beta-3}, \quad \text{when } \alpha = 4, 5, 6 \tag{6.64}$$

where the summation convention is not applied in (6.64). Then, the expression for the deflection w becomes

$$w = a_\alpha N_\alpha \tag{6.65}$$

where

$$(a_1, a_2, a_3, a_4, a_5, a_6) = \left(w_1, w_2, w_3, \left.\frac{\partial w}{\partial n}\right|_4, \left.\frac{\partial w}{\partial n}\right|_5, \left.\frac{\partial w}{\partial n}\right|_6\right)$$

and w_α and $(\partial w/\partial n)|_\alpha$ are the values of the deflection w and the normal derivative $\partial w/\partial n$ at the α node, respectively. The numbering of the nodes is the same as in the

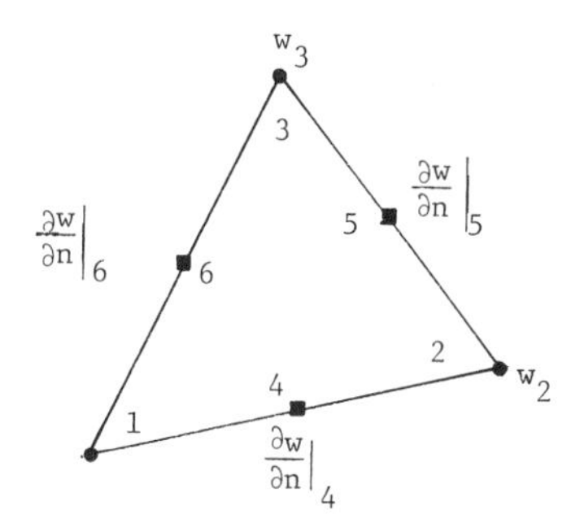

Figure 6.12 Fujino–Morley element.

six-node triangular element studied in section 2.11, as shown in Figure 6.12. It is clear that both w and its first derivatives are discontinuous across the element boundaries.

Although the Fujino–Morley elements are nonconforming in both the deflection and the normal derivative, they imply convergent finite element approximations as the size of finite element h goes to zero. Indeed, if the least-squares error of the moments is measured, it is bounded by Ch for a positive constant $C > 0$ for a smooth solution w of the plate-bending problem. That is,

$$\left[\int_{\Omega} M_{ij}(w - w_h) M_{ij}(w - w_h)\, d\Omega\right]^{1/2} = O(h)$$

where $M_{ij}(v)$ is the ij component of the bending moment $\mathbf{M}$ by the deflection v, and w_h is the finite element approximation of w by Fujino–Morley elements. It is noted that the deflection is overestimated by Fujino–Morley elements while the bending moment is approximated relatively well. The rate of convergence of the deflection is small, say, for a smooth solution w,

$$\left[\int_{\Omega} (w - w_h)^2\, d\Omega\right]^{1/2} = O(h^2)$$

Example 6.2: Let us examine the Fujino–Morley element by solving a bending problem of a circular plate, the radius of which is $R = 30$ cm. Suppose that the plate's thickness $t = 0.2$ cm, Young's modulus $E = 20.6 \times 10^6$ N/cm^2, and that Poisson's ratio $\nu = 0.3$. We shall solve the following four problems using the finite element grids shown in Figure 6.13:

1. Simply supported with the uniformly distributed load $p = 1$ N/cm^2.
2. Simply supported with the point force $p = 400$ N at the center of the plate.
3. Clamped with the uniformly distributed load $p = 1$ N/cm^2.
4. Clamped with the point force $P = 400$ N at the center of the plate.

Figure 6.14 and Table 6.1 are results of the convergence test with respect to the deflection at the center of the plate. It is clear that the convergence of the approximation is obtained. Furthermore, irregular

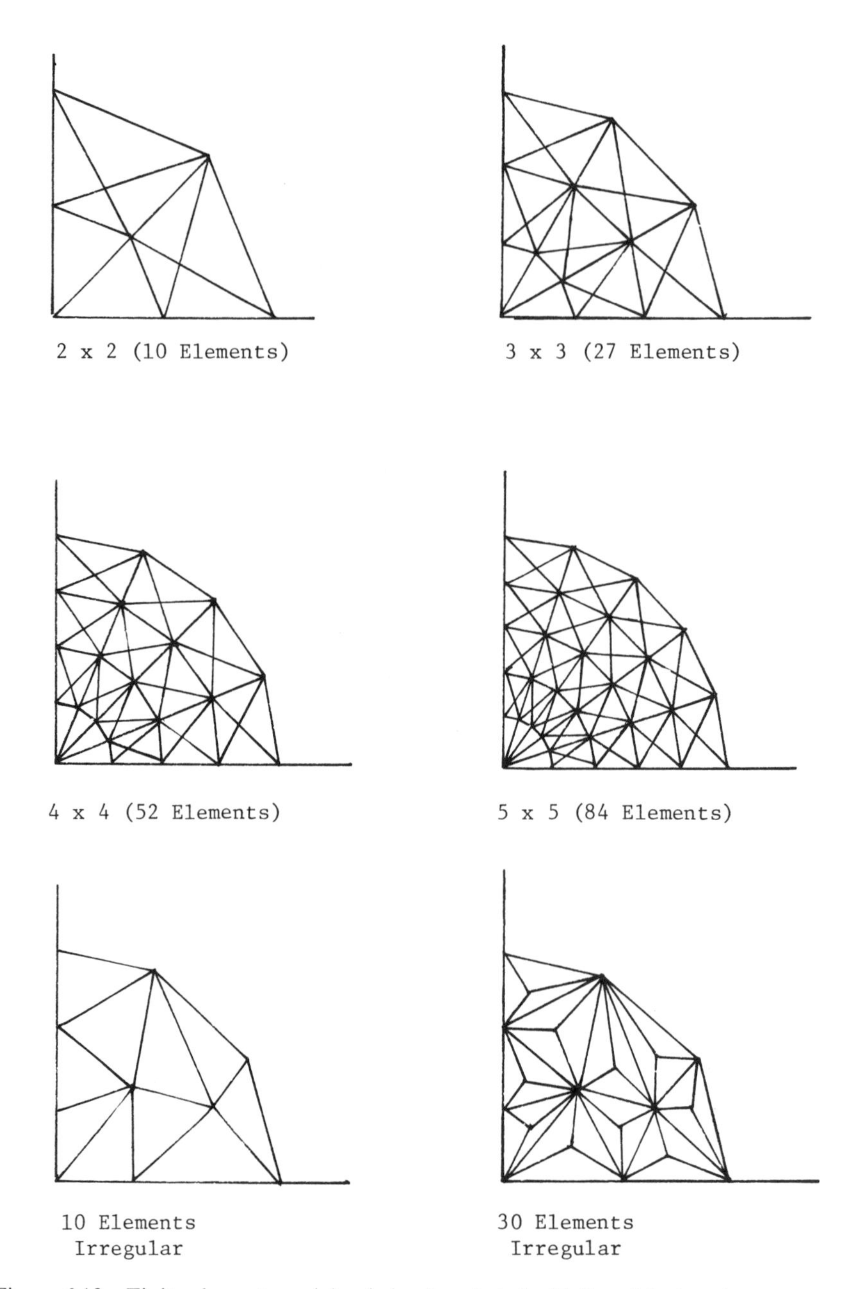

Figure 6.13 Finite element models of circular plate by Fujino–Morley elements: $E = 20.6 \times 10^6$ N/cm^2, $\nu = 0.3$, $R = 30$ cm, $t = 0.2$ cm.

finite element grids do not provide good numerical results that are very different from the case of regular finite element grids. It is also confirmed that the approximate deflection by Fujino–Morley elements is, in general, larger than the exact one.

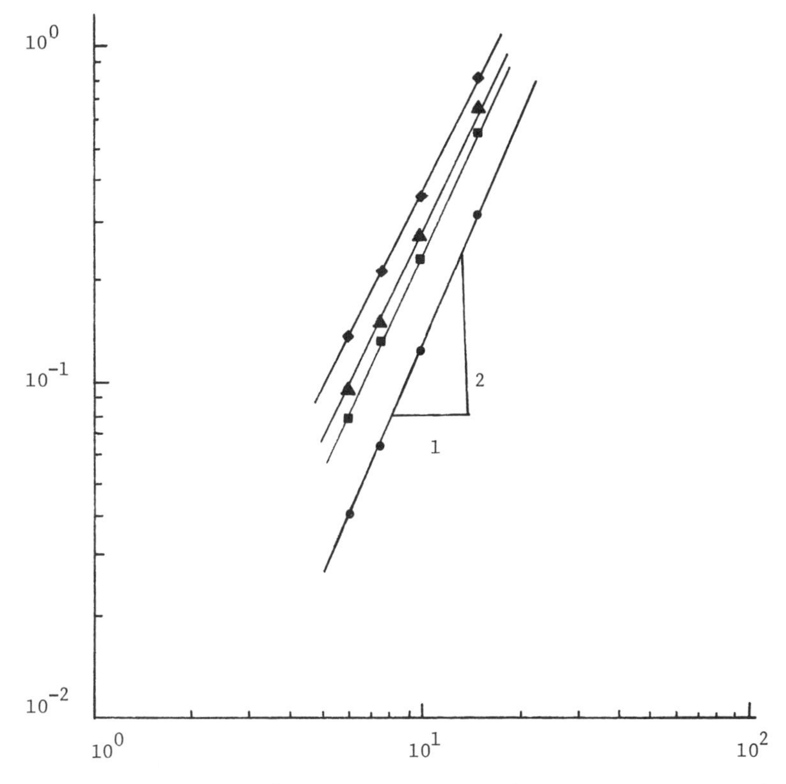

Figure 6.14 Convergence of deflection by Fujino–Morley elements: ●, case 1, simply supported, uniform load; ■, case 2, simply supported, point load; ▲, case 3, clamped, uniform load; ◆, case 4, clamped, point load.

As a final remark to this book, we shall give an example of finite element approximations of plates using three-dimensional solid (or brick) elements. As shown in Chapter 5, several beam-bending problems are treated as plane stress problems using four-node quadrilateral elements. Natural extension may be valid even for plate-bending problems, since they are special cases of general three-dimensional elasticity problems. If plate-bending problems are solvable using three-dimensional solid elements, finite element analysis will be very simplified because it may be unnecessary to introduce special elements for plates and shells. If this trial fails, careful study of the reason will clarify the essential difference between solids and plates/shells in their mechanical behavior. In any case, this will give us the opportunity to think over mechanics theory once again by studying finite element methods. It is, however, certain that this text cannot answer completely the above question. It is left to the reader as a starting point of advanced study of finite element methods in mechanics.

Suppose that the square, simply supported plate described in example 6.1 is modeled by eight-node brick elements whose shape functions are given by

$$N_\alpha(r, s, t) = \tfrac{1}{8}(1 + r_\alpha r)(1 + s_\alpha s)(1 + t_\alpha t)$$

Table 6.1 *Numerical results by Fujino–Morley elements at the center*

	Deflection and (error) for case:			
NELX	1	2	3	4
10 (2 * 2)	4.491 (0.3135)	1.867 (0.5498)	1.393 (0.661)	0.8483 (0.7875)
27 (3 * 3)	3.872 (0.1325)	1.483 (0.231)	1.068 (0.2735)	0.6449 (0.3589)
52 (4 * 4)	3.67 (0.0734)	1.36 (0.1289)	0.9629 (0.1482)	0.5729 (0.2072)
85 (5 * 5)	3.559 (0.0409)	1.296 (0.0758)	0.9192 (0.0961)	0.5395 (0.1368)
10 (irregular)	4.785 (0.3995)	1.951 (0.6195)	1.383 (0.6491)	0.8227 (0.7336)
30 (irregular)	4.385 (0.283)	1.692 (0.4045)	1.331 (0.5871)	0.7647 (0.6114)
∞	3.419 (—)	1.2047 (—)	0.8386 (—)	0.4726 (—)

Case 1: Simply supported and uniform load.
Case 2: Simply supported and point load at center.
Case 3: Clamped and uniform load.
Case 4: Clamped and point load at center.

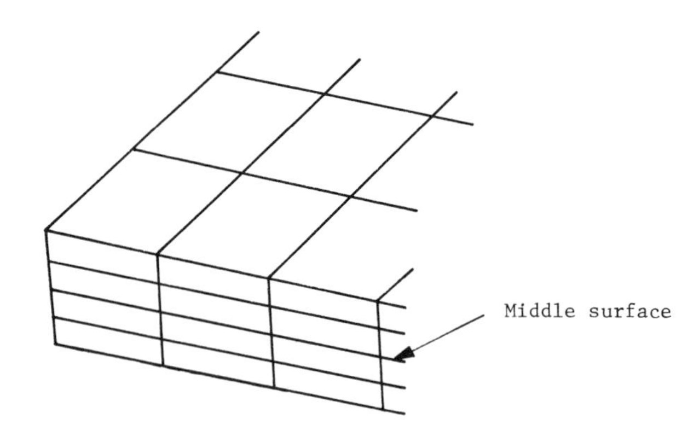

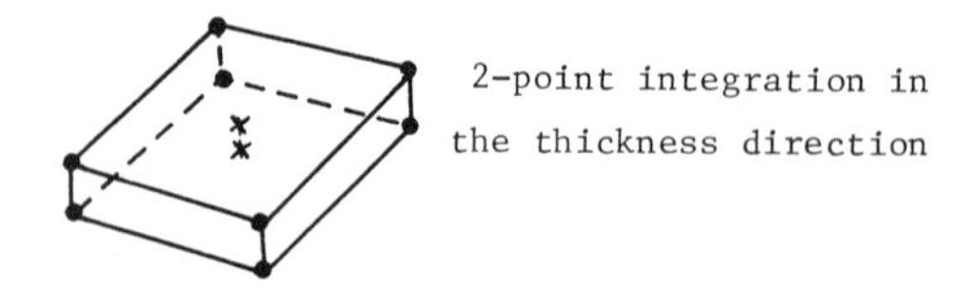

Figure 6.15 Isoparametric eight-node brick element.

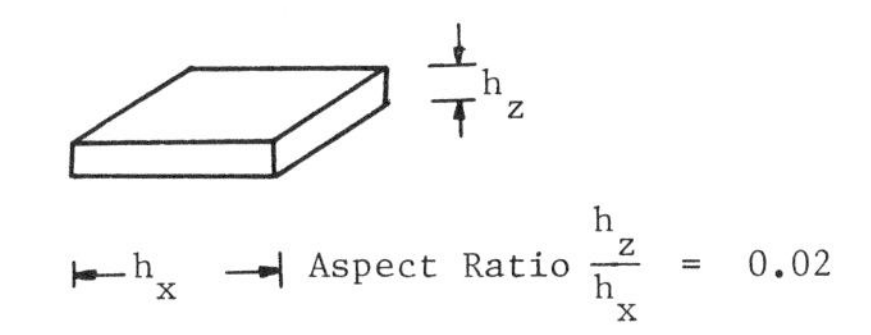

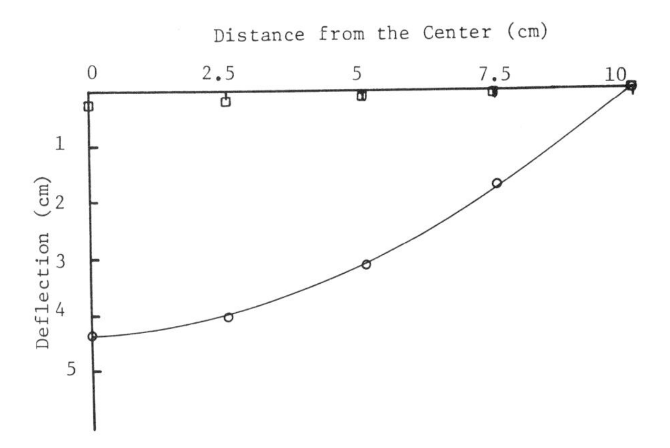

Figure 6.16 Deflection of centerline ($y = z = 0$): ○, reduced integration $1 \times 1 \times 2$ scheme; □, full integration $2 \times 2 \times 2$ scheme; ———, exact solution.

where $(r_\alpha, s_\alpha, t_\alpha)$ are the coordinates of the corner nodes in the normalized coordinate system. As shown in section 5.6.3, the reduced integration technique must be applied to eliminate unnecessary contribution to the shear stresses due to bending modes. For three-dimensional cases, we need apply reduced integration only in the plane direction, that is, the x and y directions, while ordinary integration rule is applied in the thickness direction, as shown in Figure 6.15.

Figure 6.16 shows the deflection of the plate along the centerline. It is clear that the result by three-dimensional elements with the selective reduced integration technique is fairly accurate. Indeed, only 1.4% error is observed, although the aspect ratio of elements is poorly 0.02 in the thickness direction.

As far as this example is concerned, three-dimensional solids are applicable and then plate bending can be simulated very well by three-dimensional linear elasticity. Now, is it possible to extend this idea in general?

Reference

Hughes, T. J. R., Taylor, R. L., and Kanoknukulchai, W., "A simple and efficient finite element for plate bending," *International Journal for Numerical Methods in Engineering* 11: 1529–43, 1977.

APPENDIX 1
FEM1

```
 1      C ********************************************************************
 2      C
 3      C                         PROGRAM : FEM1
 4      C
 5      C                          SPRING  1984
 6      C
 7      C ********************************************************************
 8      C
 9      C   PURPOSE:
10      C       PROGRAM FOR SOLVING HEAT CONDUCTION PROBLEMS USING
11      C       THREE-NODE LINEAR TRIANGULAR ELEMENTS
12      C
13      C   BOUNDARY VALUE PROBLEMS SOLVED:
14      C
15      C              - DIV ( K GRAD U ) = F      IN  D
16      C                            U    = G      ON S1
16      C                     K GRAD U.N = H      ON S2
18      C             K GRAD U.N = - A( U - UO ) ON S3
19      C
20      C                K  = HEAT CONDUCTIVITY
21      C                U  = TEMPERATURE
22      C                F  = HEAT SUPPLY
23      C                N  = UNIT VECTOR OUTWARD NORMAL TO THE BOUNDARY
24      C                G  = SPECIFIED BOUNDARY TEMPERATURE
25      C                H  = SPECIFIED HEAT FLUX THROUGH THE BOUNDARY
26      C                UO = OUTSIDE TEMPERATURE
27      C                A  = MATERIAL CONSTANT
28      C
29      C   NOTE:
30      C       THE PROGRAM ASSUMES THE DOUBLE PRECESSION: IMPLICIT REAL*8
31      C       (A-H,O-Z).
32      C       THE GLOBAL STIFFNESS MATRIX IS STORED IN THE ONE
33      C       DIMENSIONAL ARRAY USING THE INDEX TRANSFORMATION.
34      C       SK(I,J) --> SK(I,J-I+1) --> SK(MBAND*(I-1)+J-I+1)  FOR I<=J
35      C
36      C   CONTROL NUMBERS:
37      C       LBOUN1 = 1  IF THE 1-ST BOUNDARY CONDITION IS ASSUMED
38      C       LBOUN2 = 1  IF THE 2-ND BOUNDARY CONDITION IS ASSUMED
39      C       LBOUN3 = 1  IF THE 3-RD BOUNDARY CONDITION IS ASSUMED
40      C       LHEATG = 1  IF THE HEAT SUPPLY IS ASSUMED
41      C
42      C   STRUCTURE OF THE PROGRAM:
43      C       PRE PROCESSING
44      C          SUBROUTINE PTITLE   --------------- READ THE TITLE OF THE
45      C                                              PROBLEM
46      C          MAIN PROGRAM ---------------------- READ CONTROL NUMBERS
47      C          SUBROUTINE HRMESH   --------------- READ COORDINATES AND
48      C                                              ELEMENT CONNECTIVITIES
49      C          SUBROUTINE MATERL   --------------- READ CONDUCTIVITIES
50      C          SUBROUTINE HBANDW   --------------- COMPUTE HALF-BAND WIDTH
51      C                                              OF STIFFNESS MATRIX
52      C       FINITE ELEMENT PROCESSING
53      C          SUBROUTINE ASEMBO   --------------- ASSEMBLE ELEMENT
54      C                                              STIFFNESS MATRICES
55      C                                              AND LOAD VECTORS
56      C               SUBROUTINE ESTIF3  ----------- ELEMENT STIFFNESS
57      C                                              MATRIX
58      C                    SUBROUTINE GRADI3  ------ GRADIENT VECTOR OF
59      C                                              SHAPE FUNCTIONS
60      C          SUBROUTINE HEATGE ----------------- HEAT SOURCES/SINKS
61      C                                              ARE INPUT IN THE LOAD
62      C                                              VECTOR
63      C          SUBROUTINE BOUND1   --------------- 1-ST BOUNDARY CONDITION
64      C          SUBROUTINE BOUND2   --------------- 2-ND BOUNDARY CONDITION
65      C          SUBROUTINE BOUND3   --------------- 3-RD BOUNDARY CONDITION
66      C          SUBROUTINE BANDSL   --------------- SOLVES THE SYSTEM OF
67      C                                              LINEAR EQUATIONS
68      C       POST PROCESSING
69      C          SUBROUTINE OUTPUT   --------------- OUTPUT THE RESULTS
70      C          SUBROUTINE QUALIT ----------------- COMPUTE QUALITY INDICES
71      C                                              TO CHECK THE RESULTS
72      C   INPUT DATA TO BE READ:
73      C       1) SUBROUTINE PTITLE
74      C            1.0 FLAG(I),I=1,10                 :    10A8
75      C            1.1 TITLE(I),I=1,10                :    10A8
```

```
 76   C        2) MAIN : CONTROL NUMBERS
 77   C             2.0 FLAG(I),I=1,10                     :    10A8
 78   C             2.1 LBOUN1,REM(I),I=1,9                :    I5,9A8
 79   C             2.2 LBOUN2,REM(I),I=1,9                :    I5,9A8
 80   C             2.3 LBOUN3,REM(I),I=1,9                :    I5,9A8
 81   C             2.4 LHEATG,REM(I),I=1,9                :    I5,9A8
 82   C        3) SUBROUTINE HRMESH
 83   C             3.0 FLAG(I),I=1,10                     :    10A8
 84   C             3.1 NODADD,NELADD                      :    2I5
 85   C             3.2 II,X(II),Y(II),I=1,NODADD          :    I5,2F10.4
 86   C             3.3 NN,MPE(NN),(IJK(J,NN),J=1,3),
 87   C                 N=1,NELADD                         :    5I5
 88   C        4) SUBROUTINE MATERL
 89   C             4.0 FLAG(I),I=1,10                     :    10A8
 90   C             4.1 MATX                               :    16I5
 91   C             4.2 I,(XYK(J,I),J=1,3),I=1,MATX        :    I5,3F10.4
 92   C        5) SUBROUTINE HEATGE   IF   LHEATG =1
 93   C             5.0 FLAG(I),I=1,10                     :    10A8
 94   C             5.1 NHEATG                             :    16I5
 95   C             IF(NHEATG.LT.0) GO TO 5.3
 96   C             DO 100 I=1,NHEATG
 97   C             5.2 CXY(I,J),J=1,NHEATG                :    8F10.4
 98   C             100 CONTINUE
 99   C             GO TO 7)
100   C             5.3 NODE,VALUE,I=1,-NHEATG             :    I5,F10.4
101   C        6) SUBROUTINE BOUND1   IF   LBOUN1 =1
102   C             6.0 FLAG(I),I=1,10                     :    10A8
103   C             6.1 NBX,PENALTY                        :    I5,F20.5
104   C             6.2 NODE,VALUE,I=1,NBX                 :    I5,F10.4
105   C        7) SUBROUTINE BOUND2   IF   LBOUN2 =1
106   C             7.0 FLAG(I),I=1,10                     :    10A8
107   C             7.1 NBX                                :    16I5
108   C             7.2 NODE1,NODE2,VALUE,I=1,NBX          :    2I5,F10.4
109   C        8) SUBROUTINE BOUND3   IF   LBOUN3 =1
110   C             8.0 FLAG(I),I=1,10                     :    10A8
111   C             8.1 NBLX,MPBX                          :    16I5
112   C             8.2 I,(IJB(J,I),J=1,2),MPB(I),
113   C                 I=1,NBLX                           :    4I5
114   C             8.3 I,XKB(I),UO(I),I=1,MPBX            :    I5,2F10.4
115   C
116   C        *** LOOK AT EACH SUBROUTINE BEFORE MAKING INPUT DATA
117   C
118   C    DIMENSIONING:
119   C        SK(NX*MBAND),F(NX),U(NX),X(NX),Y(NX),CQX(NX),CQY(NX)
120   C           NX    = TOTAL NUMBER OF NODAL POINTS IN THE DOMAIN
121   C           MBAND = (HALF-)BAND WIDTH OF THE STIFFNESS MATRIX
122   C           SK    = BANDED STIFFNESS MATRIX
123   C           F     = GENERALIZED LOAD VECTOR
124   C           U     = SOLUTION VECTOR
125   C           X,Y   = COORDINATES OF NODAL POINTS
126   C           CQX   = QUALITY INDEX AT NODAL POINTS (TEMPERATURE)
127   C           CQY   = QUALITY INDEX AT NODAL POINTS (HEAT FLUX)
128   C        IJK(NODE,NELX),MP(NELX)
129   C           NELX  = TOTAL NUMBER OF FINITE ELEMENTS
130   C           NODE  = TOTAL NUMBER OF NODAL POINTS IN AN ELEMENT
131   C           IJK   = ELEMENT CONNECTIVITIES
132   C           MP    = MATERIAL GROUP OF THE FINITE ELEMENTS
133   C        XYK(3,MPX)
134   C           MPX   = TOTAL NUMBER OF MATERIAL GROUPS
135   C           XYK   = MATERIAL PROPERTIES (KXX,KYY,KXY)
136   C        IJB(NODE3,NBX),MPB(NBX),XKB(MBX),UO(MBX)
137   C           NODE3 = TOTAL NUMBER OF NODAL POINTS ON A LINE ELEMENT
138   C           NBX   = TOTAL NUMBER OF LINE ELEMENTS
139   C           MBX   = TOTAL NUMBER OF DIFFERENT MATERIAL GROUPS ALONG
140   C                   THE 3-RD BOUNDARY
141   C           IJB   = ELEMENT CONNECTIVITIES ON THE 3-RD BOUNDARY
142   C           MPB   = GROUP NUMBER OF MATERIALS
143   C           XKB   = COEFFIENTS OF HEAT TRANSFER
144   C           UO    = OUTSIDE TEMPERATURE SPECIFIED
145   C        P(4,NELX)
146   C           P     = STORES THE HEAT FLUX VECTORS AND QUALITY INDICES
147   C        FLAG(10),REM(9)
148   C           FLAG  = DESCRIPTION OF THE DATA BLOCKS
149   C                   ( ANY COMMENTS WITHIN 10A8 CAN BE INPUT IN ORDER TO
150   C                   MEMORIZE THE INPUT DATA IN EACH SUBROUTINE.  IF YOU
151   C                   DO NOT WANT IT, THEN SKIP ONE LINE IN THE DATA FILE )
152   C           REM   = REMARK TO THE INPUT DATA
153   C
154   C    DATA STATEMENT: < IMPORTANT >
```

```
155     C        LR      = DISK NUMBER FOR READ STATEMENTS   ( STD IS 5 )
156     C        LW      = DISK NUMBER FOR WRITE STATEMENTS ( STD IS 6 )
157     C        LP      = DISK NUMBER FOR WRITE STATEMENTS FOR PLOTTING
158     C                  ( STD IS 8 )
159     C        NDIM    > MBAND*NX*NDF
160     C        NODE    = TOTAL NUMBER OF NODAL POINTS IN AN ELEMENT ( = 3 )
161     C        NDF     = NUMBER OF DEGREES OF FREEDOM PER NODAL POINT ( = 1 )
162     C
163     C ********************************************************************
164     C
165           IMPLICIT REAL*8(A-H,O-Z)
166           DIMENSION SK(50000),F(300),U(300),X(300),Y(300),
167          1          IJK(3,500),MP(500),
168          3          XYK(3,10),
169          4          IJB(2,100),MPB(100),XKB(30),UO(30),
170          5          P(4,500),CQX(300),CQY(300),
171          6          FLAG(10),REM(10)
172           DATA LR,LW,LP,NDIM,NODE,NDF/5,6,8,50000,3,1/
173     C
174     C *****************    PRE-PROCESSING    ********************
175     C
176     C     ( READ THE TITLE OF THE PROBLEM AND CONTROL NUMBERS )
177     C
178                 CALL PTITLE(LR,LW)
179     C
180           READ(LR,502) (FLAG(I),I=1,10)
181       502 FORMAT(10A8)
182           WRITE(LW,602)
183       602 FORMAT(/////10X,'XXXXXXXXXX CONTROL NUMBERS XXXXXXXXXX',//)
184       500 FORMAT(I5,9A8)
185       600 FORMAT(/10X,I3,3X,9A8)
186           READ(LR,500)  LBOUN1,(REM(I),I=1,9)
187           WRITE(LW,600) LBOUN1,(REM(I),I=1,9)
188           READ(LR,500)  LBOUN2,(REM(I),I=1,9)
189           WRITE(LW,600) LBOUN2,(REM(I),I=1,9)
190           READ(LR,500)  LBOUN3,(REM(I),I=1,9)
191           WRITE(LW,600) LBOUN3,(REM(I),I=1,9)
192           READ(LR,500)  LHEATG,(REM(I),I=1,9)
193           WRITE(LW,600) LHEATG,(REM(I),I=1,9)
194     C
195     C     ( READ THE FINITE ELEMENT MODEL )
196     C
197                 CALL HRMESH(NX,NELX,X,Y,IJK,NODE,MP,LR,LW)
198     C
199     C     ( READ THE MATERIAL PROPERTIES )
200     C
201                 CALL MATERL(XYK,LR,LW)
202     C
203     C     ( FIND THE HALF-BAND WIDTH OF THE STIFFNESS MATRIX )
204     C
205                 CALL HBANDW(MBAND,IJK,NELX,NODE,NDF,LW)
206     C
207           IF(NX*MBAND.LE.NDIM) GO TO 100
208           WRITE(LW,644)
209       644 FORMAT(/////,'ARRAY FOR THE STIFFNESS MATRIX IS TOO SMALL')
210           STOP
211       100 CONTINUE
212     C
213     C ************    FINITE ELEMENT-PROCESSING    ***************
214     C
215     C     ( FORM THE STIFFNESS MATRIX AND THE LOAD VECTOR )
216     C
217                 CALL ASEMBO(SK,F,MBAND,NX,NELX,IJK,NODE,NDF,MP,X.Y.XYK,LW)
218     C
219     C     ( ROUTINE FOR THE HEAT SUPPLY TERM )
220     C
221           IF(LHEATG.EQ.1) CALL HEATGE(F,NX,NELX,IJK,X,Y,LR,LW)
222     C
223     C     ( ROUTINES FOR THE BOUNDARY CONDITIONS )
224     C
225           IF(LBOUN1.EQ.1) CALL BOUND1(MBAND,SK,F,LR,LW)
226           IF(LBOUN2.EQ.1) CALL BOUND2(F,X,Y,LR.LW)
227           IF(LBOUN3.EQ.1) CALL BOUND3(MBAND,SK,F,X,Y,IJB,MPB,XKB,UO,LR,LW)
228     C
229     C     ( SOLVE THE SYSTEM OF LINEAR EQUATIONS )
230     C
231                 CALL BANDSL(SK,F,NX,MBAND,0)
232     C
```

```
233     C ****************   POST-PROCESSING   ********************
234     C
235     C     ( PRINT OUT THE RESULTS )
236     C
237               CALL OUTPUT(F,P,NX,NELX,X,Y,IJK,MP,XYK,LW)
238     C
239               CALL QUALIT(NELX,NX,MBAND,IJK,X,Y,P,SK,CQX,CQY,NODE,LW)
240     C
241     C     ( STORE THE RESULTS FOR THE PLOTTING ROUTINE )
242     C
243           WRITE(LP,810)
244       810 FORMAT('FINITE ELEMENT MODEL')
245           WRITE(LP,800) NX,NELX
246       800 FORMAT(2I5)
247           WRITE(LP,820)
248       820 FORMAT(' < NODAL COORDINATES >')
249           WRITE(LP,802) (X(I),Y(I),I=1,NX)
250       802 FORMAT(8F10.4)
251           WRITE(LP,812)
252       812 FORMAT(' < ELEMENT CONNECTIVITIES >')
253           WRITE(LP,804) ((IJK(J,I),J=1,NODE),I=1,NELX)
254       804 FORMAT(16I5)
255           WRITE(LP,814)
256       814 FORMAT('TEMPERATURE')
257           WRITE(LP,806) (F(I),I=1,NX)
258       806 FORMAT(8E10.3)
259           WRITE(LP,816)
260       816 FORMAT('HEAT FLUX AND QUALITY INDICES')
261           WRITE(LP,808) ((P(J,I),J=1,4),I=1,NELX)
262       808 FORMAT(8E10.3)
263           WRITE(LP,818)
264       818 FORMAT('COMFORMING QUALITY INDICES')
265           WRITE(LP,808) (CQX(I),CQY(I),I=1,NX)
266     C
267           STOP
268           END
269     C ---------------------------------------------------------------
270     C
271           SUBROUTINE PTITLE(LR,LW)
272     C
273     C   PURPOSE:
274     C       PRINT OUT THE TITLE OF THE PROBLEM TO BE SOLVED.
275     C
276     C   INPUT DATA TO BE READ:
277     C       0. FLAG(I),I=1,10                       :    10A8
278     C       1. TITLE(I)                             :    10A8
279     C
280     C ---------------------------------------------------------------
281     C
282           IMPLICIT REAL*8(A-H,O-Z)
283           DIMENSION TITLE(10),FLAG(10)
284     C
285           READ(LR,500) (FLAG(I),I=1,10)
286           READ(LR,500) (TITLE(I),I=1,10)
287       500 FORMAT(10A8)
288     C
289           WRITE(LW,600) (TITLE(I),I=1,10)
290       600 FORMAT(1H1,///////////////5X,75('*'),/5X,75('*'),////,15X,
291          1 10A8,////5X,75('*'),/5X,75('*'))
292     C
293           RETURN
294           END
295     C ---------------------------------------------------------------
296     C
297           SUBROUTINE HRMESH(NX,NELX,X,Y,IJK,NODE,MPE,LR,LW)
298     C
299     C   PURPOSE:
300     C       READ THE DATA OF THE FINITE ELEMENT MODEL:  COORDINATES OF
301     C       NODAL POINTS, ELEMENT CONNECTIVITIES.
302     C
303     C   ARGUMENTS:
304     C       NX    = TOTAL NUMBER OF NODAL POINTS
305     C       NELX  = TOTAL NUMBER OF FINITE ELEMENTS
306     C       X,Y   = COORDINATES OF NODAL POINTS
307     C       IJK   = ELEMENT CONNECTIVITIES
308     C       NODE  = TOTAL NUMBER OF NODAL POINTS IN AN ELEMENT
309     C       MPE   = GROUP NUMBER OF THE MATERIALS FOR EACH ELEMENT
310     C       LR    = CONTROL NUMBER OF THE READ STATEMENT
```

```
311    C        LW    = CONTROL NUMBER OF THE WRITE STATEMENT
312    C
313    C   LOCAL VARIABLES:
314    C       NODES  = FIRST NODE NUMBER ADDED
315    C       NELMS  = FIRST ELEMENT NUMBER ADDED
316    C       NODADD = TOTAL NUMBER OF ADDITIONAL NODAL POINTS
317    C       NELADD = TOTAL NUMBER OF ADDITIONAL ELEMENTS
318    C
319    C   EXAMPLE:
320    C
321    C                         N3
322    C                       /  I            IJK(1,NEL)=N1
323    C                     /    I            IJK(2,NEL)=N2
324    C                   /      I            IJK(3,NEL)=N3
325    C                 /   NEL  I
326    C               /          I
327    C             N1----------N2
328    C
329    C   INPUT DATA TO BE READ:
330    C       0. FLAG(I),I=1,10                     :    10A8
331    C       1. NODADD,NELADD                      :    16I5
332    C       2. II,X(II),Y(II),I=1,NODADD          :    I5,2F10.4
333    C       3. NN,MPE(NN),(IJK(J,NN),J=1,NODE),
334    C          N=1,NELADD                         :    5I5
335    C
336    C ---------------------------------------------------------------------
337    C
338          IMPLICIT REAL*8(A-H,O-Z)
339          DIMENSION X(1),Y(1),IJK(NODE,1),MPE(1),FLAG(10)
340    C
341          READ(LR,598) (FLAG(I),I=1,10)
342      598 FORMAT(10A8)
343          READ(LR,500) NODADD,NELADD
344      500 FORMAT(16I5)
345          READ(LR,598) (FLAG(I),I=1,10)
346          DO 100 I=1,NODADD
347          READ(LR,502) II,X(II),Y(II)
348      502 FORMAT(I5,2F10.4)
349          IF(II.GT.NX) NX=II
350      100 CONTINUE
351          READ(LR,598) (FLAG(I),I=1,10)
352          DO 102 N=1,NELADD
353          READ(LR,500) NN,MPE(NN),(IJK(J,NN),J=1,NODE)
354          IF(NN.GT.NELX) NELX=NN
355      102 CONTINUE
356    C
357          WRITE(LW,600)
358      600 FORMAT(/////10X,'XXXXXXXXXX HRMESH : ELEMENTS AND NODES ',
359         1 'XXXXXXXXXX',//15X,
360         2 '< ELEMENT CONNECTIVITY >',/)
361    C
362          DO 104 I=1,NELX
363      104 WRITE(LW,602) I,MPE(I),(IJK(J,I),J=1,NODE)
364      602 FORMAT(10X,I5,3X,I5,5X,'<IJK>',9I5)
365          WRITE(LW,604)
366      604 FORMAT(////15X,'< COORDINATES >',/)
367          WRITE(LW,606) (I,X(I),Y(I),I=1,NX)
368      606 FORMAT(5X,I5,3X,'<',2F10.4,1X,'>',5X,I5,3X,'<',2F10.4,
369         1 2X,'>')
370    C
371          RETURN
372          END
373    C ---------------------------------------------------------------------
374    C
375          SUBROUTINE MATERL(XYK,LR,LW)
376    C
377    C   PURPOSE:
378    C       READ MATERIAL CONSTANTS KXX,KYY, AND KXY.
379    C
380    C   ARGUMENTS:
381    C       XYK = MATERIAL CONSTANTS IN THE MATRIX FORM
382    C             XYK(1,N) = KXX IN THE N-TH GROUP
383    C             XYK(2,N) = KYY IN THE N-TH GROUP
384    C             XYK(3,N) = KXY IN THE N-TH GROUP
385    C
386    C   LOCAL VARIABLES:
387    C       FLAG = ARRAY FOR THE REMARK
388    C       MATX = TOTAL NUMBER OF KINDS OF MATERIALS
389    C
```

```
390      C    INPUT DATA TO BE READ:
391      C        0. FLAG(I),I=1,10                  :    10A8
392      C        1. MATX                            :    16I5
393      C        2. I,(XYK(J,I),J=1,3),I=1,MATX :    I5,3F10.4
394      C
395      C -------------------------------------------------------------------
396      C
397             IMPLICIT REAL*8(A-H,O-Z)
398             DIMENSION XYK(3,1),FLAG(10)
399      C
400             READ(LR,500) (FLAG(I),I=1,10)
401         500 FORMAT(10A8)
402             READ(LR,502) MATX
403         502 FORMAT(16I5)
404             READ(LR,504) (I,(XYK(J,I),J=1,3),I=1,MATX)
405         504 FORMAT(I5,3F10.4)
406      C
407             WRITE(LW,600) MATX
408         600 FORMAT(/////10X,'XXXXXXXXXX MATERIAL CONSTANTS XXXXXXXXXX',//13X,
409            1 'NUMBER OF DIFFERENT MATERIALS =',I3,///,10X,
410            2 '< MATERIAL CONSTANTS >',/)
411             WRITE(LW,602) (I,(XYK(J,I),J=1,3),I=1,MATX)
412         602 FORMAT(13X,I3,2X,'KX=',E10.3,2X,'KY=',E10.3,2X,'KXY=',E10.3)
413      C
414             RETURN
415             END
416      C -------------------------------------------------------------------
417      C
418             SUBROUTINE HBANDW(MBAND,IJK,NELX,NODE,NDF,LW)
419      C
420      C    PURPOSE:
421      C        COMPUTE THE HALF-BAND WIDTH OF THE STIFFNESS MATRIX.
422      C
423      C    ARGUMENTS:
424      C        MBAND = HALF-BAND WIDTH OF THE STIFFNESS MATRIX
425      C        IJK   = ELEMENT CONNECTIVITIES
426      C        NELX  = TOTAL NUMBER OF FINITE ELEMENTS
427      C        NODE  = TOTAL NUMBER OF NODAL POINTS IN AN ELEMENT
428      C        NDF   = NUMBER OF DEGREES OF FREEDOM PER NODAL POINT
429      C
430      C -------------------------------------------------------------------
431      C
432             DIMENSION IJK(NODE,1)
433      C
434             MBAND=1
435             DO 100 NEL=1,NELX
436             DO 100 IA=1,NODE
437             IJKIA=IJK(IA,NEL)
438             DO 100  I=1,NDF
439             IJKIAI=NDF*(IJKIA-1)+I
440             DO 100 JB=1,NODE
441             IJKJB=IJK(JB,NEL)
442             DO 100  J=1,NDF
443             IJKJBJ=NDF*(IJKJB-1)+J
444             IAIJBJ=IJKJBJ-IJKIAI+1
445             IF(IAIJBJ.GT.MBAND) MBAND=IAIJBJ
446         100 CONTINUE
447      C
448             WRITE(LW,600) MBAND
449         600 FORMAT(/////10X,'XXXXXXXXXX BAND WIDTH XXXXXXXXXX',/15X,
450            1  'MBAND = ',I8)
451      C
452             RETURN
453             END
454      C -------------------------------------------------------------------
455      C
456             SUBROUTINE ASEMBO(SK,F,MBAND,NX,NELX,IJK,NODE,NDF,MP,X,Y,XYK,LW)
457      C
458      C    PURPOSE:
459      C        CONSTRUCTS THE GLOBAL STIFFNESS MATRIX AND LOAD VECTOR.
460      C
461      C    ARGUMENTS:
462      C        SK    = GLOBAL STIFFNESS MATRIX
463      C        F     = GLOBAL LOAD VECTOR
464      C        MBAND = HALF-BAND WIDTH OF THE STIFFNESS MATRIX
465      C        NX    = TOTAL NUMBER OF NODAL POINTS
466      C        NELX  = TOTAL NUMBER OF FINITE ELEMENTS
467      C        IJK   = ELEMENT CONNECTIVITIES
468      C        NODE  = TOTAL NUMBER OF NODAL POINTS IN AN ELEMENT
```

```
469   C        NDF    = NUMBER OF DEGREES OF FREEDOM PER NODAL POINT
470   C        MP     = GROUP NUMBER OF THE MATERIALS
471   C        X,Y    = COORDINATES OF NODAL POINTS
472   C        XYK    = MATERIAL CONSTANTS (KXX,KYY,KXY)
473   C
474   C   LOCAL VARIABLES:
475   C        XE,YE = COORDINATES OF NODAL POINTS IN AN ELEMENT
476   C        SKE   = ELEMENT STIFFNESS MATRIX
477   C
478   C ------------------------------------------------------------------
479   C
480          IMPLICIT REAL*8(A-H,O-Z)
481          DIMENSION SK(1),IJK(NODE,1),MP(1),X(1),Y(1),XYK(3,1),F(1)
482          DIMENSION SKE(3,3),FE(3),XE(3),YE(3),XYKE(2,2)
483   C
484   C      ( INITIALIZATION )
485   C
486          IMAX=NX*NDF
487          DO 100 I=1,IMAX
488      100 F(I)=0.DO
489          IMAX=MBAND*NX*NDF
490          DO 110 I=1,IMAX
491      110 SK(I)=0.DO
492   C
493          DO 102 NEL=1,NELX
494   C
495   C      ( SET COORDINATES OF NODAL POINTS IN AN ELEMENT )
496   C
497          DO 104 IA=1,NODE
498          IJKIA=IJK(IA,NEL)
499          XE(IA)=X(IJKIA)
500      104 YE(IA)=Y(IJKIA)
501   C
502   C      ( SET THE MATERIAL PROPERTIES OF THE ELEMENT )
503   C
504          KKK=MP(NEL)
505          XYKE(1,1)=XYK(1,KKK)
506          XYKE(1,2)=XYK(3,KKK)
507          XYKE(2,1)=XYKE(1,2)
508          XYKE(2,2)=XYK(2,KKK)
509   C
510   C      ( CONSTRUCT THE ELEMENT STIFFNESS MATRIX )
511   C
512                CALL ESTIF3(NEL,SKE,FE,AREA,XE,YE,XYKE,LW)
513   C
514   C      ( ADD THE ELEMENT STIFFNESS MATRIX TO THE GLOBAL ONE )
515   C
516          DO 108 IA=1,NODE
517          IJKIA=IJK(IA,NEL)
518          DO 108  I=1,NDF
519          IAI=NDF*(IA-1)+I
520          IJKIAI=NDF*(IJKIA-1)+I
521          F(IJKIAI)=F(IJKIAI)+FE(IAI)
522          DO 108 JB=1,NODE
523          IJKJB=IJK(JB,NEL)
524          DO 108  J=1,NDF
525          JBJ=NDF*(JB-1)+J
526          IJKJBJ=NDF*(IJKJB-1)+J
527          IAIJBJ=IJKJBJ-IJKIAI+1
528          IF(IAIJBJ.LE.O) GO TO 108
529          LOCAT=MBAND*(IJKIAI-1)+IAIJBJ
530          SK(LOCAT)=SK(LOCAT)+SKE(IAI,JBJ)
531      108 CONTINUE
532   C
533      102 CONTINUE
534   C
535          RETURN
536          END
537   C ------------------------------------------------------------------
538   C
539          SUBROUTINE ESTIF3(NEL,SKE,FE,AREA,XE,YE,XYKE,LW)
540   C
541   C   PURPOSE:
542   C       CONSTRUCT THE ELEMENT STIFFNESS MATRIX BY USING THREE-NODE
543   C       TRIANGULAR ELEMENT.
544   C
545   C   ARGUMENTS:
546   C        NEL    = ELEMENT NUMBER
```

```
547      C         SKE    = ELEMENT STIFFNESS MATRIX
548      C         FE     = ELEMENT LOAD VECTOR
549      C         AREA   = AREA OF THE TRIANGLE
550      C         XE,YE  = COORDINATES OF THREE NODAL POINTS IN AN ELEMENT
551      C         XYKE   = MATERIAL CONSTANTS (KXX,KXY,KYX,KYY)
552      C
553      C ----------------------------------------------------------------------
554      C
555            IMPLICIT REAL*8(A-H,O-Z)
556            DIMENSION SKE(3,3),FE(3),XE(3),YE(3),DN(3,2),XYKE(2,2)
557      C
558      C     ( INITIALIZATION )
559      C
560            DO 200 I=1,3
561            FE(I)=0.DO
562            DO 200 J=1,3
563        200 SKE(I,J)=0.DO
564      C
565      C     ( COMPUTE THE FIRST GRADIENT OF THE SHAPE FUNCTIONS )
566      C
567                  CALL GRADI3(NEL,XE,YE,DN,AREA,LW)
568      C
569      C     ( CONSTRUCT THE STIFFNESS MATRIX )
570      C
571            DO 100 I=1,3
572            DO 100 J=1,3
573            SKEIJ=0.
574            DO 102 K=1,2
575            DO 102 L=1,2
576        102 SKEIJ=SKEIJ+XYKE(K,L)*DN(I,K)*DN(J,L)
577        100 SKE(I,J)=SKEIJ*AREA
578      C
579            RETURN
580            END
581      C ----------------------------------------------------------------------
582      C
583            SUBROUTINE GRADI3(NEL,XE,YE,DN,AREA,LW)
584      C
585      C   PURPOSE:
586      C       CONSTRUCT THE FIRST DERIVATIVES OF THE SHAPE FUNCTIONS.
587      C
588      C   ARGUMENTS:
589      C       NEL    = ELEMENT NUMBER
590      C       XE,YE  = COORDINATES OF NODAL POINTS IN AN ELEMENT
591      C       DN     = FIRST DERIVATIVES OF THE SHAPE FUNCTIONS
592      C       AREA   = AREA OF THE TRIANGLE
593      C
594      C ----------------------------------------------------------------------
595      C
596            IMPLICIT REAL*8(A-H,O-Z)
597            DIMENSION XE(3),YE(3),DN(3,2)
598      C
599            DET=XE(2)*(YE(3)-YE(1))+XE(3)*(YE(1)-YE(2))+XE(1)*(YE(2)-YE(3))
600            AREA=0.5DO*DET
601            IF(DET.LE.1.E-30) GO TO 100
602      C
603            DN(1,1)=(YE(2)-YE(3))/DET
604            DN(2,1)=(YE(3)-YE(1))/DET
605            DN(3,1)=(YE(1)-YE(2))/DET
606            DN(1,2)=(XE(3)-XE(2))/DET
607            DN(2,2)=(XE(1)-XE(3))/DET
608            DN(3,2)=(XE(2)-XE(1))/DET
609      C
610            RETURN
611      C
612        100 WRITE(LW,600) NEL, (XE(I),YE(I),I=1,3)
613        600 FORMAT(///10X,'******** DETERMINANT IS LESS THAN ZERO ********',
614           1 /,15X,I5,5X,3('<',2F8.3,'>'))
615            STOP
616      C
617            END
618      C ----------------------------------------------------------------------
619      C
620            SUBROUTINE HEATGE(F,NX,NELX,IJK,X,Y,LR,LW)
621      C
622      C   PURPOSE:
623      C       CONSTRUCT THE GENERALIZED LOAD VECTOR DUE TO THE HEAT
624      C       GENERATION.
```

```
625   C
626   C   ARGUMENTS:
627   C       F    = GENERALIZED LOAD VECTOR
628   C       NX   = TOTAL NUMBER OF NODAL POINTS
629   C       NELX = TOTAL NUMBER OF FINITE ELEMENTS
630   C       IJK  = ELEMENT CONNECTIVITIES
631   C       X,Y  = COORDINATES OF NODAL POINTS
632   C
633   C   LOCAL VARIABLES:
634   C       NHEATG = TOTAL NUMBER OF TERMS OF POLYNOMIALS
635   C              = 0  IF THERE ARE NO HEAT SOURCES/SINKS
636   C              = NEGATIVE   IF POINT SOURCES/SINKS EXIST.  IN THIS CASE
637   C                -NHEATG BECOMES THE NUMBER OF SOURCES/SINKS
638   C       CXY    = COEFFICIENTS OF THE POLYNOMIAL BASIS
639   C       XE,YE  = COORDINATES OF NODAL POINTS IN AN ELEMENT
640   C       XC,YC  = COORDINATES OF THE CENTROID OF AN ELEMENT
641   C       AREA   = AREA OF AN ELEMENT
642   C
643   C   DIMENSIONING:
644   C       CXY(M,M)    M >= NHEATG
645   C
646   C   NOTE:
647   C       M = 11  AND  8 >= NHEATG  ARE ASSUMED IN THIS ROUTINE.
648   C
649   C   INPUT DATA TO BE READ:
650   C       0. FLAG(I),I=1,10                 :   10A8
651   C       1. NHEATG                         :   16I5
652   C       IF(NHEATG.LT.0) GO TO 3
653   C       DO 100 I=1,NHEATG
654   C       2. CXY(I,J),J=1,NHEATG            :   8F10.4
655   C       100 CONTINUE
656   C       RETURN
657   C       3. NODE,VALUE,I=1,-NHEATG         :   I5,F10.4
658   C
659   C ------------------------------------------------------------------
660   C
661         IMPLICIT REAL*8(A-H,O-Z)
662         DIMENSION F(1),IJK(3,1),X(1),Y(1),XE(3),YE(3),CXY(11,11),
663        1          FLAG(10)
664   C
665   C     ( READ INPUT DATA FOR THE BODY FORCE )
666   C
667         READ(LR,504) (FLAG(I),I=1,10)
668     504 FORMAT(10A8)
669         READ(LR,500) NHEATG
670     500 FORMAT(16I5)
671         IF(NHEATG.LE.0) GO TO 300
672   C
673   C     ( DISTRIBUTED HEAT SOURCE/SINK )
674   C
675         DO 100 I=1,NHEATG
676     100 READ(LR,502) (CXY(I,J),J=1,NHEATG)
677     502 FORMAT(8F10.4)
678   C
679         DO 102 NEL=1,NELX
680   C
681   C     ( SET AREA AND COORDINATES OF THE CENTROID )
682   C
683         DO 104 I=1,3
684         IJKI=IJK(I,NEL)
685         XE(I)=X(IJKI)
686     104 YE(I)=Y(IJKI)
687         DET=XE(2)*(YE(3)-YE(1))+XE(3)*(YE(1)-YE(2))+XE(1)*(YE(2)-YE(3))
688         AREA=0.5D0*DET
689         XC=(XE(1)+XE(2)+XE(3))/3.
690         YC=(YE(1)+YE(2)+YE(3))/3.
691   C
692   C     ( COMPUTE THE BODY FORCE )
693   C
694         FNEL=0.
695         DO 106 I=1,NHEATG
696         I1=I-1
697         DO 106 J=1,NHEATG
698         J1=J-1
699     106 FNEL=FNEL+CXY(I,J)*(XC**I1)*(YC**J1)
700   C
701         DO 108 I=1,3
702         IJKI=IJK(I,NEL)
```

```
703          108 F(IJKI)=F(IJKI)+FNEL*AREA/3.
704      C
705          102 CONTINUE
706              WRITE(LW,600)
707          600 FORMAT(/////10X,'XXXXXXXXXX BODY FORCES XXXXXXXXXX',//15X,
708            1 '< COEFFECIENTS  CXY  >',/)
709              DO 110 I=1,NHEATG
710          110 WRITE(LW,602) (CXY(I,J),J=1,NHEATG)
711          602 FORMAT(15X,10F8.3)
712              GO TO 304
713      C
714      C       ( POINT HEAT SOURCES/SINKS )
715      C
716          300 IF(NHEATG.EQ.0) RETURN
717              NHEATG=-NHEATG
718              DO 302 I=1,NHEATG
719              READ(LR,530) NODE,VALUE
720          530 FORMAT(I5,F10.4)
721          302 F(NODE)=F(NODE)+VALUE
722      C
723          304 WRITE(LW,604)
724          604 FORMAT(//10X,'< LOAD VECTOR DUE TO HEAT GENERATION >',/)
725              WRITE(LW,606) (I,F(I),I=1,NX)
726          606 FORMAT(5X,I4,1X,E10.3,3X,I4,1X,E10.3,3X,I4,1X,E10.3)
727      C
728              RETURN
729              END
730      C -----------------------------------------------------------------------
731      C
732              SUBROUTINE BOUND1(MBAND,SK,F,LR,LW)
733      C
734      C    PURPOSE:
735      C        READ THE 1-ST BOUNDARY CONDITION AND MODIFY THE STIFFNESS
736      C        MATRIX AND THE LOAD VECTOR USING PENALTY METHODS.
737      C
738      C    ARGUMENTS:
739      C        MBAND = HALF-BAND WIDTH OF THE STIFFNESS MATRIX
740      C        SK    = STIFFNESS MATRIX
741      C        F     = LOAD VECTOR
742      C
743      C    LOCAL VARIABLES:
744      C        NBX   = TOTAL NUMBER OF NODAL POINTS ON WHICH 1-ST BOUNDARY
745      C                CONDITION IS APPLIED
746      C        PENALT= PENALTY PARAMETER  (TO BE A SUFFICIENTLY LARGE NUMBER )
747      C                10000.*MAX(XK,YK) IS RECOMMENDED
748      C        XK,YK = HEAT CONDUCTIVITY IN THE X AND Y DIRECTIONS
749      C        NODE  = NODE NUMBER ON WHICH BOUNDARY CONDITION IS GIVEN
750      C        VALUE = BOUNDARY VALUE SPECIFIED
751      C
752      C    INPUT DATA TO BE READ:
753      C        0. FLAG(I),I=1,10              :    10A8
754      C        1. NBX,PENALT                  :    I5,F20.5
755      C        2. NODE,VALUE,I=1,NBX          :    I5,F10.4
756      C
757      C -----------------------------------------------------------------------
758      C
759              IMPLICIT REAL*8(A-H,O-Z)
760              DIMENSION SK(1),F(1),FLAG(10)
761      C
762      C       ( READ INPUT DATA )
763      C
764              READ(LR,504) (FLAG(I),I=1,10)
765          504 FORMAT(10A8)
766              READ(LR,500) NBX,PENALT
767          500 FORMAT(I5,F20.5)
768              WRITE(LW,600) PENALT
769          600 FORMAT(/////10X,'XXXXXXXXXX 1-ST BOUNDARY CONDITION XXXXXXXXXX',/
770            1 15X,'PENALTY PARAMETER =',E10.3,//)
771      C
772              DO 100 I=1,NBX
773              READ(LR,502) NODE,VALUE
774          502 FORMAT(I5,F10.4)
775              WRITE(LW,602) I,NODE,VALUE
776          602 FORMAT(15X,'<',I3,'>',4X,I4,3X,F10.4)
777      C
778              SK(MBAND*(NODE-1)+1)=SK(MBAND*(NODE-1)+1)+PENALT
779              F(NODE)=F(NODE)+PENALT*VALUE
780      C
```

```
781           100 CONTINUE
782      C
783              RETURN
784              END
785      C ---------------------------------------------------------------------
786      C
787              SUBROUTINE BOUND2(F,X,Y,LR,LW)
788      C
789      C    PURPOSE:
790      C       READ THE 2-ND BOUNDARY CONDITION AND FORM THE LOAD VECTOR.
791      C       <  DU/DN = H  ON  S2  >
792      C
793      C    ARGUMENTS:
794      C        F   = GENRALIZED LOAD VECTOR
795      C        X,Y = COORDINATES OF NODAL POINTS
796      C
797      C    LOCAL VARIABLES:
798      C        FLAG  = REMARK FOR THE DATA
799      C        NBX   = TOTAL NUMBER OF LINE ELEMENTS ON THIS BOUNDARY
800      C        NODE1 = NUMBER OF THE LEFT NODAL POINT IN A LINE ELEMENT
801      C        NODE2 = NUMBER OF THE RIGHT NODAL POINT IN A LINE ELEMENT
802      C        VALUE = VALUE OF THE FLUX FROM THE BOUNDARY
803      C
804      C    INPUT DATA TO BE READ:
805      C        0. FLAG(I),I=1,10                       :    10A8
806      C        1. NBX                                  :    16I5
807      C        2. NODE1,NODE2,VALUE,I=1,NBX            :    2I5,F10.4
808      C
809      C ---------------------------------------------------------------------
810      C
811              IMPLICIT REAL*8(A-H,O-Z)
812              DIMENSION FLAG(10),F(1),X(1),Y(1)
813      C
814              READ(LR,500) (FLAG(I),I=1,10)
815          500 FORMAT(10A8)
816              READ(LR,502) NBX
817          502 FORMAT(16I5)
818              IF(NBX.LE.0) RETURN
819              WRITE(LW,600)
820          600 FORMAT(/////10X,'XXXXXXXXXX 2-ND BOUNDARY CONDITION XXXXXXXXXX',
821             1 //15X,29('-'),/15X,' NODE I     VALUE   ',/15X,29('-'))
822      C
823              DO 100 I=1,NBX
824              READ(LR,504) NODE1,NODE2,VALUE
825          504 FORMAT(2I5,F10.4)
826              WRITE(LW,602) NODE1,NODE2,VALUE
827          602 FORMAT(15X,2I5,' I ',E10.3)
828              DH=DSQRT((X(NODE1)-X(NODE2))**2+(Y(NODE1)-Y(NODE2))**2)
829              F(NODE1)=F(NODE1)+0.5D0*VALUE*DH
830              F(NODE2)=F(NODE2)+0.5D0*VALUE*DH
831          100 CONTINUE
832      C
833              WRITE(LW,604)
834          604 FORMAT(15X,29('-'))
835      C
836              RETURN
837              END
838      C ---------------------------------------------------------------------
839      C
840              SUBROUTINE BOUND3(MBAND,SK,F,X,Y,IJB,MPB,XKB,U0,LR,LW)
841      C
842      C    PURPOSE:
843      C        READ THE 3-RD BOUNDARY CONDITION AND MODIFY THE STIFFNESS
844      C        MATRIX AND THE LOAD VECTOR.
845      C        < DU/DN = -K( U - U0 )  ON  S3  >
846      C
847      C    ARGUMENTS:
848      C        MBAND = HALF-BAND WIDTH OF THE STIFFNESS MATRIX
849      C        SK    = GLOBAL STIFFNESS MATRIX
850      C        F     = LOAD VECTOR
851      C        X,Y   = COORDINATES OF THE NODAL POINTS
852      C        IJB   = ELEMENT CONNECTIVITY ON THE BOUNDARY
853      C        MPB   = GROUP NUMBER OF THE MATERIALS ON THE BOUNDARY
854      C        XKB   = CONSTANTS  "K"  ON THE BOUNDARY
855      C        U0    = CONSTANTS  "U0"  ON THE BOUNDARY
856      C
857      C    LOCAL VARIABLES:
858      C        FLAG   = REMARK TO THE INPUT DATA
```

```
859    C         NBLX    = TOTAL NUMBER OF BOUNDARY ELEMENTS
860    C         MPBX    = TOTAL NUMBER OF MATERIALS ON THE BOUNDARY
861    C
862    C    INPUT DATA TO BE READ:
863    C         0. FLAG(I),I=1,10                          :    10A8
864    C         1. NBLX,MPBX                               :    16I5
865    C         2. I,MPB(I),(IJB(J,I),J=1,2),I=1,NBLX      :    4I5
866    C         3  I,XKB(I),UO(I),I=1,MPBX                 :    I5,2F10.4
867    C
868    C -----------------------------------------------------------------
869    C
870          IMPLICIT REAL*8(A-H,O-Z)
871          DIMENSION SK(1),F(1),X(1),Y(1),IJB(2,1),MPB(1),XKB(1),UO(1)
872          DIMENSION SKE(2,2),FE(2),FLAG(10)
873    C
874    C      ( INPUT DATA )
875    C
876          READ(LR,500) (FLAG(I),I=1,10)
877      500 FORMAT(10A8)
878          READ(LR,502) NBLX,MPBX
879      502 FORMAT(16I5)
880          IF(NBLX.LE.0) RETURN
881          READ(LR,504) (I,MPB(I),(IJB(J,I),J=1,2),I=1,NBLX)
882      504 FORMAT(4I5)
883          READ(LR,506) (I,XKB(I),UO(I),I=1,MPBX)
884      506 FORMAT(I5,2F10.4)
885    C
886          WRITE(LW,600)
887      600 FORMAT(/////10X,'XXXXXXXXXX 3-RD BOUNDARY CONDITION XXXXXXXXXX',
888         1 //10X,'----- BOUNDARY ELEMENT CONNECTIVITY -----',/)
889          WRITE(LW,602) (I,MPB(I),(IJB(J,I),J=1,2),I=1,NBLX)
890      602 FORMAT(2(5X,I5,3X,'<MPB>',I3,5X,'<IJB>',2I5))
891          WRITE(LW,604)
892      604 FORMAT(///10X,'----- MATERIAL CONSTANTS ON THE BOUNDARY -----',
893         1 /)
894          WRITE(LW,606) (I,XKB(I),UO(I),I=1,MPBX)
895      606 FORMAT(10X,I5,3X,'K=',E10.3,2X,'UO=',E10.3)
896    C
897    C      ( FOR STIFFNESS MATRIX AND LOAD VECTOR )
898    C
899          DO 100 NEL=1,NBLX
900    C
901    C      ( SET VARIABLES )
902    C
903          N1=IJB(1,NEL)
904          N2=IJB(2,NEL)
905          X1=X(N1)
906          Y1=Y(N1)
907          X2=X(N2)
908          Y2=Y(N2)
909          HR=DSQRT((X2-X1)**2+(Y2-Y1)**2)
910          KKK=MPB(NEL)
911          XKBE=XKB(KKK)
912          UOE=UO(KKK)
913    C
914    C      ( ELEMENT STIFFNESS MATRIX )
915    C
916          SKE(1,1)=XKBE*HR/3.D0
917          SKE(2,2)=SKE(1,1)
918          SKE(1,2)=XKBE*HR/6.D0
919          SKE(2,1)=SKE(1,2)
920    C
921    C      ( ELEMENT LOAD VECTOR )
922    C
923          FE(1)=XKBE*UOE*HR/2.D0
924          FE(2)=FE(1)
925    C
926    C      ( ASSEMBLING : BAND SOLVER )
927    C
928          DO 102 I=1,2
929          NI=IJB(I,NEL)
930          F(NI)=F(NI)+FE(I)
931          DO 102 J=1,2
932          NJ=IJB(J,NEL)
933          KI=NJ-NI+1
934          IF(KI.LE.0) GO TO 102
935          SK(MBAND*(NI-1)+KI)=SK(MBAND*(NI-1)+KI)+SKE(I,J)
936      102 CONTINUE
```

```
 937      C
 938        100 CONTINUE
 939      C
 940            RETURN
 941            END
 942      C -------------------------------------------------------------------
 943      C
 944            SUBROUTINE BANDSL(AK,R,NEQ,MBAND,KKK)
 945      C
 946      C   PURPOSE:
 947      C       SOLVES A SYSTEM OF LINEAR EQUATIONS.
 948      C       ( BANDED SYMMETRIC COEFFICIENT MATRIX )
 949      C
 950      C   CONTROL NUMBERS:
 951      C       KKK = 0   BAND SOLVER
 952      C       KKK = 1   TRIANGULATION
 953      C       KKK = 2   REDUCTION AND BACK SUBSTITUTION
 954      C
 955      C   ARGUMENTS:
 956      C       AK    = BANDED SYMMETRIC MATRIX TO BE SOLVED
 957      C       R     = RIGHT HAND SIDE ( LOAD VECTOR )
 958      C       NEQ   = TOTAL NUMBER OF DEGREES OF FREEDOM
 959      C       MBAND = HALF-WIDTH OF BAND
 960      C       KKK   = CONTROL NUMBER OF THE FLOW OF THE ROUTINE
 961      C
 962      C   NOTE:
 963      C       HOW TO STORE THE STIFFNESS MATRIX
 964      C
 965      C         ( K11 K12 K13     K15         )
 966      C         I     K22 K23 K24 K25         I
 967      C         I         K33 K34 K35 K36   O I
 968      C         I   SMY       K44 K45 K46 K47 I
 969      C         I               K55 K56 K57 I
 970      C         I                   K66 K67 I
 971      C         (                       K77 )
 972      C
 973      C       BANDED STIFFNESS MATRIX
 974      C
 975      C         ( K11 K12 K13   O K15 )
 976      C         I K22 K23 K24 K25   O I
 977      C         I K33 K34 K35 K36   O I
 978      C         I K44 K45 K46 K47   O I
 979      C         I K55 K56 K57   O   O I
 980      C         I K66 K67   O   O   O I
 981      C         ( K77   O   O   O   O )
 982      C
 983      C          NEQ = 7      MBAND = 5
 984      C
 985      C       ONE-DIMENSIONAL ARRAY OF THE STIFFNESS MATRIX "AK"
 986      C
 987      C       AK/K11,K12,K13,O,K15,K22,K23,K24,K25,O,K33,K34,K35,K36,O,
 988      C          K44,K45,K46,K47,O,K55,K56,K57,O,O,K66,K67,O,O,O,K77,O,O,O,O
 989      C
 990      C -------------------------------------------------------------------
 991      C
 992            IMPLICIT REAL*8(A-H,O-Z)
 993            DIMENSION AK(1),R(1)
 994      C
 995            NRS=NEQ-1
 996            NR=NEQ
 997            IF(KKK.EQ.2) GO TO 200
 998            DO 120 N=1,NRS
 999            M=N-1
1000            MR=MINO(MBAND,NR-M)
1001            PIVOT=AK(MBAND*(N-1)+1)
1002            DO 120 L=2,MR
1003            CP=AK(MBAND*(N-1)+L)/PIVOT
1004            IF(CP) 130,120,130
1005        130 I=M+L
1006            J=0
1007            DO 110 K=L,MR
1008            J=J+1
1009            LOCAT=MBAND*(I-1)+J
1010        110 AK(LOCAT)=AK(LOCAT)-CP*AK(MBAND*(N-1)+K)
1011        120 AK(MBAND*(N-1)+L)=CP
1012            IF(KKK.EQ.1) RETURN
1013        200 DO 230 N=1,NRS
1014            M=N-1
```

```
1015            MR=MINO(MBAND,NR-M)
1016            CP=R(N)
1017            IF(CP)210,230,210
1018        210 R(N)=CP/AK(MBAND*(N-1)+1)
1019            DO 220 L=2,MR
1020            I=M+L
1021        220 R(I)=R(I)-AK(MBAND*(N-1)+L)*CP
1022        230 CONTINUE
1023            R(NR)=R(NR)/AK(MBAND*(NR-1)+1)
1024            DO 320 I=1,NRS
1025            N=NR-I
1026            M=N-1
1027            MR=MINO(MBAND,NR-M)
1028            DO 320 K=2,MR
1029            L=M+K
1030        320 R(N)=R(N)-AK(MBAND*(N-1)+K)*R(L)
1031      C
1032            RETURN
1033            END
1034      C -------------------------------------------------------------------
1035      C
1036            SUBROUTINE OUTPUT(F,P,NX,NELX,X,Y,IJK,MP,XYK,LW)
1037      C
1038      C   PURPOSE:
1039      C       OUTPUTS THE TEMPERATURE AND THE HEAT FLUX VECTORS.
1040      C
1041      C   ARGUMENTS:
1042      C       F    = SOLUTION OBTAINED BY THE F.E. ANALYSIS
1043      C       P    = STORES THE HEAT FLUX VECTORS AND QUALITY INDICES
1044      C       NX   = TOTAL NUMBER OF NODAL POINTS
1045      C       NELX = TOTAL NUMBER OF FINITE ELEMENTS
1046      C       X,Y  = COORDINATES OF NODAL POINTS
1047      C       IJK  = ELEMENT CONNECTIVITIES
1048      C       MP   = GROUP NUMBER OF THE MATERIALS
1049      C       XYK  = MATERIAL CONSTANT IN THE MATRIX FORM
1050      C
1051      C   LOCAL VARIABLES:
1052      C       DFX,DFY = FIRST DERIVATIVES OF THE TEMPERATURE
1053      C       QX,QY   = COMPONENTS OF THE HEAT FLUX VECTOR
1054      C
1055      C -------------------------------------------------------------------
1056      C
1057            IMPLICIT REAL*8(A-H,O-Z)
1058            DIMENSION F(1),P(4,1),IJK(3,1),X(1),Y(1),XYK(3,1),MP(1)
1059            DIMENSION XE(3),YE(3),FE(3),DN(3,2)
1060      C
1061      C     ( OUTPUT THE SOLUTION VECTOR )
1062      C
1063             WRITE(LW,600)
1064        600 FORMAT(1H1,///////////////,5X,70('*'),/5X,70('*'),///20X,
1065           1 'THE SOLUTION VECTORS',///5X,70('*'),/5X,70('*'))
1066            WRITE(LW,602)
1067        602 FORMAT(1H1,//10X,43('-'),/10X,' NODE I      X         Y     ',
1068           1 'I     F',/10X,43('-'))
1069            WRITE(LW,604) (I,X(I),Y(I),F(I),I=1,NX)
1070        604 FORMAT(10X,I4,2X,'I',1X,F8.3,F9.3,2X,'I',1X,E12.5)
1071            WRITE(LW,606)
1072        606 FORMAT(10X,43('-'))
1073      C
1074      C     ( OUTPUT THE GRADIENT )
1075      C
1076            WRITE(LW,608)
1077        608 FORMAT(1H1,//////////5X,70('*'),/5X,70('*'),///20X,
1078           1 ' THE GRADIENTS OF THE SOLUTION',///5X,70('*'),/5X,70('*'))
1079            WRITE(LW,610)
1080        610 FORMAT(1H1,//5X,62('-'),/5X,' NEL I      X         Y     ',
1081           1 'I       F       I      FX         FY',/5X,62('-'))
1082      C
1083            DO 100 NEL=1,NELX
1084      C
1085      C     ( SET THE COORDINATES )
1086      C
1087            DO 102 I=1,3
1088            IJKI=IJK(I,NEL)
1089            XE(I)=X(IJKI)
1090            YE(I)=Y(IJKI)
1091        102 FE(I)=F(IJKI)
1092      C
```

```
1093             XC=(XE(1)+XE(2)+XE(3))/3.DO
1094             YC=(YE(1)+YE(2)+YE(3))/3.DO
1095             FC=(FE(1)+FE(2)+FE(3))/3.DO
1096       C
1097       C     ( CALL THE GRADIENT ROUTINE )
1098       C
1099                   CALL GRADI3(NEL,XE,YE,DN,AREA,LW)
1100       C
1101       C     ( GRADIENT OF THE SOLUTION )
1102       C
1103             KK=MP(NEL)
1104             XKNEL=XYK(1,KK)
1105             YKNEL=XYK(2,KK)
1106             XYKNEL=XYK(3,KK)
1107             DFX=0.DO
1108             DFY=0.DO
1109             DO 104 I=1,3
1110             DFX=DFX+FE(I)*DN(I,1)
1111         104 DFY=DFY+FE(I)*DN(I,2)
1112             QX=-(XKNEL*DFX+XYKNEL*DFY)
1113             QY=-(XYKNEL*DFX+YKNEL*DFY)
1114             P(1,NEL)=QX
1115             P(2,NEL)=QY
1116       C
1117           . WRITE(LW,620) NEL,XC,YC,FC,QX,QY
1118         620 FORMAT(5X,I4,' I ',F8.3,F9.3,' I ',E10.3,' I ',2E10.3)
1119       C
1120         100 CONTINUE
1121             WRITE(LW,612)
1122         612 FORMAT(5X,62('-'))
1123       C
1124             RETURN
1125             END
1126       C --------------------------------------------------------------------
1127       C
1128             SUBROUTINE QUALIT(NELX,NX,MBAND,IJK,X,Y,P,SK,CQX,CQY,NODE,LW)
1129       C
1130       C   PURPOSE :
1131       C       COMPUT QUALTY INDICES OF THE FINITE ELEMENT APPROXIMATION.
1132       C
1133       C   ARGUMENTS :
1134       C       NELX   = TOTAL NUMBER OF ELEMENTS
1135       C       NX     = TOTAL NUMBER OF NODAL POINTS
1136       C       MBAND  = HALF BAND WIDTH OF THE MASS MATRIX
1137       C       IJK    = ELEMENT CONNECTIVITIES
1138       C       X,Y    = COORDINATES OF NODAL POINTS
1139       C       P(1,N) = X COMPONENT OF THE HEAT FLUX IN THE N TH ELEMENT
1140       C       P(2,N) = Y COMPONENT OF THE HEAT FLUX IN THE N TH ELEMENT
1141       C       P(3,N) = QUALITY INDEX IN THE N TH ELEMENT (TEMPERATURE)
1142       C       P(4,N) = QUALITY INDEX IN THE N TH ELEMENT (HEAT FLUX)
1143       C       SK     = MASS MATRIX FOR THE LEAST SQUARES METHOD
1144       C       CQX    = QUALITY INDEX AT NODAL POINTS (TEMPERATURE)
1145       C       CQY    = QUALITY INDEX AT NODAL POINTS (HEAT FLUX)
1146       C       NODE   = TOTAL NUMBER OF NODAL POINTS IN AN ELEMENT
1147       C
1148       C   LOCAL VARIABLES:
1149       C       HX    = SIZE OF FINITE ELEMENT IN THE X DIRECTION
1150       C       HY    = SIZE OF FINITE ELEMENT IN THE Y DIRECTION
1151       C       QIOAV = AVERAGE QUALITY INDEX FOR TEMPERATURE
1152       C       QI1AV = AVERAGE QUALITY INDEX FOR HEAT FLUX
1153       C
1154       C   NOTE:
1155       C       ARRAYS "CQX" AND "CQY" ARE USED FOR THE CONFORMING HEAT FLUX
1156       C       VECTORS FIRST, THEN STORE THE QUALITY INDICES AT NODAL POINTS.
1157       C
1158       C --------------------------------------------------------------------
1159       C
1160             IMPLICIT REAL*8(A-H,O-Z)
1161             DIMENSION IJK(NODE,1),P(4,1),CQX(1),CQY(1),X(1),Y(1),SK(1)
1162             DIMENSION SME(3,3)
1163       C
1164             DATA SME/2.DO,1.DO,1.DO,1.DO,2.DO,1.DO,1.DO,1.DO,2.DO/
1165       C
1166       C     ( INITIAL SET )
1167       C
1168             IMAX=MBAND*NX
1169             DO 100 I=1,IMAX
1170         100 SK(I)=0.DO
1171             DO 102 I=1,NX
```

```
1172           CQX(I)=0.D0
1173       102 CQY(I)=0.D0
1174     C
1175     C     ( LEAST SQUARES METHODS )
1176     C
1177           DO 200 NEL=1,NELX
1178     C
1179           IJK1=IJK(1,NEL)
1180           IJK2=IJK(2,NEL)
1181           IJK3=IJK(3,NEL)
1182           X1=X(IJK1)
1183           X2=X(IJK2)
1184           X3=X(IJK3)
1185           Y1=Y(IJK1)
1186           Y2=Y(IJK2)
1187           Y3=Y(IJK3)
1188           AREA=0.5D0*(X1*(Y2-Y3)+X2*(Y3-Y1)+X3*(Y1-Y2))
1189           DQX=P(1,NEL)
1190           DQY=P(2,NEL)
1191     C
1192           DO 202 IA=1,NODE
1193           IJKIA=IJK(IA,NEL)
1194           CQX(IJKIA)=CQX(IJKIA)+DQX*AREA/3.D0
1195           CQY(IJKIA)=CQY(IJKIA)+DQY*AREA/3.D0
1196           DO 202 JB=1,NODE
1197           IJKJB=IJK(JB,NEL)
1198           IAJB=IJKJB-IJKIA
1199           IF(IAJB.LT.0) GO TO 202
1200           ND=MBAND*(IJKIA-1)+IAJB+1
1201           SK(ND)=SK(ND)+SME(IA,JB)*AREA/12.D0
1202       202 CONTINUE
1203     C
1204       200 CONTINUE
1205     C
1206     C     ( SOLVE THE SYSTEMS OF LINEAR EQUATIONS )
1207     C
1208                  CALL BANDSL(SK,CQX,NX,MBAND,0)
1209     C
1210                  CALL BANDSL(SK,CQY,NX,MBAND,2)
1211     C
1212           WRITE(LW,600)
1213       600 FORMAT(/////5X,70('*'),//10X,
1214          1 'CONFORMING HEAT FLUX VECTORS AT NODES',//5X,70('*'),/)
1215           WRITE(LW,602)
1216       602 FORMAT(5X,35('-'),/5X,' NODE I       CQX       I       CQY',
1217          1 /5X,35('-'))
1218           WRITE(LW,604) (I,CQX(I),CQY(I),I=1,NX)
1219       604 FORMAT(5X,I5,' I ',E11.4,' I ',E11.4)
1220           WRITE(LW,606)
1221       606 FORMAT(5X,35('-'))
1222     C
1223     C     ( COMPUTE THE QUALITY INDEX  QI0 AND QI1 )
1224     C
1225           QI0TL=0.D0
1226           QI1TL=0.D0
1227           QI0MAX=-1.D20
1228           QI1MAX=-1.D20
1229     C
1230           DO 300 NEL=1,NELX
1231     C
1232           IJK1=IJK(1,NEL)
1233           IJK2=IJK(2,NEL)
1234           IJK3=IJK(3,NEL)
1235           X1=X(IJK1)
1236           X2=X(IJK2)
1237           X3=X(IJK3)
1238           Y1=Y(IJK1)
1239           Y2=Y(IJK2)
1240           Y3=Y(IJK3)
1241           DET=X1*(Y2-Y3)+X2*(Y3-Y1)+X3*(Y1-Y2)
1242           AREA=0.5D0*DET
1243     C
1244           DQX=DABS(P(1,NEL))
1245           DQY=DABS(P(2,NEL))
1246           HX=DMAX1(DABS(X1-X2),DABS(X2-X3),DABS(X3-X1))
1247           HY=DMAX1(DABS(Y1-Y2),DABS(Y2-Y3),DABS(Y3-Y1))
1248           QI0XX=HX*DQX
1249           QI0YY=HY*DQY
1250           QI0NEL=DMAX1(QI0XX,QI0YY)*DSQRT(AREA)
```

```
1251            P(3,NEL)=QIONEL
1252            QIOTL=QIOTL+QIONEL
1253            IF(QIOMAX.LT.QIONEL) QIOMAX=QIONEL
1254      C
1255            B1=(Y2-Y3)/DET
1256            B2=(Y3-Y1)/DET
1257            B3=(Y1-Y2)/DET
1258            C1=(X3-X2)/DET
1259            C2=(X1-X3)/DET
1260            C3=(X2-X1)/DET
1261            DQXX=CQX(IJK1)*B1+CQX(IJK2)*B2+CQX(IJK3)*B3
1262            DQXY=CQX(IJK1)*C1+CQX(IJK2)*C2+CQX(IJK3)*C3
1263            DQYX=CQY(IJK1)*B1+CQY(IJK2)*B2+CQY(IJK3)*B3
1264            DQYY=CQY(IJK1)*C1+CQY(IJK2)*C2+CQY(IJK3)*C3
1265            QI1XXX=HX*DABS(DQXX)
1266            QI1XYX=HX*DABS(DQYX)
1267            QI1YXY=HY*DABS(DQXY)
1268            QI1YYY=HY*DABS(DQYY)
1269            QI1NEL=DMAX1(QI1XXX,QI1XYX,QI1YXY,QI1YYY)*DSQRT(AREA)
1270            P(4,NEL)=QI1NEL
1271            QI1TL=QI1TL+QI1NEL
1272            IF(QI1MAX.LT.QI1NEL) QI1MAX=QI1NEL
1273      C
1274        300 CONTINUE
1275      C
1276            QIOAV=QIOTL/NELX
1277            QI1AV=QI1TL/NELX
1278      C
1279            WRITE(LW,620) QIOAV,QI1AV
1280        620 FORMAT(/////5X,70('*'),//15X,
1281           1 'QUALITY INDICES FOR TEMPERATURE AND HEAT FLUX',//5X,70('*'),
1282           2 //5X,'AVERAGE QUALITY INDEX (TEMPERATURE) = ',E12.5,/5X,
1283           3 'AVERAGE QUALITY INDEX (HEAT FLUX) = ',E12.5,//,
1284           4 5X,45('-'),/5X,' NEL  I  LO L1  I    QIO     I      QI1',
1285           5 /5X,45('-'))
1286            DO 310 NEL=1,NELX
1287            QIONEL=P(3,NEL)
1288            QI1NEL=P(4,NEL)
1289            LO=10*(QIONEL+1.D-10)/QIOMAX
1290            L1=10*(QI1NEL+1.D-10)/QI1MAX
1291            WRITE(LW,622) NEL,LO,L1,QIONEL,QI1NEL
1292        622 FORMAT(5X,I5,' I ',2I3,' I ',E11.4,' I ',E11.4)
1293        310 CONTINUE
1294            WRITE(LW,626)
1295        626 FORMAT(5X,45('-'))
1296      C
1297      C     ( COMPUTE NODAL VALUES OF QUALITY INDICES )
1298      C
1299            DO 400 I=1,NX
1300            CQX(I)=0.DO
1301        400 CQY(I)=0.DO
1302      C
1303            DO 402 NEL=1,NELX
1304            IJK1=IJK(1,NEL)
1305            IJK2=IJK(2,NEL)
1306            IJK3=IJK(3,NEL)
1307            X1=X(IJK1)
1308            X2=X(IJK2)
1309            X3=X(IJK3)
1310            Y1=Y(IJK1)
1311            Y2=Y(IJK2)
1312            Y3=Y(IJK3)
1313            AREA=0.5DO*(X1*(Y2-Y3)+X2*(Y3-Y1)+X3*(Y1-Y2))
1314            CQX(IJK1)=CQX(IJK1)+P(3,NEL)*AREA/3.DO
1315            CQX(IJK2)=CQX(IJK2)+P(3,NEL)*AREA/3.DO
1316            CQX(IJK3)=CQX(IJK3)+P(3,NEL)*AREA/3.DO
1317            CQY(IJK1)=CQY(IJK1)+P(4,NEL)*AREA/3.DO
1318            CQY(IJK2)=CQY(IJK2)+P(4,NEL)*AREA/3.DO
1319            CQY(IJK3)=CQY(IJK3)+P(4,NEL)*AREA/3.DO
1320        402 CONTINUE
1321      C
1322                  CALL BANDSL(SK,CQX,NX,MBAND,2)
1323      C
1324                  CALL BANDSL(SK,CQY,NX,MBAND,2)
1325      C
1326            RETURN
1327            END
```

APPENDIX 2
PRE-FEM1: PREPROCESSING TO FEM1

This program generates a finite element model for plane problems using three-node triangular elements. It comprises three parts: (1) automatic mesh generator, (2) modification and addition of geometric data such as nodal coordinates and element connectivities, and (3) renumbering of node numbers to "minimize" the bandwidth of the global stiffness matrix.

The automatic mesh generator put in this problem follows the idea of finite element method. Indeed a given domain is decomposed into possibly curved quadrilateral blocks in which much smaller finite elements are generated by specifying number of division and mesh gradient. These quadrilateral blocks are then identified with 4-, 8-, or 12-node serendipity elements.

Connection of two blocks are performed to search nodes which have the same coordinate of a node on the common line within a given tolerance EPXY. To make this algorithm possible, data check routine is executed right after inputting, e.g., number of division and mesh gradient to each block. In many cases, it is necessary to modify nodal coordinates or create few number of additional elements without using the automatic mesh generator. To do this, the second block of the program is applicable. Because of the automatic mesh generator, numbering of nodes is, in general, far from the optimum of having the smallest bandwidth of the global stiffness matrix. The last part of the program may reduce the bandwidth by renumbering of node numbers.

Graphic results can be obtained using POST-FEM1 that plots a finite element model with node and element numbers. "Blow-up" capability is recommended in a graphic terminal or plotting system in computers.

```
  1   C **********************************************************************
  2   C
  3   C                          PROGRAM : PRE-FEM1
  4   C
  5   C                             SPRING  1984
  6   C
  7   C **********************************************************************
  8   C
  9   C   PURPOSE:
 10   C       AUTOMATIC MESH GENERATOR FOR FINITE ELEMENT ANALYSIS USING
 11   C       3-NODE TRIANGULAR ELEMENTS FOR PLANE AND AXISYMMETRIC
 12   C       PROBLEMS IN MECHANICS
 13   C
 14   C   VARIABLES: < IMPORTANT >
 15   C       NX     = TOTAL NUMBER OF NODAL POINTS
 16   C       NELX   = TOTAL NUMBER OF FINITE ELEMENTS
 17   C       IJK    = ELEMENT CONNECTIVITIES
 18   C       X,Y    = COORDINATES OF NODAL POINTS
 19   C       MPE    = THE GROUP NUMBER OF MATERIALS ASSIGNED EACH ELEMENT
 20   C       NODE   = TOTAL NUMBER OF NODAL POINTS IN AN ELEMENT
 21   C       NBX    = TOTAL NUMBER OF "NODES" THAT CHARACTERIZE BLOCKS
 22   C       XB,YB  = COORDINATES OF "NODES" OF THE BLOCKS
 23   C       XO,YO  = COORDINATES OF NODAL POINTS IN THE OLD NUMBERING SYSTEM
```

```
 24     C        JOP1    = STORES THE NUMBER OF ELEMENTS CONNECTED TO A NODAL
 25     C                  POINT
 26     C        JOP2    = STORES ELEMENT NUMBERS CONNECTED TO A NODAL POINT
 27     C        JOP3    = CONTAINS NODE NUMBERS IN THE OLD SYSTEM
 28     C        JOP4    = DUMMY STORAGE
 29     C        JNEW    = STORES NODE NUMBERS IN THE NEW NUMBERING SYSTEM
 30     C        LLXY    = TOTAL NUMBER OF "NODES" IN A BLOCK ( 4, 8 OR 12 )
 31     C        IIJB    = BLOCK CONNECTIVITIES
 32     C        MATPR   = GROUP NUMBER OF MATERIALS FOR THE BLOCK
 33     C        MMX     = NUMBER OF DIVISIONS IN THE 1-ST DIRECTION
 34     C        MMY     = NUMBER OF DIVISIONS IN THE 2-ND DIRECTION
 35     C        PPX     = MESH GRADIENT IN THE 1-ST DIRECTION
 36     C        PPY     = MESH GRADIENT IN THE 2-ND DIRECTION
 37     C        LLMN    = NODE CONNECTIVITIES
 38     C        LBLOCK  = TOTAL NUMBER OF BLOCKS USED FOR MESH GENERATION
 39     C        LHANDR  = CONTROL NUMBER FOR MANUAL INPUT DATA
 40     C        LR,LW   = CONTROL NUMBERS FOR READ AND WRITE STATEMENTS
 41     C        LP      = CONTROL NUMBER TO STORE THE DATA
 42     C
 43     C    INPUT DATA TO BE READ:
 44     C        1. MAIN ROUTINE
 45     C           1.1 LBLOCK                          :    I5
 46     C           IF(LBLOCK.EQ.0) GOTO THE 3-RD STEP
 47     C           1.2 NBX                             :    I5
 48     C           1.3 XB(I),YB(I),I=1,NBX             :    2F10.4
 49     C        2. SUBROUTINE READBL
 50     C           DO 100 N=1,LBLOCK
 51     C           2.1 MATPR(N)                        :    I5
 52     C           2.2 MMX(N),MMY(N)                   :    2I5
 53     C           2.3 LLXY(N)                         :    I5
 54     C           2.4 IIJB(I,N),I=1,LLXY(N)           :    I5
 55     C           2.5 PPX(N),PPY(N)                   :    2F10.4
 56     C           100 CONTINUE
 57     C        3. MAIN ROUTINE
 58     C           3.1 LHANDR                          :    A4
 59     C           IF(LHANDR.NE.'YES') GO TO THE THE 5-TH STEP
 60     C        4. SUBROUTINE HRMESH
 61     C           4.1 NODADD                          :    I5
 62     C           4.2 NELADD                          :    I5
 63     C           4.3 N,X(N),Y(N),I=1,NODADD          :    I5,2F10.4
 64     C           4.4 N,MPE(N),(IJK(J,N),J=1,NODE),
 65     C               I=1,NELADD                      :    2I5,NODE*I5
 66     C        5. MAIN ROUTINE
 67     C           5.1 LOPTIM                          :    A4
 68     C
 69     C    DIMENSIONING:
 70     C        X(NX),Y(NX),XO(NX),YO(NX)
 71     C           NX    = TOTAL NUMBER OF NODAL POINTS CREATED BEFORE CONNECTION
 72     C        IJK(NODE,NELX),MPE(NELX),P(3,NELX)
 73     C           NODE = TOTAL NUMBER OF NODAL POINTS IN AN ELEMENT
 74     C           NELX = TOTAL NUMBER OF ELEMENTS IN A MODEL
 75     C        JOP1(NX),JOP2(8*NX),JOP3(NX),JOP4(NX),JNEW(NX)
 76     C        XB(NBX),YB(NBX)
 77     C           NBX   = TOTAL NUMBER OF "NODES" THAT CHARACTERIZE BLOCKS
 78     C        MATPR(LBLOCK),MMX(LBLOCK),MMY(LBLOCK),PPX(LBLOCK),PPY(LBLOCK)
 79     C        IIJB(LXY,LBLOCK),LLMN(4,LBLOCK),LLXY(LBLOCK)
 80     C           LBLOCK = TOTAL NUMBER OF BLOCKS
 81     C           LXY    = TOTAL NUMBER OF "NODES" THAT DEFINE THE BLOCK
 82     C        IJB(LXY)
 83     C
 84     C *************************************************************************
 85     C
 86           DIMENSION X(500),Y(500),XO(500),YO(500),
 87          1          IJK(3,1000),MPE(1000),
 88          2          JOP1(500),JOP2(4000),JOP3(500),JOP4(500),JNEW(500),
 89          3          XB(100),YB(100),
 90          4          MATPR(100),MMX(100),MMY(100),PPX(100),PPY(100),
 91          5          IIJB(12,100),LLMN(4,100),LLXY(100)
 92           DIMENSION IJB(12)
 93     C
 94           DATA LR,LW,LP,NODE,NDF/5,6,8,3,1/
 95           DATA IYES,INO/'YES','NO'/
 96     C
 97     C     ( READ CONTROL NUMBERS )
 98     C
 99           WRITE(LW,500)
100       500 FORMAT(//,'AUTOMATIC MESH GENERATOR')
101           WRITE(LW,504)
```

```
102         504 FORMAT(//,'HOW MANY BLOCKS ARE USED TO CREATE A FINITE ELEMENT',
103            1  ' MODEL?   INPUT THE NUMBER OF BLOCKS.')
104             READ(LR,502) LBLOCK
105         502 FORMAT(16I5)
106       C
107       C       ( READ THE COORDINATES OF "NODES" CHARACTERIZING BLOCKS )
108       C
109             IF(LBLOCK.LE.0) GO TO 112
110             WRITE(LW,520)
111         520 FORMAT(/,'INPUT COORDINATES OF NODES CHARACTERIZING BLOCKS',
112            1 //5X,'NUMBER OF NODES')
113             READ(LR,502) NBX
114             WRITE(LW,522)
115         522 FORMAT(/,'< COORDINATES : XB(I),YB(I) >')
116             DO 524 I=1,NBX
117             WRITE(LW,526) I
118         526 FORMAT('NODE ',I4)
119             READ(LR,508) XB(I),YB(I)
120         508 FORMAT(2F10.4)
121         524 CONTINUE
122       C
123       C       ( READ DATA FOR BLOCKS )
124       C
125                   CALL READBL(LBLOCK,MATPR,MMX,MMY,LLXY,IIJB,PPX,PPY,LLMN,
126            1                   LR,LW)
127       C
128             NX=0
129             NELX=0
130             DO 100 NBLOCK=1,LBLOCK
131       C
132             MATPRO=MATPR(NBLOCK)
133             MX=MMX(NBLOCK)
134             MY=MMY(NBLOCK)
135             PX=PPX(NBLOCK)
136             PY=PPY(NBLOCK)
137             LXY=LLXY(NBLOCK)
138             DO 116 I=1,LXY
139         116 IJB(I)=IIJB(I,NBLOCK)
140       C
141                   CALL ATMESH(NBLOCK,NX,NELX,X,Y,XB,YB,IJK,MPE,NODE,
142            1                   MATPRO,MX,MY,PX,PY,LXY,IJB,LR,LW)
143       C
144         100 CONTINUE
145       C
146       C       ( DATA READ MANUALLY I.E. WITHOUT THE GENERATOR )
147       C
148         112 WRITE(LW,506)
149         506 FORMAT(/,'DO YOU NEED TO READ DATA MANUALLY ?  (YES/NO)',
150            1  '  DO NOT PUT COMMA AT THE END OF YES/NO.')
151             READ(LR.510) LHANDR
152         510 FORMAT(A4)
153       C
154             IF(LHANDR.NE.IYES) GO TO 102
155       C
156                   CALL HRMESH(NX,NELX,X,Y,IJK,MPE,NODE,LR,LW)
157       C
158       C       ( CONNECTION OF TWO DISJOINT BLOCKS IF NECESSARY )
159       C
160         102       CALL CONECT(NX,NELX,NODE,X,Y,IJK,JOP1,JOP3,JOP4,LR,LW)
161       C
162       C       ( CONDENSATION OF THE ELEMENTS WHOSE AREA IS ALMOST ZERO )
163       C
164                   CALL CONDEN(NELX,IJK,X,Y,3,LW)
165       C
166       C       ( OPTIMIZATION OF THE NUMBERING )
167       C
168             WRITE(LW,530)
169         530 FORMAT(/,'DO YOU OPTIMIZE THE NUMBERING TO HAVE THE MINIMUM',
170            1  ' BAND WIDTH ?   (YES/NO)',/,
171            2  'DO NOT PUT COMMA AT THE END OF YES/NO.')
172             READ(LR,532) LOPTIM
173         532 FORMAT(A4)
174       C
175             IF(LOPTIM.NE.IYES) GO TO 104
176       C
177                   CALL OPTIMN(NX,NELX,3,IJK,JOP1,JOP2,JOP3,JNEW,JOP4,
178            1                   LR,LW)
179       C
```

```
180     C       ( CHANGE THE COORDINATES )
181     C
182             DO 108 I=1,NX
183             XO(I)=X(I)
184         108 YO(I)=Y(I)
185             DO 110 I=1,NX
186             II=JNEW(I)
187             X(II)=XO(I)
188         110 Y(II)=YO(I)
189     C
190     C       ( PRINT OUT )
191     C
192         104 WRITE(LW,602)
193         602 FORMAT(/////5X,'********** FINITE ELEMENT MODEL **********',//)
194             WRITE(LW,604) NX,NELX
195         604 FORMAT(10X,'THE TOTAL NUMBER OF NODAL POINTS =',I5,/10X,
196            1 'THE TOTAL NUMBER OF FINITE ELEMENTS =',I5,////10X,
197            2 '< COORDINATES >',/)
198             WRITE(LW,606) (I,X(I),Y(I),I=1,NX)
199         606 FORMAT(2(5X,I5,3X,'(',2F10.4,2X,')'))
200             WRITE(LW,608)
201         608 FORMAT(/////10X,'< ELEMENT CONNECTIVITY >',/)
202             DO 106 I=1,NELX
203         106 WRITE(LW,610) I,MPE(I),(IJK(J,I),J=1,NODE)
204         610 FORMAT(15X,I5,5X,'<MPE>',I4,5X,'< IJK >',12I5)
205     C
206     C       ( STORE THE DATA )
207     C
208        1000 WRITE(LP,814)
209         814 FORMAT('FINITE ELEMENT MODEL')
210             WRITE(LP,800) NX,NELX
211         800 FORMAT(2I5)
212             WRITE(LP,816)
213         816 FORMAT(' < COORDINATES >')
214             DO 802 I=1,NX
215         802 WRITE(LP,804) I,X(I),Y(I)
216         804 FORMAT(I5,2F10.4)
217             WRITE(LP,818)
218         818 FORMAT(' < ELEMENT CONNECTIVITIES >')
219             DO 806 I=1,NELX
220         806 WRITE(LP,808) I,MPE(I),(IJK(J,I),J=1,NODE)
221         808 FORMAT(16I5)
222     C
223             STOP
224             END
225     C ----------------------------------------------------------------------
226     C
227             SUBROUTINE READBL(LBLOCK,MATPR,MMX,MMY,LLXY,IIJB,PPX,PPY,LLMN,
228            1                  LR,LW)
229     C
230     C   PURPOSE:
231     C       READ DATA OF BLOCKS AND CHECK THE DATA.
232     C
233     C   ARGUMENTS:
234     C       LBLOCK = TOTAL NUMBER OF BLOCKS
235     C       MATPR  = GROUP NUMBER OF MATERIALS
236     C       MMX    = NUMBER OF DIVISIONS IN THE 1-ST DIRECTION
237     C       MMY    = NUMBER OF DIVISIONS IN THE 2-ND DIRECTION
238     C       LLXY   = TOTAL NUMBER OF "NODES" IN A BLOCK ( 4, 8 OR 12 )
239     C       IIJB   = BLOCK CONNECTIVITIES
240     C       PPX    = MESH GRADIENT IN THE 1-ST DIRECTION
241     C       PPY    = MESH GRADIENT IN THE 2-ND DIRECTION
242     C       LLMN   = NODE CONNECTIVITIES
243     C
244     C   INPUT DATA TO BE READ:
245     C       DO 100 N=1,LBLOCK
246     C       1. MATPR(N)                    :   I5
247     C       2. MMX(N),MMY(N)               :   2I5
248     C       3. LLXY(N)                     :   I5
249     C       4. IIJB(I,N),I=1,LLXY(N)       :   I5
250     C       5. PPX(N),PPY(N)               :   2F10.4
251     C       100 CONTINUE
252     C
253     C ----------------------------------------------------------------------
254     C
255             DIMENSION MATPR(1),MMX(1),MMY(1),LLXY(1),PPX(1),PPY(1),
256            1          IIJB(12,1),LLMN(4,1)
257     C
```

```
258     C       ( READ INPUT DATA )
259     C
260             DO 100 N=1,LBLOCK
261     C
262             WRITE(LW,500) N
263       500 FORMAT(////,'----- INPUT DATA FOR THE BLOCK',I4,' -----',/)
264             WRITE(LW,502)
265       502 FORMAT(/,'INPUT THE GROUP NUMBER(1-99) OF MATERIALS')
266             READ(LR,504) MATPR(N)
267       504 FORMAT(I5)
268             WRITE(LW,506)
269       506 FORMAT(/,'INPUT THE NUMBER OF DIVISION IN THE 1-ST AND 2-ND',
270          1  ' DIRECTIONS ; MMX,MMY')
271             READ(LR,508) MMX(N),MMY(N)
272       508 FORMAT(2I5)
273             WRITE(LW,510)
274       510 FORMAT(/,'INPUT THE NUMBER OF NODES CHARACTERIZING THE BLOCK',
275          1  ' ; 4, 8, OR 12')
276             READ(LR,512) LLXY(N)
277       512 FORMAT(I5)
278             WRITE(LW,514)
279       514 FORMAT(/,'INPUT THE BLOCK CONNECTIVITY')
280             IMAX=LLXY(N)
281             DO 112 I=1,IMAX
282             WRITE(LW,516) I
283       516 FORMAT('IJB(',I3,' )')
284             READ(LR,518) IIJB(I,N)
285       518 FORMAT(I5)
286       112 CONTINUE
287             WRITE(LW,520)
288       520 FORMAT(/,'INPUT REAL PARAMETERS OF MESH GRADIENT IN THE 1-ST',
289          1  ' AND 2-ND DIRECTION.',/10X,
290          2  'IF PPX = 2. THEN THE GRID GRADUALLY BECOMES LARGE.',/10X,
291          3  'IF PPX = 1. THEN THE GRID IS UNIFORM.',/10X,
292          4  'IF PPX = 0.5THEN THE GRID GRADUALLY BECOMES SMALL.',/)
293             READ(LR,522) PPX(N),PPY(N)
294       522 FORMAT(2F10.4)
295     C
296       100 CONTINUE
297     C
298     C       ( CHECK THE DATA )
299     C
300             DO 114 N=1,LBLOCK
301             DO 114 J=1,4
302       114 LLMN(J,N)=0
303     C
304             DO 116 N=1,LBLOCK
305             DO 116 I=1,4
306             I1=IIJB(I,N)
307             I3=I+1
308             IF(I.EQ.4) I3=1
309             I2=IIJB(I3,N)
310             NN=0
311       118 NN=NN+1
312             IF(NN.GT.LBLOCK) GO TO 116
313             DO 122 J=1,4
314             J1=IIJB(J,NN)
315             J3=J+1
316             IF(J.EQ.4) J3=1
317             J2=IIJB(J3,NN)
318             IF(I1.EQ.J2.AND.I2.EQ.J1) LLMN(I,N)=10*NN+J
319       122 CONTINUE
320             GO TO 118
321       116 CONTINUE
322     C
323             LBLOC1=LBLOCK-1
324             DO 180 NES=1,LBLOC1
325             NMAX=LBLOCK-NES+1
326             DO 130 N=1,NMAX
327             DO 130 I=1,4
328             M=LLMN(I,N)/10
329             IF(M.GE.N) GO TO 130
330             L=LLMN(I,N)-10*M
331             IF(I.NE.1.OR.L.NE.1) GO TO 132
332             PPX(M)=1./PPX(N)
333             MMX(M)=MMX(N)
334             GO TO 130
335       132 IF(I.NE.1.OR.L.NE.2) GO TO 134
```

```
336            PPY(M)=1./PPX(N)
337            MMY(M)=MMX(N)
338            GO TO 130
339        134 IF(I.NE.1.OR.L.NE.3) GO TO 136
340            PPX(M)=PPX(N)
341            MMX(M)=MMX(N)
342            GO TO 130
343        136 IF(I.NE.1.OR.L.NE.4) GO TO 138
344            PPY(M)=PPX(N)
345            MMY(M)=MMX(N)
346            GO TO 130
347        138 IF(I.NE.2.OR.L.NE.1) GO TO 140
348            PPX(M)=1./PPY(N)
349            MMX(M)=MMY(N)
350            GO TO 130
351        140 IF(I.NE.2.OR.L.NE.2) GO TO 142
352            PPY(M)=1./PPY(N)
353            MMY(M)=MMY(N)
354            GO TO 130
355        142 IF(I.NE.2.OR.L.NE.3) GO TO 144
356            PPX(M)=PPY(N)
357            MMX(M)=MMY(N)
358            GO TO 130
359        144 IF(I.NE.2.OR.L.NE.4) GO TO 146
360            PPY(M)=PPY(N)
361            MMY(M)=MMY(N)
362            GO TO 130
363        146 IF(I.NE.3.OR.L.NE.1) GO TO 148
364            PPX(M)=PPX(N)
365            MMX(M)=MMX(N)
366            GO TO 130
367        148 IF(I.NE.3.OR.L.NE.2) GO TO 150
368            PPY(M)=PPX(N)
369            MMY(M)=MMX(N)
370            GO TO 130
371        150 IF(I.NE.3.OR.L.NE.3) GO TO 152
372            PPX(M)=1./PPX(N)
373            MMX(M)=MMX(N)
374            GO TO 130
375        152 IF(I.NE.3.OR.L.NE.4) GO TO 154
376            PPY(M)=1./PPX(N)
377            MMY(M)=MMX(N)
378            GO TO 130
379        154 IF(I.NE.4.OR.L.NE.1) GO TO 156
380            PPX(M)=PPY(N)
381            MMX(M)=MMY(N)
382            GO TO 130
383        156 IF(I.NE.4.OR.L.NE.2) GO TO 158
384            PPY(M)=PPY(N)
385            MMY(M)=MMY(N)
386            GO TO 130
387        158 IF(I.NE.4.OR.L.NE.3) GO TO 160
388            PPX(M)=1./PPY(N)
389            MMX(M)=MMY(N)
390            GO TO 130
391        160 PPY(M)=1./PPY(N)
392            MMY(M)=MMY(N)
393        130 CONTINUE
394        180 CONTINUE
395      C
396      C     ( OUTPUT )
397      C
398            WRITE(LW,606)
399        606 FORMAT(/////10X,'---------- FINAL INPUT DATA FOR BLOCKS',
400           1  ' ----------')
401            DO 162 N=1,LBLOCK
402            WRITE(LW,600) N
403        600 FORMAT(//10X,'----- BLOCK : ',I3,' -----',/)
404            WRITE(LW,602) MMX(N),MMY(N),PPX(N),PPY(N)
405        602 FORMAT(/,10X,'MX =',I4,5X,'MY =',I4,/10X,'PX =',F10.4,5X
406           1 ,'PY =',F10.4,/)
407            IMAX=LLXY(N)
408            WRITE(LW,604) (IIJB(I,N),I=1,IMAX)
409        604 FORMAT(10X,'<IJB>',2X,15I5)
410        162 CONTINUE
411      C
412            RETURN
413            END
```

```
414     C -------------------------------------------------------------------
415     C
416           SUBROUTINE ATMESH(NBLOCK,NX,NELX,X,Y,XC,YC,IJK,MPE,NODE,
417          1                  MATPRO,MX,MY,PX,PY,LXY,IJB,LR,LW)
418     C
419     C   PURPOSE:
420     C       CREATE THE FINITE ELEMENT MODEL USING THREE-NODE TRIANGULAR
421     C       ELEMENTS.   THE MESH IS ARRANGED AS THE UNION-JACK PATTERN.
422     C       THAT IS,  A QUADRILATERAL ELEMENT CONSISTS OF
423     C       FOUR THREE-NODE-TRIANGULAR ELEMENTS.
424     C
425     C   ARGUMENTS:
426     C       NBLOCK = BLOCK NUMBER
427     C       NX     = TOTAL NUMBER OF NODAL POINTS
428     C       NELX   = TOTAL NUMBER OF FINITE ELEMENTS
429     C       X,Y    = COORDINATES OF NODAL POINTS
430     C       XC,YC  = COORDINATES OF "NODES" OF THE BLOCKS
431     C       IJK    = ELEMENT CONNECTIVITIES
432     C       MPE    = GROUP NUMBER OF THE MATERIALS FOR EACH ELEMENT
433     C       NODE   = TOTAL NUMBER OF NODAL POINTS IN AN ELEMENT
434     C       MATPRO = GROUP NUMBER OF THE MATERIALS FOR THE BLOCK
435     C       MX     = NUMBER OF DIVISION IN THE FIRST DIRECTION
436     C       MY     = NUMBER OF DIVISION IN THE SECOND DIRECTION
437     C       PX     = PARAMETER OF THE DIVISION FOR THE 1-ST DIRECTION
438     C       PY     = PARAMETER OF THE DIVISION FOR THE 2-ND DIRECTION
439     C        (EXAMPLE)   IF  PX=1 ,   THEN MESH IS EQUALLY DIVIDED
440     C                    IF  PX=2 ,   THEN MESH IS GRADUALLY LARGER
441     C                    IF  PX=0.5 , THEN MESH IS GRADUALLY SMALLER
442     C       LXY    = TOTAL NUMBER OF "NODES" THAT DEFINES THE BLOCK
443     C                4, 8 OR 12 IN THIS PROGRAM
444     C       IJB    = BLOCK CONNECTIVITY
445     C
446     C   LOCAL VARIABLES:
447     C       NODES  = NUMBER OF THE FIRST NODE OF THE BLOCK
448     C       NELMS  = NUMBER OF THE FIRST ELEMENT OF THE BLOCK
449     C       XB,YB  = COORDINATES OF "NODES" IN THE BLOCK
450     C
451     C   DIMENSIONING:
452     C       XB,YB,SH   MUST BE DIMENSIONED FOR THE CASE THAT MORE
453     C       SOPHISTICATED CURVED BLOCKS ARE ASSUMED THAN 12-NODE
454     C       SERENDIPITY BLOCK.
455     C
456     C   NOTES:
457     C       THE NUMBERING SYSTEM FOR EACH ELEMENT IS
458     C
459     C              L2 *---------------* L4
460     C                 I      -N3-     I
461     C                 I               I
462     C                 I -N4-   *  -N2- I
463     C                 I      L5       I
464     C                 I      -N1-     I
465     C              L1 *---------------* L3
466     C
467     C
468     C -------------------------------------------------------------------
469     C
470           DIMENSION X(1),Y(1),IJK(NODE,1),MPE(1),XC(1),YC(1)
471           DIMENSION XB(12),YB(12),SH(12),IJB(12)
472     C
473     C     ( READ INPUT DATA FOR A QUADRILATERAL BLOCK )
474     C
475           WRITE(LW,598) NBLOCK
476       598 FORMAT(/////,' ---------- BLOCK ',I3,' ----------',//)
477     C
478     C     ( SET COORDINATES OF A BLOCK )
479     C
480           DO 514 I=1,LXY
481           IJBI=IJB(I)
482           XB(I)=XC(IJBI)
483       514 YB(I)=YC(IJBI)
484     C
485           MX1=MX+1
486           MY1=MY+1
487     C
488     C     ( SET THE LAST NODE AND ELEMENT NUMBER )
489     C
490           NODES=NX+1
491           NELMS=NELX+1
```

```
492            NX=NODES-1+MX1*MY1+MX*MY
493            NELX=NELMS-1+4*MX*MY
494            NEQ=2*NX
495      C
496            DO 100 IX=1,MX
497            DO 100 IY=1,MY
498      C
499      C     ( CONSTRUCT THE ELEMENT CONNECTIVITY )
500      C
501            L1=NODES-1+(IX-1)*MY1+(IX-1)*MY+IY
502            L2=L1+1
503            L3=L1+MY1+MY
504            L4=L3+1
505            L5=L1+MY1
506            N1=NELMS+4*((IX-1)*MY+(IY-1))
507            N2=N1+1
508            N3=N1+2
509            N4=N1+3
510      C
511            IJK(1,N1)=L1
512            IJK(2,N1)=L3
513            IJK(3,N1)=L5
514            IJK(1,N2)=L3
515            IJK(2,N2)=L4
516            IJK(3,N2)=L5
517            IJK(1,N3)=L4
518            IJK(2,N3)=L2
519            IJK(3,N3)=L5
520            IJK(1,N4)=L2
521            IJK(2,N4)=L1
522            IJK(3,N4)=L5
523      C
524      C     ( SPECIFY THE NUMBER OF MATERIAL GROUP )
525      C
526            MPE(N1)=MATPRO
527            MPE(N2)=MATPRO
528            MPE(N3)=MATPRO
529            MPE(N4)=MATPRO
530      C
531      C     ( COORDINATES )
532      C
533            XL1=FUN1(IX,MX,PX)
534            XL2=XL1
535            XL3=FUN1(IX+1,MX,PX)
536            XL4=XL3
537            YL1=FUN1(IY,MY,PY)
538            YL2=FUN1(IY+1,MY,PY)
539            YL3=YL1
540            YL4=YL2
541      C
542                  CALL SEREND(SH,XL1,YL1,LXY)
543      C
544            X(L1)=DOT(XB,SH,LXY)
545            Y(L1)=DOT(YB,SH,LXY)
546                  CALL SEREND(SH,XL2,YL2,LXY)
547            X(L2)=DOT(XB,SH,LXY)
548            Y(L2)=DOT(YB,SH,LXY)
549                  CALL SEREND(SH,XL3,YL3,LXY)
550            X(L3)=DOT(XB,SH,LXY)
551            Y(L3)=DOT(YB,SH,LXY)
552                  CALL SEREND(SH,XL4,YL4,LXY)
553            X(L4)=DOT(XB,SH,LXY)
554            Y(L4)=DOT(YB,SH,LXY)
555            X(L5)=0.25*(X(L1)+X(L2)+X(L3)+X(L4))
556            Y(L5)=0.25*(Y(L1)+Y(L2)+Y(L3)+Y(L4))
557      C
558        100 CONTINUE
559      C
560      C     ( PRINT OUT THE MESH DATA FOR THE PRESENT BLOCK )
561      C
562            WRITE(LW,606)
563        606 FORMAT(////10X,'< ELEMENT CONNECTIVITIES >',/)
564            WRITE(LW,608) (I,MPE(I),(IJK(J,I),J=1,3),I=NELMS,NELX)
565        608 FORMAT(5X,I5,3X,I2,4X,'<IJK>',3I5,6X,I5,3X,I2,4X,'<IJK>',3I5)
566            WRITE(LW,610)
567        610 FORMAT(///15X,'< COORDINATES >',/)
568            WRITE(LW,612) (I,X(I),Y(I),I=NODES,NX)
569        612 FORMAT(5X,I5,2X,'<',2F10.4,1X,'>',5X,I5,2X,'<',2F10.4,
570           1 1X,'>')
```

```
571      C
572            RETURN
573            END
574      C ------------------------------------------------------------------
575      C
576            FUNCTION DOT(A,B,N)
577      C
578      C   PURPOSE:
579      C       TAKE DOT PRODUCT OF TWO VECTORS  "A"  AND  "B".
580      C
581      C ------------------------------------------------------------------
582      C
583            DIMENSION A(1),B(1)
584      C
585            DOT=0.
586            DO 100 I=1,N
587        100 DOT=DOT+A(I)*B(I)
588      C
589            RETURN
590            END
591      C ------------------------------------------------------------------
592      C
593            FUNCTION FUN1(I,M,P)
594      C
595      C   PURPOSE:
596      C       DEFINE THE SIZE OF THE ELEMENT.
597      C
598      C   ARGUMENTS:
599      C       I    = I-TH ELEMENT
600      C       M    = TOTAL NUMBER OF ELEMENTS IN EACH DIRECTION
601      C       P    = 2 ,   IF THE SIZE GRADUALLY BECOMES LARGER
602      C            = 1 ,   IF THE SIZE IS UNIFORM
603      C            = 0.5 , IF THE SIZE GRADUALLY BECOMES SMALLER
604      C
605      C ------------------------------------------------------------------
606      C
607            RI=I
608            RM=M
609      C
610            IF(P.LE.1.) GO TO 100
611            FUN1=-1.+2.*(RI-1.)*RI/(RM*(RM+1.))
612            RETURN
613      C
614        100 IF(P.LT.1.) GO TO 102
615            FUN1=-1.+2.*(RI-1.)/RM
616            RETURN
617      C
618        102 FUN1=-1.+2.*(RI-1.)*(2.*RM-RI+2.)/(RM*(RM+1.))
619      C
620            RETURN
621            END
622      C ------------------------------------------------------------------
623      C
624            SUBROUTINE SEREND(SH,S,T,N)
625      C
626      C   PURPOSE:
627      C       CONSTRUCT THE SHAPE FUNCTIONS OF A SERENDIPITY ELEMENT
628      C       FOR THE MESH GENERATOR.  ( N=4 , 8  OR 12 )
629      C
630      C   ARGUMENTS:
631      C       SH   = SHAPE FUNCTIONS
632      C       S    = FIRST LOCAL COORDINATE  ( -1, 1 )
633      C       T    = SECOND LOCAL COORDINATE ( -1, 1 )
634      C       N    = TOTAL NUMBER OF "NODES"
635      C            = 4  FOR THE QUADRILATERAL BLOCK
636      C            = 8 OR 12  FOR CURVED QUADRILATERAL BLOCKS
637      C
638      C   NUMBERING SYSTEM:
639      C
640      C
641      C            4-------7-------3          4---10----9----3
642      C            I               I          I              I
643      C            I               I         11              8
644      C            8               6          I              I
645      C            I               I         12              7
646      C            I               I          I              I
647      C            1-------5-------2          1----5----6----2
648      C
649      C ------------------------------------------------------------------
```

```
650     C
651           DIMENSION SH(12)
652     C
653           KKK=N/4
654           GO TO (100,200,300), KKK
655     C
656     C     ( FOR 4-NODE CASE )
657     C
658       100 SH(1)=0.25*(1.-S)*(1.-T)
659           SH(2)=0.25*(1.+S)*(1.-T)
660           SH(3)=0.25*(1.+S)*(1.+T)
661           SH(4)=0.25*(1.-S)*(1.+T)
662           RETURN
663     C
664     C     ( FOR 8-NODE CASE )
665     C
666       200 SS=S**2
667           TT=T**2
668           ST=S*T
669           SST=SS*T
670           STT=S*TT
671           SH(1)=(-1.+ST+SS+TT-SST-STT)/4.
672           SH(2)=(-1.-ST+SS+TT-SST+STT)/4.
673           SH(3)=(-1.+ST+SS+TT+SST+STT)/4.
674           SH(4)=(-1.-ST+SS+TT+SST-STT)/4.
675           SH(5)=(1.-T-SS+SST)/2.
676           SH(6)=(1.+S-TT-STT)/2.
677           SH(7)=(1.+T-SS-SST)/2.
678           SH(8)=(1.-S-TT+STT)/2.
679           RETURN
680     C
681     C     ( FOR 12-NODE CASE )
682     C
683       300 S1=1.-S
684           S2=1.+S
685           T1=1.-T
686           T2=1.+T
687           ST=-10.+9.*(S*S+T*T)
688           SH(1)=S1*T1*ST/32.
689           SH(2)=S2*T1*ST/32.
690           SH(3)=S2*T2*ST/32.
691           SH(4)=S1*T2*ST/32.
692           SH(5)=9.*(1.-3.*S)*S1*S2*T1/32.
693           SH(6)=9.*(1.+3.*S)*S1*S2*T1/32.
694           SH(7)=9.*S2*(1.-3.*T)*T1*T2/32.
695           SH(8)=9.*S2*(1.+3.*T)*T1*T2/32.
696           SH(9)=9.*(1.+3.*S)*S1*S2*T2/32.
697           SH(10)=9.*(1.-3.*S)*S1*S2*T2/32.
698           SH(11)=9.*S1*(1.+3.*T)*T1*T2/32.
699           SH(12)=9.*S1*(1.-3.*T)*T1*T2/32.
700     C
701           RETURN
702           END
703     C -----------------------------------------------------------------------
704     C
705           SUBROUTINE HRMESH(NX,NELX,X,Y,IJK,MPE,NODE,LR,LW)
706     C
707     C   PURPOSE:
708     C       READ THE MESH DATA ( ELEMENT CONNECTIVITIES AND COORDINATES )
709     C       WITHOUT USING THE AUTOMATIC MESH GENERATOR.
710     C
711     C   ARGUMENTS:
712     C       NX    = TOTAL NUMBER OF NODAL POINTS
713     C       NELX  = TOTAL NUMBER OF FINITE ELEMENTS
714     C       X,Y   = COORDINATES OF NODAL POINTS
715     C       IJK   = ELEMENT CONNECTIVITIES
716     C       MPE   = GROUP NUMBER OF THE MATERIALS FOR EACH ELEMENT
717     C       NODE  = TOTAL NUMBER OF NODAL POINTS IN AN ELEMENT
718     C
719     C   LOCAL VARIABLES:
720     C       NODES  = NUMBER OF THE FIRST NODAL POINT ADDED
721     C       NELMS  = NUMBER OF THE FIRST ELEMENT ADDED
722     C       NODADD = TOTAL NUMBER OF ADDITIONAL NODAL POINTS
723     C       NELADD = TOTAL NUMBER OF ADDITIONAL ELEMENTS
724     C
725     C   INPUT DATA TO BE READ:
726     C       1. NODADD                       :   I5
727     C       2. NELADD                       :   I5
```

```
728      C         3. I,X(I),Y(I),J=1,NODADD        :    I5,2F10.4
729      C         4. I,MPE(I),(IJK(J,I),J=1,3)
730      C            K=1,NELADD                    :    5I5
731      C
732      C -----------------------------------------------------------------
733      C
734            DIMENSION X(1),Y(1),IJK(NODE,1),MPE(1)
735      C
736            WRITE(LW,500)
737        500 FORMAT(////,'----- (MANUAL) INPUT DATA -----')
738            WRITE(LW,502)
739        502 FORMAT(/,'INPUT THE TOTAL NUMBER OF NODES INPUT.')
740            READ(LR,504) NODADD
741        504 FORMAT(16I5)
742            WRITE(LW,506)
743        506 FORMAT(/,'INPUT THE TOTAL NUMBER OF ELEMENTS INPUT')
744            READ(LR,504) NELADD
745      C
746            IF(NODADD.LE.0) GO TO 110
747            WRITE(LW,508)
748        508 FORMAT(//,'INPUT COORDINATES OF NODES MODIFIED/ADDED',/)
749            DO 100 I=1,NODADD
750            WRITE(LW,510)
751        510 FORMAT('NODE NUMBER,X-COORDINATE,Y-COORDINATE')
752            READ(LR,512) II,X(II),Y(II)
753        512 FORMAT(I5,2F10.4)
754            IF(II.GT.NX) NX=II
755        100 CONTINUE
756      C
757        110 IF(NELADD.LE.0) GOTO 112
758            WRITE(LW,514)
759        514 FORMAT(//,'INPUT THE GROUP NUMBER AND ELEMENT CONNECTIVITY.',/)
760            DO 102 I=1,NELADD
761            WRITE(LW,516)
762        516 FORMAT('ELEMENT NUMBER,GROUP NUMBER,<IJK>')
763            READ(LR,518) II,MPE(II),(IJK(J,II),J=1,NODE)
764        518 FORMAT(16I5)
765            IF(II.GT.NELX) NELX=II
766        102 CONTINUE
767      C
768        112 CONTINUE
769      C
770            RETURN
771            END
772      C -----------------------------------------------------------------
773      C
774            SUBROUTINE CONECT(NX,NELX,NODE,X,Y,IJK,JOP,IBLO1,IBLO2,LR,LW)
775      C
776      C   PURPOSE:
777      C       CONNECT TWO-DISJOINT BLOCKS BY RENUMBERING THE NODE NUMBERS.
778      C
779      C   ARGUMENTS:
780      C       NX    = TOTAL NUMBER OF NODAL POINTS
781      C       NELX  = TOTAL NUMBER OF ELEMENTS
782      C       NODE  = TOTAL NUMBER OF NODAL POINTS IN AN ELEMENT
783      C       X,Y   = COORDINATES OF NODAL POINTS
784      C       IJK   = ELEMENT CONNECTIVITIES
785      C       JOP   = ARRAY TO STORE NODE NUMBERS
786      C       IBLO1 = NODE NUMBERS OF THE 1-ST BLOCK
787      C       IBLO2 = NODE NUMBERS OF THE 2-ND BLOCK
788      C
789      C   LOCAL VARIABLES:
790      C       NCX   = TOTAL NUMBER OF NODAL POINTS CONNECTED
791      C
792      C -----------------------------------------------------------------
793      C
794            DIMENSION X(1),Y(1),IJK(NODE,1),JOP(1),IBLO1(1),IBLO2(1)
795            DATA TOR/1.E-5/
796      C
797      C     ( FIND THE NODES OVERLAPPED )
798      C
799            NXO=NX
800            NCX=0
801            DO 300 I=1,NX
802            JOP(I)=I
803            XI=X(I)
804            YI=Y(I)
805            J1=I+1
```

```
806            DO 302 J=J1,NX
807            XIJ=ABS(XI-X(J))
808            YIJ=ABS(YI-Y(J))
809            IF(XIJ.GT.TOR.OR.YIJ.GT.TOR) GO TO 302
810            NCX=NCX+1
811            IBLO1(NCX)=I
812            IBLO2(NCX)=J
813        302 CONTINUE
814        300 CONTINUE
815            IF(NCX.LE.0) RETURN
816      C
817            WRITE(LW,600) NCX
818        600 FORMAT(/////10X,'----- CONNECTION -----',5X,'NCX=',I5,//)
819            WRITE(LW,602) (I,IBLO1(I),IBLO2(I),I=1,NCX)
820        602 FORMAT(10X,I5,5X,'(',2I5,'  )')
821      C
822      C     ( RENUMBERING )
823      C
824            NC=0
825        210 NC=NC+1
826            I1=IBLO1(NC)
827            I2=IBLO2(NC)
828            IF(I1-I2) 202,200,204
829        202 IBLO2(NC)=I1
830            NNODE=I1
831            NPIVOT=I2
832            GO TO 206
833        204 IBLO1(NC)=I2
834            NNODE=I2
835            NPIVOT=I1
836        206 NC1=NC+1
837            DO 208 I=NC1,NCX
838            IA=IBLO1(I)
839            IB=IBLO2(I)
840            IF(IA.EQ.NPIVOT) IBLO1(I)=NNODE
841            IF(IB.EQ.NPIVOT) IBLO2(I)=NNODE
842            IF(IA.GT.NPIVOT) IBLO1(I)=IA-1
843            IF(IB.GT.NPIVOT) IBLO2(I)=IB-1
844        208 CONTINUE
845            DO 120 NEL=1,NELX
846            DO 120   I=1,NODE
847            IJKI=IJK(I,NEL)
848            IF(IJKI.EQ.NPIVOT) IJK(I,NEL)=NNODE
849            IF(IJKI.GT.NPIVOT) IJK(I,NEL)=IJKI-1
850        120 CONTINUE
851            DO 114 I=1,NXO
852            JOPI=JOP(I)
853            IF(JOPI.EQ.NPIVOT) JOP(I)=NNODE
854            IF(JOPI.GT.NPIVOT) JOP(I)=JOPI-1
855        114 CONTINUE
856            NOD1=NPIVOT+1
857            DO 112 I=NOD1,NX
858            X(I-1)=X(I)
859        112 Y(I-1)=Y(I)
860            NX=NX-1
861      C
862            NC2=NC+2
863            IF(NC2.GT.NCX) GO TO 200
864            DO 212 N=NC1,NCX
865            I1=IBLO1(N)
866            I2=IBLO2(N)
867            N1=N+1
868            NCOUNT=0
869            DO 214 M=N1,NCX
870            IF(IBLO1(M).NE.I1.OR.IBLO2(M).NE.I2) GO TO 214
871            NCOUNT=NCOUNT+1
872            DO 216 L=M,NCX
873            IBLO1(L)=IBLO1(L+1)
874        216 IBLO2(L)=IBLO2(L+1)
875        214 CONTINUE
876            NCX=NCX-NCOUNT
877        212 CONTINUE
878        200 IF(NC.LT.NCX) GO TO 210
879      C
880      C     ( PRINT OUT )
881      C
882            WRITE(LW,620)
883        620 FORMAT(/////10X,'----- CONDENSED NODE NUMBERS -----',/)
```

```
884           WRITE(LW,622) (I,JOP(I),I=1,NXO)
885       622 FORMAT(5(5X,'(',I5,'  -',I5,' )'))
886     C
887           RETURN
888           END
889     C -----------------------------------------------------------------
890     C
891           SUBROUTINE CONDEN(NELX,IJK,X,Y,NODE,LW)
892     C
893     C   PURPOSE:
894     C       TAKE OUT THE ELEMENTS IF THEY HAVE NON-POSITIVE DETERMINANT.
895     C
896     C   ARGUMENTS:
897     C       NELX  = TOTAL NUMBER OF ELEMENTS
898     C       IJK   = ELEMENT CONNECTIVITIES
899     C       X,Y   = COORDINATES OF NODAL POINTS
900     C       NODE  = TOTAL NUMBER OF NODAL POINTS IN AN ELEMENT
901     C
902     C   LOCAL VARIABLES:
903     C       RMIN  = RADIUS OF CIRCLE IN A TRIANGLE
904     C       RMAX  = RADIUS OF CIRCLE CIRCUMSCRIBING A TRIANGLE
905     C       RATIO = RMIN/RMAX
906     C       TOR1  = TOLERANCE TO ZERO AREA (1.E-4)
907     C       TOR2  = MINIMUM VALUE OF RATIO ALLOWABLE
908     C
909     C ----------------------------------------------------------------
910     C
911           DIMENSION IJK(NODE,1),X(1),Y(1)
912     C
913           DATA TOR1,TOR2/1.E-4,0.01/
914     C
915           NEL=0
916       100 NEL=NEL+1
917           I1=IJK(1,NEL)
918           I2=IJK(2,NEL)
919           I3=IJK(3,NEL)
920           X1=X(I1)
921           Y1=Y(I1)
922           X2=X(I2)
923           Y2=Y(I2)
924           X3=X(I3)
925           Y3=Y(I3)
926           A=SQRT((X1-X2)*(X1-X2)+(Y1-Y2)*(Y1-Y2))
927           B=SQRT((X2-X3)*(X2-X3)+(Y2-Y3)*(Y2-Y3))
928           C=SQRT((X3-X1)*(X3-X1)+(Y3-Y1)*(Y3-Y1))
929           S=0.5*(A+B+C)
930           IF(A.LE.TOR1) GO TO 102
931           IF(B.LE.TOR1) GO TO 102
932           IF(C.LE.TOR1) GO TO 102
933     C
934     C     ( IF AREA IS POSITIVE )
935     C
936           AREA=0.5*(X1*(Y2-Y3)+X2*(Y3-Y1)+X3*(Y1-Y2))
937           RMIN=AREA/S
938           RMAX=0.25*A*B*C/AREA
939           RATIO=RMIN/RMAX
940           IF(RATIO.LE.TOR2) WRITE(LW,600) NEL,RMIN,RMAX,AREA
941       600 FORMAT(//'***** INADEQUATE TRIANGULAR ELEMENT',5X,'NEL = ',I5,
942          1 5X,'RMIN = ',E10.3,5X,'RMAX = ',E10.3,5X,'AREA = ',E10.3)
943           GO TO 108
944     C
945     C     ( IF AREA IS CLOSE TO ZERO )
946     C
947       102 NEL1=NEL+1
948           IF(NEL1.GT.NELX) GO TO 106
949           DO 104 LEL=NEL1,NELX
950           DO 104 IA=1,NODE
951       104 IJK(IA,LEL-1)=IJK(IA,LEL)
952       106 NELX=NELX-1
953           NEL=NEL-1
954     C
955       108 IF(NEL.LT.NELX) GO TO 100
956     C
957           RETURN
958           END
959     C -----------------------------------------------------------------
960     C
961           SUBROUTINE OPTIMN(NX,NELX,NODE,IJK,JM,MT,JO,JP,NW,LR,LW)
```

```
 962      C
 963      C   PURPOSE:
 964      C       OPTIMIZE THE NUMBERING TO HAVE THE MINIMUM BAND WIDTH
 965      C       OF A FINITE ELEMENT MODEL BY 3-NODE ELEMENTS.
 966      C
 967      C   ARGUMENTS:
 968      C       NX    = TOTAL NUMBER OF NODAL POINTS
 969      C       NELX  = TOTAL NUMBER OF ELEMENTS
 970      C       NODE  = TOTAL NUMBER OF NODAL POINTS IN AN ELEMENT
 971      C       IJK   = ELEMENT CONNECTIVITIES
 972      C       JM    = STORES THE NUMBER OF ELEMENTS RELATED TO AN ELEMENT
 973      C       MT    = CONTAINS THE ELEMENT NUMBERS RELATED TO A NODAL POINT
 974      C       JO    = CONTAINS THE OLD NUMBERING OF NODAL POINTS
 975      C       JP    = CONTAINS THE NEW NUMBERING OF NODAL POINTS
 976      C       NW    = DUMMY ARRAY
 977      C
 978      C -----------------------------------------------------------------
 979      C
 980            DIMENSION IJK(NODE,1),JM(1),MT(1),JO(1),JP(1),NW(1)
 981      C
 982      C     ( RENUMBERING )
 983      C
 984            IMAX=8*NX
 985            DO 132 I=1,IMAX
 986        132 MT(I)=0
 987            MB=0
 988            DO 100 J=1,NX
 989        100 JM(J)=0
 990            DO 102 J=1,NELX
 991            DO 104 I=1,NODE
 992            JN=IJK(I,J)
 993            JS=8*(JN-1)
 994            DO 106 II=1,NODE
 995            IF(I.EQ.II) GO TO 106
 996            JJ=IJK(II,J)
 997            ME=JM(JN)
 998            IF(ME.EQ.0) GO TO 108
 999            DO 110 I3=1,ME
1000            IF(MT(JS+I3).EQ.JJ) GO TO 106
1001        110 CONTINUE
1002        108 JM(JN)=JM(JN)+1
1003            MT(JS+JM(JN))=JJ
1004            IF(IABS(JN-JJ).GT.MB) MB=IABS(JN-JJ)
1005        106 CONTINUE
1006        104 CONTINUE
1007        102 CONTINUE
1008      C
1009            WRITE(LW,600) MB
1010        600 FORMAT(/////5X,'---------- OPTIMIZATION OF THE NUMBERING',
1011          1 ' ----------',//10X,'BAND WIDTH BEFORE THE OPTIMIZATION = ',I5)
1012      C
1013            NJ=NX
1014            MI=MB
1015            DO 112 IK=1,NX
1016            DO 114 J=1,NX
1017            JO(J)=0
1018        114 NW(J)=0
1019            MA=0
1020            I=1
1021            NW(1)=IK
1022            JO(IK)=1
1023            K=1
1024        122 K4=JM(NW(I))
1025            IF(K4.LE.0) GO TO 116
1026            JS=8*(NW(I)-1)
1027            DO 118 JJ=1,K4
1028            K5=MT(JS+JJ)
1029            IF(JO(K5).GT.0) GO TO 118
1030            K=K+1
1031            NW(K)=K5
1032            JO(K5)=K
1033            ND=IABS(I-K)
1034            IF(ND.GE.MI) GO TO 112
1035            IF(MA.LT.ND) MA=ND
1036        118 CONTINUE
1037            IF(K.EQ.NJ) GO TO 120
1038        116 I=I+1
1039            IF(I.GT.NX) GO TO 120
```

```
1040          GO TO 122
1041      120 MI=MA
1042          DO 124 J=1,NX
1043      124 JP(J)=JO(J)
1044      112 CONTINUE
1045    C
1046          INDEX=0
1047          DO 134 I=1,NX
1048      134 IF(JP(I).LE.0) INDEX=INDEX+1
1049          IF(INDEX.EQ.0) GO TO 138
1050          WRITE(LW,612)
1051      612 FORMAT(///,'***** THE OPTIMIZER DID NOT WORK FOR THIS CASE *****',//)
1052          DO 140 I=1,NX
1053      140 JP(I)=I
1054          GO TO 136
1055    C
1056    C       ( OUTPUT OF THE NEW NUMBERING )
1057    C
1058      138 WRITE(LW,608)
1059      608 FORMAT(///10X,'< THE NEW NUMBERING SYSTEM >',/)
1060          WRITE(LW,610) (I,JP(I),I=1,NX)
1061      610 FORMAT(3(5X,'(',I5,'  - ',I5,' )'))
1062    C
1063    C       ( NEW CONNECTIVITIES )
1064    C
1065          DO 126 NEL=1,NELX
1066          DO 126 IA=1,NODE
1067          IJKIA=IJK(IA,NEL)
1068      126 IJK(IA,NEL)=JP(IJKIA)
1069    C
1070      136 WRITE(LW,602)
1071      602 FORMAT(///10X,'< OPTIMIZED ELEMENT CONNECTIVITIES >',/)
1072          WRITE(LW,604) (NEL,(IJK(IA,NEL),IA=1,3),NEL=1,NELX)
1073      604 FORMAT(10X,I5,3X,'<IJK>',3I5,10X,I5,3X,'<IJK>',3I5)
1074    C
1075    C       ( NEW BAND WIDTH )
1076    C
1077          MB=0
1078          DO 130 NEL=1,NELX
1079          DO 130 IA=1,NODE
1080          IJKIA=IJK(IA,NEL)
1081          DO 130 JB=1,NODE
1082          IJKJB=IJK(JB,NEL)
1083          IAJB=IJKJB-IJKIA+1
1084          IF(IAJB.GT.MB) MB=IAJB
1085      130 CONTINUE
1086    C
1087          WRITE(LW,606) MB
1088      606 FORMAT(///10X,'NEW BAND WIDTH = ',I5)
1089    C
1090          RETURN
1091          END
```

APPENDIX 3
POST-FEM1: POSTPROCESSING TO FEM1

This program is basically for graphics to display a finite element model generated by PRE-FEM1 and computed results by FEM1 such as temperature, heat flux vector, and quality indices. This program is developed for the MTS system at The University of Michigan. The parts system dependent are the following statements.

CALL PLTEND
CALL PLINE
CALL PNUMBR
CALL PARROW

Brief description of these subroutines are given in the list of the program POST-FEM1. These should be easily replaced by similar subroutines in users' computer systems.

```
 1  C ***********************************************************************
 2  C
 3  C                        PROGRAM : POST-FEM1
 4  C
 5  C                          SPRING  1984
 6  C
 7  C ***********************************************************************
 8  C
 9  C   PURPOSE:
10  C       PLOT THE FINITE ELEMENT MODEL, THE DISTRIBUTION OF THE
11  C       TEMPERATURE, THE ISO-THERMAL LINES, THE HEAT FLUX VECTORS
12  C       AND THE CONTOUR LINES OF THE QUALITY INDICES
13  C
14  C   INPUT DATA TO BE READ:
15  C       1.  LPRN  (YES/NO)                            :   A4
16  C       2.  LPRE  (YES/NO)                            :   A4
17  C       3.  NELES                                     :   I5
18  C       4.  NELE                                      :   I5
19  C       5.  THETA1,THETA2                             :   2F10.4
20  C       6.  NCONT                                     :   I5
21  C       7.  I,FCONT(I),I=1,NCONT                      :   I5,F10.4
22  C       8.  NTL (1-3)                                 :   I5
23  C       9.  NRM  (1-10)                               :   I5
24  C      10.  (TITLE(I,J),J=1,5),I=1,NTL                :   5A4
25  C      11.  (REMARK(I,J),J=1,6),I=1,NRM               :   6A4
26  C      12.  NCONT  (FOR QUALITY INDICES)              :   I5
27  C
28  C   NOTES:
29  C       BESIDES ABOVE INPUT DATA,  THIS PROGRAM READS THE FINITE
30  C       ELEMENT MODEL AND THE RESULTS FROM A FILE8.
31  C
32  C   DIMENSIONING:
33  C       X(NX),Y(NX),F(NX),CQX(NX),CQY(NX)
34  C          NX     = TOTAL NUMBER OF NODAL POINTS IN A MODEL
35  C       IJK(NODE,NELX),P(4,NELX)
36  C          NELX  = TOTAL NUMBER OF ELEMENTS IN A MODEL
37  C          NODE  = TOTAL NUMBER OF NODAL POINTS IN AN ELEMENT
38  C                = 3 FOR THE 3-NODE TRIANGULAR ELEMENT
39  C       IDEL(2*IDELX)
40  C          IDELX = TOTAL NUMBER OF NODAL POINTS ON THE BOUNDARY
41  C
42  C ***********************************************************************
```

```
 43      C
 44            DIMENSION X(500),Y(500),IJK(3,1000),IDEL(1000),F(500),P(4,1000)
 45            DIMENSION CQX(500),CQY(500),FLAG(20)
 46      C
 47            DATA LR,LW,LP,NODE/5,6,8,3/
 48      C
 49      C     ( READ THE MODEL AND RESULTS FROM A FILE8 )
 50      C
 51            READ(LP,510) (FLAG(I),I=1,20)
 52        510 FORMAT(20A4)
 53            READ(LP,500) NX,NELX
 54        500 FORMAT(2I5)
 55            READ(LP,510) (FLAG(I),I=1,20)
 56            READ(LP,502) (X(I),Y(I),I=1,NX)
 57        502 FORMAT(8F10.4)
 58            READ(LP,510) (FLAG(I),I=1,20)
 59            READ(LP,504) ((IJK(J,I),J=1,NODE),I=1,NELX)
 60        504 FORMAT(16I5)
 61            READ(LP,510) (FLAG(I),I=1,20)
 62            READ(LP,506) (F(I),I=1,NX)
 63        506 FORMAT(8E10.3)
 64            READ(LP,510) (FLAG(I),I=1,20)
 65            READ(LP,508) ((P(J,I),J=1,4),I=1,NELX)
 66        508 FORMAT(8E10.3)
 67            READ(LP,510) (FLAG(I),I=1,20)
 68            READ(LP,508) (CQX(I),CQY(I),I=1,NX)
 69      C
 70      C     ( FIND THE NODES ON THE BOUNDARY )
 71      C
 72                  CALL DELINE(NELX,IJK,IDEL,IDELX)
 73      C
 74      C     ( PLOTTING ROUTINES )
 75      C
 76                  CALL PLOTPR(X,Y,IJK,IDEL,NX,NELX,IDELX,F,P,1,1,CQX,CQY,3,
 77           1                     LR,LW)
 78      C
 79            STOP
 80            END
 81      C -----------------------------------------------------------------------
 82      C
 83            SUBROUTINE DELINE(NELX,IJK,IDEL,IM)
 84      C
 85      C   PURPOSE:
 86      C      FIND OUT THE NODAL POINTS ON THE OUTSIDE BOUNDARY.
 87      C
 88      C   ARGUMENTS:
 89      C      NELX = TOTAL NUMBER OF ELEMENTS
 90      C      IJK  = ELEMENT CONNECTIVITIES
 91      C      IDEL = ARRAY TO KEEP NODE NUMBERS ON THE BOUNDARY
 92      C      IM   = TOTAL NUMBER OF NODAL POINTS ON THE BOUNDARY
 93      C
 94      C -----------------------------------------------------------------------
 95      C
 96            DIMENSION IJK(3,1),IDEL(1)
 97      C
 98            IM=0
 99            DO 100 NEL=1,NELX
100            DO 100 I=1,3
101            I1=IJK(I,NEL)
102            I3=I+1
103            IF(I3.GE.4) I3=1
104            I2=IJK(I3,NEL)
105            N=0
106            ID=1
107        102 N=N+1
108            IF(N.GT.NELX) GO TO 106
109            DO 104 J=1,3
110            J1=IJK(J,N)
111            J3=J+1
112            IF(J3.GE.4) J3=1
113            J2=IJK(J3,N)
114            IF(I1.EQ.J2.AND.I2.EQ.J1) ID=0
115        104 CONTINUE
116            IF(ID.EQ.0) GO TO 100
117            GO TO 102
118        106 IM=IM+1
119            IDEL(2*IM-1)=I1
120            IDEL(2*IM)  =I2
```

```
121         100 CONTINUE
122   C
123           RETURN
124           END
125   C -----------------------------------------------------------------
126   C
127           SUBROUTINE PLOTPR(X,Y,IJK,IDEL,NX,NELX,IDELX,F,P,KKK,MPR,CQX,CQY,
128          1                  NODE,LR,LW)
129   C
130   C  PURPOSE:
131   C      PLOT THE FINITE ELEMENT MODEL AND THE RESULTS COMPUTED.
132   C      THERE ARE FOUR CAPABILITIES IN THIS ROUTINE.   THAT IS,
133   C      (1) THE F.E.DOMAIN, (2) CONTOUR LINES FOR THE SOLUTION
134   C      F(.) AND QUALITY INDICES AT NODAL POINTS CQX(.), CQY(.),
135   C      (3) CURVED SURFACE FORMED BY THE FUNCTION  Z=F(.)(X,Y)
136   C      IN THE THREE DIMENSIONAL FIELD (X,Y,Z),  AND
137   C      (4) THE VELOCITY FIELD IN THE ARRAY  P(.,.) CAN BE PLOTTED
138   C      BY USING THE MTS-SYSTEM IN THE UNIVERSITY OF MICHIGAN.
139   C      DETAILS OF THE SUBROUTINES USED IN THIS PROGRAM ARE FOUND
140   C      IN THE MTS-MANUAL VOL.11.  SPECIFICALLY, SUBROUTINES PLTBGN,
141   C      PLINE,PSYM,PARROW,AND PLTEND ARE EXPLAINED THERE.
142   C
143   C  ARGUMENTS:
144   C      X,Y    = COORDINATES OF THE NODAL POINTS
145   C      IJK    = ELEMENT CONNECTIVITIES
146   C      IDEL   = ARRAY TO KEEP NODE NUMBERS ON THE BOUNDARY
147   C      NX     = TOTAL NUMBER OF NODAL POINTS
148   C      NELX   = TOTAL NUMBER OF ELEMENTS
149   C      IDELX  = TOTAL NUMBER OF NODAL POINTS ON THE BOUNDARY
150   C      F      = FUNCTION PLOTTED LIKE THE TEMPERATURE FIELD
151   C      P      = FUNCTIONS OF THE PRINCIPAL VALUES
152   C      KKK    = 1 ,  IF THE INPUT DATA HAVE TO BE READ IN THIS ROUTINE
153   C      MPR    = INTEGER PARAMETER TO CONTROL ROUTINES
154   C               1 IS ASSUMED IN THIS PROGRAM
155   C      CQX    = QUALITY INDEX AT NODAL POINTS (TEMPERATURE)
156   C      CQY    = QUALITY INDEX AT NODAL POINTS (HEAT FLUX)
157   C      NODE   = TOTAL NUMBER OF NODAL POINTS IN AN ELEMENT
158   C
159   C  LOCAL VARIABLES:
160   C      LPRN    = 1 , IF THE NODE NUMBERS ARE PLOTTED
161   C      LPRE    = 1 , IF THE ELEMENT NUMBERS ARE PLOTTED
162   C      LPRF    = 1 , IF THE ELEMENT IS PLOTTED FOR THE VELOCITY
163   C      NELS    = FIRST ELEMENT NUMBER PLOTTED
164   C      NELE    = LAST ELEMENT NUMBER PLOTTED
165   C      THETA1 = ANGLE OF THE 1-ST COORDINATE AXIS
166   C      THETA2 = ANGLE OF THE 2-ND COORDINATE AXIS
167   C      NCONT   = TOTAL NUMBER OF THE CONTOUR LINES
168   C      NTL     = TOTAL NUMBER OF LINES FOR THE TITLE REMARKS
169   C      NRM     = TOTAL NUMBER OF LINES FOR THE REMARKS
170   C      TITLE   = TITLE REMARKS
171   C      REMARK = REMARKS ON THE PLOTTING
172   C      SCALEG = SCALING FACTOR TO THE GEOMETRY OF THE MODEL
173   C      SCALE   = SCALING FACTOR TO THE CURVED SURFACE
174   C      XMIN    = MINIMUM VALUE OF THE 1-ST COORDINATE
175   C      XMAX    = MAXIMUM VALUE OF THE 1-ST COORDINATE
176   C      YMIN    = MINIMUM VALUE OF THE 2-ND COORDINATE
177   C      YMAX    = MAXIMUM VALUE OF THE 2-ND COORDINATE
178   C      FMIN    = MINIMUM VALUE OF THE FUNCTION  F
179   C      FMAX    = MAXIMUM VALUE OF THE FUNCTION  F
180   C      XORG    = 1-ST COORDINATE OF THE ORIGIN IN THE PLOTTING
181   C      YORG    = 2-ND COORDINATE OF THE ORIGIN IN THE PLOTTING
182   C      PMAX    = MAXIMUM VELOCITY
183   C      STDLTH = STANDARD SIZE OF THE PLOTTING ( 7-IN )
184   C      STDARW = STANDARD SIZE OF THE ARROW
185   C      HIG1,2 = SIZE OF THE CHARACTERS USED IN THE PLOTTING
186   C
187   C  INPUT DATA TO BE READ:
188   C      1.  LPRN  (YES/NO)                                 :   A4
189   C      2.  LPRE  (YES/NO)                                 :   A4
190   C      3.  NELES                                          :   I5
191   C      4.  NELE                                           :   I5
192   C      5.  THETA1,THETA2                                  :   2F10.4
193   C      6.  NCONT                                          :   I5
194   C      7.  I,FCONT(I),I=1,NCONT                           :   I5,F10.4
195   C      8.  NTL (1-3)                                      :   I5
196   C      9.  NRM  (1-10)                                    :   I5
197   C     10.  (TITLE(I,J),J=1,5),I=1,NTL                     :   5A4
198   C     11.  (REMARK(I,J),J=1,6),I=1,NRM                    :   6A4
```

```
199     C        12.  NCONT  (FOR QUALITY INDICES)               :    I5
200     C
201     C -----------------------------------------------------------------
202     C
203           DIMENSION IJK(NODE,1),X(1),Y(1),F(1),P(4,1),IDEL(1),CQX(1),CQY(1)
204           DIMENSION REMARK(10,6),TITLE(3,5),FCONT(99)
205     C
206           DATA STDLTH,STDARW,HIG1,HIG2,XORG,YORG/7.,0.7,0.13,0.11,3.,2./
207     C
208     C     ( READ INPUT DATA )
209     C
210           WRITE(LW,620)
211       620 FORMAT(/////,60('*'),/,'            INPUT DATA   :  PLOTPR-FEM1',
212          1  /,60('*'))
213           WRITE(LW,622)
214       622 FORMAT(/,'DO YOU PLOT THE NODE NUMBERS ?  (YES/NO)',
215          1  '    DO NOT PUT COMMA AT THE END OF YES/NO.')
216           READ(LR,500) LPRN
217       500 FORMAT(A4)
218           WRITE(LW,624)
219       624 FORMAT(/,'DO YOU PLOT THE ELEMENT NUMBERS ?  (YES/NO)',
220          1  '    DO NOT PUT COMMA AT THE END OF YES/NO.')
221           READ(LR,500) LPRE
222           WRITE(LW,626)
223       626 FORMAT(/,'INPUT THE NUMBER OF THE 1-ST ELEMENT PLOTTED')
224           READ(LR,502) NELS
225       502 FORMAT(I5)
226           WRITE(LW,628) NELX
227       628 FORMAT(//,'* THE LAST ELEMENT NUMBER OF THE MODEL IS',I5,
228          1  /,'INPUT THE NUMBER OF THE LAST ELEMENT PLOTTED',/)
229           READ(LR,502) NELE
230           WRITE(LW,630)
231       630 FORMAT(//,'INPUT THE ANGLES OF THE X AND Y AXES TO THE',
232          1  ' HORIZONTAL LINE.',/,'IF YOU DO NOT HAVE YOUR CHOICE,',
233          2   INPUT  15.,AND 20., FOR THESE ANGLES.')
234           READ(LR,504) THETA1,THETA2
235       504 FORMAT(2F10.4)
236           WRITE(LW,632)
237       632 FORMAT(/,'INPUT THE TOTAL NUMBER OF CONTOUR LINES PLOTTED')
238           READ(LR,502) NCONT
239           WRITE(LW,634)
240       634 FORMAT(//,'INPUT THE VALUES OF CONTOUR LINES')
241           DO 200 I=1,NCONT
242           WRITE(LW,636) I
243       636 FORMAT('CONTOUR LINE ',I3,'  = ')
244           READ(LR,506) FCONT(I)
245       506 FORMAT(F20.7)
246       200 CONTINUE
247           WRITE(LW,638)
248       638 FORMAT(//,'INPUT THE NUMBER OF LINES FOR THE TITLE (0-3)')
249           READ(LR,502) NTL
250           IF(NTL.LE.0) GO TO 202
251           DO 204 I=1,NTL
252           WRITE(LW,640) I
253       640 FORMAT('LINE ',I1)
254       204 READ(LR,508) (TITLE(I,J),J=1,5)
255       508 FORMAT(5A4)
256       202 WRITE(LW,642)
257       642 FORMAT(/,'INPUT THE NUMBER OF LINES FOR THE REMARK (0-10)')
258           READ(LR,502) NRM
259           IF(NRM.LE.0) GO TO 206
260           DO 208 I=1,NRM
261           WRITE(LW,644) I
262       644 FORMAT('LINE ',I2)
263       208 READ(LR,510) (REMARK(I,J),J=1,6)
264       510 FORMAT(6A4)
265       206 CONTINUE
266           WRITE(LW,600) NELS,NELE,THETA1,THETA2
267       600 FORMAT(/////10X,'----- PLOTPR : READ -----',
268          1 10X,'FIRST ELEMENT PLOTTED = ',I5,/,
269          2 'LAST ELEMENT PLOTTED = ',I5,
270          3 /10X,'ANGLE OF THE 1-ST COORDINATE =',F10.4,/10X,
271          4 'ANGLE OF THE 2-ND COORDINATE =',F10.4)
272           IF(NTL.GT.0) WRITE(LW,612) ((TITLE(N,J),J=1,4),N=1,NTL)
273       612 FORMAT(10X,'< TITLE TO BE PLOTTED >',5X,4A4)
274           IF(NRM.GT.0) WRITE(LW,614) ((REMARK(N,J),J=1,5),N=1,NRM)
275       614 FORMAT(10X,'< REMARK TO BE PLOTTED >',4X,5A4)
276     C
```

```
277             THETA1=3.141592654*THETA1/180.
278             THETA2=3.141592654*THETA2/180.
279       C
280       C     ( FIND THE SCALING FACTOR OF THE GEOMETRY )
281       C
282                   CALL GSCALE(NX,X,Y,XMIN,YMIN,XMAX,YMAX,SCALEG,LW)
283            1
284       C
285       C     ( PLOT THE FINITE ELEMENT MODEL )
286       C
287                   CALL LAYOUT(NTL,NRM,TITLE,REMARK)
288       C
289                   CALL DOMAIN(NELS,NELE,IJK,X,Y,3,XMIN,YMIN,SCALEG,
290            1                  LPRE,LPRN)
291       C
292                   CALL PLTEND
293       C
294       C     ( PLOT THE FUNCTION )
295       C
296                   CALL LAYOUT(NTL,NRM,TITLE,REMARK)
297       C
298                   CALL FUNC3D(NELS,NELE,NX,F,X,Y,IJK,THETA1,THETA2,STDLTH,
299            1                  XMIN,YMIN,XMAX,YMAX,XORG,YORG,IDELX,IDEL,3,LW)
300       C
301                   CALL PLTEND
302       C
303       C     ( PLOT THE CONTOUR LINES )
304       C
305                   CALL LAYOUT(NTL,NRM,TITLE,REMARK)
306       C
307                   CALL CONTR3(NELS,NELE,NX,F,X,Y,IJK,IDELX,IDEL,NCONT,FCONT,
308            1                  3,XMIN,YMIN,SCALEG,XORG,YORG,LW)
309       C
310                   CALL PLTEND
311       C
312       C     ( PLOT THE VELOCITY VECTOR )
313       C
314                   CALL LAYOUT(NTL,NRM,TITLE,REMARK)
315       C
316                   CALL VELOCI(NELS,NELE,P,X,Y,IJK,IDELX,IDEL,
317            1                  STDARW,XMIN,YMIN,XORG,YORG,SCALEG,3,LW)
318       C
319                   CALL PLTEND
320       C
321       C     ( PLOT THE CONTOUR LINES OF THE QUALITY INDEX )
322       C
323             WRITE(LW,680)
324         680 FORMAT(//,'INPUT THE NUMBER OF CONTOUR LINES FOR THE QUALITY',
325            1  ' INDEX')
326             READ(LR,580) NCONT
327         580 FORMAT(I5)
328       C
329                   CALL VCONT(NX,CQX,FCONT,NCONT)
330       C
331                   CALL LAYOUT(NTL,NRM,TITLE,REMARK)
332       C
333                   CALL CONTR3(NELS,NELE,NX,CQX,X,Y,IJK,IDELX,IDEL,NCONT,
334            1                  FCONT,3,XMIN,YMIN,SCALEG,XORG,YORG,LW)
335       C
336                   CALL PLTEND
337       C
338                   CALL VCONT(NX,CQY,FCONT,NCONT)
339       C
340                   CALL LAYOUT(NTL,NRM,TITLE,REMARK)
341       C
342                   CALL CONTR3(NELS,NELE,NX,CQY,X,Y,IJK,IDELX,IDEL,NCONT,
343            1                  FCONT,3,XMIN,YMIN,SCALEG,XORG,YORG,LW)
344       C
345                   CALL PLTEND
346       C
347             RETURN
348             END
349       C -----------------------------------------------------------------
350       C
351             SUBROUTINE GSCALE(NX,X,Y,XMIN,YMIN,XMAX,YMAX,SCALEG,LW)
352       C
353       C   PURPOSE:
354       C      FIND THE SCALING FACTOR OF THE GEOMETRY TO PLOT.
```

```
355      C
356      C   ARGUMENTS:
357      C       NX      = TOTAL NUMBER OF NODAL POINTS
358      C       X,Y     = COORDINATES OF NODAL POINTS
359      C       XMIN    = MINIMUM VALUE OF THE X COORDINATE
360      C       YMIN    = MINIMUM VALUE OF THE Y COORDINATE
361      C       XMAX    = MAXIMUM VALUE OF THE X COORDINATE
362      C       YMAX    = MAXIMUM VALUE OF THE Y COORDINATE
363      C       SCALEG = SCALING FACTOR OF THE GEOMETRY
364      C
365      C ------------------------------------------------------------------
366      C
367            DIMENSION X(1),Y(1)
368            DATA STDLTH/7./
369      C
370      C     ( SCALING OF THE GEOMETRY )
371      C
372            XMIN= 1.E20
373            YMIN= 1.E20
374            XMAX=-1.E20
375            YMAX=-1.E20
376            DO 102 I=1,NX
377            XI=X(I)
378            XMIN=AMIN1(XMIN,XI)
379            XMAX=AMAX1(XMAX,XI)
380            YI=Y(I)
381            YMIN=AMIN1(YMIN,YI)
382        102 YMAX=AMAX1(YMAX,YI)
383            XSIZE=XMAX-XMIN
384            YSIZE=YMAX-YMIN
385            ASIZE=AMAX1(XSIZE,YSIZE)
386            SCALEG=STDLTH/ASIZE
387            WRITE(LW,602) XMIN,XMAX,YMIN,YMAX,SCALEG
388        602 FORMAT(///10X,'< SCALING THE GEOMETRY >',/10X,
389          1 'XMIN=',E10.3,5X,'XMAX=',E10.3,/10X,'YMIN=',E10.3,5X,
390          2 'YMAX=',E10.3,/10X,'SCALE =',E10.3)
391      C
392            RETURN
393            END
394      C ------------------------------------------------------------------
395      C
396            SUBROUTINE LAYOUT(NTL,NRM,TITLE,REMARK)
397      C
398      C   PURPOSE:
399      C       PLOT THE LAYOUT OF GRAPHS IN THE HARD COPY.
400      C
401      C   ARGUMENTS:
402      C       NTL     = NUMBER OF LINES FOR TITLES OF THE PLOT
403      C       NRM     = NUMBER OF LINES FOR REMARKS OF THE PLOT
404      C       TITLE   = TITLES OF THE PLOT
405      C       REMARK = REMARKS OF THE PLOT
406      C
407      C ------------------------------------------------------------------
408      C
409            DIMENSION TITLE(3,5),REMARK(10,6),TIT(5),REM(6),EX(99),EY(99)
410      C
411      C     ( INITIALIZATION )
412      C
413        100       CALL PLTBGN
414      C
415      C     ( PLOT THE BOX )
416      C
417            EX(1)=2.
418            EY(1)=1.
419            EX(2)=14.
420            EY(2)=1.
421            EX(3)=14.
422            EY(3)=10.
423            EX(4)=2.
424            EY(4)=10.
425            EX(5)=2.
426            EY(5)=1.
427      C
428                  CALL PLINE(EX,EY,5,1,0,0,0)
429      C
430            EX(1)=10.5
431            EY(1)=1.
432            EX(2)=10.5
```

```
433           EY(2)=2.2
434           EX(3)=14.
435           EY(3)=2.2
436           EX(4)=14.
437           EY(4)=1.8
438           EX(5)=10.5
439           EY(5)=1.8
440           EX(6)=10.5
441           EY(6)=1.4
442           EX(7)=14.
443           EY(7)=1.4
444     C
445                 CALL PLINE(EX,EY,7,1,0,0,0)
446     C
447     C     ( PLOT THE TITLE )
448     C
449           IF(NTL.LE.0) GO TO 20
450           DO 10 I=1,NTL
451           DO 12 J=1,5
452        12 TIT(J)=TITLE(I,J)
453           SX=10.6
454           SY=1.05+(I-1)*0.4
455     C
456                 CALL PSYM(SX,SY,0.25,TIT,0.,20,0)
457     C
458        10 CONTINUE
459        20 IF(NRM.LE.0) GO TO 22
460           DO 14 I=1,NRM
461           DO 16 J=1,6
462        16 REM(J)=REMARK(I,J)
463           SX=10.
464           SY=5.0+0.4*(I-1)
465     C
466                 CALL PSYM(SX,SY,0.2,REM,0.,24,0)
467     C
468        14 CONTINUE
469        22 CONTINUE
470     C
471           RETURN
472           END
473     C -----------------------------------------------------------------
474     C
475           SUBROUTINE DOMAIN(NELS,NELE,IJK,X,Y,NODE,XMIN,YMIN,SCALEG,
476          1                  LPRE,LPRN)
477     C
478     C   PURPOSE:
479     C       PLOT THE FINITE ELEMENT MODEL CREATED ON THE DEFORMED DOMAIN.
480     C
481     C   ARGUMENTS:
482     C       NELS   = 1-ST ELEMENT NUMBER OF ELEMENTS PLOTTED
483     C       NELE   = LAST ELEMENT NUMBER OF ELEMENTS PLOTTED
484     C       IJK    = ELEMENT CONNECTIVITIES
485     C       X,Y    = COORDINATES OF NODAL POINTS
486     C       NODE   = TOTAL NUMBER OF NODAL POINTS IN AN ELEMENT
487     C       XMIN   = MINIMUM VALUE OF THE X COORDINATE
488     C       YMIN   = MINIMUM VALUE OF THE Y COORDINATE
489     C       SCALEG = SCALING FACTOR OF THE GEOMETRY
490     C       LPRE   = 1 , IF THE ELEMENT NUMBERS ARE PLOTTED
491     C       LPRN   = 1 , IF THE NODE NUMBERS ARE PLOTTED
492     C
493     C -----------------------------------------------------------------
494     C
495           DIMENSION IJK(NODE,1),X(1),Y(1),EX(99),EY(99)
496           DATA HIG1,HIG2,XORG,YORG/0.13,0.11,3.,2./
497           DATA IYES,INO/'YES','NO'/
498     C
499     C     ( PLOT THE ELEMENTS )
500     C
501           DO 106 NEL=NELS,NELE
502           DO 108 I=1,NODE
503           IJKI=IJK(I,NEL)
504           EX(I)=SCALEG*(X(IJKI)-XMIN)+XORG
505       108 EY(I)=SCALEG*(Y(IJKI)-YMIN)+YORG
506           EX(4)=EX(1)
507           EY(4)=EY(1)
508     C
509                 CALL PLINE(EX,EY,4,1,0,0,0)
510     C
```

```
511         106 CONTINUE
512   C
513   C         ( PLOT THE ELEMENT NUMBERS )
514   C
515             IF(LPRE.NE.IYES) GO TO 110
516             DO 112 NEL=NELS,NELE
517             XC=0.
518             YC=0.
519             DO 114 I=1,NODE
520             IJKI=IJK(I,NEL)
521             XC=XC+X(IJKI)
522         114 YC=YC+Y(IJKI)
523             XC=SCALEG*(XC/3.-XMIN)+XORG-0.8*HIG1
524             YC=SCALEG*(YC/3.-YMIN)+YORG-0.5*HIG1
525   C
526                   CALL PNUMBR(XC,YC,HIG1,NEL,0.,'I3*',0)
527   C
528         112 CONTINUE
529   C
530   C         ( PLOT THE NODE NUMBERS )
531   C
532         110 IF(LPRN.NE.IYES) RETURN
533             DO 118 NEL=NELS,NELE
534             DO 118 I=1,NODE
535             IJKI=IJK(I,NEL)
536             XC=SCALEG*(X(IJKI)-XMIN)+XORG-0.8*HIG2
537             YC=SCALEG*(Y(IJKI)-YMIN)+YORG-0.5*HIG2
538   C
539                   CALL PNUMBR(XC,YC,HIG2,IJKI,0.,'I3*3',0)
540   C
541         118 CONTINUE
542   C
543             RETURN
544             END
545   C ---------------------------------------------------------------------
546   C
547             SUBROUTINE CONTR3(NELS,NELE,NX,F,X,Y,IJK,IDELX,IDEL,NCONT,FCONT,
548            1                  NODE,XMIN,YMIN,SCALEG,XORG,YORG,LW)
549   C
550   C     PURPOSE:
551   C         PLOT CONTOUR LINES BASED ON A TRIANGULAR ELEMENT.
552   C
553   C     ARGUMENTS:
554   C         NELS   = 1-ST ELEMENT NUMBER OF ELEMENTS PLOTTED
555   C         NELE   = LAST ELEMENT NUMBER OF ELEMENTS PLOTTED
556   C         NX     = TOTAL NUMBER OF NODAL POINTS IN A MODEL
557   C         F      = STORES THE NODAL VALUES OF A FUNCTION PLOTTED
558   C         X,Y    = COORDINATES OF NODAL POINTS
559   C         IJK    = ELEMENT CONNECTIVITIES
560   C         IDELX  = TOTAL NUMBER OF NODAL POINTS ON THE BOUNDARY
561   C         IDEL   = ARRAY TO KEEP NODE NUMBERS ON THE BOUNDARY
562   C         NCONT  = TOTAL NUMBER OF CONTOUR LINES
563   C         FCONT  = VALUES OF THE FUNCTION FOR CONTOUR LINES
564   C         NODE   = TOTAL NUMBER OF NODAL POINTS IN AN ELEMENT
565   C         XMIN   = MINIMUM VALUE OF THE X COORDINATE
566   C         YMIN   = MINIMUM VALUE OF THE Y COORDINATE
567   C         SCALEG = SCALING FACTOR OF THE GEOMETRY
568   C         XORG   = ORIGIN OF THE X COORDINATE
569   C         YORG   = ORIGIN OF THE Y COORDINATE
570   C
571   C ---------------------------------------------------------------------
572   C
573             DIMENSION X(1),Y(1),IJK(NODE,1),F(1),IDEL(1),FCONT(1)
574             DIMENSION FE(3),XE(3),YE(3),IARY1(6),EX(99),EY(99)
575   C
576             DATA IARY1/1,2,2,3,3,1/
577   C
578   C         ( DETERMINE THE INTERVAL OF THE CONTOUR LINES )
579   C
580             FMIN= 1.E20
581             FMAX=-1.E20
582             DO 202 I=1,NX
583             FI=F(I)
584             IF(FI.GT.FMAX) FMAX=FI
585         202 IF(FI.LT.FMIN) FMIN=FI
586             WRITE(LW,600) FMIN,FMAX
587         600 FORMAT(/////10X,'----- PLOT : CONTOUR LINES -----',//15X,
588            1  'THE MINIMUM VALUE = ',E12.5,/15X,
```

```
589          2  'THE MAXIMUM VALUE = ',E12.5,//10X,
590          3  '< CONTOUR LINES >',/)
591             WRITE(LW,602) (I,FCONT(I),I=1,NCONT)
592         602 FORMAT(10X,'CONTOUR(',I3,' )',5X,E12.5)
593             EPP=0.0001*(FMAX-FMIN)
594       C
595       C     ( PLOT THE CONTOUR LINES )
596       C
597             DO 206 NEL=NELS,NELE
598       C
599             DO 208 I=1,NODE
600             IJKI=IJK(I,NEL)
601             XE(I)=X(IJKI)
602             YE(I)=Y(IJKI)
603         208 FE(I)=F(IJKI)
604       C
605             DO 210 N=1,NCONT
606       C
607             FSI=FCONT(N)
608             LIN=1
609             DO 212 J=1,NODE
610             J1=2*(J-1)+1
611             J2=J1+1
612             J1A=IARY1(J1)
613             J2A=IARY1(J2)
614             XE1=XE(J1A)
615             YE1=YE(J1A)
616             XE2=XE(J2A)
617             YE2=YE(J2A)
618             FE1=FE(J1A)
619             FE2=FE(J2A)
620             IF(FE2-FE1.GT.EPP) GO TO 214
621             IF(FE1-FE2.GT.EPP) GO TO 218
622             GO TO 216
623         214 IF(FSI.GT.FE2.OR.FSI.LT.FE1) GO TO 212
624             GO TO 220
625         218 IF(FSI.GT.FE1.OR.FSI.LT.FE2) GO TO 212
626         220 TA=(FSI-FE2)/(FE1-FE2)
627             EX(LIN)=SCALEG*(XE2+TA*(XE1-XE2)-XMIN)+XORG
628             EY(LIN)=SCALEG*(YE2+TA*(YE1-YE2)-YMIN)+YORG
629             LIN=LIN+1
630             GO TO 212
631         216 IF(ABS(FSI-FE1).GT.EPP) GO TO 212
632             EX(1)=SCALEG*(XE1-XMIN)+XORG
633             EY(1)=SCALEG*(YE1-YMIN)+YORG
634             EX(2)=SCALEG*(XE2-XMIN)+XORG
635             EY(2)=SCALEG*(YE2-YMIN)+YORG
636             LIN=3
637         212 CONTINUE
638       C
639             LIN1=LIN-1
640             IF(LIN.GE.3) CALL PLINE(EX,EY,LIN1,1,0,0,0)
641       C
642         210 CONTINUE
643       C
644         206 CONTINUE
645       C
646       C     ( PLOT THE DELINEATION )
647       C
648             DO 224 I=1,IDELX
649             I1=IDEL(2*I-1)
650             I2=IDEL(2*I)
651             EX(1)=SCALEG*(X(I1)-XMIN)+XORG
652             EY(1)=SCALEG*(Y(I1)-YMIN)+YORG
653             EX(2)=SCALEG*(X(I2)-XMIN)+XORG
654             EY(2)=SCALEG*(Y(I2)-YMIN)+YORG
655       C
656                   CALL PLINE(EX,EY,2,1,0,0,0)
657       C
658         224 CONTINUE
659       C
660             RETURN
661             END
662       C ---------------------------------------------------------------
663       C
664             SUBROUTINE FUNC3D(NELS,NELE,NX,F,X,Y,IJK,THETA1,THETA2,STDLTH,
665            1                  XMIN,YMIN,XMAX,YMAX,XORG,YORG,IDELX,IDEL,NODE,LW)
666       C
```

```
667       C    PURPOSE:
668       C        PLOT THE FUNCTION PROFILE ON THE SKEW COORDINATE SYSTEM.
669       C
670       C    ARGUMENTS:
671       C        NELS    = 1-ST ELEMENT NUMBER OF ELEMENTS PLOTTED
672       C        NELE    = LAST ELEMENT NUMBER OF ELEMENTS PLOTTED
673       C        NX      = TOTAL NUMBER OF NODAL POINTS IN THE MODEL
674       C        F       = STORES THE NODAL VALUES OF A FUNCTION PLOTTED
675       C        X,Y     = COORDINATES OF NODAL POINTS
676       C        IJK     = ELEMENT CONNECTIVITIES
677       C        THETA1 = ANGLE OF THE 1-ST AXIS FROM THE HORIZONTAL LINE
678       C                 15.-30. ARE RECOMMENDED
679       C        THETA2 = ANGLE OF THE 2-ND AXIS; 20.-35. ARE RECOMMENDED
680       C        STDLTH = STANDARD SIZE OF THE PLOTTING ( 7-IN )
681       C        XMIN    = MINIMUM VALUE OF THE X COORDINATE
682       C        YMIN    = MINIMUM VALUE OF THE Y COORDINATE
683       C        XMAX    = MAXIMUM VALUE OF THE X COORDINATE
684       C        YMAX    = MAXIMUM VALUE OF THE Y COORDINATE
685       C        XORG    = ORIGIN OF THE X COORDINATE
686       C        YORG    = ORIGIN OF THE Y COORDINATE
687       C        IDELX   = TOTAL NUMBER OF NODAL POINTS ON THE BOUNDARY
688       C        IDEL    = ARRAY TO KEEP NODE NUMBERS ON THE BOUNDARY
689       C        NODE    = TOTAL NUMBER'OF NODAL POINTS IN AN ELEMENT
690       C
691       C ----------------------------------------------------------------------
692       C
693             DIMENSION F(1),X(1),Y(1),IJK(NODE,1),IDEL(1)
694             DIMENSION EX(99),EY(99)
695       C
696       C      ( SCALING THE FUNCTION  F  )
697       C
698             FMIN= 1.E20
699             FMAX=-1.E20
700             DO 302 I=1,NX
701             FI=F(I)
702             IF(FI.LT.FMIN) FMIN=FI
703         302 IF(FI.GT.FMAX) FMAX=FI
704             WRITE(LW,666) FMIN,FMAX
705         666 FORMAT(/////10X,'----- PLOTPR : FUNC3D -----',//10X,
706            1  'THE MINIMUM VALUE = ',E12.5,/10X,
707            2  'THE MAXIMUM VALUE = ',E12.5)
708             IF(FMIN.GT.0.) FMIN=0.
709             FSIZE=FMAX-FMIN
710             ASIZE=AMAX1(XMAX-XMIN,YMAX-YMIN)
711             SXMIN= XMIN*COS(THETA1)+YMIN*COS(THETA2)
712             SYMIN=-XMAX*SIN(THETA1)+YMIN*SIN(THETA2)+FMIN*ASIZE/FSIZE
713             SXMAX= XMAX*COS(THETA1)+YMAX*COS(THETA2)
714             SYMAX=-XMIN*SIN(THETA1)+YMAX*SIN(THETA2)+FMAX*ASIZE/FSIZE
715             SIZEFX=AMAX1(SXMAX-SXMIN,SYMAX-SYMIN)
716             SCALE=STDLTH/SIZEFX
717       C
718       C      ( PLOT THE DELINEATION OF THE DOMAIN )
719       C
720             DO 310 I=1,IDELX
721             I1=IDEL(2*I-1)
722             I2=IDEL(2*I)
723             EX(1)=SCALE*( X(I1)*COS(THETA1)+Y(I1)*COS(THETA2)-SXMIN)+XORG
724             EY(1)=SCALE*(-X(I1)*SIN(THETA1)+Y(I1)*SIN(THETA2)-SYMIN)+YORG
725             EX(2)=SCALE*( X(I2)*COS(THETA1)+Y(I2)*COS(THETA2)-SXMIN)+XORG
726             EY(2)=SCALE*(-X(I2)*SIN(THETA1)+Y(I2)*SIN(THETA2)-SYMIN)+YORG
727       C
728                   CALL PLINE(EX,EY,2,1,0,0,0)
729       C
730         310 CONTINUE
731       C
732       C      ( PLOT THE FUNCTION )
733       C
734             NODE1=NODE+1
735             DO 314 NEL=NELS,NELE
736             DO 316 N=1,NODE
737             IJKN=IJK(N,NEL)
738             EX(N)=SCALE*( X(IJKN)*COS(THETA1)+Y(IJKN)*COS(THETA2)-SXMIN)+XORG
739             EY(N)=SCALE*(-X(IJKN)*SIN(THETA1)+Y(IJKN)*SIN(THETA2)+
740            1         F(IJKN)*ASIZE/FSIZE-SYMIN)+YORG
741         316 CONTINUE
742             EX(NODE1)=EX(1)
743             EY(NODE1)=EY(1)
744       C
```

```
745                 CALL PLINE(EX,EY,NODE1,1,0,0,0)
746     C
747       314 CONTINUE
748     C
749           EX(1)=2.5
750           EY(1)=1.5
751           EX(2)=2.5+0.5*COS(THETA1)
752           EY(2)=1.5-0.5*SIN(THETA1)
753                 CALL PARROW(EX,EY,2,1,0.1,0,0)
754           EX(2)=2.5+0.5*COS(THETA2)
755           EY(2)=1.5+0.5*SIN(THETA2)
756                 CALL PARROW(EX,EY,2,1,0.1,0,0)
757           EX(2)=2.5
758           EY(2)=2.
759                 CALL PARROW(EX,EY,2,1,0.1,0,0)
760     C
761           RETURN
762           END
763     C -----------------------------------------------------------------
764     C
765           SUBROUTINE VELOCI(NELS,NELE,P,X,Y,IJK,IDELX,IDEL,
766          1                  STDARW,XMIN,YMIN,XORG,YORG,SCALEG,NODE,LW)
767     C
768     C   PURPOSE:
769     C       PLOT THE VELOCITY VECTOR ON TRIANGULAR ELEMENTS.
770     C
771     C   ARGUMENTS:
772     C       NELS   = 1-ST ELEMENT NUMBER OF ELEMENTS PLOTTED
773     C       NELE   = LAST ELEMENT NUMBER OF ELEMENTS PLOTTED
774     C       P      = STORES THE VELOCITY COMPONENTS
775     C       X,Y    = COORDINATES OF NODAL POINTS
776     C       IJK    = ELEMENT CONNECTIVITIES
777     C       IDELX  = TOTAL NUMBER OF NODAL POINTS ON THE BOUNDARY
778     C       IDEL   = ARRAY TO KEEP NODE NUMBERS ON THE BOUNDARY
779     C       STDARW = STANDARD SIZE OF THE ARROW
780     C       XMIN   = MINIMUM VALUE OF THE X COORDINATE
781     C       YMIN   = MINIMUM VALUE OF THE Y COORDINATE
782     C       XORG   = ORIGIN OF THE X COORDINATE
783     C       YORG   = ORIGIN OF THE Y COORDINATE
784     C       SCALEG = SCALING FACTOR OF THE GEOMETRY
785     C       NODE   = TOTAL NUMBER OF NODAL POINTS IN AN ELEMENT
786     C
787     C -----------------------------------------------------------------
788     C
789           DIMENSION P(4,1),X(1),Y(1),IJK(NODE,1),IDEL(1)
790           DIMENSION EX(99),EY(99)
791     C
792     C     ( FIND THE MAXIMUM VELOCITY )
793     C
794           PMAX=0.
795           DO 402 NEL=NELS,NELE
796           P1=P(1,NEL)
797           P2=P(2,NEL)
798           AP=SQRT(P1*P1+P2*P2)
799       402 IF(AP.GT.PMAX) PMAX=AP
800           WRITE(LW,668) PMAX
801       668 FORMAT(/////10X,'----- PLOTPR : VELOCITY -----',//10X,
802          2 '0.7 INCH CORRESPONDS TO THE SPEED  SP = ',E10.3)
803     C
804           SCALEA=STDARW/PMAX
805           ARWLTH=0.15*STDARW
806     C
807           DO 404 NEL=NELS,NELE
808     C
809     C     ( PLOT THE ELEMENTS )
810     C
811           DO 406 I=1,NODE
812           IJKI=IJK(I,NEL)
813           EX(I)=SCALEG*(X(IJKI)-XMIN)+XORG
814       406 EY(I)=SCALEG*(Y(IJKI)-YMIN)+YORG
815           EX(4)=EX(1)
816           EY(4)=EY(1)
817     C
818           IF(LPRF.EQ.1) CALL PLINE(EX,EY,4,1,0,0,0)
819     C
820     C     ( PLOT THE VELOCITY )
821     C
822           ECX=(EX(1)+EX(2)+EX(3))/3.
```

```
823              ECY=(EY(1)+EY(2)+EY(3))/3.
824              ADX=SCALEA*P(1,NEL)
825              ADY=SCALEA*P(2,NEL)
826              EX(1)=ECX-0.5*ADX
827              EY(1)=ECY-0.5*ADY
828              EX(2)=ECX+0.5*ADX
829              EY(2)=ECY+0.5*ADY
830        C
831                    CALL PARROW(EX,EY,2,1,ARWLTH,0,0)
832        C
833          404 CONTINUE
834        C
835        C      ( PLOT THE DELINEATION )
836        C
837              DO 408 I=1,IDELX
838              I1=IDEL(2*I-1)
839              I2=IDEL(2*I)
840              EX(1)=SCALEG*(X(I1)-XMIN)+XORG
841              EY(1)=SCALEG*(Y(I1)-YMIN)+YORG
842              EX(2)=SCALEG*(X(I2)-XMIN)+XORG
843              EY(2)=SCALEG*(Y(I2)-YMIN)+YORG
844        C
845                    CALL PLINE(EX,EY,2,1,0,0,0)
846        C
847          408 CONTINUE
848        C
849              RETURN
850              END
851        C -----------------------------------------------------------------------
852        C
853              SUBROUTINE VCONT(NX,CQ,FCONT,NCONT)
854        C
855        C   PURPOSE:
856        C       COMPUTE THE VALUES OF THE CONTOUR LINES.
857        C
858        C   ARGUMENTS:
859        C       NX     = TOTAL NUMBER OF NODAL POINTS
860        C       CQ     = QUALITY INDEX AT NODAL POINTS
861        C       FCON   = VALUES OF THE FUNCTION FOR CONTOUR LINES
862        C       NCONT  = TOTAL NUMBER OF CONTOUR LINES
863        C
864        C -----------------------------------------------------------------------
865        C
866              DIMENSION CQ(1),FCONT(1)
867        C
868        C      ( OBTAIN MAXIMUM AND MINIMUM VALUES )
869        C
870              QCIMAX=-1.E20
871              QCIMIN=1.E20
872              DO 100 I=1,NX
873              QCIMAX=AMAX1(QCIMAX,CQ(I))
874          100 QCIMIN=AMIN1(QCIMIN,CQ(I))
875        C
876        C      ( SET THE VALUES OF THE CONTOUR LINES )
877        C
878              DO 102 I=1,NCONT
879          102 FCONT(I)=QCIMIN+(QCIMAX-QCIMIN)*(I-1)/(NCONT-1)
880        C
881              RETURN
882              END
```

APPENDIX 4
FEM2

```
  1   C ***********************************************************************
  2   C
  3   C                          PROGRAM : FEM2
  4   C
  5   C                           SPRING  1984
  6   C
  7   C ***********************************************************************
  8   C
  9   C   PURPOSE:
 10   C       PROGRAM OF STRESS ANALYSIS FOR LINEARLY ELASTIC STRUCTURES IN
 11   C       A PLANE USING 4-NODE ISOPARAMETRIC ELEMENTS
 12   C
 13   C   BOUNDARY VALUE PROBLEMS SOLVED:
 14   C
 15   C          - DS(I,J)/DX(J) = F(I)    IN  G
 16   C                     U(I) = G(I)    ON  S1
 17   C              S(I,J)*N(J) = T(I)    ON  S2
 18   C              S(I,J)*N(J) = K(I,J)*(U(J)-G(J))+T(I)   ON  S3
 19   C
 20   C       S(I,J)      = (I,J) COMPONENT OF THE STRESS TENSOR
 21   C                   = D(I,J,K,L)*E(K,L)
 22   C       E(K,L)      = (K,L) COMPONENT OF THE LINEARIZED STRAIN TENSOR
 23   C                   = ( DU(K)/DX(L) + DU(L)/DX(K) )/2
 24   C       D(I,J,K,L)= TANGENT MODULUS OF THE MATERIAL
 25   C       F(I)        = I-TH COMPONENT OF THE BODY FORCE
 26   C       T(I)        = I-TH COMPONENT OF THE SURFACE TRACTION
 27   C       G(I)        = I-TH COMPONENT OF THE CONSTRAINED DISPLACEMENT ON
 28   C                     THE BOUNDARY  S1
 29   C       K(I,J)      = GENERALIZED SPRING CONSTANTS
 30   C
 31   C   NOTES:
 32   C       THIS PROGRAM IS SET UP FOR NONHOMOGENEOUS ISOTROPIC MATERIALS.
 33   C       IF ANISOTROPIC MATERIALS MUST BE CONSIDERED, WE NEED TO
 34   C       MODIFY THE SUBROUTINES MATERL AND TANMOD IN ORDER TO CONSTRUCT
 35   C       THE ELASTICITY CONSTANTS  "D(I,J,K,L)"  FOR THOSE MATERIALS.
 36   C
 37   C   CONTROL NUMBERS READ IN THE MAIN PROGRAM:
 38   C       MBOUN1 = 1  IF DISPLACEMENTS ARE SPECIFIED ON NODAL POINTS
 39   C              = 0  OTHERWISE
 40   C       MBOUN2 = 1  IF APPLIED LOADS ARE GIVEN ON NODAL POINTS
 41   C              = 0  OTHERWISE
 42   C       MBOUN3 = 1  IF GENERALIZED 3-RD BOUNDARY CONDITION EXISTS
 43   C              = 0  OTHERWISE
 44   C       MBOUN4 = 1  IF DISPLACEMENTS ARE SPECIFIED ALONG THE SLOPES
 45   C              = 0  OTHERWISE
 46   C       MBODYF = 1  IF THERE ARE BODY FORCES
 47   C              = 0  OTHERWISE
 48   C       MTEMPE = 1  IF THERMAL STRESSES ARE CONSIDERED
 49   C              = 0  OTHERWISE
 50   C       MPREPR = 1  IF ONLY PRE-PROCESSING WILL BE EXECUTED
 51   C              = 0  IF ALL PROCESSINGS WILL BE EXECUTED
 52   C
 53   C   INPUT DATA TO BE READ:
 54   C           ALL VARIABLES ARE EXPLAINED IN EACH SUBROUTINE
 55   C
 56   C           1 SUBROUTINE PTITLE
 57   C                1.0 FLAG(I),I=1,10                   :    10A8
 58   C                1.1 TITLE(I),I=1,10                  :    10A8
 59   C           2 MAIN ROUTINE
 60   C                2.0 FLAG(I),I=1,10                   :    10A8
 61   C                2.1 MBOUN1,REM(I),I=1,9              :    I5,9A8
 62   C                2.2 MBOUN2,REM(I),I=1,9              :    I5,9A8
 63   C                2.3 MBOUN3,REM(I),I=1,9              :    I5,9A8
 64   C                2.4 MBOUN4,REM(I),I=1,9              :    I5,9A8
 65   C                2.5 MBODYF,REM(I),I=1,9              :    I5,9A8
 66   C                2.6 MTEMPE,REM(I),I=1,9              :    I5,9A8
 67   C                2.7 MPREPR,REM(I),I=1,9              :    I5,9A8
 68   C           3 SUBROUTINE HRMESH
 69   C                3.0 FLAG(I),I=1,10                   :    10A8
 70   C                3.1 NODADD,NELADD                    :    16I5
 71   C                3.2 N,X(N),Y(N),I=1,NODADD           :    I5,2F10.4
 72   C                3.3 N,MPE(N),(IJK(NODE*(N-1)+J),J=1,4),
 73   C                    I=1,NELADD                       :    6I5
 74   C           4 SUBROUTINE MATERL
```

```
 75   C               4.0 FLAG(I),I=1,10                    :     10A8
 76   C               4.1 LMAT,KK1,KK2                      :     16I5
 77   C               4.2 (PROMAT(J,I),J=1,8),I=1,LMAT      :     8F10.4
 78   C            IF(MBOUN1.NE.1) GO TO 100
 79   C          5 SUBROUTINE RBOUN1
 80   C               5.0 FLAG(I),I=1,10                    :     10A8
 81   C               5.1 LFIX,SPRING                       :     I5,F20.4
 82   C               5.2 (NDFIX(J,I),J=1,2),(VFIXED(J,I),J=1,2),
 83   C                   I=1,LFIX                          :     2I5,2F10.4
 84   C        100 IF(MBOUN2.NE.1) GO TO 102
 85   C          6 SUBROUTINE RBOUN2
 86   C               6.0 FLAG(I),I=1,10                    :     10A8
 87   C               6.1 LFOR                              :     I5
 88   C               6.2 NLOAD(I),(VLOAD(J,I),J=1,2),
 89   C                   I=1,LFOR                          :     I5,2F10.4
 90   C        102 IF(MBOUN3.NE.1) GO TO 104
 91   C          7 SUBROUTINE RBOUN3
 92   C               7.0 FLAG(I),I=1,10                    :     10A8
 93   C               7.1 NB3X                              :     I5
 94   C               DO 200 MNEL=1,NB3X
 95   C               7.2 NEL,(IJ3(J,NEL),J=1,2)            :     16I5
 96   C               7.3 SPTN(1,NEL),GTN(1,NEL),TRN(1,NEL)
 97   C                                                     :     F20.5,2F10.4
 98   C               7.4 SPTN(2,NEL),GTN(2,NEL),TRN(2,NEL)
 99   C                                                     :     F20.5,2F10.4
100   C               200 CONTINUE
101   C        104 IF(MBOUN4.NE.1) GO TO 106
102   C          8 SUBROUTINE RBOUN4
103   C               8.0 FLAG(I),I=1,10                    :     10A8
104   C               8.1 LSLID,SPRING                      :     I5,F20.4
105   C               8.2 NSLIDE(I),(VSLIDE(J,I),J=1,2),
106   C                   I=1,LSLID                         :     I5,2F10.4
107   C        106 IF(MBODYF.NE.1) GO TO 108
108   C          9 SUBROUTINE RBODYF
109   C               9.0 FLAG(I),I=1,10                    :     10A8
110   C               9.1 LBODY                             :     16I5
111   C               9.2 (C1(I,J),J=1,8),I=1,LBODY         :     8F10.4
112   C               9.3 (C2(I,J),J=1,8),I=1,LBODY         :     8F10.4
113   C        108 IF(MTEMPE.NE.1) STOP
114   C         10 SUBROUTINE RTEMPE
115   C              10.0 FLAG(I),I=1,10                    :     10A8
116   C              10.1 LTEMP                             :     16I5
117   C              10.2 (C1(I,J),J=1,8),I=1,LTEMP         :     8F10.4
118   C
119   C    DIMENSIONING:
120   C         X(NX),Y(NX)
121   C            NX      = TOTAL NUMBER OF NODAL POINTS
122   C         F(NEQ),U(NEQ),JDIAG(NEQ)
123   C            NEQ     = TOTAL NUMBER OF EQUATIONS SOLVED ( = 2*NX )
124   C         IJK(NMAX),MPE(NELX)
125   C            NMAX    = NODE*NELX
126   C            NODE    = TOTAL NUMBER OF NODAL POINTS IN AN ELEMENT
127   C            NELX    = TOTAL NUMBER OF ELEMENTS
128   C         PROMAT(16,LMAT)
129   C            LMAT    = TOTAL NUMBER OF GROUPS OF MATERIAL
130   C         SK(MSIZE1)
131   C            MSIZE1 = MAXIMUM SIZE OF THE ARRAY FOR STIFFNESS MATRIX
132   C         NDFIX(2,LFIX),VFIXED(3,LFIX)
133   C            LFIX    = TOTAL NUMBER OF NODAL POINTS CONSTRAINED
134   C         NLOAD(LFOR),VLOAD(2,LFOR)
135   C            LFOR    = TOTAL NUMBER OF NODAL POINTS APPLIED FORCES
136   C         IJ3(2,NE3),MPBE(NE3),UJ(2,NDMX),TJ(2,NDMX)
137   C            NE3     = TOTAL NUMBER OF LINE ELEMENTS ON THE 3-RD BOUNDARY
138   C            NDMX    = NE3+1
139   C         XK(3,MPBX),AL(3,MPBX)
140   C            MPBX    = NUMBER OF DIFFERENT MATERIALS ON THE 3-RD BOUNDARY
141   C         NSLIDE(LSLID),VSLIDE(2,LSLID)
142   C            LSLID   = TOTAL NUMBER OF NODAL POINTS SPECIFIED THE SLOPE
143   C         TEMP(NX),BFR(2,NX)
144   C            NX      = TOTAL NUMBER OF NODAL POINTS
145   C
146   C    DATA STATEMENTS BE SPECIFIED:
147   C         LR      = NUMBER FOR THE READ STATEMENT ( =5 )
148   C         LW      = NUMBER FOR THE WRITE STATEMENT ( =6 )
149   C         LP      = NUMBER FOR THE WRITE STATEMENT TO THE PLOT FILE ( =8 )
150   C         MSIZE1 = MAXIMUM NUMBER OF ARRAY FOR THE STIFFNESS MATRIX
151   C         NDF     = NUMBER OF DEGREES OF FREEDOM PER NODAL POINT
152   C         NODE    = TOTAL NUMBER OF NODAL POINTS IN AN ELEMENT
```

```
153   C        INTXY  = TOTAL NUMBER OF INTEGRATION POINTS IN AN ELEMENT
154   C
155   C ***********************************************************************
156   C
157         IMPLICIT REAL*8(A-H,O-Z)
158         DIMENSION SK(10000),JDIAG(600),F(600),U(600),X(300),Y(300),
159        1          IJK(2000),MPE(500),
160        2          PROMAT(16,10),
161        3          NDFIX(2,50),VFIXED(3,50),
162        4          NLOAD(50),VLOAD(2,50),
163        5          IJ3(200),SPTN(2,100),GTN(2,100),TRN(2,100),
164        6          NSLIDE(50),VSLIDE(2,50),
165        7          TEMP(300),BFR(2,300)
166         DIMENSION FLAG(10),REM(9)
167   C
168         DATA LR,LW,LP,MSIZE1,NDF,NODE,INTXY/5,6,8,10000,2,4,4/
169   C
170   C *******************   PRE-PROCESSING   *******************
171   C
172   C
173   C       ( READ THE TITLE OF THE PROBLEM )
174   C
175               CALL PTITLE(LR,LW)
176   C
177   C       ( READ NUMBERS WHICH CONTROL THE FLOW OF ROUTINES )
178   C
179         READ(LR,502) (FLAG(I),I=1,10)
180     502 FORMAT(10A8)
181         WRITE(LW,600)
182     600 FORMAT(/////30X,'----- CONTROL NUMBERS -----',//)
183         READ(LR,500) MBOUN1,(REM(J),J=1,9)
184     500 FORMAT(I5,9A8)
185         WRITE(LW,612) MBOUN1,(REM(I),I=1,9)
186     612 FORMAT(25X,I5,5X,9A8)
187         READ(LR,500) MBOUN2,(REM(J),J=1,9)
188         WRITE(LW,612) MBOUN2,(REM(I),I=1,9)
189         READ(LR,500) MBOUN3,(REM(J),J=1,9)
190         WRITE(LW,612) MBOUN3,(REM(I),I=1,9)
191         READ(LR,500) MBOUN4,(REM(J),J=1,9)
192         WRITE(LW,612) MBOUN4,(REM(I),I=1,9)
193         READ(LR,500) MBODYF,(REM(J),J=1,9)
194         WRITE(LW,612) MBODYF,(REM(I),I=1,9)
195         READ(LR,500) MTEMPE,(REM(J),J=1,9)
196         WRITE(LW,612) MTEMPE,(REM(I),I=1,9)
197         READ(LR,500) MPREPR,(REM(J),J=1,9)
198         WRITE(LW,612) MPREPR,(REM(I),I=1,9)
199         WRITE(LW,602)
200     602 FORMAT(1H1,///10X,'XXXXXXXXXX DATA OF THE FINITE ELEMENT',
201        1 ' MODEL XXXXXXXXXX',/)
202   C
203   C       ( READ THE FINITE ELEMENT MESH )
204   C
205         NX=0
206         NELX=0
207   C
208               CALL HRMESH(NX,NELX,4,X,Y,IJK,MPE,LR,LW)
209   C
210   C       ( SET THE TOTAL NUMBER OF DEGREES OF FREEDOM )
211   C
212         NEQ=NDF*NX
213   C
214   C       ( KEEP THE DATA ON THE FILE8 FOR PLOTTING )
215   C
216         WRITE(LP,800) NX,NELX,NEQ,NODE,NDF,INTXY
217     800 FORMAT(6I5)
218         WRITE(LP,802) (X(I),Y(I),I=1,NX)
219     802 FORMAT(8F10.4)
220         IE=NODE*NELX
221         WRITE(LP,804) (IJK(I),I=1,IE)
222     804 FORMAT(16I5)
223   C
224   C       ( OUTPUT OF COORDINATES AND ELEMENT CONNECTIVITIES )
225   C
226         WRITE(LW,604)
227     604 FORMAT(/////10X,'---------- < COORDINATES OF NODES > ----------',/)
228         WRITE(LW,606) (I,X(I),Y(I),I=1,NX)
229     606 FORMAT(10X,I5,5X,'( ',2F10.4,' )',10X,I5,5X,'( ',2F10.4,' )')
230         WRITE(LW,608)
```

```
231         608 FORMAT(/////10X,'---------- < ELEMENT CONNECTIVITIES >',
232            1   ' ----------',/)
233              DO 118 I=1,NELX
234              JS=NODE*(I-1)+1
235              JE=NODE*I
236         118 WRITE(LW,610) I,MPE(I),(IJK(J),J=JS,JE)
237         610 FORMAT(10X,I5,5X,'MPE ',I3,5X,'<IJK>',20I5)
238              IF(MPREPR.EQ.1) STOP
239       C
240       C      ( PRE-PROCESSING TO THE SKYLINE SOLVER )
241       C
242                    CALL PRESKY(JDIAG,IJK,NELX,NEQ,4,2,MSIZE1,LW)
243       C
244       C      ( MATERIAL CONSTANTS )
245       C
246                    CALL MATERL(PROMAT,KK1,KK2,LR,LW)
247       C
248       C      ( READ INPUT DATA FOR BOUNDARY CONDITIONS )
249       C
250              IF(MBOUN1.EQ.1) CALL RBOUN1(NDFIX,VFIXED,SPRIN1,2,LFIX,LR,LW)
251              IF(MBOUN2.EQ.1) CALL RBOUN2(NLOAD,VLOAD,2,LFOR,LR,LW)
252              IF(MBOUN3.EQ.1) CALL RBOUN3(NB3X,IJ3,SPTN,GTN,TRN,2,LR,LW)
253              IF(MBOUN4.EQ.1) CALL RBOUN4(NSLIDE,VSLIDE,SPRIN3,LSLID,LR,LW)
254              IF(MBODYF.EQ.1) CALL RBODYF(BFR,X,Y,NX,LR,LW)
255              IF(MTEMPE.EQ.1) CALL RTEMPE(TEMP,X,Y,NX,LR,LW)
256       C
257       C    *******************   F.E.-PROCESSING   *******************
258       C
259       C      ( CONSTRUCT THE GLOBAL STIFFNESS MATRIX AND THE LOAD VECTOR )
260       C
261                    CALL ASSEMB(SK,F,JDIAG,IJK,X,Y,MPE,PROMAT,TEMP,BFR,
262            1                   NX,NELX,KK1,KK2,NEQ,4,2,MBOUN1,LR,LW)
263       C
264       C      ( BOUNDARY CONDITIONS )
265       C
266              IF(MBOUN1.EQ.1) CALL BOUND1(SK,JDIAG,F,NDFIX,VFIXED,SPRIN1,2,LFIX)
267              IF(MBOUN2.EQ.1) CALL BOUND2(F,NLOAD,VLOAD,2,LFOR)
268              IF(MBOUN3.EQ.1) CALL BOUND3(SK,F,JDIAG,NB3X,SPTN,GTN,TRN,2,2,
269            1                                 IJ3,X,Y)
270              IF(MBOUN4.EQ.1) CALL BOUND4(SK,JDIAG,F,NSLIDE,VSLIDE,SPRIN3,LSLID)
271       C
272       C      ( SOLVE THE SYSTEM OF LINEAR EQUATIONS )
273       C
274                    CALL SKYLIN(SK,F,JDIAG,NEQ,0)
275       C
276       C    *******************   POST-PROCESSING   *******************
277       C
278       C      ( DEFORMATION AND STRESS FIELDS )
279       C
280                    CALL DEFORM(F,U,X,Y,NX,LW,LP)
281                    CALL STRESS(X,Y,U,IJK,NELX,PROMAT,MPE,4,2,KK1,KK2,TEMP,
282            1                   NX,SK,JDIAG,MSIZE1,LW,LP)
283       C
284               STOP
285               END
286       C -----------------------------------------------------------------
287       C
288              SUBROUTINE PTITLE(LR,LW)
289       C
290       C    PURPOSE:
291       C        PRINT OUT THE TITLE OF THE PROBLEM TO BE SOLVED.
292       C
293       C    LOCAL VARIABLES:
294       C        FLAG  = ARRAY TO PUT REMARKS OR DATA BLOCK SPECIFICATION
295       C        TITLE = TITLE OF THE PROBLEM TO BE SOLVED
296       C
297       C    INPUT DATA TO BE READ:
298       C        0. FLAG(I),I=1,10             :     10A8
299       C        1. TITLE(I)                   :     10A8
300       C
301       C -----------------------------------------------------------------
302       C
303              IMPLICIT REAL*8(A-H,O-Z)
304              DIMENSION TITLE(10),FLAG(10)
305       C
306              READ(LR,500) (FLAG(I),I=1,10)
307              READ(LR,500) (TITLE(I),I=1,10)
308         500 FORMAT(10A8)
```

```
309      C
310            WRITE(LW,600) (TITLE(I),I=1,10)
311        600 FORMAT(1H1,///////////////10X,80('*'),/10X,80('*'),////,20X,
312           1 10A8,////10X,80('*'),/10X,80('*'))
313      C
314            RETURN
315            END
316      C ---------------------------------------------------------------------
317      C
318            SUBROUTINE HRMESH(NX,NELX,NODE,X,Y,IJK,MPE,LR,LW)
319      C
320      C   PURPOSE:
321      C       READ THE MESH DATA ( ELEMENT CONNECTIVITIES AND COORDINATES )
322      C       WITHOUT USING THE AUTOMATIC MESH GENERATOR.
323      C
324      C   ARGUMENTS:
325      C       NX   = TOTAL NUMBER OF NODAL POINTS
326      C       NELX = TOTAL NUMBER OF FINITE ELEMENTS
327      C       NODE = TOTAL NUMBER OF NODAL POINTS IN AN ELEMENT
328      C       X,Y  = COORDINATES OF NODAL POINTS
329      C       IJK  = ELEMENT CONNECTIVITIES
330      C       MPE  = GROUP NUMBER OF THE MATERIALS FOR EACH ELEMENT
331      C
332      C   LOCAL VARIABLES:
333      C       NODADD = TOTAL NUMBER OF ADDITIONAL NODAL POINTS
334      C       NELADD = TOTAL NUMBER OF ADDITIONAL ELEMENTS
335      C       IJKD   = DUMMY ARRAY TO READ THE ELEMENT CONECTIVITY
336      C
337      C   DIMENSIONING:
338      C       IJKD IS RANGED FROM 1 TO 27.  IF THERE ARE MORE THAN 27 NODAL
339      C       POINTS IN AN ELEMENT, WE HAVE TO INCREASE THE DIMENSION.
340      C
341      C   INPUT DATA TO BE READ:
342      C       0. FLAG(I),I=1,10                          :    10A8
343      C       1. NODADD,NELADD                           :    16I5
344      C       2. N,X(N),Y(N),I=1,NODADD                  :    I5,2F10.4
345      C       3. N,MPE(N),(IJK(NODE*(N-1)+J),J=1,4),
346      C          I=1,NELADD                              :    6I5
347      C
348      C ---------------------------------------------------------------------
349      C
350            IMPLICIT REAL*8(A-H,O-Z)
351            DIMENSION X(1),Y(1),IJK(1),MPE(1),FLAG(10),IJKD(27)
352      C
353            READ(LR,598) (FLAG(I),I=1,10)
354            WRITE(LW,606)
355        606 FORMAT(/////10X,'---------- MODIFICATION/ADDITION OF NODES',
356           1  ' AND ELEMENT CONNECTIVITIES ----------',//,10X,
357           2  '< COORDINATES OF NODES >',/)
358        598 FORMAT(10A8)
359            READ(LR,500) NODADD,NELADD
360        500 FORMAT(16I5)
361            DO 106 I=1,NODADD
362            READ(LR,502) N,X(N),Y(N)
363        502 FORMAT(I5,2F10.4)
364            IF(N.GT.NX) NX=NX+1
365            WRITE(LW,600) N,X(N),Y(N),NX
366        600 FORMAT(15X,'NODE ',I4,5X,'( ',2F12.4,' )',5X,'NEW NX = ',I5)
367        106 CONTINUE
368            WRITE(LW,604)
369        604 FORMAT(/////10X,'< ELEMENT CONNECTIVITIES >',/)
370            DO 100 I=1,NELADD
371            READ(LR,504) NEL,MPE(NEL),(IJKD(J),J=1,NODE)
372            DO 102 J=1,NODE
373        102 IJK(NODE*(NEL-1)+J)=IJKD(J)
374        504 FORMAT(16I5)
375            IF(NEL.GT.NELX) NELX=NELX+1
376            WRITE(LW,602) NEL,MPE(NEL),(IJKD(J),J=1,NODE)
377        602 FORMAT(15X,'NEL ',I5,5X,'MPE ',I3,5X,'<IJK>',20I5)
378        100 CONTINUE
379      C
380            RETURN
381            END
382      C ---------------------------------------------------------------------
383      C
384            SUBROUTINE PRESKY(JDIAG,IJK,NELX,NEQ,NODE,NDF,MSIZE1,LW)
385      C
386      C   PURPOSE:
387      C       COMPUTE THE HEIGHT OF THE SKYLINE AND CONSTRUCT THE DIAGONAL
```

```
388     C        POINTERS  "JDIAG"  FOR THE SKYLINE SOLVER.  THIS ROUTINE MUST
389     C        BE APPLIED AT THE VERY END OF THE PRE-PROCESSOR.
390     C
391     C   ARGUMENTS:
392     C        JDIAG   = DIAGONAL POINTERS FOR THE SKYLINE SOLVER
393     C        IJK     = ELEMENT CONNECTIVITIES
394     C        NELX    = TOTAL NUMBER OF FINITE ELEMENTS IN THE MODEL
395     C        NEQ     = NUMBER OF UNKNOWN ( DEGREES OF FREEDOM ) IN THE
396     C                  SYSTEM WHICH IS EQUAL TO  "NDF*NX"
397     C        NODE    = TOTAL NUMBER OF NODAL POINTS IN AN ELEMENT
398     C        NDF     = NUMBER OF DEGREES OF FREEDOM PER NODAL POINTS
399     C        MSIZE1 = MAXIMUM SIZE OF THE STIFFNESS MATRIX
400     C
401     C ------------------------------------------------------------------
402     C
403           IMPLICIT REAL*8(A-H,O-Z)
404           DIMENSION JDIAG(1),IJK(1)
405     C
406     C     ( INITIALIZATION )
407     C
408           DO 106 I=1,NEQ
409       106 JDIAG(I)=0
410     C
411     C     ( COMPUTE THE SKYLINE HEIGHT )
412     C
413           DO 100 NEL=1,NELX
414           DO 102 IA=1,NODE
415           IJKIA=IJK(NODE*(NEL-1)+IA)
416           DO 102  I=1,NDF
417           NIAI=NDF*(IJKIA-1)+I
418           DO 102 JB=1,NODE
419           IJKJB=IJK(NODE*(NEL-1)+JB)
420           DO 102  J=1,NDF
421           NJBJ=NDF*(IJKJB-1)+J
422           KIJ=NIAI-NJBJ+1
423           IF(KIJ.LT.1) GO TO 102
424           JDIAG(NIAI)=MAX0(JDIAG(NIAI),KIJ)
425       102 CONTINUE
426       100 CONTINUE
427     C
428     C     ( DEFINE THE DIAGONAL POINTER )
429     C
430           DO 104 I=2,NEQ
431       104 JDIAG(I)=JDIAG(I-1)+JDIAG(I)
432     C
433     C     ( PRINT OUT THE SIZE OF THE STIFFNESS MATRIX )
434     C
435           WRITE(LW,600) JDIAG(NEQ)
436       600 FORMAT(/////10X,'----- THE SIZE OF THE STIFFNESS MATRIX -----',
437          1 /20X,'JDIAG(NEQ)=',I8)
438           NX=NEQ/2
439           WRITE(LW,602) NX,NELX
440       602 FORMAT(/////10X,'----- SIZE OF THE FINITE ELEMENT MODEL',
441          1 ' -----',/15X,'THE TOTAL NUMBER OF THE NODES=',I5,
442          2 /15X,'THE TOTAL NUMBER OF ELEMENTS =',I5)
443     C
444           IF(JDIAG(NEQ).GT.MSIZE1) STOP
445     C
446           RETURN
447           END
448     C ------------------------------------------------------------------
449     C
450           SUBROUTINE MATERL(PROMAT,KK1,KK2,LR,LW)
451     C
452     C   PURPOSE:
453     C        READ THE MATERIAL PROPERTIES.
454     C
455     C   ARGUMENTS:
456     C        PROMAT = MATERIAL PROPERTIES
457     C        KK1     = CONTROL NUMBER OF THE CONSTITUTIVE EQUATIONS
458     C        KK2     = CONTROL NUMBER OF THE PLANE STRAIN/STRESS PROBLEM
459     C
460     C   LOCAL VARIABLES:
461     C        LMAT    = TOTAL NUMBER OF GROUPS OF MATERIAL
462     C
463     C   RESTRICTION:
464     C        IN THE ARRAY OF "PROMAT" WE ASSUME THAT 16 DIFFERENT
465     C        PARAMETERS CAN BE SPECIFIED.   HOWEVER IN THIS PROGRAM
466     C        ONLY 8 PARAMETERS ARE READ.   IF NECESSARY, MODIFY THE
```

```
467     C        PROGRAM PROPERLY.
468     C
469     C   CONTENTS IN THE ARRAY PROMAT:
470     C        1.  YOUNG'S MODULUS
471     C        2.  POISSON'S RATIO
472     C        3.  THICKNESS OF THE BODY
473     C        4.  COEFFICIENT OF THERMAL EXPANSION
474     C
475     C   INPUT DATA TO BE READ:
476     C        0. FLAG(I),I=1,10                     :    10A8
477     C        1. LMAT,KK1,KK2                       :    16I5
478     C        2. (PROMAT(J,I),J=1,8),I=1,LMAT       :    8F10.4
479     C
480     C ----------------------------------------------------------------------
481     C
482           IMPLICIT REAL*8(A-H,O-Z)
483           DIMENSION PROMAT(16,1),FLAG(10)
484     C
485           READ(LR,598) (FLAG(I),I=1,10)
486       598 FORMAT(10A8)
487           READ(LR,500) LMAT,KK1,KK2
488       500 FORMAT(16I5)
489           READ(LR,502) ((PROMAT(J,I),J=1,8),I=1,LMAT)
490       502 FORMAT(8F10.4)
491     C
492           WRITE(LW,600) KK1,KK2
493       600 FORMAT(/////10X,'----- MATERIAL PROPERTIES -----',//15X,
494          1 'CONTROL NUMBER <KK1> =',I3,//15X,
495          2 'CONTROL NUMBER <KK2> =',I3,//5X,88('-'),/5X,
496          3 'GROUP I      P1          P2          P3          P4          P5
497          5 '     P6          P7          P8',/5X,88('-'))
498           WRITE(LW,602) (I,(PROMAT(J,I),J=1,8),I=1,LMAT)
499       602 FORMAT(7X,I2,'  I ',8E10.3)
500           WRITE(LW,604)
501       604 FORMAT(5X,88('-'))
502     C
503           RETURN
504           END
505     C ----------------------------------------------------------------------
506     C
507           SUBROUTINE RBOUN1(NDFIX,VFIXED,SPRING,NDF,LFIX,LR,LW)
508     C
509     C   PURPOSE:
510     C       READ INPUT DATA FOR THE 1-ST ( DISPLACEMENT ) BOUNDARY
511     C       CONDITION IN THE X- AND Y-DIRECTIONS AT THE NODAL POINTS.
512     C
513     C   ARGUMENTS:
514     C       NDIFIX = NUMBER OF THE NODAL POINT AT WHICH THE 1-ST BOUNDARY
515     C                CONDITION IS APPLIED
516     C          NDFIX(1,I) = STORES THE NODE NUMBER CONSTRAINED
517     C          NDFIX(2,I) = STORES THE INDEX SPECIFYING THE DIRECTION OF
518     C                       CONSTRAINED DISPLACEMENTS
519     C       VFIXED = VALUE OF THE BOUNDARY CONDITIONS
520     C       SPRING = SPRING CONSTANT FOR THE SUPPORT
521     C       NDF    = NUMBER OF DEGREES OF FREEDOM PER NODAL POINT
522     C       LFIX   = TOTAL NUMBER OF NODAL POINTS ON THE 1-ST BOUNDARY
523     C
524     C   NOTE:
525     C       NDFIX(2,N) = 11 , IF TWO DEGREES OF FREEDOM ARE CONSTRAINED
526     C       NDFIX(2,N) = 10 , IF THE FIRST DEGREE OF FREEDOM IS CONSTRAINED
527     C       NDFIX(2,N) =  1 , IF THE SECOND DEGREE OF FREEDOM IS CONSTRAINED
528     C
529     C   INPUT DATA TO BE READ:
530     C       0. FLAG(I),I=1,10                              :    10A8
531     C       1. LFIX,SPRING                                 :    I5,F20.5
532     C       2. (NDFIX(J,I),J=1,2),(VFIXED(J,I),J=1,2),
533     C          I=1,LFIX                                    :    2I5,2F10.4
534     C
535     C ----------------------------------------------------------------------
536     C
537           IMPLICIT REAL*8(A-H,O-Z)
538           DIMENSION FLAG(10),NDFIX(2,1),VFIXED(3,1)
539     C
540           READ(LR,500) (FLAG(I),I=1,10)
541       500 FORMAT(10A8)
542           READ(LR,502) LFIX,SPRING
543       502 FORMAT(I5,F20.5)
544     C
```

```
545          DO 100 I=1,LFIX
546          READ(LR,504) (NDFIX(J,I),J=1,2),(VFIXED(J,I),J=1,NDF)
547      504 FORMAT(2I5,7F10.4)
548      100 CONTINUE
549    C
550          WRITE(LW,600) SPRING
551      600 FORMAT(/////10X,'---------- 1-ST BOUNDARY CONDITION ----------',//
552         1  15X,'SPRING CONSTANT = ',E10.3,//10X,50('-'),/10X,
553         2  ' NODE I  KIND I             VALUE',/10X,50('-'))
554          DO 102 I=1,LFIX
555      102 WRITE(LW,602) (NDFIX(J,I),J=1,2),(VFIXED(J,I),J=1,NDF)
556      602 FORMAT(10X,I5,' I ',I5,' I ',7E10.3)
557          WRITE(LW,604)
558      604 FORMAT(10X,50('-'))
559    C
560          RETURN
561          END
562    C --------------------------------------------------------------------
563    C
564          SUBROUTINE RBOUN2(NLOAD,VLOAD,NDF,LFOR,LR,LW)
565    C
566    C   PURPOSE:
567    C       READ INPUT DATA FOR THE NON-HOMOGENEOUS 2-ND ( TRACTION )
568    C       BOUNDARY CONDITON.  APPLIED TRACTIONS MUST BE TRANSFERRED
569    C       TO EQUIVALENT NODAL FORCES DEFINED AT THE NODAL POINTS.
570    C
571    C   ARGUMENTS:
572    C       NLOAD = NODE NUMBER AT WHICH THE TRACTION IS APPLIED
573    C       VLOAD = VALUES OF THE TRACTION IN X- AND Y- DIRECTION
574    C       NDF   = NUMBER OF DEGREES OF FREEDOM PER NODAL POINT
575    C       LFOR  = TOTAL NUMBER OF NODAL POINTS ON THE TRACTION BOUNDARY
576    C
577    C   INPUT DATA TO BE READ:
578    C       0.  FLAG(I),I=1,10                         :    10A8
579    C       1.  LFOR                                   :    16I5
580    C       2.  NLOAD(I),(VLOAD(J,I),J=1,2),I=1,LFOR   :    I5,2F10.4
581    C
582    C --------------------------------------------------------------------
583    C
584          IMPLICIT REAL*8(A-H,O-Z)
585          DIMENSION FLAG(10),NLOAD(1),VLOAD(NDF,1)
586    C
587          READ(LR,500) (FLAG(I),I=1,10)
588      500 FORMAT(10A8)
589    C
590          READ(LR,502) LFOR
591      502 FORMAT(16I5)
592          DO 100 I=1,LFOR
593      100 READ(LR,504) NLOAD(I),(VLOAD(J,I),J=1,NDF)
594      504 FORMAT(I5,7F10.4)
595    C
596          WRITE(LW,600)
597      600 FORMAT(/////10X,'---------- 2-ND BOUNDARY CONDITION ----------',
598         1 //15X,31('-'),/15X,'NODE I      TX          TY',/15X,31('-'))
599          DO 102 I=1,LFOR
600      102 WRITE(LW,602) NLOAD(I),(VLOAD(J,I),J=1,NDF)
601      602 FORMAT(13X,I5,2X,'I',2X,7E10.3)
602          WRITE(LW,604)
603      604 FORMAT(15X,31('-'))
604    C
605          RETURN
606          END
607    C --------------------------------------------------------------------
608    C
609          SUBROUTINE RBOUN3(NB3X,IJ3,SPTN,GTN,TRN,NODE,LR,LW)
610    C
611    C   PURPOSE:
612    C       READ INPUT DATA FOR THE GENERALIZED 3-RD BOUNDARY CONDITION
613    C       IN THE NORMAL AND TANGENTIAL DIRECTIONS.
614    C
615    C   ARGUMENTS:
616    C       NB3X      = TOTAL NUMBER OF LINE ELEMENTS ON THE 3-RD BOUNDARY
617    C       IJ3       = LINE ELEMENT CONNECTIVITIES
618    C       SPTN(1,I) = TANGENTIAL SPRING CONSTANT OF THE I-TH LINE ELEMENT
619    C       SPTN(2,I) = NORMAL SPRING CONSTANT   ......
620    C       GTN(1,I)  = CONSTRAINED TANGENTIAL DISPLACEMENT ......
621    C       GTN(2,I)  = CONSTRAINED NORMAL DISPLACEMENT ......
622    C       TRN(1,I)  = APPLIED TANGENTIAL TRACTION ......
```

```
623     C        TRN(2,N)   = APPLIED NORMAL TRACTION ......
624     C        NODE       = TOTAL NUMBER OF NODAL POINTS ON A LINE ELEMENT
625     C
626     C   NUMBERING ORDER OF IJ3:
627     C
628     C              N2
629     C                  I
630     C                     I        OUTSIDE OF THE DOMAIN
631     C        INSIDE          I
632     C                           N1
633     C
634     C                     IJ3(1,NEL)=N1    IJ3(2,NEL)=N2
635     C
636     C   INPUT DATA TO BE READ:
637     C        0. FLAG(I),I=1,10                           :    10A8
638     C        1. NB3X                                     :    I5
639     C        DO 200 MNEL=1,NB3X
640     C        2. NEL,(IJ3(J,NEL),J=1,2)                   :    16I5
641     C        3. SPTN(1,NEL),GTN(1,NEL),TRN(1,NEL)        :    F20.5,2F10.4
642     C        4. SPTN(2,NEL),GTN(2,NEL),TRN(2,NEL)        :    F20.5,2F10.4
643     C        200 CONTINUE
644     C
645     C -----------------------------------------------------------------
646     C
647             IMPLICIT REAL*8(A-H,O-Z)
648             DIMENSION SPTN(2,1),GTN(2,1),TRN(2,1),IJ3(1),FLAG(10)
649     C
650             READ(LR,500) (FLAG(I),I=1,10)
651         500 FORMAT(10A8)
652     C
653             READ(LR,502) NB3X
654         502 FORMAT(I5)
655             DO 100 MNEL=1,NB3X
656             READ(LR,506) NEL,(IJ3(NODE*(NEL-1)+J),J=1,NODE)
657         506 FORMAT(16I5)
658             READ(LR,504) SPTN(1,NEL),GTN(1,NEL),TRN(1,NEL)
659         504 FORMAT(F20.5,F10.4,F10.4)
660             READ(LR,504) SPTN(2,NEL),GTN(2,NEL),TRN(2,NEL)
661         100 CONTINUE
662     C
663             WRITE(LW,600) NB3X
664         600 FORMAT(/////5X,'---------------- 3-RD BOUNDARY CONDITION ',
665            1  '---------------',//,5X,'NUMBER OF LINE ELEMENTS = ',I5,//5X,
666            2 68('-'),/5X,' NEL  I  CONNECT  I DIR I   SPRING   I  ',
667            3 'DISPLACEMENT  I  TRACTION',/5X,68('-'))
668             DO 102 NEL=1,NB3X
669             WRITE(LW,602) NEL,(IJ3(NODE*(NEL-1)+J),J=1,NODE),SPTN(1,NEL),
670            1               GTN(1,NEL),TRN(1,NEL)
671         602 FORMAT(5X,I4,'  I ',2I4,'  I  T  I ',E10.3,' I    ',E10.3,'  I ',
672            1  E10.3)
673             WRITE(LW,604) SPTN(2,NEL),GTN(2,NEL),TRN(2,NEL)
674         604 FORMAT(26X,'N  I ',E10.3,' I    ',E10.3,'  I ',E10.3)
675         102 CONTINUE
676             WRITE(LW,606)
677         606 FORMAT(5X,68('-'))
678     C
679             RETURN
680             END
681     C -----------------------------------------------------------------
682     C
683             SUBROUTINE RBOUN4(NSLIDE,VSLIDE,SPRING,LSLID,LR,LW)
684     C
685     C   PURPOSE:
686     C        READ INPUT DATA FOR THE BOUNDARY CONDITION WHICH YIELDS
687     C        SLIDING ALONG AN INCLINED SURFACE.  THIS CONDITION IS
688     C        SPECIFIED AT THE NODAL POINTS.
689     C
690     C   ARGUMENTS:
691     C        NSLIDE = NODE NUMBER AT WHICH THIS CONDITION IS ASSUMED
692     C        VSLIDE = ANGLE OF THE SLOPE AND THE NORMAL DISPLACEMENT
693     C        SPRING = SPRING CONSTANT
694     C        LSLID  = TOTAL NUMBER OF THE NODAL POINTS ON THIS BOUNDARY
695     C
696     C   INPUT DATA TO BE READ:
697     C        0.  FLAG(I),I=1,10                              :    10A8
698     C        1.  LSLID,SPRING                                :    I5,F20.5
699     C        2.  NSLIDE(I),(VSLIDE(J,I),J=1,2),I=1,LSLID     :    I5,2F10.4
700     C
```

```
701     C ----------------------------------------------------------------------
702     C
703           IMPLICIT REAL*8(A-H,O-Z)
704           DIMENSION NSLIDE(1),VSLIDE(2,1),FLAG(10)
705     C
706           READ(LR,500) (FLAG(I),I=1,10)
707       500 FORMAT(10A8)
708           READ(LR,502) LSLID,SPRING
709       502 FORMAT(I5,F20.4)
710           READ(LR,504) (NSLIDE(I),(VSLIDE(J,I),J=1,2),I=1,LSLID)
711       504 FORMAT(I5,2F10.4)
712     C
713           WRITE(LW,600)
714       600 FORMAT(/////10X,'--------------- SLIDING BOUNDARY CONDITION',
715          1  ' ---------------',//15X,27('-'),/15X,
716          2  'NODE   I  THETA   I     GN',/15X,27('-'))
717           WRITE(LW,602) (NSLIDE(I),(VSLIDE(J,I),J=1,2),I=1,LSLID)
718       602 FORMAT(15X,I3,3X,'I',2X,F6.2,'I',F10.4)
719           WRITE(LW,604)
720       604 FORMAT(15X,27('-'))
721     C
722           RETURN
723           END
724     C ----------------------------------------------------------------------
725     C
726           SUBROUTINE RBODYF(BFR,X,Y,NX,LR,LW)
727     C
728     C   PURPOSE:
729     C       FORM THE BODY FORCE VECTORS AT EACH NODAL POINT.
730     C
731     C   ARGUMENTS:
732     C       BFR    = BODY FORCE VECTORS
733     C       X,Y    = COORDINATES OF THE NODAL POINTS
734     C       NX     = TOTAL NUMBER OF NODAL POINTS
735     C
736     C   LOCAL VARIABLES:
737     C       LBODY = ORDER OF THE POLYNOMIAL ASSUMED
738     C             = 0 ,   IF NO BODY FORCES ARE ASSUMED
739     C       XG,YG = COORDINATES OF NODAL POINTS IN AN ELEMENT
740     C       C1    = PARAMETERS OF THE BODY FORCE IN THE 1-ST DIRECTION
741     C       C2    = PARAMETERS OF THE BODY FORCE IN THE 2-ND DIRECTION
742     C       PXY   = POLYNOMIAL BASIS SUCH AS (1,X,Y,X**2,X*Y,Y**2,...)
743     C
744     C   DIMENSIONING:
745     C       C1,C2,PXY  MUST BE SPECIFIED ACCORDING TO THE PROBLEM.   IN
746     C       THIS PROGRAM,  8X8  ARRAYS ARE ASSUMED.
747     C
748     C   NOTES:
749     C       THE BODY FORCES ARE REPRESENTED BY THE POLYNOMIAL EXPANSIONS.
750     C       IF ANOTHER EXPRESSION IS PREFERABLE, PLEASE CHANGE THE ROUTINE.
751     C
752     C   INPUT DATA TO BE READ:
753     C       0. FLAG(I),I=1,10                  :    10A8
754     C       1. LBODY                           :    16I5
755     C       2. (C1(I,J),J=1,8),I=1,LBODY       :    8F10.4
756     C       3. (C2(I,J),J=1,8),I=1,LBODY       :    8F10.4
757     C
758     C ----------------------------------------------------------------------
759     C
760           IMPLICIT REAL*8(A-H,O-Z)
761           DIMENSION BFR(2,1),X(1),Y(1)
762           DIMENSION PXY(8,8),C1(8,8),C2(8,8),FLAG(10)
763     C
764           READ(LR,598) (FLAG(I),I=1,10)
765       598 FORMAT(10A8)
766           READ(LR,500) LBODY
767       500 FORMAT(16I5)
768     C
769           IF(LBODY.EQ.0) RETURN
770     C
771           DO 100 I=1,LBODY
772       100 READ(LR,502) (C1(I,J),J=1,LBODY)
773       502 FORMAT(8F10.4)
774           DO 102 I=1,LBODY
775       102 READ(LR,502) (C2(I,J),J=1,LBODY)
776           WRITE(LW,600) LBODY
777       600 FORMAT(/////10X,'---------- DATA OF THE BODY FORCES ----------',
778          1 //15X,'LBODY=',I5,/15X,'< TO THE 1-ST DIRECTION >',/)
```

```
779             DO 104 I=1,LBODY
780         104 WRITE(LW,602) I,(C1(I,J),J=1,LBODY)
781         602 FORMAT(10X,'X**(',I2,1X,'-1)',3X,8(E10.3,1X))
782             WRITE(LW,604)
783         604 FORMAT(//15X,'< TO THE 2-ND DIRECTION >',/)
784             DO 106 I=1,LBODY
785         106 WRITE(LW,606) I,(C2(I,J),J=1,LBODY)
786         606 FORMAT(10X,'Y**(',I2,1X,'-1)',3X,8(E10.3,1X))
787             WRITE(LW,608)
788         608 FORMAT(////10X,'---------- BODY FORCES (NODEWISE) ----------',
789            1 /15X,35('-'),/15X,' NODE  I    FX    I    FY    ',/15X,
790            2 35('-'))
791       C
792             DO 200 N=1,NX
793       C
794             XG=X(N)
795             YG=Y(N)
796       C
797       C     ( FOR THE POLYNOMIAL BASIS )
798       C
799             DO 204 I=1,LBODY
800             IF(I.EQ.1) PXI=1.
801             IF(I.GT.1) PXI=XG**(I-1)
802             DO 204 J=1,LBODY
803             IF(J.EQ.1) PYJ=1.
804             IF(J.GT.1) PYJ=YG**(J-1)
805         204 PXY(I,J)=PXI*PYJ
806       C
807       C     ( COMPUTE THE BODY FORCES )
808       C
809             FX=0.
810             FY=0.
811             DO 206 I=1,LBODY
812             DO 206 J=1,LBODY
813             FX=FX+C1(I,J)*PXY(I,J)
814         206 FY=FY+C2(I,J)*PXY(I,J)
815             BFR(1,N)=FX
816             BFR(2,N)=FY
817             WRITE(LW,610) N,FX,FY
818         610 FORMAT(15X,I5,'I',E10.3,'I',E10.3)
819       C
820         200 CONTINUE
821             WRITE(LW,612)
822         612 FORMAT(15X,35('-'))
823       C
824             RETURN
825             END
826       C ----------------------------------------------------------------------
827       C
828             SUBROUTINE RTEMPE(TEMP,X,Y,NX,LR,LW)
829       C
830       C   PURPOSE:
831       C       FORM THE TEMPERATURES AT EACH NODAL POINT.
832       C
833       C   ARGUMENTS:
834       C       TEMP  = TEMPERATURES AT EACH NODAL POINT
835       C       X,Y   = COORDINATES OF NODAL POINTS
836       C       NX    = TOTAL NUMBER OF NODAL POINTS
837       C
838       C   LOCAL VARIABLES:
839       C       LTEMP = THE ORDER OF THE POLYNOMIAL ASSUMED
840       C             =  0 ,  IF THERMAL STRESSES NEED NOT TO BE CONSIDERED
841       C             = -1 ,  IF TEMPERATURES ARE GIVEN AT EACH NODAL POINT
842       C       XG,YG = COORDINATES OF NODAL POINTS IN AN ELEMENT
843       C       C1    = PARAMETERS OF THE TEMPERATURE DISTRIBUTION
844       C       PXY   = POLYNOMIAL BASIS SUCH AS (1,X,Y,X**2,X*Y,Y**2,...)
845       C
846       C   DIMENSIONING:
847       C       C1,PXY  MUST BE SPECIFIED ACCORDING TO THE PROBLEM.   IN
848       C       THIS PROGRAM,  8X8  ARRAYS ARE ASSUMED.
849       C
850       C   NOTES:
851       C       THE TEMPERATURES ARE REPRESENTED BY A POLYNOMIAL EXPRESSION.
852       C       IF ANOTHER EXPRESSION IS PREFERABLE, PLEASE CHANGE THE ROUTINE.
853       C
854       C   INPUT DATA TO BE READ:
855       C       0. FLAG(I),I=1,10                :    10A8
```

```
856      C        1. LTEMP                          :    I5
857      C        2. (C1(I,J),J=1,8),I=1,LTEMP      :    8F10.4
858      C
859      C ---------------------------------------------------------------------
860      C
861            IMPLICIT REAL*8(A-H,O-Z)
862            DIMENSION TEMP(1),X(1),Y(1)
863            DIMENSION PXY(8,8),C1(8,8),FLAG(10)
864      C
865            READ(LR,598) (FLAG(I),I=1,10)
866        598 FORMAT(10A8)
867            READ(LR,500) LTEMP
868        500 FORMAT(16I5)
869      C
870            IF(LTEMP.EQ.0) RETURN
871      C
872            IF(LTEMP.LT.0) GO TO 300
873      C
874            DO 100 I=1,LTEMP
875        100 READ(LR,502) (C1(I,J),J=1,LTEMP)
876        502 FORMAT(8F10.4)
877            WRITE(LW,600) LTEMP
878        600 FORMAT(/////10X,'---------- DATA OF THE TEMPERATURE ----------',
879           1 //15X,'LTEMP = ',I5,//)
880            DO 104 I=1,LTEMP
881        104 WRITE(LW,602) I,(C1(I,J),J=1,LTEMP)
882        602 FORMAT(10X,'X**(',I2,1X,'-1)',3X,8(E10.3,1X))
883            WRITE(LW,608)
884        608 FORMAT(/////10X,'---------- TEMPERATURE DISTRIBUTION ----------',
885           1 //15X,45('-'),/15X,' NODE I  TEMPERATURE I NODE I TEMPERATURE',
886           2 /15X,45('-'))
887      C
888            DO 200 N=1,NX
889      C
890            XG=X(N)
891            YG=Y(N)
892      C
893      C     ( FOR THE POLYNOMIAL BASIS )
894      C
895            DO 204 I=1,LTEMP
896            IF(I.EQ.1) PXI=1.
897            IF(I.GT.1) PXI=XG**(I-1)
898            DO 204 J=1,LTEMP
899            IF(J.EQ.1) PYJ=1.
900            IF(J.GT.1) PYJ=YG**(J-1)
901        204 PXY(I,J)=PXI*PYJ
902      C
903      C     ( COMPUTE THE TEMPERATURE DISTRIBUTION )
904      C
905            TEM=0.
906            DO 206 I=1,LTEMP
907            DO 206 J=1,LTEMP
908        206 TEM=TEM+C1(I,J)*PXY(I,J)
909            TEMP(N)=TEM
910        610 FORMAT(15X,I5,' I ',E11.3,' I ',I5,' I ',E11.3)
911      C
912        200 CONTINUE
913      C
914            WRITE(LW,610) (I,TEMP(I),I=1,NX)
915            WRITE(LW,612)
916        612 FORMAT(15X,45('-'))
917      C
918            RETURN
919      C
920        300 READ(LR,504) (TEMP(I),I=1,NX)
921        504 FORMAT(8E10.3)
922            WRITE(LW,608)
923            WRITE(LW,610) (I,TEMP(I),I=1,NX)
924            WRITE(LW,612)
925      C
926            RETURN
927            END
928      C ---------------------------------------------------------------------
929      C
930            SUBROUTINE ASSEMB(SK,F,JDIAG,IJK,X,Y,MPE,PROMAT,TEMP,BFR,
931           1                 NX,NELX,KK1,KK2,NEQ,NODE,NDF,MBOUN1,LR,LW)
932      C
```

```
 933      C   PURPOSE:
 934      C       ASSEMBLE EACH ELEMENT STIFFNESS MATRIX AND ELEMENT LOAD
 935      C       VECTOR TO THE SKYLINE METHOD SOLVING THE SYSTEM OF LINEAR
 936      C       EQUATIONS.
 937      C
 938      C   ARGUMENTS:
 939      C       SK      = GLOBAL STIFFNESS MATRIX ASSEMBLED TO THE 1-DIMENSIONAL
 940      C                 ARRAY WHOSE SIZE MUST BE LESS THAN OR EQUAL TO THE
 941      C                 INPUT DATA "MSIZE1"
 942      C       F       = GLOBAL LOAD VECTOR
 943      C       JDIAG   = DIAGONAL POINTERS
 944      C       IJK     = ELEMENT CONNECTIVITIES
 945      C       X,Y     = COORDINATES OF NODAL POINTS
 946      C       MPE     = GROUP NUMBER OF THE MATERIALS FOR EACH ELEMENT
 947      C       PROMAT  = MATERIAL PROPERTIES
 948      C       TEMP    = TEMPERATURES AT EACH NODAL POINT
 949      C       BFR     = BODY FORCES AT EACH NODAL POINT
 950      C       NX      = TOTAL NUMBER OF NODAL POINTS
 951      C       NELX    = TOTAL NUMBER OF FINITE ELEMENTS
 952      C       KK1     = CONTROL NUMBER FOR THE CONSTITUTIVE EQUATIONS
 953      C       KK2     = CONTROL NUMBER FOR SPECIFICATION OF PLANE STRAIN/STRESS
 954      C                 PROBLEM SOLVED
 955      C       NEQ     = NUMBER OF UNKNOWN ( DEGREES OF FREEDOM ) IN THE
 956      C                 SYSTEM WHICH IS EQUAL TO  "NDF*NX"
 957      C       NODE    = TOTAL NUMBER OF NODAL POINTS IN AN ELEMENT
 958      C       NDF     = NUMBER OF DEGREES OF FREEDOM PER NODAL POINT
 959      C       MBOUN1  = CONTROL NUMBER FOR THE 1-ST BOUNDARY CONDITION
 960      C
 961      C   DIMENSIONING:
 962      C       SKE(NODMX,NODMX) AND FE(NODMX) MUST BE SPECIFIED ACCORDING TO
 963      C       FINITE ELEMENTS CHOSEN. ( NODMX=NODE*NDF )
 964      C
 965      C ----------------------------------------------------------------------
 966      C
 967             IMPLICIT REAL*8(A-H,O-Z)
 968             DIMENSION SK(1),F(1),X(1),Y(1),IJK(1),MPE(1),PROMAT(16,1),
 969            1          JDIAG(1),TEMP(1),BFR(2,1)
 970             DIMENSION SKE(8,8),FE(8)
 971      C
 972      C      ( INITIALIZATION )
 973      C
 974             DO 122 I=1,NEQ
 975         122 F(I)=0.
 976             MAX=JDIAG(NEQ)
 977             DO 120 I=1,MAX
 978         120 SK(I)=0.
 979      C
 980      C      ( CONSTRUCT THE GLOBAL STIFFNESS MATRIX AND THE LOAD VECTOR )
 981      C
 982             DO 100 NEL=1,NELX
 983      C
 984                   CALL ESTIFO(NEL,SKE,FE,IJK,X,Y,PROMAT,MPE,TEMP,BFR,
 985            1                  NODE,NDF,MBOUN1,KK1,KK2,LW)
 986      C
 987             DO 110 IA=1,NODE
 988             IJKIA=IJK(NODE*(NEL-1)+IA)
 989             DO 110  I=1,NDF
 990             NIAI=NDF*(IJKIA-1)+I
 991             IAI=NDF*(IA-1)+I
 992             F(NIAI)=F(NIAI)+FE(IAI)
 993             DO 110 JB=1,NODE
 994             IJKJB=IJK(NODE*(NEL-1)+JB)
 995             DO 110  J=1,NDF
 996             NJBJ=NDF*(IJKJB-1)+J
 997             JBJ=NDF*(JB-1)+J
 998             KIJ=NJBJ-NIAI
 999             IF(KIJ.LT.O) GO TO 110
1000             ND=JDIAG(NJBJ)-KIJ
1001             SK(ND)=SK(ND)+SKE(IAI,JBJ)
1002         110 CONTINUE
1003      C
1004         100 CONTINUE
1005      C
1006             RETURN
1007             END
1008      C ----------------------------------------------------------------------
1009      C
1010             SUBROUTINE ESTIFO(NEL,SKE,FE,IJK,X,Y,PROMAT,MPE,TEMP,BFR,
```

```
1011          1                     NODE,NDF,MBOUN1,KK1,KK2,LW)
1012    C
1013    C   PURPOSE:
1014    C       CONSTRUCT THE ELEMENT STIFFNESS MATRIX AND THE ELEMENT LOAD
1015    C       VECTOR BY USING 4-NODE ELEMENTS AND 2X2 GAUSSIAN INTEGRATION.
1016    C
1017    C   ARGUMENTS:
1018    C       NEL    = ELEMENT NUMBER
1019    C       SKE    = ELEMENT STIFFNESS MATRIX
1020    C       FE     = ELEMENT LOAD VECTOR
1021    C       IJK    = ELEMENT CONNECTIVITIES
1022    C       X,Y    = COORDINATES OF NODAL POINTS
1023    C       PROMAT = MATERIAL PROPERTIES FOR THE MODEL
1024    C       MPE    = GROUP NUMBER OF THE MATERIALS
1025    C       TEMP   = TEMPERATURES AT EACH NODAL POINT
1026    C       BFR    = BODY FORCES AT EACH NODAL POINT
1027    C       NODE   = TOTAL NUMBER OF NODAL POINTS IN AN ELEMENT
1028    C       NDF    = NUMBER OF DEGREES OF FREEDOM PER NODAL POINT
1029    C       MBOUN1 = CONTROL NUMBER FOR THE 1-ST BOUNDARY CONDITION
1030    C       KK1    = CONTROL NUMBER WHICH SPECIFIES THE CONSTITUTIVE
1031    C                EQUATIONS
1032    C              = 1 , THEN ISOTROPIC LINEAR ELASTICITY
1033    C              = 2 , THEN ORTHOTROPIC LINEAR ELASTICITY
1034    C              = 3 , THEN ISOTROPIC ELASTO-PLASTICITY
1035    C       KK2    = CONTROL NUMBER WHICH SPECIFIES THE PROBLEM
1036    C              = 0 , THEN PLANE STRAIN PROBLEMS ARE ASSUMED
1037    C              = 1 , THEN PLANE STRESS PROBLEMS ARE ASSUMED
1038    C
1039    C   LOCAL VARIABLES:
1040    C       XE     = COORDINATES OF THE NODAL POINTS IN AN ELEMENT
1041    C       ELMMAT = MATERIAL PROPERTIES OF AN ELEMENT
1042    C       DET    = DETERMINANT OF THE TRANSFORMATION
1043    C       D      = TANGENT MODULUS
1044    C       SH     = SHAPE FUNCTION OF THE 4-NODE ELEMENT
1045    C       GDN    = GRADIENT VECTOR IN THE GLOBAL COORDINATE SYSTEM
1046    C       ITX    = TOTAL NUMBER OF INTEGRATION POINTS IN THE 1-ST
1047    C                DIRECTION
1048    C       ITY    = TOTAL NUMBER OF INTEGRATION POINTS IN THE 2-ND
1049    C                DIRECTION
1050    C       GX     = INTEGRATION POINTS IN THE QUADRATURE RULE
1051    C       GW     = WEIGHTS OF THE INTEGRATION RULE
1052    C       BF     = BODY FORCES AT THE NODAL POINTS IN AN ELEMENT
1053    C       TM     = TEMPERATURES AT THE NODAL POINTS IN AN ELEMENT
1054    C       XL,YL  = COORDINATES IN THE MASTER ELEMENT
1055    C       EPI    = REGULARIZATION PARAMETER
1056    C
1057    C   NOTE:
1058    C       IF THERE ARE NO 1-ST/3-RD BOUNDARY CONDITIONS, THE STIFFNESS
1059    C       MATRIX BECOMES SINGULAR.  TO AVOID THIS, REGULARIZATION METHOD
1060    C       IS APPLIED IN THIS PROGRAM.  SEE EXERCISE 5.12.
1061    C
1062    C
1063    C -----------------------------------------------------------------
1064    C
1065          IMPLICIT REAL*8(A-H,O-Z)
1066          DIMENSION SKE(8,8),FE(8),XE(4,2),ELMMAT(16),D(2,2,2,2),
1067         1          SH(4),GDN(4,2),BF(2,4),TM(4),DELTA(2,2)
1068          DIMENSION X(1),Y(1),IJK(1),MPE(1),PROMAT(16,1),TEMP(1),
1069         1          BFR(NDF,1)
1070          DIMENSION GX(10),GW(10)
1071    C
1072          DATA DELTA/1.,0.,0.,1./
1073    C
1074    C     ( SET THE INTEGRATION POINTS AND WEIGHTS )
1075    C
1076          DATA GX/0.,-0.577350269189626,0.57735026918926,
1077         1        -0.774596669241483,0.,0.774596669241483/
1078          DATA GW/2.,1.,1.,0.5555555555555556,0.8888888888888889,
1079         1        0.5555555555555556/
1080          DATA ITX,ITY/2,2/
1081    C
1082    C     ( COORDINATES, BODY FORCES AND TEMPERATURES )
1083    C
1084          DO 102 I=1,NODE
1085          II=IJK(NODE*(NEL-1)+I)
1086          BF(1,I)=BFR(1,II)
1087          BF(2,I)=BFR(2,II)
1088          TM(I)=TEMP(II)
```

```
1089                XE(I,1)=X(II)
1090            102 XE(I,2)=Y(II)
1091          C
1092          C     ( SET THE MATERIAL CONSTANTS )
1093          C
1094                MNEL=MPE(NEL)
1095                DO 104 I=1,16
1096            104 ELMMAT(I)=PROMAT(I,MNEL)
1097          C
1098                EPI=1.D-5*ELMMAT(1)
1099          C
1100                      CALL TANMOD(KK1,KK2,D,BT,ELMMAT)
1101          C
1102                NODNDF=NDF*NODE
1103                DO 106 I=1,NODNDF
1104                FE(I)=0.
1105                DO 106 J=1,NODNDF
1106            106 SKE(I,J)=0.
1107          C
1108                DO 100 IX=1,ITX
1109                I1=(ITX*(ITX-1))/2+IX
1110                XL=GX(I1)
1111                WX=GW(I1)
1112                DO 100 IY=1,ITY
1113                I2=(ITY*(ITY-1))/2+IY
1114                YL=GX(I2)
1115                WY=GW(I2)
1116          C
1117          C     ( SHAPE FUNCTIONS AND THEIR GLOBAL GRADIENTS )
1118          C
1119                      CALL GGRAD4(NEL,XL,YL,SH,DET,GDN,XE,LW)
1120          C
1121          C     ( BODY FORCES AND TEMPERATURE DISTRIBUTION )
1122          C
1123                DO 212 IA=1,NODE
1124                DO 212  I=1,NDF
1125                IAI=NDF*(IA-1)+I
1126                BFI=0.
1127                TEP=0.
1128                DO 216 JB=1,NODE
1129                BFI=BFI+BF(I,JB)*SH(JB)
1130            216 TEP=TEP+TM(JB)*SH(JB)
1131            212 FE(IAI)=FE(IAI)+(BT*TEP*GDN(IA,I)+BFI*SH(IA))*WX*WY*DET
1132          C
1133          C     ( STIFFNESS MATRIX )
1134          C
1135            210 DO 206 IA=1,NODE
1136                DO 206  I=1,NDF
1137                IAI=NDF*(IA-1)+I
1138                DO 206 JB=IA,NODE
1139                DO 206  J=1,NDF
1140                JBJ=NDF*(JB-1)+J
1141                SUM=0.
1142                DO 208 K=1,NDF
1143                DO 208 L=1,NDF
1144            208 SUM=SUM+D(I,K,J,L)*GDN(IA,K)*GDN(JB,L)
1145                IF(MBOUN1.EQ.0) SUM=SUM+EPI*DELTA(I,J)*SH(IA)*SH(JB)
1146            206 SKE(IAI,JBJ)=SKE(IAI,JBJ)+SUM*DET*WX*WY
1147          C
1148            100 CONTINUE
1149          C
1150          C     ( THICKNESS OF THE "PLATE" )
1151          C
1152                TH=ELMMAT(3)
1153          C
1154                DO 214 I=1,8
1155                FE(I)=TH*FE(I)
1156                DO 214 J=I,8
1157                SKE(I,J)=TH*SKE(I,J)
1158            214 SKE(J,I)=SKE(I,J)
1159          C
1160                RETURN
1161                END
1162          C -----------------------------------------------------------------
1163          C
1164                SUBROUTINE GGRAD4(NEL,XL,YL,SH,DET,GDN,XE,LW)
1165          C
```

```
1166      C   PURPOSE:
1167      C       CONSTRUCT THE SHAPE FUNCTIONS AND THEIR DERIVATIVES WITH
1168      C       RESPECT TO THE GLOBAL COORDINATES.
1169      C
1170      C   ARGUMENTS:
1171      C       NEL   = ELEMENT NUMBER
1172      C       XL,YL = LOCAL COORDINATES FOR AN ISOPARAMETRIC ELEMENT
1173      C       SH    = SHAPE FUNCTIONS
1174      C       DET   = DETERMINANT OF THE JACOBIAN MATRIX
1175      C       GDN   = GLOBAL GRADIENT OF THE SHAPE FUNCTIONS
1176      C       XE    = COORDINATES OF THE NODAL POINTS IN AN ELEMENT
1177      C
1178      C   LOCAL VARIABLES:
1179      C       DN    = 1-ST DERIVATIVES OF THE SHAPE FUNCTIONS IN LOCAL
1180      C               COORDINATE SYSTEM
1181      C       DJ    = JACOBIAN MATRIX
1182      C
1183      C -----------------------------------------------------------------------
1184      C
1185            IMPLICIT REAL*8(A-H,O-Z)
1186            DIMENSION SH(4),GDN(4,2),DN(4,2),DJ(2,2),XE(4,2)
1187      C
1188      C     ( SET THE SHAPE FUNCTIONS FOR THE 4-NODE ELEMENT )
1189      C
1190            SH(1)=0.25*(1.-XL)*(1.-YL)
1191            SH(2)=0.25*(1.+XL)*(1.-YL)
1192            SH(3)=0.25*(1.+XL)*(1.+YL)
1193            SH(4)=0.25*(1.-XL)*(1.+YL)
1194      C
1195            DN(1,1)=-0.25*(1.-YL)
1196            DN(2,1)= 0.25*(1.-YL)
1197            DN(3,1)= 0.25*(1.+YL)
1198            DN(4,1)=-0.25*(1.+YL)
1199            DN(1,2)=-0.25*(1.-XL)
1200            DN(2,2)=-0.25*(1.+XL)
1201            DN(3,2)= 0.25*(1.+XL)
1202            DN(4,2)= 0.25*(1.-XL)
1203      C
1204      C     ( COMPUTE THE JACOBIAN MATRIX )
1205      C
1206            DO 200 I=1,2
1207            DO 200 J=1,2
1208            DJIJ=0.
1209            DO 202 IA=1,4
1210        202 DJIJ=DJIJ+XE(IA,J)*DN(IA,I)
1211        200 DJ(I,J)=DJIJ
1212            DET=DJ(1,1)*DJ(2,2)-DJ(1,2)*DJ(2,1)
1213            IF(DET.LE.0.) GO TO 110
1214            DJ11=DJ(1,1)
1215            DJ22=DJ(2,2)
1216            DJ(1,1)=DJ22/DET
1217            DJ(1,2)=-DJ(1,2)/DET
1218            DJ(2,1)=-DJ(2,1)/DET
1219            DJ(2,2)=DJ11/DET
1220      C
1221      C     ( GLOBAL GRADIENT OF THE SHAPE FUNCTIONS )
1222      C
1223            DO 204 IA=1,4
1224            GDN(IA,1)=DJ(1,1)*DN(IA,1)+DJ(1,2)*DN(IA,2)
1225        204 GDN(IA,2)=DJ(2,1)*DN(IA,1)+DJ(2,2)*DN(IA,2)
1226      C
1227            RETURN
1228      C
1229        110 WRITE(LW,600) NEL,XL,YL
1230        600 FORMAT(/////2X,'STOP STOP !  THE DETERMINANT BECOMES ZERO',/
1231           1 /10X,'ELEMENT ',I4,5X,'(',2F10.4,' )')
1232            STOP
1233            END
1234      C -----------------------------------------------------------------------
1235      C
1236            SUBROUTINE TANMOD(KK1,KK2,D,BT,ELMMAT)
1237      C
1238      C   PURPOSE:
1239      C       CONSTRUCT THE TANGENT MODULUS OF THE MATERIAL.
1240      C
1241      C   ARGUMENTS:
1242      C       KK1    = CONTROL NUMBER WHICH DEFINES THE MATERIAL
```

```
1243     C        KK2    = CONTROL NUMBER WHICH DEFINES THE PROBLEM
1244     C        D      = TANGENT MODULUS OF THE MATERIAL
1245     C        BT     = ADJUSTED COEFFICIENT OF THERMAL EXPANSION
1246     C        ELMMAT = MATERIAL PROPERTIES OF AN ELEMENT
1247     C
1248     C    LOCAL VARIABLES:
1249     C        RAM  = FIRST LAME CONSTANT FOR ISOTROPIC MATERIAL
1250     C        GNU  = SECOND LAME CONSTANT FOR ISOTROPIC MATERIAL
1251     C
1252     C    NOTES:
1253     C        FOR THE ISOTROPIC MATERIAL,
1254     C        ELMMAT(1) = YOUNG'S MODULUS
1255     C        ELMMAT(2) = POISSON'S RATIO
1256     C        ELMMAT(3) = THICKNESS OF THE PLATE FOR PLANE PROBLEMS
1257     C        ELMMAT(4) = COEFFICIENT OF THERMAL EXPANSION
1258     C
1259     C ----------------------------------------------------------------------
1260     C
1261           IMPLICIT REAL*8(A-H,O-Z)
1262           DIMENSION D(2,2,2,2),ELMMAT(16),DEL(2,2)
1263           DATA DEL/1.,0.,0.,1./
1264     C
1265           GO TO (102,104,106), KK1
1266     C
1267     C     ( ISOTROPIC LINEARLY ELASTIC MATERIALS )
1268     C
1269       102 CONTINUE
1270           E1=ELMMAT(1)
1271           E2=ELMMAT(2)
1272           E4=ELMMAT(4)
1273           IF(KK2.EQ.1) GO TO 202
1274           RAM=E1*E2/((1.+E2)*(1.-2.*E2))
1275           GNU=E1/(2.*(1.+E2))
1276           BT=E1*E4/(1.-2.*E2)
1277           GO TO 204
1278       202 RAM=E1*E2/(1.-E2*E2)
1279           GNU=E1/(2.*(1.+E2))
1280           BT=E1*E4/(1.-E2)
1281       204 DO 206 I=1,2
1282           DO 206 J=1,2
1283           DO 206 K=1,2
1284           DO 206 L=1,2
1285       206 D(I,J,K,L)=RAM*DEL(I,J)*DEL(K,L)+
1286          1            GNU*(DEL(I,K)*DEL(J,L)+DEL(I,L)*DEL(J,K))
1287     C
1288       104 CONTINUE
1289     C
1290       106 CONTINUE
1291     C
1292           RETURN
1293           END
1294     C ----------------------------------------------------------------------
1295     C
1296           SUBROUTINE BOUND1(SK,JDIAG,F,NDFIX,VFIXED,SPRING,NDF,LFIX)
1297     C
1298     C    PURPOSE:
1299     C        ADJUST THE STIFFNESS MATRIX AND LOAD VECTOR BY PENALTY METHOD
1300     C        ACCORDING TO THE GIVEN BOUNDARY CONDITIONS ON THE DISPLACEMENT
1301     C        FIELDS. ( THE 1-ST BOUNDARY CONDITION )
1302     C
1303     C    ARGUMENTS:
1304     C        SK     = STIFFNESS MATRIX
1305     C        JDIAG  = DIAGONAL POINTERS
1306     C        F      = LOAD VECTOR
1307     C        NDFIX  = NODE NUMBERS AND CONTROL NUMBERS
1308     C           NDFIX(1,I) = STORES THE NODE NUMBERS CONSTRAINED
1309     C           NDFIX(2,I) = 11 ,  IF BOTH DIRECTIONS ARE CONSTRAINED
1310     C           NDFIX(2,I) = 10 ,  IF THE 1-ST DIRECTION IS CONSTRAINED
1311     C           NDFIX(2,I) =  1 ,  IF THE 2-ND DIRECTION IS CONSTRAINED
1312     C        VFIXED = CONSTRAINED DISPLACEMENTS
1313     C        SPRING = SPRING CONSTANT FOR THE PENALTY METHOD
1314     C        NDF    = NUMBER OF DEGREES OF FREEDOM PER NODAL POINT
1315     C        LFIX   = TOTAL NUMBER OF THE NODAL POINTS CONSTRAINED
1316     C
1317     C ----------------------------------------------------------------------
1318     C
1319           IMPLICIT REAL*8(A-H,O-Z)
1320           DIMENSION SK(1),JDIAG(1),F(1),NDFIX(2,1),VFIXED(3,1)
```

```
1321      C
1322            DO 100 I=1,LFIX
1323      C
1324            NNODE=NDFIX(1,I)
1325            NFIX2=NDFIX(2,I)
1326            DO 100 J=1,NDF
1327            IF(NFIX2.LT.10**(NDF-J)) GO TO 100
1328            NC=NDF*(NNODE-1)+J
1329            ND=JDIAG(NC)
1330            SK(ND)=SK(ND)+SPRING
1331            F(NC)=F(NC)+SPRING*VFIXED(J,I)
1332            NFIX2=NFIX2-10**(NDF-J)
1333        100 CONTINUE
1334      C
1335            RETURN
1336            END
1337      C -----------------------------------------------------------------
1338      C
1339            SUBROUTINE BOUND2(F,NLOAD,VLOAD,NDF,LFOR)
1340      C
1341      C   PURPOSE:
1342      C       MODIFY THE LOAD VECTOR BY THE EQUIVALENT NODAL FORCES.
1343      C       ( THE 2-ND BOUNDARY CONDITION )
1344      C
1345      C   ARGUMENTS:
1346      C       F     = LOAD VECTOR
1347      C       NLOAD = NODE NUMBER AT WHICH THE TRACTION IS APPLIED
1348      C       VLOAD = TRACTION FORCE
1349      C       NDF   = NUMBER OF DEGREES OF FREEDOM PER NODAL POINT
1350      C       LFOR  = TOTAL NUMBER OF NODAL POINTS ON THE BOUNDARY
1351      C
1352      C -----------------------------------------------------------------
1353      C
1354            IMPLICIT REAL*8(A-H,O-Z)
1355            DIMENSION F(1),NLOAD(1),VLOAD(NDF,1)
1356      C
1357            DO 100 I=1,LFOR
1358      C
1359            NODE=NLOAD(I)
1360            DO 100 J=1,NDF
1361            NCJ=NDF*(NODE-1)+J
1362            F(NCJ)=F(NCJ)+VLOAD(J,I)
1363      C
1364        100 CONTINUE
1365      C
1366            RETURN
1367            END
1368      C -----------------------------------------------------------------
1369      C
1370            SUBROUTINE BOUND3(SK,F,JDIAG,NB3X,SPTN,GTN,TRN,NODE,NDF,IJ3,X,Y)
1371      C
1372      C   PURPOSE:
1373      C       MODIFY THE STIFFNESS MATRIX AND LOAD VECTOR DUE TO THE
1374      C       GENERALIZED 3-RD BOUNDARY CONDITION.
1375      C
1376      C          ( TANGENTIAL DIRECTION )
1377      C               ST(I) = -KT(UT(I)-GT(I)) + HT(I)    ON  S3
1378      C
1379      C          ( NORMAL DIRECTION )
1380      C               SN = = -KN(UN-GN) + HN    ON S3
1381      C
1382      C       WHERE
1383      C          ST(I) = I-TH COMPONENT OF TANGENTIAL TRACTION
1384      C          SN    = NORMAL TRACTION
1385      C          KT    = SPRING CONSTANT IN THE TANGENTIAL DIRECTION
1386      C          KN    = SPRING CONSTANT IN THE NORMAL DIRECTION
1387      C          UT(I) = I-TH COMPONENT OF THE TANGENTIAL DISPLACEMENT
1388      C          UN    = NORMAL DISPLACEMENT
1389      C          GT(I) = I-TH COMPONENT OF THE CONSTRAINED TANGENTIAL
1390      C                  DISPLACEMENT
1391      C          GN    = CONSTRAINED NORMAL DISPLACEMENT
1392      C          HT(I) = I-TH COMPONENT OF THE APPLIED TANGENTIAL TRACTION
1393      C          HN    = APPLIED NORMAL TRACTION
1394      C
1395      C   ARGUMENTS:
1396      C       SK    = STIFFNESS MATRIX
1397      C       F     = LOAD VECTOR
1398      C       JDIAG = DIAGONAL POINTERS
```

```
1399   C        NB3X  = TOTAL NUMBER OF LINE ELEMENTS ON THE 3-RD BOUNDARY
1400   C        SPTN  = STORES  KT  AND  KN  FOR EACH ELEMENT
1401   C        GTN   = STORES  GT  AND  GN  FOR EACH ELEMENT
1402   C        TRN   = STORES  HT  AND  HN  FOR EACH ELEMENT
1403   C        NODE  = TOTAL NUMBER OF NODAL POINTS IN A LINE ELEMENT
1404   C        NDF   = NUMBER OF DEGREES OF FREEDOM PER NODAL POINT
1405   C        IJ3   = LINE ELEMENT CONNECTIVITIES
1406   C        X,Y   = COORDINATES OF NODAL POINTS
1407   C
1408   C    LOCAL VARIABLES:
1409   C        VTX,VTY = UNIT TANGENTIAL VECTOR
1410   C        VNX,VNY = UNIT NORMAL VECTOR
1411   C        HE      = LENGTH OF A LINE ELEMENT
1412   C
1413   C -------------------------------------------------------------------
1414   C
1415         IMPLICIT REAL*8(A-H,O-Z)
1416         DIMENSION SK(1),F(1),JDIAG(1),X(1),Y(1),SPTN(2,1),GTN(2,1),
1417        1          TRN(2,1),IJ3(1)
1418         DIMENSION SKE(4,4),FE(4)
1419   C
1420         DO 100 NEL=1,NB3X
1421   C
1422         IJ31=IJ3(NODE*(NEL-1)+1)
1423         IJ32=IJ3(NODE*(NEL-1)+2)
1424         X1=X(IJ31)
1425         X2=X(IJ32)
1426         Y1=Y(IJ31)
1427         Y2=Y(IJ32)
1428         HE=DSQRT((X2-X1)**2+(Y2-Y1)**2)
1429         VNX=(Y2-Y1)/HE
1430         VNY=(X1-X2)/HE
1431         VTX=(X2-X1)/HE
1432         VTY=(Y2-Y1)/HE
1433         VT11=VTX*VTX
1434         VT12=VTX*VTY
1435         VT22=VTY*VTY
1436         VN11=VNX*VNX
1437         VN12=VNX*VNY
1438         VN22=VNY*VNY
1439   C
1440         SH11=HE/3.DO
1441         SH12=HE/6.DO
1442         SH22=SH11
1443   C
1444         SPT=SPTN(1,NEL)
1445         SPN=SPTN(2,NEL)
1446         HT=SPT*GTN(1,NEL)+TRN(1,NEL)
1447         HN=SPN*GTN(2,NEL)+TRN(2,NEL)
1448   C
1449   C     ( ELEMENT STIFFNESS MATRIX )
1450   C
1451         SKE(1,1)=(SPT*VT11+SPN*VN11)*SH11
1452         SKE(1,2)=(SPT*VT12+SPN*VN12)*SH11
1453         SKE(1,3)=(SPT*VT11+SPN*VN11)*SH12
1454         SKE(1,4)=(SPT*VT12+SPN*VN12)*SH12
1455         SKE(2,2)=(SPT*VT22+SPN*VN22)*SH11
1456         SKE(2,3)=(SPT*VT12+SPN*VN12)*SH12
1457         SKE(2,4)=(SPT*VT22+SPN*VN22)*SH12
1458         SKE(3,3)=(SPT*VT11+SPN*VN11)*SH22
1459         SKE(3,4)=(SPT*VT12+SPN*VN12)*SH22
1460         SKE(4,4)=(SPT*VT22+SPN*VN22)*SH22
1461         DO 102 I=1,4
1462         DO 102 J=I,4
1463     102 SKE(J,I)=SKE(I,J)
1464   C
1465         FE(1)=(HT*VTX+HN*VNX)*HE*O.5DO
1466         FE(2)=(HT*VTY+HN*VNY)*HE*O.5DO
1467         FE(3)=FE(1)
1468         FE(4)=FE(2)
1469   C
1470   C     ( ASSEMBLE TO THE GLOBAL )
1471   C
1472         DO 104 IA=1,NODE
1473         IJ3IA=IJ3(NODE*(NEL-1)+IA)
1474         DO 104 I=1,NDF
1475         IJ3IAI=NDF*(IJ3IA-1)+I
1476         IAI=NDF*(IA-1)+I
```

```
1477            F(IJ3IAI)=F(IJ3IAI)+FE(IAI)
1478            DO 104 JB=1,NODE
1479            IJ3JB=IJ3(NODE*(NEL-1)+JB)
1480            DO 104 J=1,NDF
1481            IJ3JBJ=NDF*(IJ3JB-1)+J
1482            JBJ=NDF*(JB-1)+J
1483            IAIJBJ=IJ3JBJ-IJ3IAI
1484            IF(IAIJBJ.LT.0) GO TO 104
1485            LOCAT=JDIAG(IJ3JBJ)-IAIJBJ
1486            SK(LOCAT)=SK(LOCAT)+SKE(IAI,JBJ)
1487        104 CONTINUE
1488      C
1489        100 CONTINUE
1490      C
1491            RETURN
1492            END
1493      C ---------------------------------------------------------------------
1494      C
1495            SUBROUTINE BOUND4(SK,JDIAG,F,NSLIDE,VSLIDE,SPRING,LSLID)
1496      C
1497      C   PURPOSE:
1498      C       ADJUST THE STIFFNESS MATRIX AND LOAD VECTOR BY THE BOUNDARY
1499      C       CONDITION WHICH YIELDS SLIDING ALONG AN INCLINED SURFACE.
1500      C       THIS CONDITION IS SPECIFIED AT THE NODAL POINTS.
1501      C
1502      C   ARGUMENTS:
1503      C       SK     = STIFFNESS MATRIX
1504      C       JDIAG  = DIAGONAL POINTERS
1505      C       F      = LOAD VECTOR
1506      C       NSLIDE = NODE NUMBER AT WHICH THE SLIDING IS CONSIDERED
1507      C       VSLIDE = DATA FOR THE SLIDING
1508      C       SPRING = SPRING CONSTANT
1509      C       LSLID  = TOTAL NUMBER OF NODAL POINTS ON THE BOUNDARY
1510      C
1511      C ---------------------------------------------------------------------
1512      C
1513            IMPLICIT REAL*8(A-H,O-Z)
1514            DIMENSION SK(1),JDIAG(1),F(1),NSLIDE(1),VSLIDE(2,1)
1515      C
1516            DO 100 I=1,LSLID
1517      C
1518            NODE=NSLIDE(I)
1519            THETA=VSLIDE(1,I)
1520            GN=VSLIDE(2,I)
1521      C
1522            RAD=3.141592653*THETA/180.
1523            XSIN=DSIN(RAD)
1524            XCOS=DCOS(RAD)
1525            NCX=2*NODE-1
1526            NCY=2*NODE
1527            NDX=JDIAG(NCX)
1528            NDY=JDIAG(NCY)
1529            SK(NDX)=SK(NDX)+SPRING*XSIN*XSIN
1530            SK(NDY)=SK(NDY)+SPRING*XCOS*XCOS
1531            SK(NDY-1)=SK(NDY-1)-SPRING*XSIN*XCOS
1532            F(NCX)=F(NCX)-SPRING*XSIN*GN
1533            F(NCY)=F(NCY)+SPRING*XCOS*GN
1534      C
1535        100 CONTINUE
1536      C
1537            RETURN
1538            END
1539      C ---------------------------------------------------------------------
1540      C
1541            SUBROUTINE SKYLIN(A,B,JDIAG,NEQ,KKK)
1542      C
1543      C   PURPOSE:
1544      C       SOLVE THE SYMMETRIC SYSTEM OF LINEAR EQUATIONS. ( R.L.TAYLOR )
1545      C
1546      C   ARGUMENTS:
1547      C       A     = COEFFECIENT MATRIX
1548      C       B     = LOAD VECTOR AT THE TIME OF INPUT, AND THE SOLUTION
1549      C               AT THE END OF THE ROUTINE
1550      C       JDIAG = DIAGONAL POINTERS
1551      C       NEQ   = TOTAL NUMBER OF EQUATIONS TO BE SOLVED
1552      C       KKK   = 0 , SOLVER
1553      C             = 1 , LU DECOMPOSITION
1554      C             = 2 , REDUCTION OF THE LOAD VECTOR AND BACK SUBSTITUTION
```

```
1555      C
1556      C -----------------------------------------------------------------------
1557      C
1558            IMPLICIT REAL*8(A-H,O-Z)
1559            DIMENSION A(1),B(1),JDIAG(1)
1560      C
1561      C     ( FACTOR THE MATRIX "A" TO "UT*D*U" AND REDUCE THE VECTOR "B" )
1562      C
1563            JR=0
1564            DO 600 J=1,NEQ
1565            JD=JDIAG(J)
1566            JH=JD-JR
1567            IS=J-JH+2
1568            IF(JH-2) 600,300,100
1569        100 IF(KKK.EQ.2) GO TO 500
1570            IE=J-1
1571            K=JR+2
1572            ID=JDIAG(IS-1)
1573      C
1574      C     ( REDUCE ALL EQUATIONS EXCEPT DIAGONAL )
1575      C
1576            DO 200 I=IS,IE
1577            IR=ID
1578            ID=JDIAG(I)
1579            IH=MIN0(ID-IR-1,I-IS+1)
1580            IF(IH.GT.0) A(K)=A(K)-DOT(A(K-IH),A(ID-IH),IH)
1581        200 K=K+1
1582      C
1583      C     ( REDUCE THE DIAGONAL )
1584      C
1585        300 IF(KKK.EQ.2) GO TO 500
1586            IR=JR+1
1587            IE=JD-1
1588            K=J-JD
1589            DO 400 I=IR,IE
1590            ID=JDIAG(K+I)
1591            IF(A(ID).EQ.0.) GO TO 400
1592            D=A(I)
1593            A(I)=A(I)/A(ID)
1594            A(JD)=A(JD)-D*A(I)
1595        400 CONTINUE
1596      C
1597      C     ( REDUCE THE LOAD VECTOR )
1598      C
1599        500 IF(KKK.NE.1) B(J)=B(J)-DOT(A(JR+1),B(IS-1),JH-1)
1600        600 JR=JD
1601            IF(KKK.EQ.1) RETURN
1602      C
1603      C     ( DIVIDE BY THE DIAGONAL PIVOTS )
1604      C
1605            DO 700 I=1,NEQ
1606            ID=JDIAG(I)
1607            IF(A(ID).NE.0.) B(I)=B(I)/A(ID)
1608        700 CONTINUE
1609      C
1610      C     ( BACK SUBSTITUTION )
1611      C
1612       1100 J=NEQ
1613            JD=JDIAG(J)
1614        800 D=B(J)
1615            J=J-1
1616            IF(J.LE.0) RETURN
1617            JR=JDIAG(J)
1618            IF(JD-JR.LE.1) GO TO 1000
1619            IS=J-JD+JR+2
1620            K=JR-IS+1
1621            DO 900 I=IS,J
1622        900 B(I)=B(I)-A(I+K)*D
1623       1000 JD=JR
1624            GO TO 800
1625      C
1626            END
1627      C -----------------------------------------------------------------------
1628      C
1629            FUNCTION DOT(A,B,N)
1630      C
1631      C   PURPOSE:
1632      C       TAKE THE DOT PRODUCT OF TWO VECTORS  "A" AND "B" .
```

```
1633      C
1634      C -----------------------------------------------------------------
1635      C
1636            IMPLICIT REAL*8(A-H,O-Z)
1637            DIMENSION A(1),B(1)
1638      C
1639            DOT=0.
1640            DO 100 I=1,N
1641        100 DOT=DOT+A(I)*B(I)
1642      C
1643            RETURN
1644            END
1645      C -----------------------------------------------------------------
1646      C
1647            SUBROUTINE DEFORM(F,U,X,Y,NX,LW,LP)
1648      C
1649      C   PURPOSE:
1650      C       OUTPUT THE RESULTS OF DEFORMATION.
1651      C
1652      C   ARGUMENTS:
1653      C       F   = (INCREMENTAL) DISPLACEMENT VECTOR
1654      C       U   = "CURRENT" DISPLACEMENT VECTOR
1655      C       X,Y = COORDINATES OF NODAL POINTS AT THE INITIAL CONFIGURATION
1656      C       NX  = TOTAL NUMBER OF NODAL POINTS
1657      C
1658      C   NOTE:
1659      C       STORE THE "CURRENT" DISPLACEMENTS ON THE FILE8 FOR PLOTTING.
1660      C
1661      C -----------------------------------------------------------------
1662      C
1663            IMPLICIT REAL*8(A-H,O-Z)
1664            DIMENSION F(1),U(1),X(1),Y(1)
1665      C
1666            WRITE(LW,600)
1667        600 FORMAT(1H1,//////////////////10X,80('*'),///30X,
1668           1 'DEFORMATIONS OF THE BODY',///10X,80('*'))
1669      C
1670            NPAGE=NX/55.00001+1
1671            DO 200 NP=1,NPAGE
1672      C
1673            WRITE(LW,602)
1674        602 FORMAT(1H1,5X,102('-'),/6X,
1675           1 'NODE I    DU(1)        DU(2)    I   U(1)     ',
1676           2 '    U(2)      I    X(1)        X(2)     I     CX(1) ',
1677           3 '      CX(2)',/5X,102('-'))
1678      C
1679            IS=55*(NP-1)+1
1680            IE=55*NP
1681            IF(IE.GT.NX) IE=NX
1682            DO 202 I=IS,IE
1683            IX=2*I-1
1684            IY=2*I
1685            DU1=F(IX)
1686            DU2=F(IY)
1687            U1=DU1+U(IX)
1688            U2=DU2+U(IY)
1689            U(IX)=U1
1690            U(IY)=U2
1691            X1=X(I)
1692            X2=Y(I)
1693            CX1=X1+DU1
1694            CX2=X2+DU2
1695            WRITE(LW,604) I,DU1,DU2,U1,U2,X1,X2,CX1,CX2
1696        604 FORMAT(5X,I5,' I ',E10.3,1X,E10.3,' I ',E10.3,1X,E10.3,' I ',
1697           1 E10.3,1X,E10.3,' I ',E10.3,1X,E10.3)
1698        202 CONTINUE
1699      C
1700            WRITE(LW,606)
1701        606 FORMAT(5X,102('-'))
1702      C
1703        200 CONTINUE
1704      C
1705      C     ( KEEP THE RESULTS ON FILE8 FOR PLOTTING )
1706      C
1707            NEQ=2*NX
1708            WRITE(LP,800) (U(I),I=1,NEQ)
1709        800 FORMAT(8E10.3)
1710      C
```

```
1711              RETURN
1712              END
1713        C -----------------------------------------------------------------------
1714        C
1715              SUBROUTINE STRESS(X,Y,U,IJK,NELX,PROMAT,MPE,NODE,NDF,KK1,KK2,TEMP,
1716             1                  NX,SK,JDIAG,MSIZE1,LW,LP)
1717        C
1718        C   PURPOSE:
1719        C       COMPUTE THE STRESS TENSORS IN EACH FINITE ELEMENT.  APPLYING
1720        C       THE LEAST SQUARE METHOD, CONTINUOUS STRESS FIELDS ARE OBTAINED
1721        C       AT EACH NODAL POINT.
1722        C
1723        C   ARGUMENTS:
1724        C       X,Y     = COORDINATES OF NODAL POINTS
1725        C       U       = "CURRENT" DISPLACEMENT FIELDS
1726        C       IJK     = ELEMENT CONNECTIVITIES
1727        C       NELX    = TOTAL NUMBER OF ELEMENTS
1728        C       PROMAT  = MATERIAL PROPERTIES
1729        C       MPE     = GROUP NUMBER OF MATERIALS
1730        C       NODE    = TOTAL NUMBER OF NODAL POINTS IN AN ELEMENT
1731        C       NDF     = NUMBER OF DEGREES OF FREEDOM PER NODAL POINT
1732        C       KK1     = CONTROL NUMBER OF THE CONSTITUTIVE EQUATION
1733        C       KK2     = CONTROL NUMBER FOR SPECIFICATION OF PLANE STRAIN/STRESS
1734        C                 PROBLEM SOLVED
1735        C       TEMP    = TEMPERATURES AT EACH NODAL POINT
1736        C       NX      = TOTAL NUMBER OF NODAL POINTS
1737        C       SK      = STIFFNESS MATRIX
1738        C       JDIAG   = DIAGONAL POINTERS
1739        C       MSIZE1  = MAXIMUM SIZE OF THE STIFFNESS MATRIX
1740        C
1741        C   LOCAL VARIABLES:
1742        C       XE,YE   = COORDINATES OF NODAL POINTS IN AN ELEMENT
1743        C       UE      = DISPLACEMENTS OF NODAL POINTS IN AN ELEMENT
1744        C       E       = STRAIN TENSOR
1745        C       S       = STRESS TENSOR
1746        C       SD      = DEVIATORIAL STRESS TENSOR
1747        C       DIV     = DIVERGENCE
1748        C       PRES    = HYDROSTATIC-PRESSURE, NOT THE TRACE OF THE
1749        C                 STRESS TENSOR  "S"
1750        C       D       = TANGENT MODULUS OF THE MATERIAL
1751        C       ENERGY  = STRAIN ENERGY DENSITY WITHIN AN ELEMENT
1752        C       DNORM   = L-2 NORM OF THE DIVERGENCE
1753        C       PNORM   = L-2 NORM OF THE HYDROSTATIC PRESSURE
1754        C       TENERG  = TOTAL STRAIN ENERGY OF THE ELASTIC BODY
1755        C       D1D2    = SUMMATION OF THE DIVERGENCE OF ELEMENTS 1 AND 3
1756        C       D2D4    = SUMMATION OF THE DIVERGENCE OF ELEMENTS 2 AND 4
1757        C       AVPR    = AVERAGE PRESSURE ON ELEMENTS 1,2,3,AND 4
1758        C       PS1     = MAXIMUM PRINCIPAL STRESS
1759        C       PS2     = MINIMUM PRINCIPAL STRESS
1760        C       THETA   = ANGLE OF THE PRINCIPAL AXIS TO THE X-DIRECTION
1761        C       PS3     = MAXIMUM SHEAR STRESS
1762        C       ESTR    = VON MISES'S EQUIVALENT STRESS
1763        C
1764        C   NOTE:
1765        C       STORE THE COMPUTED RESULTS, THE PRINCIPAL STRESSES(PS1,PS2,
1766        C       THETA), THE MAXIMUM SHEAR STRESS(PS3), THE EQUIVALENT STRESS
1767        C       (ESTR), AND THE SPECIFIC STRAIN ENERGY(ENERGY), ON THE FILE8
1768        C
1769        C -----------------------------------------------------------------------
1770        C
1771              IMPLICIT REAL*8(A-H,O-Z)
1772              DIMENSION X(1),Y(1),U(1),IJK(1),PROMAT(16,1),MPE(1),TEMP(1)
1773              DIMENSION XE(4,2),UE(2,4),ELMMAT(16),E(2,2),S(3,3),D(2,2,2,2),
1774             1          SD(3,3),DEL(3,3),SH(4),DN(4,2),GDN(4,2),DJ(2,2),TEMPE(4)
1775              DIMENSION SE(4,4),FE(4,4),FEE(4,4),FS(4,4),SI(4),SK(1),JDIAG(1)
1776              DIMENSION GX(10),GW(10)
1777        C
1778        C      ( SET THE INTEGRATION POINTS AND WEIGHTS )
1779        C
1780              DATA GX/0.,-0.577350269189626,0.57735026918926,
1781             1        -0.774596669241483,0.,0.774596669241483/
1782              DATA GW/2.,1.,1.,0.5555555555555556,0.8888888888888889,
1783             1        0.5555555555555556/
1784              DATA ITX,ITY/2,2/
1785        C
1786              DATA DEL/1.,0.,0.,0.,1.,0.,0.,0.,1./
1787        C
```

```
1788      C        ( PRE SKYLINE )
1789      C
1790                       CALL PRESKY(JDIAG,IJK,NELX,NX,NODE,1,MSIZE1,LW)
1791      C
1792      C        ( INITIALIZATION )
1793      C
1794               MAX=JDIAG(NX)+3*NX
1795               DO 312 I=1,MAX
1796           312 SK(I)=0.
1797               WRITE(LW,600)
1798           600 FORMAT(1H1,///////////////10X,80('*'),///30X,
1799              1 'STRESSES IN EACH FINITE ELEMENT',///10X,80('*'))
1800               WRITE(LW,602)
1801           602 FORMAT(1H1)
1802               AA=0.57735026918926
1803               DNORM=0.
1804               PNORM=0.
1805               TENERG=0.D0
1806               NELPR=53/(ITX*ITY+1)
1807               NELP=55
1808               DO 100 NEL=1,NELX
1809               IF(NELP.LE.NELPR) GO TO 1000
1810               NELP=0
1811               WRITE(LW,610)
1812           610 FORMAT(1H1,121('-'),/,'  NEL I    X       Y    I      SXX    ',
1813              1 '     SYY         SXY     I      S1          S2         TH      ',
1814              2 ' I     MISES     MSHEAR     ENERGY',/1X,120('-'))
1815          1000 NELP=NELP+1
1816      C
1817      C        ( DEFINE ELEMENT LEVEL DATA )
1818      C
1819               DO 202 IA=1,NODE
1820               IJKIA=IJK(NODE*(NEL-1)+IA)
1821               TEMPE(IA)=TEMP(IJKIA)
1822               XE(IA,1)=X(IJKIA)
1823               XE(IA,2)=Y(IJKIA)
1824               UE(1,IA)=U(2*IJKIA-1)
1825           202 UE(2,IA)=U(2*IJKIA)
1826      C
1827      C        ( SET THE MATERIAL CONSTANTS IN AN ELEMENT )
1828      C
1829               MNEL=MPE(NEL)
1830               DO 204 I=1,16
1831           204 ELMMAT(I)=PROMAT(I,MNEL)
1832      C
1833                       CALL TANMOD(KK1,KK2,D,BT,ELMMAT)
1834      C
1835      C        ( INITIALIZATION )
1836      C
1837               DO 308 IA=1,NODE
1838               DO 308 JB=1,NODE
1839               FE(IA,JB)=0.
1840           308 SE(IA,JB)=0.
1841      C
1842      C        ( GAUSSIAN INTEGRATION POINTS )
1843      C
1844               DO 200 IX=1,ITX
1845               I1=(ITX*(ITX-1))/2+IX
1846               XL=GX(I1)
1847               WX=GW(I1)
1848               DO 200 IY=1,ITY
1849               I2=(ITY*(ITY-1))/2+IY
1850               YL=GX(I2)
1851               WY=GW(I2)
1852               IXY=ITY*(IX-1)+IY
1853      C
1854      C        ( SHAPE FUNCTIONS AND THEIR GLOBAL GRADIENTS )
1855      C
1856                       CALL GGRAD4(NEL,XL,YL,SH,DET,GDN,XE,LW)
1857      C
1858      C        ( COORDINATES OF THE INTEGRATION POINT )
1859      C
1860               XIT=0.D0
1861               YIT=0.D0
1862               TEP=0.D0
1863               DO 230 IA=1,NODE
1864               TEP=TEP+TEMPE(IA)*SH(IA)
```

```
1865              XIT=XIT+XE(IA,1)*SH(IA)
1866          230 YIT=YIT+XE(IA,2)*SH(IA)
1867        C
1868        C     ( COMPUTE STRAIN TENSOR )
1869        C
1870              DO 214 I=1,2
1871              DO 214 J=1,2
1872              EIJ=0.
1873              DO 216 IA=1,NODE
1874          216 EIJ=EIJ+0.5*(UE(I,IA)*GDN(IA,J)+UE(J,IA)*GDN(IA,I))
1875          214 E(I,J)=EIJ
1876              DIV=E(1,1)+E(2,2)
1877        C
1878        C     ( COMPUTE STRESS TENSOR )
1879        C
1880              DO 218 I=1,2
1881              DO 218 J=1,2
1882              SIJ=0.
1883              DO 220 K=1,2
1884              DO 220 L=1,2
1885          220 SIJ=SIJ+D(I,J,K,L)*E(K,L)
1886          218 S(I,J)=SIJ-BT*TEP*DEL(I,J)
1887        C
1888              DO 228 I=1,3
1889              S(I,3)=0.
1890          228 S(3,I)=0.
1891              IF(KK2.NE.1) S(3,3)=ELMMAT(2)*(S(1,1)+S(2,2))
1892        C
1893              PRES=-D(1,1,2,2)*DIV
1894              DNORM=DNORM+DIV*DIV*DET*WX*WY
1895              PNORM=PNORM+PRES*PRES*DET*WX*WY
1896        C
1897        C     ( DEVIATORIAL STRESS TENSOR :  SD(I,J) )
1898        C
1899              SDII=(S(1,1)+S(2,2)+S(3,3))/3.
1900              DO 224 I=1,3
1901              DO 224 J=1,3
1902          224 SD(I,J)=S(I,J)-SDII*DEL(I,J)
1903        C
1904        C     ( VON MISES'S STRESS )
1905        C
1906              ESTR=0.
1907              DO 226 I=1,3
1908              DO 226 J=1,3
1909          226 ESTR=ESTR+SD(I,J)*SD(I,J)
1910              ESTR=DSQRT(1.5*ESTR)
1911        C
1912        C     ( PRINCIPAL STRESSES )
1913        C
1914                    CALL PRSTRS(S,PS1,PS2,THETA)
1915        C
1916              PS3=0.D0
1917              IF(KK2.NE.1) PS3=ELMMAT(2)*(PS1+PS2)
1918              PS12=0.5*(PS1-PS2)
1919              PS23=0.5*DABS(PS2-PS3)
1920              PS31=0.5*DABS(PS3-PS1)
1921              PS3=DMAX1(PS12,PS23,PS31)
1922        C
1923        C     ( COMPUTE THE DENSITY OF STRAIN ENERGY )
1924        C
1925              ENERGY=0.
1926              DO 222 I=1,2
1927              DO 222 J=1,2
1928          222 ENERGY=ENERGY+0.5*S(I,J)*E(I,J)
1929              TENERG=TENERG+ENERGY*DET*WX*WY
1930        C
1931        C     ( PRINT OUT )
1932        C
1933              WRITE(LW,612) NEL,XIT,YIT,S(1,1),S(2,2),S(1,2),PS1,PS2,THETA,
1934             1  ESTR,PS3,ENERGY
1935          612 FORMAT(I5,' I ',2F6.2,'I ',3E10.3,' I ',3E10.3,' I ',3E10.3)
1936        C
1937        C     ( KEEP THE RESULTS ON THE FILE8 FOR PLOTTING )
1938        C
1939              WRITE(LP,800) PS1,PS2,THETA
1940          800 FORMAT(6E10.3)
1941        C
1942        C     ( LEAST SQUARE )
```

```
1943   C
1944         SI(1)=0.25*(1.-XL/AA)*(1.-YL/AA)
1945         SI(2)=0.25*(1.-XL/AA)*(1.+YL/AA)
1946         SI(3)=0.25*(1.+XL/AA)*(1.-YL/AA)
1947         SI(4)=0.25*(1.+XL/AA)*(1.+YL/AA)
1948   C
1949         DO 300 IA=1,NODE
1950         DO 300 JB=1,NODE
1951         FE(IA,JB)=FE(IA,JB)+SH(IA)*SI(JB)*WX*WY*DET
1952         SE(IA,JB)=SE(IA,JB)+SH(IA)*SH(JB)*WX*WY*DET
1953     300 CONTINUE
1954   C
1955         FS(1,IXY)=PS3
1956         FS(2,IXY)=ESTR
1957         FS(3,IXY)=ENERGY
1958   C
1959     200 CONTINUE
1960   C
1961         WRITE(LW,614)
1962     614 FORMAT(1X,120('-'))
1963   C
1964   C     ( RIGHT HAND SIDE )
1965   C
1966         DO 306 IA=1,NODE
1967         DO 306 II=1,3
1968         SUM=0.
1969         DO 314 JB=1,NODE
1970     314 SUM=SUM+FE(IA,JB)*FS(II,JB)
1971     306 FEE(IA,II)=SUM
1972   C
1973   C     ( ASSEMBLING )
1974   C
1975         DO 302 IA=1,NODE
1976         IJKIA=IJK(NODE*(NEL-1)+IA)
1977         SK(JDIAG(NX)+IJKIA)=SK(JDIAG(NX)+IJKIA)+FEE(IA,1)
1978         SK(JDIAG(NX)+NX+IJKIA)=SK(JDIAG(NX)+NX+IJKIA)+FEE(IA,2)
1979         SK(JDIAG(NX)+2*NX+IJKIA)=SK(JDIAG(NX)+2*NX+IJKIA)+FEE(IA,3)
1980         DO 302 JB=1,NODE
1981         IJKJB=IJK(NODE*(NEL-1)+JB)
1982         IAJB=IJKJB-IJKIA
1983         IF(IAJB.LT.0) GO TO 302
1984         ND=JDIAG(IJKJB)-IAJB
1985         SK(ND)=SK(ND)+SE(IA,JB)
1986     302 CONTINUE
1987   C
1988     100 CONTINUE
1989         DNORM=DSQRT(DNORM)
1990         PNORM=DSQRT(PNORM)
1991         WRITE(LW,616) DNORM,PNORM,TENERG
1992     616 FORMAT(//////5X,'----- NORMS : CONVERGENCE -----',
1993        1 //10X,'L2-NORM OF THE DIVERGENCE = ',E12.4,
1994        2  /10X,'L2-NORM OF THE HYDROSTATIC PRESSURE = ',E12.4,
1995        3  /10X,'TOTAL STRAIN ENERGY OF THE BODY = ',E12.4)
1996   C
1997   C     ( NODAL VALUES OF STRESSES )
1998   C
1999               CALL SKYLIN(SK,SK(JDIAG(NX)+1),JDIAG,NX,0)
2000               CALL SKYLIN(SK,SK(JDIAG(NX)+NX+1),JDIAG,NX,2)
2001               CALL SKYLIN(SK,SK(JDIAG(NX)+2*NX+1),JDIAG,NX,2)
2002   C
2003         WRITE(LW,624)
2004     624 FORMAT(1H1,//10X,'++++++++++ MAXIMUM SHEAR AND EQUIVALENT',
2005        1  ' STRESSES AND DENSITY OF THE STRAIN ENERGY +++++',//10X,
2006        2  64('-'),/10X,' NODE I  MAXIMUM SHEAR  I  MISES EQUIVALENT  I
2007        3  'STRAIN ENERGY',/10X,64('-'),/)
2008         WRITE(LW,626) (I,SK(JDIAG(NX)+I),SK(JDIAG(NX)+NX+I),
2009        1  SK(JDIAG(NX)+2*NX+I),I=1,NX)
2010     626 FORMAT(10X,I5,' I  ',E13.5,'  I    ',E13.5,'   I  ',E13.5)
2011         WRITE(LW,628)
2012     628 FORMAT(10X,64('-'))
2013   C
2014         WRITE(LP,802) (SK(JDIAG(NX)+I),SK(JDIAG(NX)+NX+I),
2015        1  SK(JDIAG(NX)+2*NX+I),I=1,NX)
2016     802 FORMAT(6E10.3)
2017   C
2018         RETURN
2019         END
2020   C -----------------------------------------------------------------
```

```
2021      C
2022            SUBROUTINE PRSTRS(S,S1,S2,THETA)
2023      C
2024      C   PURPOSE:
2025      C       COMPUTE THE PRINCIPAL STRESSES  "S1"  AND  "S2".
2026      C
2027      C   ARGUMENTS:
2028      C       S     = STRESS TENSOR AT A POINT
2029      C       S1    = FIRST PRINCIPAL STRESS
2030      C       S2    = SECOND PRINCIPAL STRESS
2031      C       THETA = ANGLE OF THE PRINCIPAL AXIS TO THE X-COORDINATE
2032      C
2033      C ---------------------------------------------------------------------
2034      C
2035            IMPLICIT REAL*8(A-H,O-Z)
2036            DIMENSION S(3,3)
2037      C
2038            SX=S(1,1)
2039            SY=S(2,2)
2040            SXY=S(1,2)
2041            A1=0.5*(SX+SY)
2042            A2=0.5*(SX-SY)
2043            A3=DSQRT(A2*A2+SXY*SXY)
2044      C
2045            S1=A1+A3
2046            S2=A1-A3
2047            THETA=0.5*57.29577952*DATAN2(SXY,A2)
2048      C
2049            RETURN
2050            END
```

APPENDIX 5
FINITE ELEMENT PROGRAMS

Many powerful and large-scale, general-purpose finite element codes have been developed and are commercially available. The success of finite element methods is due to these programs, and their applicability to design and analysis is proven. Especially MRC-NASTRAN, ANSYS, PAFEC, SAP, ASKA, MARC, ADINA, ABAQUS, and other professional commercial codes provide strong confidence of applicability of finite element methods to solve realistic and practical engineering problems. Furthermore, recent development of computer codes for geometric modeling of solids and structures and interactive graphics accelerates popular use of finite element methods, because these pre- and postprocessings for finite element analysis give users easy access. Some recent trends are summarized by Fong (1984).

For those who wish to know more about commercially available finite element codes, the following is a list of some of these codes and their suppliers.

ABAQUS	Hibbitt and Karlsson, Inc. Providence, Rhode Island
ADINA	Mr. G. Larsson ADINA Engineering AB Stangjarnsgatan 227 S 72473 Vasteras, Sweden
ANSYS	Mr. P. Kohnke Swanson Analysis Systems, Inc. Johnson Road P. O. Box 65 Houston, PA 15342
ASAS	Atkins Research and Development Woodcote Grove Ashley Road Epsom Surrey KY18 5BW, England
ASKA	IKO Software Service GmbH Albstadtweg 10 D 7000 Stuttgart 80, Germany
BEASY	Mr. D. Danson Computational Mechanics Centre Ashurst Lodge, Ashurst Southampton SO4 2AA, England

COMET-PR	McDonnell Douglas Automation Company St. Louis, MO In Europe: Istituto Sperimentale Modelli e Strutture SpA Viale Giulio Cesare 29 24100 Bergamo, Italy
COSMOS	Structural Research and Analysis Corporation 1661 Lincoln Boulevard, Suite 100 Santa Monica, CA 90404
DIAL	Mr. Don Wong Structures Department 81-12 Building 154 Lockheed Missiles and Space Company P. O. Box 504 Sunnyvale, CA 94086
GIFTS-1100	Sperry Univac Computer Systems Sperry Univac Centre Stonebridge Park North Circular Road London NW10 8LS
MARC	Marc Analysis Research Corp. 260 Sheridan Avenue, Suite 200 Palo Alto, CA 94306
MSC/NASTRAN	The MacNeal-Schwendler Corporation 7442 North Figueroa Street Los Angeles, CA 90041 MacNeal-Schwendler GmbH 8000 Munchen 80 Prinzregentenstrasse 78 West Germany
PAFEC	PAFEC Limited Strelley Hall Main Street Strelley Nottingham NG8 6PE
PDA/PATRAN-G	Kins Development Woodcote Grove Ashley Road Epsom Surrey KT18 5BW

SERFEM	Gothenburg University Computing Centre Box 19070 S 400 12 Goteborg Sweden
STARS	Dr. V. Svalbonas Manager, Engineering Koppers Company, Inc. Mineral Processing Systems Division P. O. Box 312 York, PA 17405
STDYNL	Dr. Bulent Ovunc Department of Civil Engineering University of Southwestern Louisiana Lafayette, LA 70504

Reference

Fong, H. H. "Interactive graphics and commercial finite element codes," *Mechanical Engineering,* 106(6): 18–27, 1984.

BIBLIOGRAPHY

Excellent and very informative textbooks have been published for finite element methods in both introductory and advanced levels. The series *Finite elements* (6 vols.) by Oden, Carey, and Becker published by Prentice-Hall covers various aspects of finite element methods, whereas *The finite element method,* 3rd edition, by Zienkiewicz provides a variety of applications of finite element methods to engineering problems and gives many important references for further study. It is, however, still true that few textbooks are based on the view of applied mechanics, although many are based on structures and fluid mechanics, as shown in the list below. The present text is especially written for engineering students interested in applied mechanics applicable to solve industrial problems and is based on many works published so far. The following list was compiled from the article "List of textbooks and monographs on finite element technology" in *State-of-the-art surveys on finite element technology,* edited by Noor and Pilkey, The American Society of Mechanical Engineers, 1983, New York.

Many of these references are used in the textbook without explicit citation.

FUNDAMENTALS

Bathe, K. J., *Finite element procedures in engineering analysis,* Prentice-Hall, Englewood Cliffs, New Jersey, 1981.

Becker, E. B., Carey, G. F., and Oden, J. T., *Finite elements: an introduction,* vol. 1, Prentice-Hall, Englewood Cliffs, New Jersey, 1981.

Carey, G. F., and Oden, J. T., *Finite elements: a second course,* vol. 2, Prentice-Hall, Englewood Cliffs, New Jersey, 1983.

Cook, R. D., *Concepts and applications of finite element analysis,* 2nd ed., Wiley, New York, 1981.

Desai, C. S., and Abel, J. F., *Introduction to the finite element method. A numerical method for engineering analysis,* Van Nostrand Reinhold, New York, 1972.

Gallagher, R. H., *Finite element analysis—fundamentals,* Prentice-Hall, Englewood Cliffs, New Jersey, 1975.

Hinton, E., and Owen, D. R. J., *An introduction to finite element computations,* Pineridge Press, Swansea, United Kingdom, 1979.

Huebner, K. H., and Thornton, E. A., *The finite element method for engineers,* 2nd ed., Wiley-Interscience, New York, 1982.

Irons, B., and Ahmad, S., *Techniques of finite elements,* Wiley, New York, 1980.

Martin, H. C., and Carey, G. F., *Introduction to finite element analysis. Theory and application,* McGraw-Hill, New York, 1972.

Norrie, D. H., and de Vries, G., *An introduction to the finite element method,* 2nd ed., Academic, New York, 1978.

Oden, J. T., *Finite elements of nonlinear continua,* McGraw-Hill, New York, 1972.

Owen, D. R. J., and Hinton, E., *A simple guide to finite elements*, Pineridge Press, Swansea, United Kingdom, 1980.

Tong, P., and Rossettos, J. N., *Finite element method—basic techniques and implementation*, MIT Press, Cambridge, Massachusetts, 1977.

Zienkiewicz, O. C., *The finite element method*, 3rd ed., McGraw-Hill, New York, 1977.

Zienkiewicz, O. C., and Morgan, K., *Finite elements and approximation*, Wiley-Interscience, New York, 1982.

MATHEMATICAL FOUNDATIONS

Aziz, A. K. (ed.), *The mathematical foundations of the finite element method with applications to partial differential equations*, Academic, New York, 1972.

Ciarlet, P. G., *The finite element method for elliptic problems*, North-Holland, New York, 1978.

de Boor, C. (ed.), *Mathematical aspects of finite elements in partial differential equations*, Academic, New York, 1974.

Galligani, I., and Magenes, E. (eds.), *Mathematical aspects of finite element methods, lecture notes in mathematics*, vol. 606, Springer-Verlag, New York, 1977.

Mercier, B., *Lectures on topics in finite element solution of elliptic problems*, Springer-Verlag, Berlin, 1979.

Mitchell, A. R., and Wait, R. A., *The finite element method in partial differential equations*, Wiley, New York, 1977.

Oden, J. T., and Reddy, J. N., *An introduction to the mathematical theory of finite elements*, Wiley, New York, 1976.

Oden, J. T. (in collaboration with G. F. Carey), *Finite elements: mathematical aspects*, vol. 4, Prentice-Hall, New York, 1982.

Strang, R., and Fix, G., *An analysis of the finite element method*, Prentice-Hall, New York, 1973.

Whiteman, J. R. (ed.), *The mathematics of finite elements and applications*, vol. 1, Academic, London, 1973.

Whiteman, J. R. (ed.), *The mathematics of finite elements and applications*, vol. 2, Academic, London, 1977.

Whiteman, J. R. (ed.), *The mathematics of finite elements and applications*, vol. 3, Academic, London, 1979.

Whiteman, J. R. (ed.), *The mathematics of finite elements and applications*, vol. 4, Academic, London, 1982.

Vichnevetsky, R., *Computer methods for partial differential equations*, vol. 1, Prentice-Hall, Englewood Cliffs, New Jersey, 1981.

STRUCTURAL AND SOLID MECHANICS APPLICATIONS

Ashwell, D. G., and Gallagher, R. H. (eds.), *Finite elements for thin shells and curved members*, Wiley, London, 1976.

Belytschko, T., and Marcal, P. V. (eds.), *Finite element analysis of transient nonlinear structural behavior*, AMD vol. 14, American Society of Mechanical Engineers, New York, 1975.

Bernadou, M., and Boisserie, J. M., *The finite element method in thin shell theory: application to arch dam simulations*, Birkhauser, Boston, 1982.

Brebbia, C. A., and Connor, J. J., *Fundamentals of finite element techniques for structural engineers*, Wiley, New York, 1974.

Cheung, Y. K., *Finite strip method in structural analysis*, Pergamon, Elmsford, New York, 1976.

Hinton, E., and Owen, D. R. J (eds.), *Finite element software for plates and shells*, Pineridge Press, Swansea, United Kingdom, 1983.

Holand, I., and Bell, K. (eds.), *Finite element methods in stress analysis,* Lectures presented at the Technical University of Norway, Jan. 6–11, 1969; published by TAPIR.

Hughes, T. J. R., Pifko, A., and Jay, A. (eds.), *Nonlinear finite element analysis of plates and shells,* presented at the Winter Annual Meeting of the American Society of Mechanical Engineering, Washington, D. C., Nov. 15–20, 1981, AMD vol. 48, 1981.

Kamal, M. M., and Wolf, J. A. (eds.), *Finite element applications in vibration problems,* American Society of Mechanical Engineers, New York, 1977.

Oden, J. T., and Carey, G. F. (compliers), *Finite elements: special problems in solid mechanics,* Prentice-Hall, New York, 1983.

Owen, D. R. J., and Hinton, E., *Finite elements in plasticity—theory and practice,* Pineridge Press, Swansea, United Kingdom, 1981.

Przemieniecki, J. S., *Theory of matrix structural analysis,* McGraw-Hill, New York, 1968.

Robinson, J., *An integrated theory of finite element methods*, Wiley-Interscience, New York, 1973.

FLUID MECHANICS APPLICATIONS

Baker, A. J., *Finite element computational fluid mechanics,* McGraw-Hill, New York, 1983.

Connor, J. C., and Brebbia, C. A., *Finite element techniques for fluid flow,* Butterworth, London, 1976.

Chung, T. J., *Finite element analysis in fluid dynamics,* McGraw-Hill, New York, 1978.

Gallagher, R. H., Oden, J. T., Tayler, C., and Zienkiewicz, O. C. (eds.), *Finite elements in fluids,* vols. 1 and 3, Wiley, New York, 1975.

Gallagher, R. H., Zienkiewicz, O. C., Oden, J. T., Cecchi, M. M., and Taylor, C. (eds.), *Finite elements in fluids,* vol. 3, Wiley, New York, 1978.

Gallagher, R. H., Norrie, D. H., Oden, J. T., and Zienkiewicz, O. C. (eds.), *Finite elements in fluids,* vol. 4, Wiley, New York, 1982.

Hughes, T. J. R. (ed.), *Finite element methods for convection dominated flows,* presented at the Winter Annual Meeting of the American Society of Mechanical Engineers, New York, Dec. 2–7, 1979; sponsored by Applied Mechanics Division, ASME/AMD, New York, vol. 34, 1979.

Norrie, D. H. (ed.), *Proceedings of the international conference on finite elements in flow problems,* Third, Banff, 1980.

Oden, J. T., Gallagher, R. H., Zienkiewicz, O. C., Kawahara, M. T., and Kawai, T. (eds.), *Finite elements in fluids,* vol. 5, Wiley, London, 1983.

Raviart, P. A., *Finite element methods in fluid mechanics,* Editions Eyrolles, Paris, Electricité de France, direction des études et recherches, 1981.

Taylor, C., and Hughes, T. G., *Finite element programming of the Navier–Stokes equations,* Pineridge Press, Swansea, United Kingdom, 1981.

MIXED, HYBRID, AND PENALTY FINITE ELEMENT METHODS

Atluri, S. N., Gallagher, R. H., and Zienkiewicz, O. C. (eds.), *Hybrid and mixed finite element methods,* Wiley-Interscience, New York, 1983.

Glowinski, R., Rodin, E. Y., and Zienkiewicz, O. C. (eds.), *Energy methods in finite element analysis,* Wiley, New York, 1979.

Kardestuncer, H., *Finite element methods via tensors,* International Center for Mechanical Sciences, Udine, Italy, 1974.

Reddy, J. N. (ed.), *Penalty-finite element methods in mechanics,* presented at the Winter Annual Meeting of the American Society of Mechanical Engineers, Phoenix, Arizona, Nov. 14–19, 1982, AMD, New York, vol. 51, 1982.

BIBLIOGRAPHIES AND HANDBOOKS

Norrie, D., and de Vries, G., *Finite element bibliography,* Plenum, New York, 1976.
Whiteman, J. R., *A bibliography for finite elements,* Academic, London, 1975.
Kardestuncer, H. (ed.), *Finite element handbook,* McGraw-Hill, New York, 1984.

PROCEEDINGS OF SYMPOSIA AND CONFERENCES ON FINITE ELEMENT METHODS

Abel, J. F., Kawai, T., and Shen, S. F. (eds.), *Interdisciplinary finite element analysis,* Proceedings of the U.S.–Japan seminar held at Cornell University, Aug, 7–11, 1978; sponsored by U.S. National Science Foundation and the Japan Society for Promotion of Science, Ithaca, N.Y.

Bader, R. M. (ed.), *Proceedings of the third conference on matrix methods in structural mechanics,* Wright Patterson Air Force Base, Ohio, AFFDL-TR-71-160, Oct. 1971.

Bathe, K. J., Oden, J. T. and Wunderlich, W. (eds.), *Formulations and computational algorithms in finite element analysis,* U.S.–Germany symposium held at Massachusetts Institute of Technology, Cambridge, Massachusetts, Aug. 1976, MIT Press, Cambridge, Massachusetts, 1977.

Berke, L., and J. S. Przemieniecki (eds.). *Proceedings of the second conference on matrix methods in structural mechanics,* Wright Patterson Air Force Base, Ohio, AFFDL-TR-68-150, Oct. 15–17, 1968.

Brebbia, C. A. (ed.), *New developments in boundary element methods,* Proceedings of the second international seminar at Southampton University, March 1980.

Brebbia, C. A. (ed.), *Further developments in boundary element methods,* Proceedings of the third international seminar, July 1981, CML Publications, Computational Mechanics Center, Ashurst Lodge, Hampshire, England.

de Veubeke, Fraeijs B. (ed.), *Matrix methods of structural analysis,* Pergamon, Elmsford, New York, 1964.

Gallagher, R. H., Yamada, Y., and Oden, J. T. (eds.), *Recent advances in matrix methods in structural analysis and design,* Proceedings of the first Japan–U.S. Seminar on Matrix Methods in Structural Analysis and Design, Tokyo, Aug. 25–30, 1969, University of Alabama Press, Huntsville, 1970.

Hughes, T. J. R., Gartling, D., and Spilker, R. L. (eds.), *New concepts in finite element analysis,* presented at the Applied Mechanics Conference, Boulder, Colorado, June 22–24, 1981, ASME/AMD vol. 44, 1981, American Society of Mechanical Engineers, New York.

Kabaila, A. P., and Pulmano, V. A. (eds.), *Finite element methods in engineering,* Proceedings of the third international conference in Australia on finite element methods held at the University of New South Wales, Sydney, Australia, July 2–6, 1979.

Kardestuncer, H. (ed.), *Proceedings of the symposia on the unification of finite elements—finite differences and calculus of variations,* University of Connecticut, Storrs, 1979, 1980, and 1982, University Park.

Oden, J. T., Clough, R. W., and Yamamoto, Y. (eds.), *Advances in computational methods in structural mechanics and design,* Proceedings of the second U.S.–Japan seminar on matrix methods in structural analysis and design, Aug. 1972, University of Alabama in Huntsville.

Oden, J. T., and Oliveira, E. R. A. (eds.), *Lectures on finite element methods in continuum mechanics,* The University of Alabama in Huntsville, 1973.

Pister, K. S., Reynolds, R. R., and William, K. J. (eds.), *Proceedings of the international conference on finite elements in nonlinear mechanics* (*FENOMECH '78*), Aug. 30–Sept. 1, 1978, Institut für Statik und Dynamik der Luft-und Raumfahrtkonstruktionen, Universitat Stuttgart, West Germany, 1979, Stuttgart.

Proceedings of the workshop meeting on computational aspects of the finite element method, Sept. 17–18, 1973, University of Stuttgart, West Germany.

Przemieniecki, J. S., et al. (eds.), Matrix methods in structural mechanics—Proceedings of the first conference, Wright Patterson Air Force Base, Ohio, AFFDL-TR-66-80, Oct. 26–28, 1965.

Pulmano, V. A., and Kabaila, A. P. (eds.), Proceedings of the 1974 international conference on finite element methods in engineering, University of New South Wales, Australia, Aug. 28–30, 1974.

St. Doltsinis, J., Straub, K., and William, K. J. (eds.), *FEMOMECH '81, Proceedings of the second international conference on finite elements in nonlinear mechanics,* Institut für Statik and Dynamik der Luft-und Raumfahrtkonstruktionen, Universitat Stuttgart, West Germany, Aug. 25–28, 1981, North-Holland, Amsterdam, 1982.

Wunderlich, W., Stein, E., and Bathe, K. J. (eds.), *Nonlinear finite element analysis in structural mechanics,* Proceedings of the Europe–U.S. workshop, Ruhr Universitat Bochum, West Germany, July 28–31, 1980, Springer-Verlag, Berlin, 1981.

LIST OF NOTATION

CHAPTER 1

$\mathbf{A}$	Matrices
$\mathbf{B}$	Rotation matrix
$\mathbf{e}_I$	Base vector in IJK coordinate system
$\mathbf{f}$: f_i	Force vector: component
$\mathrm{F}(\ )$	Functionals
$\mathbf{i}_i$	Base vector in ijk coordinate system
$\mathbf{K}$: K_{ij}	Stiffness matrix: component
$L(\ ,\ )$	Lagrangian functional
$\mathbf{M}$	Mass matrix
$\mathbf{M}^{-1}$	Inverse matrix of $\mathbf{M}$
$\mathbf{P}$, $[P_{kl}]$: P_{kl}	Orthogonal matrix: component
$\mathbf{P}^{\mathrm{T}}$	Transposed matrix of P
$\mathbf{r}$	Position vector
$\mathbb{R}$	Real space
$\mathbb{R}^N$	N-Dimensional real space
$\mathbf{S}$, $\hat{\mathbf{S}}$: S_{kl}, $\hat{S}_{kl}$	Similarity transformation matrices: components
$\mathbf{T}$: T_{iI}	Transformation matrix: component
$\mathbf{u}$, $\mathbf{v}$, $\mathbf{w}$: u_i, v_i, w_i	Vectors: components
$\mathbf{a} \circ \mathbf{b}$	"Product" of two vectors
$\mathbf{a} \cdot \mathbf{b}$	Dot product of two vectors
$\mathbf{a} \times \mathbf{b}$	Cross product of two vectors
$\mathbf{i}_i\mathbf{e}_I$	Dyadic product of two base vectors
Δ	Laplacian
∇	Gradient vector
δ	Variation of
δ_{ij}	Kronecker delta
$\forall$	For all, for every
$\in$	Element of
λ	Eigenvalue
$\phi_{,i}$	Derivative of ϕ with respect to $\mathbf{i}$

CHAPTER 2

Δa	Infinitesimal area
$\{a_\alpha\}$	Set of a_α
A_e	Area of an element
ΔA_{xy}	Area element in (x, y) coordinate
$\Delta A_{\xi_1\xi_2}$	Area element in (ξ_1, ξ_2) coordinate

$\{b_\alpha\}$	Set of b_α
$\{c_\alpha\}$	Set of c_α
c_p	Heat capacity
d_0, d_1, d_2	Constants
d	Deviation from e_A^I
e_h	Error of finite element approximation for a function
e_h^I	Interpolation error for a function
e_0^I, e_1^I	Global interpolation errors
e_{i0}^I, e_{i1}^I	Interpolation errors in ith element of first and second kinds
$\hat{e}_{i0}^I, \hat{e}_{i1}^I$	Square norms of interpolation errors
$\bar{e}_{i\alpha}^I$ or $(\bar{e}_{i0}^I, \bar{e}_{i1}^I)$	Designed interpolation errors ($\alpha = 1, 2$)
$\bar{e}_{A\alpha}^I$ or $(\bar{e}_{A0}^I, \bar{e}_{A1}^I)$	Average of interpolation errors ($\alpha = 1, 2$)
E	Total head
E_0	Reference head
E_2', E_3'	Numbers of line elements on second and third boundaries
f	Applied load
$\hat{f}, \hat{f}^0, \hat{f}^1, \hat{f}^2$	A function and its coefficients
$\{f_\alpha^e\}$: f_α^e	Element load vector: component
$f_\alpha^{e,2}, f_\alpha^{e,3}$	Components of element load vectors for second and third boundaries
g	(1) A function; (2) gravity acceleration
$\{g_I\}$: g_I	Global interpolation vector of g: component
G	Shear modulus
h	(1) Mesh size; (2) normal traction; (3) thickness of film; (4) hydraulic head
h_i	Interval of a line
h_x, h_y	Sizes of an element in x, y directions
H	Arc length
H_i	Normal distance of ith vertex to its opposite edge
J	Determinant of Jacobian matrix
J_2	Equivalent stress
k	Heat conductivity in one dimension
k_0	Convection coefficient
$\mathbf{k}$: $k_{ij}(k_{xx}, k_{yy}, k_{xy}, k_{yx})$	Heat conductivity matrix: components
$[K_{\alpha\beta}^e]$: $K_{\alpha\beta}^e$	Element stiffness matrix: component
$[K_{\alpha\beta}^{e,3}]$: $K_{\alpha\beta}^{e,3}$	Element stiffness matrix for third boundary condition: component
$[K_{IJ}]$: K_{IJ}	Global stiffness matrix: components
$\mathbf{n}$: (n_x, n_y)	Outward normal vector: components
$\{N_\alpha(x)\}$	Set of shape functions
p	(1) Pressure; (2) penalty parameter
p^-	Special function
(r, θ, z)	Cylindrical coordinate
R_e	Aspect's ratio
$R_{\text{old}}, R_{\text{new}}$	Radii of before and after iteration

$\mathbf{q}$: (q_x, q_y) or (q_n, q_s)	Heat flux vector: components
q	Distributed source or sink
$\mathbf{q}_h$	Heat flux vector by finite element solution
$\hat{\mathbf{q}}_h$	"Spanned" heat flux vector
$q_{i\beta}$	Heat flux component at β node
s	(1) Arc coordinate; (2) a nonnegative parameter
S_i	Length of arc
$\mathbf{u}$: u_i or (u_x, u_y)	Velocity vector: components
u_{old}	Velocity at previous step
$u(x)$	Function to be interpolated
$u_h^I(x)$	Finite element interpolation function
$(u_h^I)'(x)$	First derivative of u_h^I
$\mathbf{v}$: v_i or (v_x, v_y)	Velocity vector of flow: components
$\mathbf{x}$: (x, y)	Position vector: components
$\hat{\mathbf{x}}$	Position vector of centroid
$\mathbf{x}_n$	Position vector of nth node
$\mathbf{x}_{nk}^c$	Position vector of centroid of kth element
$(x_i, x_i + 1)$	Subinterval in one dimension
(x^β, y^β)	Position of β node
Ω	Domain
Ω_e	Domain of an element
Ω_R	Subdomain
Γ: $\Gamma_1, \Gamma_2, \Gamma_3$	Boundary of the domain: of first, second, and third kinds
Γ_e: Γ_e^3	Boundary of an element: of third kind
Γ_s	Inner surface
ϕ	(1) Velocity potential; (2) a function
ϕ_h	Finite element solution of ϕ
ϕ_n	Value of ϕ at nth node
ψ	Stream function
η	Viscosity
ρ	Density
ξ_i	Normalized coordinate
$\bar{\sigma}$	Yield stress
$\sigma_{ij}, \sigma_{xx}, \sigma_{xy}, \sigma_{yy}$	Stress components
$\boldsymbol{\omega}$: ω_i or $(\omega_x, \omega_y, \omega_z)$	Angular velocity vector: components
$\ni$	such that

CHAPTER 3

c_p	Heat capacity
f	Heat source
f	Global heat source vector in finite element
$\{f_\beta^e\}$: f_β^e	Element heat source vector: a component at β node
g	A given function
$\{g_\beta\}$: g_β	Finite element approximation vector of g: component
h	(1) A given function; (2) representative mesh size
h_e	Length of a line element

k	Heat conductivity
k_{ad}	Artificial conductivity
k_0	Convection coefficient
$\mathbf{k}$: k_{ij}	Heat conductivity tensor: component
$\mathbf{k}^{ac}$: k_{ij}^{ac}	Artificial conductivity tensor: component
$[K_{\alpha\beta}^c]$: $K_{\alpha\beta}^c$	Stiffness matrix for convection for a line element; component
$[K_{\alpha\beta}^d]$: $K_{\alpha\beta}^d$	Stiffness matrix for diffusion for a line element; component
$[K_{\alpha\beta}^e]$: $K_{\alpha\beta}^e$	Element stiffness matrix: component
$[K_{\alpha\beta}^{ad}]$: $K_{\alpha\beta}^{ad}$	Stiffness matrix for artificial conductivity for a line element; component
$[K_{\alpha\beta}^{c,e}]$: $K_{\alpha\beta}^{c,e}$	Element stiffness matrix for convection: component
$[K_{\alpha\beta}^{d,e}]$: $K_{\alpha\beta}^{d,e}$	Element stiffness matrix for diffusion: component
$[K_{\alpha\beta}^{ac,e}]$: $K_{\alpha\beta}^{ac,e}$	Element stiffness matrix for artificial conductivity: component
$[M_{\alpha\beta}^e]$: $M_{\alpha\beta}^e$	Element mass matrix: component
$\mathbf{n}$: n_i or (n_x, n_y)	Outward normal vector: component
$\{N_\beta\}$: N_β	Set of shape functions: component
s	Length coordinate
t	Time
Δt	Time increment
t^n	nth incremental time
T	Temperature
T_0	Initial temperature
T_h	Finite element approximation of temperature
T_β	(1) Temperature at β node; (2) temperature at a position $\mathbf{x}_\beta$
$T_{0\beta}$	(1) Initial temperature at β node; (2) initial temperature at a position $\mathbf{x}_\beta$
T_β^n	nth iterative value of T at a position $\mathbf{x}_\beta$
u	Velocity in one dimension
$\mathbf{u}$: u_i or (u_x, u_y)	Velocity vector: components
$\mathbf{u}^{n+1}$	$\mathbf{u}$ at $(n+1)$th time step
$\mathbf{u}_k^{n+1}$	kth iterative value of $\mathbf{u}$ at $(n+1)$th time step
$\mathbf{x}$: x_i or (x, y)	Position vector: components
ξ	Coordinate
ξ^n	Coordinate at nth time step
ξ_i^n	ith iterative value of ξ at nth time step

CHAPTER 4

A	Cross-sectional area
$\underline{A}$	Lower bound of designed area
A_e	Cross-sectional area of a truss element
A_e^0	Initial area
A_e^i	Cross-sectional area of an element at ith step
$\mathbf{A}$, $\bar{\mathbf{A}}$: A_{ij}, $\bar{A}_{ij}$	Matrices $\mathbf{A}$, $\bar{\mathbf{A}}$: components
$\hat{\mathbf{A}}$: $\hat{A}_{ij}$	Similarity transformation of $\mathbf{A}$: components

$\mathbf{B}$: B_{ij}	Matrix $\mathbf{B}$: components
$\mathbf{C}$: C_{ij}	Damping matrix: components
$\hat{\mathbf{C}}$: $\hat{C}_{ij}$	Modified damping matrix: components
E	Young's modulus
f	Distributed load
$\mathbf{f}$	Load vector
$\{f^e_\alpha\}$: f^e_α	Element load vector: components
$\{\bar{f}^e_\alpha\}$: $\bar{f}^e_\alpha$	Transformed element load vector: components
h	Beam thickness
$\underline{h}, \bar{h}$	Lower, upper bound of thickness
h_e	Thickness of a beam element
h^0_e	Initial thickness of an element
h^i_e	Thickness of an element after ith iteration
G	Shear modulus
$I, \hat{I}$	Moments of inertia
$\mathbf{I}$	Identity matrix
$J, \hat{J}$	Polar moments of inertia
K	Torsion rigidity
$\mathbf{K}$: K_{IJ}	Global stiffness matrix: components
$\hat{\mathbf{K}}$: $\bar{K}_{IJ}$	Transformed global stiffness matrix: components
$\mathbf{K}_e$, $[K^e_{\alpha\beta}]$: $K^e_{\alpha\beta}$	Element stiffness matrix for truss: components
$[K^e_{ab}]$: K^e_{ab}	Element stiffness matrix for beam: components
$\bar{\mathbf{K}}_e$, $[\bar{K}^e_{pq}]$: $\bar{K}^e_{pq}$	Modified element stiffness matrix: components
l_e	Length of truss
(l, m, n)	Direction cosines
M_1, M_2	Bending moments
$\mathbf{M}$: M_{IJ}	Global mass matrix: components
$[M^e_{\alpha\beta}]$: $M^e_{\alpha\beta}$	Mass stiffness matrix for truss: components
$[M^e_{ab}]$: M^e_{ab}	Mass stiffness matrix for beam: components
$[\bar{M}^e_{pq}]$: $\bar{M}^e_{pq}$	Modified mass stiffness matrix: components
$N(s)$	Internal force at s
$\{N_\beta\}$: N_β or N_1, N_2	Set of shape functions: components
N_e	Total number of beams
$\mathbf{P}$: P_{ij}	Orthogonal matrix: components
$\mathbf{P}^{(ij)}$	Orthogonal matrix for (ij) transformation: components
P_e	Axial force in a truss
Q	A quantity
Q_e	A quantity for a beam element
Q^i_e	A quantity for a beam element at ith step
$\bar{Q}^i$	Average of Q^i_e
$\mathbb{R}$	Real space
s	Axial coordinate
Δs	Incremental length
Δt	Time increment
T	(1) Temperature; (2) axial force; (3) torque
T_0	(1) Initial temperature; (2) maximum tolerance
$\mathbf{T}$, $[T_{\alpha\beta}]$: $T_{\alpha\beta}$	Transformation matrix: components

$\mathbf{u}$: u_i	Displacement vector
$\mathbf{u}_{k+1}$	$\mathbf{u}$ at $(k+1)$th time step
u_{ps}	Displacement of a point P in s direction
$\mathbf{U}$: U_J	Global displacement vector: component
$\hat{\mathbf{U}}$: $\hat{U}_J$	Transformed global displacement vector: component
$\mathbf{v}_m$	mth modal vector
$\mathbf{V}$: V_{ij}	Modal matrix
V_0	Volume constraint
V_1, V_2	Shear forces
$\mathbf{w}$	A vector
$\{W_b\}$: W_b	Global vector: component
$\bar{W}$	Upper bound of W
$\mathbf{x}^0$	Initial position vector
$\{x^k\}$	Position vector after kth iteration
α	Thermal expansion coefficient
ε	Strain in truss
ε_0	Initial strain
$\Delta\varepsilon^n$	Increment of ε at nth time step
θ	Angle
θ_1, θ_2	Gradient of deflection at ends of beam
ψ	Lagrangian function
λ_{1e}, λ_{2e}	Constants
λ_i	Eigenvalue
Λ: Λ_{ii}	Eigenvalue matrix: components
δ	Variation of
μ, μ_e, $\bar{\mu}$	Constants
ρ, ρ_s	Density
ξ	Normalized coordinate
σ, σ_A	Axial stress
$\bar{\sigma}$	Upper bound of axial stress
σ_Y	Yield stress in compression
τ	Shear stress
τ_Y	Yield stress in shear

CHAPTER 5

$\mathbf{a}$	Position vector
a_α: (a_1, a_2, a_3)	Initial position coordinate
da_α: (da_1, da_2, da_3)	Differential coordinate
A	Area
ΔA_i	Area component in ith direction
$\mathbf{A}$: A_{ij}	Matrix $\mathbf{A}$: component
$\mathbf{b}_x$, $\mathbf{b}_y$	Constant vectors
$\mathbf{B}$, $[B_{i\alpha}]$: $B_{i\alpha}$	B (gradient) matrix: component
$[B^e_{\alpha i\delta}]$: $B^e_{\alpha i\delta}$	Element stiffness matrix for pressure: component
$\mathbf{C}$, $[C_{IJ}]$: C_{IJ}	Matrix of contracted elasticity tensor: component
$[\hat{C}_{IJ}]$: $\hat{C}_{IJ}$	Contracted elasticity matrix for orthotropic material: component
e	Relative error in strain energy

e_d	Relative error in displacement
e_τ	Relative error in stress
$\mathbf{e}$: e_{ij}	Strain tensor
$(\mathbf{e}_r, \mathbf{e}_\theta, \mathbf{e}_z)$	Set of base vectors
E	Young's modulus
$\hat{E}_1, \hat{E}_2$	Principal Young's moduli for orthotropy
$\mathbf{E}$: E_{ijkl}	Elasticity tensor: component
$\hat{\mathbf{E}}$: $\bar{E}_{ijkl}$	Generalized elasticity tensor: component
$[E^e_{i\alpha j\beta}]$: $E^e_{i\alpha j\beta}$	Element stiffness matrix: component
$[E^{e,1}_{i\alpha j\beta}]$: $E^{e,1}_{i\alpha j\beta}$	Part of element stiffness matrix by one-point integration: component
$[E^{e,2}_{i\alpha j\beta}]$: $E^{e,2}_{i\alpha j\beta}$	Part of element stiffness matrix by 2×2 integration: component
$[E^{1,e}_{i\alpha j\beta}]$: $E^{1,e}_{i\alpha j\beta}$	Element stiffness matrix for first boundary condition: component
$[E^{3,e}_{i\alpha j\beta}]$: $E^{3,e}_{i\alpha j\beta}$	Element stiffness matrix for third boundary condition: component
f	Function
f^I	Interpolated function
$f^e_{i\alpha}$	Component of element force vector
$f^{0,e}_{i\alpha}$	Component of element force vector by temperature difference
$f^{1,e}_{i\alpha}$	Component of element force vector by body force
$f^{2,e}_{i\alpha}$	Component of element force vector by second boundary condition
$f^{3,e}_{i\alpha}$	Component of element force vector by third boundary condition
$\mathbf{g}_s, \mathbf{g}_t$	Base vectors
G	Shear modulus
G_{IJ}	Shear modulus component in orthotropy
H^0, H^1	Hermite interpolation functions
I	Real interval
J	(1) Jacobian; (2) J integral value
k	Spring constant
k_n, k_T	Spring constants in normal, tangential direction
$\mathbf{k}$: k_{ij}	Spring constant tensor: component
K_I	Stress intensity factor of first kind
$(K_I)_a$	"Apparent" stress intensity factor of first kind
$[K^e_{pq}]$: K^e_{pq}	Element stiffness matrix: component
$[K^e_{11}], [K^e_{12}], [K^e_{21}], [K^e_{22}]$	Partial matrices of element stiffness matrix
$\mathbf{L}$: L_{ij}	Lower triangular matrix: component
$\{L_i(x)\}$	Set of Lagrangian polynomials
$\{M_\alpha(\mathbf{x})\}$	Set of shape functions
$\{\hat{M}_\alpha(s, t)\}$: $\hat{M}_\alpha(s, t)$	Set of shape functions in (s, t) coordinate: component
$\{N_\alpha(\mathbf{x})\}$ or $\{N_\alpha(x, y)\}$: $N_\alpha(x, y)$	Set of shape functions
$\{\hat{N}_\alpha(s, t)\}$: $\hat{N}_\alpha(s, t)$	Set of shape functions in (s, t) coordinate: component
p	Hydrostatic pressure

$\mathbf{r}$	Position vector
$\mathbf{r}_0$	Reference position vector
$\mathbf{r}_p$	Position vector of P
$d\mathbf{r}$	Differential vectors of $\mathbf{r}$
$d\mathbf{r}_0$	Differential vectors of $\mathbf{r}_0$
$\mathbf{R}$	Residual vector
$\mathbb{R}$	Real space
$\mathbb{R}^2$, $\mathbb{R}^3$	Two- and three-dimensional real spaces
S_0	Distance
$\mathbf{t}$: t_i	Traction vector: components
t_n	Normal component of traction vector
$\mathbf{t}_T$	Tangential traction vector
$t_{i\alpha}$	Component of traction vector at node α
$\mathbf{u}$	Displacement vector
$\mathbf{u}_P$	Displacement vector at P
$\bar{\mathbf{u}}$	Virtual displacement vector
$\mathbf{u}_h$	Displacement vector by finite element approximation
$\mathbf{u}^n$	Velocity vector at nth step in flow problem
$\Delta\mathbf{u}^n$	Increment of velocity vector during nth step
U	Total strain energy in a domain
U_0	Strain energy density
ΔU	Increment of total strain energy
$\mathbf{U}$: U_{ij}	Upper triangular matrix: component
x_0	Initial approximation of x
x_n	x at nth iteration
Δx_n	Increment of x at nth iteration
$\boldsymbol{\beta}$: β_{ij}	Thermal expansion tensor
ε	Regularization parameter
$\boldsymbol{\varepsilon}$: ε_{ij}	Cauchy strain tensor
$\boldsymbol{\varepsilon}$: ε_i	Contracted strain vector
$\boldsymbol{\sigma}$: σ_{ij}	Cauchy stress tensor
$\boldsymbol{\sigma}$: σ_i	Contracted stress vector
σ_m	Mean stress
τ	Stress component at integration point
$\tilde{\tau}$	Continuous stress component
$\hat{\tau}_\alpha$	Stress component at node α
$\bar{\tau}$	Virtual stress component
κ	Bulk modulus
κ_{ad}	Adiabatic bulk modulus

CHAPTER 6

D	Bending rigidity
D_{ijkl}	Components of elastic modulus tensor for plate
E	(1) Young's modulus; (2) total number of element
$\mathbf{f}$: f_i or (f_x, f_y)	Body force vector per unit mass: component
h	Thickness of plate
k	A "fudge" factor

$K^{1,e}_{i\alpha j\beta}$	Element stiffness matrix for bending
$K^{2,e}_{i\alpha\beta}$	Element stiffness matrix for coupling of bending and shear
$K^{3,e}_{\alpha j\beta}$	Element stiffness matrix for coupling of shear and bending
$K^{4,e}_{\alpha\beta}$	Element stiffness matrix for coupling of shear
m_x, m_y	Moments of resultants of body forces
M_{ij}	Components of bending moment matrix
M_{nn}, M_{ns}	Bending moments on an edge of plate
M_{xx}, M_{yy}	Bending moments
M_{xy}	Twisting moment
$\{M_\alpha(x, y)\}$	Set of shape functions
$\{\hat{M}_\alpha(s, t)\}$	Set of shape functions in (s, t) coordinate
$\mathbf{n}$: n_i, or (n_x, n_y, n_z)	Unit outward normal vector: components
N_{xx}, N_{yy}, N_{xy}	In-plane resultant forces
$\{N_\alpha(x, y)\}$	Set of shape functions
$\{\hat{N}_\alpha(s, t)\}$	Set of shape functions in (s, t) coordinate
q	Resultants of body forces
Q_n	Transverse shear force on an edge
Q_x, Q_y	Transverse shear forces
u, v, w	Displacement components at neutral plane
u_x, u_y, u_z	Displacement in x, y, z directions
w_α	Deflection at node α
μ	Shear modulus
ν	Poisson's ratio
θ_i, or (θ_x, θ_y)	Rotation angle in x, y direction
ε_{xx}, ε_{yy}, ε_{zz}, ε_{xy}, ε_{yz}, ε_{xz}	Strain components
σ_{xx}, σ_{yy}, σ_{zz}, σ_{xy}, σ_{yz}, σ_{xz}	Stress components

INDEX

PROGRAM MANUAL AND DISKETTE ORDERING INSTRUCTIONS
(FOR OUTSIDE U.S.A. & CANADA)

Also published by Cambridge University Press and described below are a program manual and software diskette that accompany *Finite Element Methods in Mechanics*.

PROGRAM MANUAL FOR FINITE ELEMENT METHODS IN MECHANICS
ISBN 0-521-30952-2, about 196 pages.
The program manual provides the student and teacher with additional programs written for personal computers and with practice problems for finite element methods. Each program is introduced by an explanatory comment and several examples and exercises pertaining to the program are given. All in all, there are twenty-one programs written in BASIC for the personal computer, which may be used on a variety of problems: heat conduction problems, stress analysis problems, the Gauss elimination method (unsymmetric), band method for Gauss elimination (symmetric and unsymmetric), Crout elimination method (symmetric and unsymmetric), skyline method for Crout's method (symmetric and unsymmetric), conjugate gradient method, Jacobi's method, generalized Jacobi's method, inverse iteration method, θ-method for parabolic problems, Newmark β-method for vibration problems, beam-bending problems, and plate-bending problems. The five important finite element programs found in the main book are repeated in the manual, to make it self-contained as far as programs are concerned. This manual will be a very valuable aid and resource to students and teachers studying finite element methods in mechanics.

FINITE ELEMENT METHODS IN MECHANICS DISKETTE 0-521-30953-0
5¼-inch, double-sided/double-density floppy diskette
The diskette accompanies the main book and manual, and contains the BASIC programs found in the book and the manual in machine-readable form. It operates on DOS 2.0/3.0 on the IBM PC, XT, AT, and IBM-compatible machines. The diskette can save hours of tedious keyboarding, leaving users free to experiment with and apply the programs.

To order the example books or latest version of the diskettes, complete the information below and mail this page or a copy to Cambridge University Press. RESIDENTS OF THE U.S.A. AND CANADA PLEASE USE THE FORM ON THE FOLLOWING PAGES.

Technical questions, corrections, and requests for information on other available formats should be directed in writing to Professor N. Kikuchi, Department of Mechanical Engineering and Applied Mechanics, The University of Michigan, Ann Arbor, Michigan 48109, U.S.A.

Only diskettes with manufacturing defects may be returned to the publisher for replacement (no cash refunds).

To: Customer Services Department, Cambridge University Press, Edinburgh Building, Shaftesbury Road, Cambridge CB2 2RU, U.K.

Please send me:

_______ 30952-2 *PROGRAM MANUAL FOR FINITE ELEMENT METHODS IN MECHANICS* £15.00 each

_______ 30953-0 *FINITE ELEMENT METHODS IN MECHANICS DISKETTE* £15.00 each

Name ____ (Block capitals please)
Address ____

____

____

Please accept my payment by cheque or money order in pounds sterling:
I enclose (circle one) a Cheque (made payable to Cambridge University Press)/UK Postal Order/ International Money Order/Bank Draft/Post Office Giro.

Please accept by payment by credit card:
Charge my (circle one) Barclaycard/VISA/Eurocard/Access/Mastercard/Bank Americard/any other credit card bearing the Interbank symbol (please specify).

Card No. ____ Expiry date: __________
Signature ____ Date: __________
Address as registered by card company: ______________________________________

____

____

All prices include VAT and are subject to alteration without prior notice.

PROGRAM MANUAL AND DISKETTE ORDERING INSTRUCTIONS
(FOR U.S.A. & CANADA)

Also published by Cambridge University Press and described below are a program manual and software to accompany *Finite Element Methods in Mechanics*.

PROGRAM MANUAL FOR FINITE ELEMENT METHODS IN MECHANICS
0-521-30952-2, about 196 pages.
The program manual provides the student and teacher with additional programs written for personal computers and with practice problems for finite element methods. Each program is introduced by an explanatory comment and several examples and exercises pertaining to the program are given. All in all, there are twenty-one programs written in BASIC for the personal computer, which may be used on a variety of problems: heat conduction problems, stress analysis problems, the Gauss elimination method (unsymmetric), band method for Gauss elimination (symmetric and unsymmetric), Crout elimination method (symmetric and unsymmetric), skyline method for Crout's method (symmetric and unsymmetric), conjugate gradient method, Jacobi's method, generalized Jacobi's method, inverse iteration method, θ-method for parabolic problems, Newmark β-method for vibration problems, beam-bending problems, and plate-bending problems. The five important finite element programs found in the main book are repeated in the manual, to make it self-contained as far as programs are concerned. This manual will be a very valuable aid and resource to students and teachers studying finite element methods in mechanics.

FINITE ELEMENT METHODS IN MECHANICS DISKETTE 0-521-30953-0
5¼-inch, double-sided/double-density floppy diskette
The diskette accompanies the main book and manual, and contains the BASIC programs found in the book and the manual in machine-readable form. It operates on DOS 2.0/3.0 on the IBM PC, XT, AT, and IBM-compatible machines. The diskette can save hours of tedious keyboarding, leaving users free to experiment with the programs.

To order the program manual or the latest version of the diskette, complete the information below and mail this page or a copy to Cambridge University Press. [Alternatively, customers may call the Press at 914/235-0300 (in N.Y. and Canada) or 800/431-1580 (in rest of U.S.) to place an order or verify prices.] Orders must be accompanied by payment in U.S. funds or the equivalent in Canadian funds. New York and California residents please add appropriate sales tax. Prices printed below are subject to change without notice. Ordinary postage for shipping orders is paid by the publisher. RESIDENTS OF COUNTRIES OTHER THAN THE U.S.A. AND CANADA PLEASE USE THE FORM ON THE PRECEDING PAGES.

Technical questions, corrections, and requests for information on other available formats should be directed in writing to Professor N. Kikuchi, Department of Mechanical Engineering and Applied Mechanics, The University of Michigan, Ann Arbor, Michigan 48109.

Only diskettes with manufacturing defects may be returned to the publisher for replacement (no cash refunds).

To: Cambridge University Press, Order Department, 510 North Avenue, New Rochelle, New York 10801.

Please indicate method of payment check ----- , Mastercard ----- , or Visa ----- .

Name --
Address --

Card No. ---------------------- Expiration date ------------------------
Signature --

Please send me:

------- 30952-2 *PROGRAM MANUAL FOR FINITE ELEMENT METHODS IN MECHANICS* $18.95 each
------- 30953-0 *FINITE ELEMENT METHODS IN MECHANICS DISKETTE* $19.95 each
------- Please indicate the total number of items ordered.

------- total price
------- tax, if applicable (NY and CA residents)
------- total enclosed